26

SIGNAL TRANSDUCTION

SIGNAL TRANSDUCTION

Prokaryotic and Simple Eukaryotic Systems

EDITED BY

Janet Kurjan
Department of Microbiology and Molecular Genetics and
The Vermont Cancer Center
College of Medicine and College of Life Sciences and Agriculture
University of Vermont
Burlington, Vermont

Barry L. Taylor
Department of Microbiology
School of Medicine
Loma Linda University
Loma Linda, California

ACADEMIC PRESS, INC.
A Division of Harcourt Brace & Company
San Diego New York Boston
London Sydney Tokyo Toronto

This book is printed on acid-free paper. ∞

Academic Press, Inc.
1250 Sixth Avenue, San Diego, California 92101-4311

United Kingdom Edition published by
Academic Press Limited
24–28 Oval Road, London NW1 7DX

Library of Congress Cataloging-in-Publication Data

Signal transduction : prokaryotic and simple eukaryotic systems
edited by Janet Kurjan, Barry L. Taylor.
p. cm.
Includes bibliographical references and index.
ISBN 0-12-429350-6
1. Cellular signal transdution. I. Kurjan, Janet. II. Taylor, Barry L.
[DNLM: 1. Signal Transduction--physiology. 2. Bacteria--physiology. 3. Eukaryotic Cells--physiology. QH 645 S578 1993]
QP517.C45S55 1993
DNLM/DLC
for Library of Congress 93-19086
CIP

PRINTED IN THE UNITED STATES OF AMERICA
93 94 95 96 97 98 MM 9 8 7 6 5 4 3 2 1

Contents

Part II

Eukaryotic Signaling Systems

CHAPTER 13

Biochemistry and Genetics of Sensory Transduction in *Dictyostelium*

PETER J. M. VAN HAASTERT and PETER N. DEVREOTES

CHAPTER 14

Gene Regulation by Hormone-like Signals in *Dictyostelium*

PAULINE SCHAAP, DORIEN J. M. PETERS, BODDULURI HARIBABU, and ROBERT P. DOTTIN

CHAPTER 15

G Proteins of *Drosophila melanogaster*

JAMES B. HURLEY

Contributors

Numbers in parentheses indicate the pages on which the authors' contributions begin.

Robert G. Ankenbauer (67), Laboratory of Microbial Structure and Function, Rocky Mountain Laboratories, National Institutes of Health, Hamilton, Montana 59840

Judith P. Armitage (43), Microbiology Unit, Department of Biochemistry, University of Oxford, Oxford OX1 3QU, United Kingdom

Raffi V. Aroian (391), Division of Biology, California Institute of Technology, Pasadena, California 91125

Richard Betz (239), Ruhr-Universität Bochum, Abteilung für Biochemie der Differenzierung, D-4630 Bochum, Germany

Peter N. Devreotes (329), Department of Biological Chemistry, The Johns Hopkins University School of Medicine, Baltimore, Maryland 21205-2185

Robert P. Dottin (353), Department of Biological Sciences, Hunter College, City University of New York, New York, New York 10021

Wolfgang Duntze (239), Ruhr-Universität Bochum, Abteilung für Biochemie der Differenzierung, D-4630 Bochum, Germany

Beverly Errede (189), Department of Chemistry, University of North Carolina, Chapel Hill, North Carolina 27599

Bodduluri Haribabu (353), Department of Biological Sciences, Hunter College, City University of New York, New York, New York 10021

Hartmann Harz (279), Botanisches Institut, 8000 Munich 19, Germany

Peter Hegemann (279), Institut für Biochemie I, 93040 Regensburg, Germany

David A. Hughes (123), Chester Beatty Laboratories, Institute of Cancer Research, London SW3 6JB, United Kingdom

James B. Hurley (377), Department of Biochemistry, and Howard Hughes Medical Institute, University of Washington, Seattle, Washington 98195

Mark S. Johnson (3), Department of Microbiology, School of Medicine, Loma Linda University, Loma Linda, California 92350

Janet Kurjan (107), Department of Microbiology and Molecular Genetics and The Vermont Cancer Center, College of Medicine and College of Life Sciences and Agriculture, University of Vermont, Burlington, Vermont 05405

William R. McCleary (17), Department of Molecular Biology, Princeton University, Princeton, New Jersey 08544

Eugene W. Nester (67), Department of Microbiology, University of Washington, Seattle, Washington 98195

Dorien J. M. Peters (353), Cell Biology and Genetics, Department of Biology, University of Leiden, 2311 GP Leiden, The Netherlands

Bert Pöpping (239), Ruhr-Universität Bochum, Abteilung für Biochemie der Differenzierung, D-4630 Bochum, Germany

Pauline Schaap (353), Department of Biology, University of Leiden, 2333 AL Leiden, The Netherlands

William J. Snell (255), Department of Cell Biology and Neuroscience, University of Texas Southwestern Medical Center, Dallas, Texas 75235

Paul W. Sternberg (391), Division of Biology, California Institute of Technology, Pasadena, California 91125

Jeffry B. Stock (17), Department of Molecular Biology, Princeton University, Princeton, New Jersey 08544

Kelly Tatchell (147), Department of Microbiology, North Carolina State University, Raleigh, North Carolina 27695

Barry L. Taylor (3), Department of Microbiology, School of Medicine, Loma Linda University, Loma Linda, California 92350

Peter J. M. Van Haastert (329), Department of Biochemistry, University of Groningen, 9747 AG Groningen, The Netherlands

Judith Van Houten (309), Department of Zoology, University of Vermont, Burlington, Vermont 05405

Malcolm Whiteway (189), National Research Council Canada, Biotechnology Research Institute, Montreal, Quebec, Canada H4P 2R2

Masayuki Yamamoto (123), Department of Biophysics and Biochemistry, Faculty of Science, University of Tokyo, Tokyo 113, Japan

Preface

The past several years have seen an explosion of information in the field of signal transduction. As information from prokaryotes and eukaryotes has accumulated, it has become evident that diverse sensory systems share common mechanistic themes. Extracellular signals usually act through transmembrane receptors to activate intracellular response pathways. Modification of the receptors plays a role in desensitization in both prokaryotes and eukaryotes; however, the nature of the modification is different. There are few sequence similarities between components of prokaryotic and eukaryotic signal transduction pathways, but this difference should not divert the reader from recognizing the mechanistic and conceptual similarities between diverse species. Microorganisms and other simple eukaryotes in which extensive genetic analysis is possible provide systems in which signal transduction pathways can be dissected using approaches not possible in higher organisms; molecular genetics is one of the most important of these approaches. As a result, remarkable progress has been made in elucidating sensory transduction pathways in these systems. This volume examines the current knowledge of the molecular mechanisms for sensory transduction in a variety of such systems, providing an overview of recent progress in this field.

This book is intended for a broader audience than researchers in the field of signal transduction in bacteria and simple eukaryotes. Historically, many biochemical pathways have been first discovered in microorganisms and later extended to higher organisms. Similarly, many insights into sensory phenomena were first made in microorganisms. Investigators and students of signal transduction in more complex organisms will clearly benefit from knowledge gained in these simpler systems. This compilation provides readers the opportunity to make comparisons between systems, not only with regard to the mechanisms by which signals are transmitted but also with respect to different strategies that can be used to study these processes.

We have divided this book into two major sections describing prokaryotic and simple eukaryotic signaling systems. In addition to including better known systems, we have included less well-known systems that are equally interesting. In prokaryotes, many signaling systems have been found to have similar components. These systems utilize membrane-spanning receptors whose sensitivity is adjusted by covalent modification and two-component regulatory systems composed of a histidine kinase and a response regulator. Phosphorylation transfer in the two-component systems constitutes a central sensory signal, not only in bacterial chemotaxis but also in host–pathogen interactions in crown gall tumorigenesis. In eukaryotes, many of the systems involve signaling through Ras or heterotrimeric G proteins. Some chapters describe systems in which the signaling components are less well-defined and for which the future characterization of such components is of great interest. One chapter describes a novel type of signaling molecule discovered in recent years, the prenylated peptides. So far, the lipopeptide pheromones have been discovered only in fungal systems, but researchers in other systems should be aware of the possibility of similar signaling molecules.

Robert Dottin had a seminal role in developing the concept for this volume, including choosing the authors for the chapters on sensory responses in *Dictyostelium*. We thank him for his advice and encouragement. We also thank Dr. Phyllis B. Moses, our insightful and patient editor at Academic Press.

Janet Kurjan
Barry L. Taylor

Part I

Prokaryotic Signaling Systems

Chapter 1

Universal Themes of Signal Transduction in Bacteria

BARRY L. TAYLOR and MARK S. JOHNSON

Department of Microbiology
School of Medicine
Loma Linda University
Loma Linda, California 92350

I. Introduction

Bacterial chemotaxis has been studied primarily as a model for sensory transduction in more complex organisms and not for the purpose of understanding the adaptation of bacteria to their natural environment (Koshland, 1974). Significantly, few published studies examine the role of chemotaxis in a natural habitat. Many chemotaxis experiments, notably those involving temporal stimuli, have been conducted under nonphysiological conditions. As young investigators in the 1970s, we saw bacterial chemotaxis as a first step toward a distant goal of understanding the functioning of the human mind in molecular detail.

The publication of this book comes at an opportune time to survey the rapid progress in investigations of bacterial chemotaxis and to evaluate the contribution of the bacterial studies to the current understanding of the mechanisms of sensory transduction in more complex organisms. With the discovery of the phosphorylation cascade (Hess *et al.*, 1987), a complete pathway has been elucidated in enteric bacteria for signal transduction from chemoreceptor to flagellar motors and for adaptation to changes in environmental stimuli. Future studies no doubt will describe alternative pathways and reveal unexpected complexities of integration and feedback regulation. However, the currently known pathway is likely to continue to be the central transduction pathway for chemotaxis.

In the 30 years since Julius Adler (1965) began the modern mechanistic era of bacterial chemotaxis studies, studying sensory transduction in more complex organisms has become an enormously popular endeavor. At large national meetings we are sometimes bemused to meet earnest graduate students, hurrying from session to session on mammalian signal transduction, who assume that bacterial signal transduction is irrelevant to their research. As the chapters in this section illustrate, bacterial chemotaxis not only contributed substantially to our current knowledge of sensory transduction mechanisms, but will continue to be at the forefront in advancing our knowledge of intra- and intermolecular aspects of receptor function.

McCleary and Stock (Chapter 2) present the signal transduction pathways in enteric bacteria that involve receptor methylation or demethylation in adaptation to stimuli. In these most studied chemotaxis pathways, the common paradigms for signal transduction are evident. Armitage (Chapter 3) addresses alternative bacterial strategies for sensory transduction that are less well understood mechanistically, but are probably highly important in the survival of bacterial. The treatment of bacterial chemotaxis as a model system for sensory transduction comes full circle in Chapter 4 by Ankenbauer and Nester, and also in Chapter 2. Knowledge of signal transduction mechanisms now is being used to understand survival strategies by bacteria in host–pathogen interactions and in responding to a variety of environmental stimuli. The purpose of this chapter is to give a more general introduction than is found in the specialized chapters in this section.

II. Chemotactic Behavior

The range of stimuli that elicits a behavioral response varies with the bacterial species, but the responses of the enteric bacteria *Escherichia coli* and *Salmonella typhimurium* have become the references to which the responses of other species may be compared. These bacteria are attracted to organic molecules such as many amino acids, sugars, and dipeptides (*E. coli*) and are repelled by hydrophobic amino acids, certain aliphatic alcohols, membrane permeant weak acids, indole, phenol (*S. typhimurium*), and inorganic cations (*E. coli*) such as Co^{2+} and Ni^{2+} (Macnab, 1987). Electron acceptors for the bacterial respiratory chain such as oxygen, fumarate, and nitrate are attractants

(Taylor, 1983a), although oxygen at high concentrations is a repellent (Shioi *et al.*, 1987). Changes in the internal or external pH and the ambient temperature also can elicit attractant or repellent responses (Macnab, 1987).

Bacteria are too small to detect shallow spatial gradients of chemoeffectors accurately by comparing the concentrations of effectors at the fore and aft of the bacterium. Instead, they compare concentrations over time. In an isotropic medium, the swimming of *E. coli* is interrupted at frequent intervals by brief chaotic tumbling that reorients the swimming (Macnab, 1987). The resulting swimming pattern is a three-dimensional random walk (Berg and Brown, 1972). In a gradient, the bacterium continuously samples the concentrations of chemoeffector and compares the present concentration with the concentration in the immediate past. if the present concentration is more favorable than the past concentration, that is, if the concentration of an attractant has increased or that of a repellent has decreased, the bacterium decreases the probability of tumbling so it continues to swim in the favorable direction. This response biases the random walk toward the favorable environment without requiring direct directional navigation by the bacteria.

Tumbling results from a brief reversal of the flagella that propel the bacteria. Chemotaxis is effected by modulating the probability of reversal of the flagellar motors. A simple memory is required to compare concentrations of effectors over time. In addition to sensing stimuli, the bacteria are able to integrate competing stimuli. Bacteria experiencing an increase in the concentration of an attractant (favorable stimulus) and an increase in the concentration of a repellent (unfavorable stimulus) can respond to the algebraic sum of the stimuli (Tsang *et al.*, 1973).

Escherichia coli and *S. typhimurium* cells respond to a change in chemoeffectors in their environment by altering the frequency of tumbling. However, over an interval of time, the bacteria adapt to the new conditions and return to a normal probability of tumbling. This adaptation enables the bacteria to respond optimally to future changes in their environment. Adaptation to attractants such as aspartate and serine involves methylation of the chemoreceptor (Springer *et al.*, 1979). The sensory transduction pathways for such stimuli, known as methylation-dependent pathways, are the most studied chemotaxis pathways and are the subject of Chapter 2 by McCleary and Stock. Adaptation to other stimuli such as oxygen and phosphotransferase substrates is independent of receptor methylation and is reviewed in Chapter 3 by Armitage.

III. Chemotaxis Methodology

Mechanistic studies of chemotaxis were feasible only after quantitative assays of behavior were developed. The capillary assay measures the response of bacteria to a spatial gradient of chemoattractant (Adler, 1973). A capillary is filled with a chemical to be tested and inserted into a pond of bacteria. The chemical diffuses out of the capillary, forming a gradient. Bacteria attracted by the test chemical swim up the gradient and enter the capillary. At the end of the incubation period, the capillary is removed and the number of bacteria that entered the capillary is counted. By adding additional chemicals to the pond, the effect of competition on the chemotactic response can be measured.

The temporal assay for chemotaxis can quantitate responses to both attractants and repellents (Macnab and Koshland, 1972; Brown and Berg, 1974). The chemical to be tested is mixed instantaneously with bacteria and the change in tumbling frequency is observed through a microscope. The time required for 50% of the bacteria to return to the prestimulus tumbling frequency is recorded. Various approaches have been used to mechanize the analysis of tumbling frequency, but these have been superseded largely by computerized motion analysis.

By far the most important methods in elucidating the mechanisms of bacterial chemotaxis are genetic. Because *E. coli* and *S. typhimurium* have the most advanced systems for genetic analysis, bacterial chemotaxis in these bacteria yielded to the genetic approach in a manner that has not been achieved with any other behavioral system (Parkinson, 1981). More recent successes in studying sensory transduction in simple eukaryotic organisms frequently have followed the development of a system of genetic analysis in those organisms.

An early strategy used in studying bacterial chemotaxis was to select a large number of mutants of *E. coli* or *S. typhimurium* that were deficient in chemotaxis and to group the mutants into complementation groups (Armstrong *et al.*, 1967; Armstrong and Adler, 1969; Aswad and Koshland, 1975; Parkinson, 1976). The behavior of the mutants in each complementation group was analyzed to determine the precise nature of the behavioral deficit. When possible, parallel studies were made of accompanying biochemical deficits. Soon assigning the complementation groups to hypothetical segments of the sensory transduction pathway became possible (Figure 1.1). Mutants deficient in a periplasmic binding protein lacked chemotaxis to one attractant or, at most, several structurally related attractants. Those defective in a chemotaxis receptor (transducer) had lost the ability to respond to a subset of diverse chemoeffectors and were deficient in chemoeffector-stimulated methylation of a specific protein. (The receptors initially were known as methyl-accepting chemotaxis proteins.) Other mutants that were motile but totally defective in chemotaxis in spatial gradients had either a constantly tumbling phenotype or a smooth swimming (no tumbling) phenotype. Such mutants were assumed to be defective in signal processing (Fig. 1.1).

Subsequently, all chemotaxis genes were cloned and sequenced. The gene products were identified, isolated, and characterized. This procedure often preceded discovery of the function of the gene product and was a powerful aid

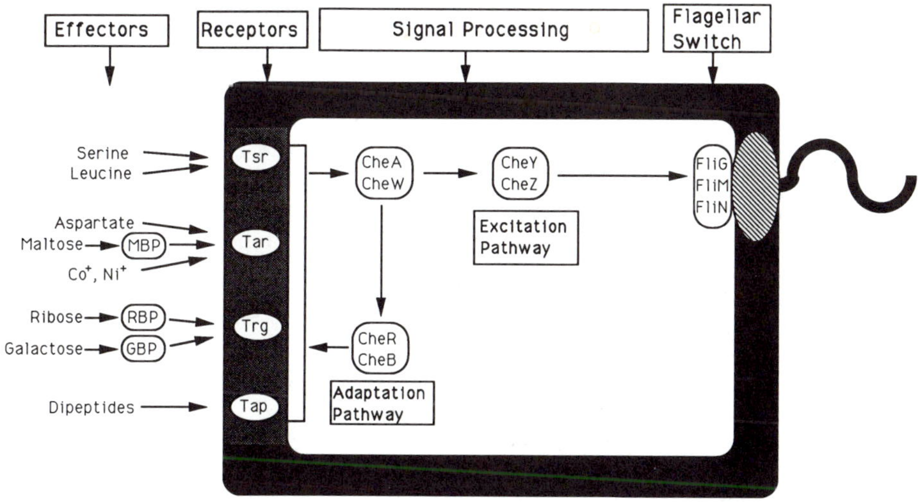

Figure 1.1 Chemotaxis gene products arranged in a scheme for sensory transduction that explains the observed classes of chemotaxis mutations (see text). Abbreviations: MBP, maltose binding protein; RBP, ribose binding protein; GBP, galactose binding protein.

in focusing the investigation of the transduction mechanism on the correct intermediates in the pathway. Since additional mutant hunts failed to identify new genes that were required for signal processing, researchers assumed that most or all of the essential genes had been identified. The copy number of the individual proteins has been determined. Figure 1.2 summarizes current knowledge of the constituents for chemotaxis in *E. coli*.

Chemotaxis investigations benefitted greatly from analysis of allele-specific second-site suppressor mutations. This technique, used extensively by Parkinson and his colleagues (Parkinson, 1981), provided information about components of the transduction pathways that came into physical contact with one another. Site-directed mutagenesis also has been valuable as a probe of protein structure and function.

More recently, as discussed by McCleary and Stock (Chapter 2) and by Ankenbauer and Nester (Chapter 4), molecular genetics has made another powerful contribution. Comparison of chemotaxis gene sequences with the sequences of all genes stored in the major gene database revealed that the *cheA* (Stock *et al.*, 1988) and *cheY* (Stock *et al.*, 1985) genes are members of a class of two-component regulatory systems and that the N-terminal region of the *cheB* (Stock *et al.*, 1985) gene also contained consensus sequences found in such systems (Stock *et al.*, 1989; Parkinson and Kofoid, 1992). This discovery was an important contribution to the elucidation of the phosphorylation cascade in bacterial chemotaxis. In turn, the discovery of the signal transduction

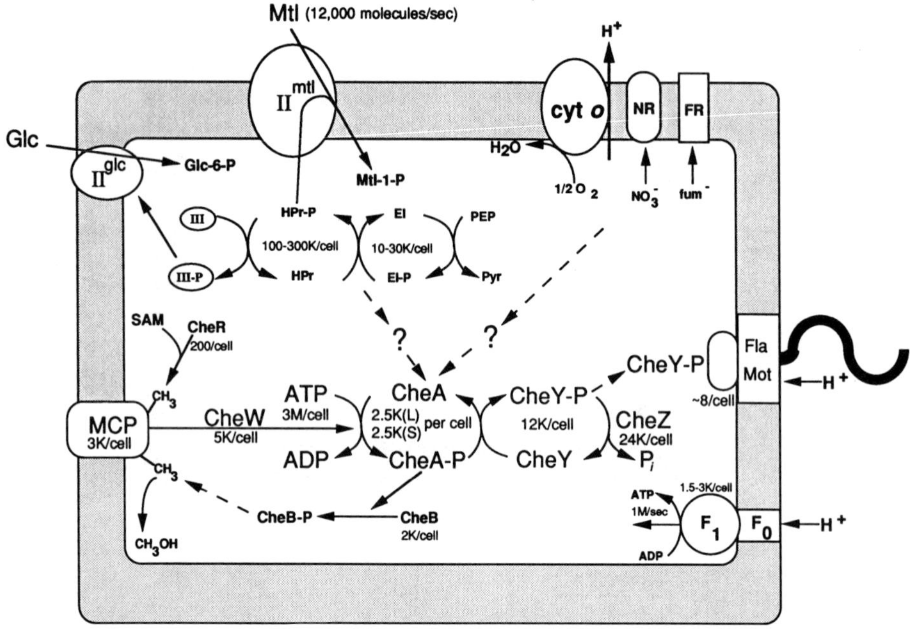

Figure 1.2 General scheme for bacterial chemotaxis in *Escherichia coli* and *Salmonella typhimurium* depicting the central excitation pathway from transmembrane receptor (MCP) to flagellar motor and the enigmatic chemotaxis pathways from the phosphotransferase system and the proton motive force generators. External chemoattractants or repellents bind to the MCP, inducing a conformational change in the MCP cytoplasmic region that affects CheA protein kinase activity through physical interactions with CheW. CheZ is a phosphatase that decreases the lifetime of the tumble generator CheY–P. Signal attenuation or adaptation occurs after changing the steady state level of MCP methylation (controlled by relative CheR and CheB activities). Phosphotransferase system chemoattractants such as glucose or mannitol, and electron acceptors such as oxygen, nitrate, and fumarate (fum) signal to the central chemotactic pathway by unknown mechanisms. The copy numbers per cell of selective proteins (monomers) are shown if available. Values given are from the following references: Enzyme I and HPr, Mattoo and Waygood (1983); MCPs, Hazelbauer and Engstrom (1981); CheW, $CheA_L$, $CheA_S$, Gegner and Dahlquist (1991); CheR, CheB, CheY, and CheZ, Stock *et al.* (1991); ATPase, Maloney (1987); total rate of mannitol transport per cell, Lengeler (1975); total rate of ATP synthesis per cell, Maloney (1987).

pathway in bacteria has aided in demonstrating the role of histidine kinases and phosphorylation in other two-component regulatory systems.

Wolfe and Berg (Wolfe *et al.*, 1987, 1988; Conley *et al.*, 1989) constructed so-called gutted strains in which all four methylation-dependent receptors and the cytoplasmic signal processing proteins (CheAWRBYZ) were deleted from *E. coli*. The missing genes then were restored selectively to the gutted strain, singly or in combination. The constructs were analyzed for recovery of any chemotactic behavior. A surprising discovery arising from these reconstitution studies was that, when CheY was restored to the gutted strain, the cells responded to the repellent acetate in the absence of all other signal processing components (Wolfe *et al.*, 1988). McCleary and Stock (Chapter 2) discuss the resulting investigations of a role of acetyl phosphate in chemotaxis and suggest a model to explain the acetate effect. Chemotaxis reconstitution studies in *E. coli* cell envelopes (Barak *et al.*, 1992) suggest that acetyl adenylate is required for chemotaxis. The discrepancy remains to be explained.

IV. Universal Themes

Some sensory transduction pathways presented in this book are similar although they are found in different organisms. Collectively they contribute to an understanding of the nuances of a particular signal transduction mechanism. Other signaling pathways involve diverse strategies. Our thesis is that discovering underlying principles in the seemingly diverse sensory systems is also likely to yield insights into fundamental principles of signal transduction. Bacterial chemotaxis is expected to be the first complete behavioral system to be understood as a sequence of molecular events. Individuals attuned to universal themes will continue to find that the unfolding chemotaxis mechanism will suggest strategies for investigating signal transduction in eukaryotes.

A. Receptors

In *E. coli*, the methylated receptors (methyl-accepting chemotaxis proteins, MCP) for chemotaxis—Tsr, Tar, Trg, and Tap—are members of a class of transmembrane receptors in which receptors are composed of globular extracellular and cytoplasmic domains connected by one or two membrane-spanning regions (see Chapter 2). The receptors are environmental sensors that transduce a signal and transmit it across the membrane to the cytoplasm where it is processed. Mammalian members of this class include receptors for epidermal growth factor, insulin, erythropoietin, interleukin 2, and nerve growth factor (Jancarik *et al.*, 1991). This class is distinct from that of receptors with seven transmembrane regions, such as rhodopsin and the β-adrenergic receptor.

The mechanism of signal transduction in the chemotaxis receptor has not been established and is currently the subject of intense investigation using genetic, biochemical, and crystallographic techniques. The mechanism presumably involves a conformational change, induced by ligand binding to the extracellular (periplasmic) domain, that is transmitted perpendicular to the plane of the membrane to the cytoplasmic signaling domain. The mechanism of transmembrane signaling is likely to be similar in bacterial and mammalian receptors from this class. A chimera formed from the N-terminal domain of the *E. coli* aspartate chemotaxis receptor (Tar) and the cytosolic tyrosine phosphotransferase domain of the human insulin receptor exhibited tyrosine kinase activity that was stimulated by aspartate (Moe *et al.*, 1989).

Another reason for interest in the chemotaxis receptors is the complexity of the different reactions in which they participate. For example, the Tsr receptor in *E. coli* responds to changes in the level of serine, NH_4^+, leucine, temperature, and pH; transmits the signal across the membrane; initiates excitation signals; and is modulated and desensitized by adaptation signals (Bourret *et al.*, 1991). The cytoplasmic signaling domain has three signaling states: positive, negative, and nonsignaling. Clearly, in terms of function, the chemotaxis receptors are among the most complex proteins known.

B. Signal Processing

When the biochemical mechanism of signal processing in bacterial chemotaxis first was investigated in the early 1970s, behavioral responses in other organisms were associated with membrane depolarization and ion gradients. Second messenger at that time was synonymous with cyclic AMP. Evidence was obtained that indicated that chemotactic signaling was not via membrane potential (Miller and Koshland, 1977), except for responses to oxygen, electron acceptors, and a few other effectors (Taylor, 1983b). This evidence initiated a search for a unique second messenger system (Black *et al.*, 1980; Tribhuwan *et al.*, 1986) and eventually to discovery of the phosphorylation cascade as the transduction pathway for chemotaxis (Hess *et al.*, 1987).

The so-called two-component (Nixon *et al.*, 1986) histidine kinase or phosphorylation cascade paradigm for signal transduction is now known to be ubiquitous in prokaryotes and involved in regulating, in addition to chemotaxis, such diverse systems as nitrogen assimilation, outer membrane protein expression, *Agrobacterium* virulence, sporulation, and repression of oxidative metabolism in anaerobiosis (Stock *et al.*, 1989; Chapter 2; Chapter 4). The common components of these regulatory systems are (1) a detector of environmental stimuli that modulates phosphorylation of a response regulator, (2) an autophosphorylating histidine kinase, (3) a response regulator that is phosphorylated by the histidine kinase, (4) a target that is regulated by the response regulator, and (5) a specific phosphatase that dephosphorylates the response regulator (Bourret *et al.*, 1991). These components appear to be arranged in modules appropriate to each signaling system. In chemotaxis, multiple detectors (receptors) regulate autophosphorylation of a common histidine kinase, the CheA protein, integrating different chemotactic stimuli. Interaction of a detector or kinase with more than one regulator distributes the environmental stimulus to regulate more than one pathway. McCleary and Stock (Chapter 2) and Ankenbauer and Nester (Chapter 4) review the rapidly increasing knowledge about the molecular and intramolecular events in signal transduction in bacterial chemotaxis and *Agrobacter* crown gall tumorigenesis. In the discussion of signal processing in methylation-independent behavioral systems, Armitage (Chapter 3) presents new evidence that a histidine kinase also may be involved in these systems.

In addition to the obvious similarity between chemotaxis and prokaryote histidine kinase-dependent signal transduction pathways, parallels exist between chemotaxis and the eukaryotic sensory transduction pathways involving *ras* and heterotrimeric G proteins that are discussed in the second section of this book. In each system, a component detects environmental stimuli and activates a G protein that modulates an effector, typically a second messenger system. An intrinsic catalytic activity deactivates the G protein. The second messenger system substantially amplifies the environmental signal. Like the prokaryote signal transduction pathways, the eukaryote pathways are modular and are observed with numerous different combinations of modules as well as diverse second messenger systems. Some eukaryote signal transduction pathways are themselves the subject of complex regulation. Similar feedback regulation probably remains to be discovered in bacterial chemotaxis.

C. Adaptation and Signal Attenuation

Behavioral adaptation by organisms to continuous or repetitive stimuli is observed universally. Prior to studies of chemotaxis in enteric bacteria, adaptation was believed to be solely a property of neural networks. Adaptation in bacterial chemotaxis was the first demonstration of behavioral adaptation in a single cell. In *E. coli*, the process of adaptation to a new concentration of chemoeffector effectively resets the signal transduction machinery to a new zero point, allowing the bacteria to respond to a small change in chemoeffector concentration over a 10^5-fold range of background intensity (Adler, 1969). Adaptation in single cells now has been demonstrated in retinal rod cells, in which it is quantitatively more important than adaptation occurring elsewhere in the mammalian visual system (Stryer, 1986). Adaptation (habituation) also occurs in neuronally differentiated PC12 cells, in which decreasing amounts of norepinephrine are secreted in response to repetitive stimulation by depolarizing concentrations of potassium (McFadden and Koshland, 1990).

The mechanism of cellular adaptation that has been elucidated in enteric bacteria involves methylation of glutamyl residues on the cytosolic face of the chemotaxis receptors. This covalent modification modulates the signaling properties of the receptor and has little effect on the affinity of the periplasmic ligand-binding domain (Dunten and Koshland, 1991). In retinal rod cells, a comparable covalent modification of the light receptor occurs: a steady stimulation of the rod cells, at an intensity appreciably higher than background, results in increased phosphorylation of rhodopsin (Shichi and Somers, 1978). However, adaptation by covalent modification of the receptors is proving to be the exception rather than the rule in eukaryotes.

Armitage (Chapter 3) reviews bacterial chemotaxis in which adaptation is independent of receptor methylation. These methylation-independent chemosensory pathways are widespread in bacteria; in some bacteria they are the only chemotaxis pathways present. These alternative pathways may be as important as the methylation-dependent pathways, since they involve responses to oxygen and sugars in *E. coli*, but the mechanism of methylation-independent adaptation is unknown.

Increasing evidence suggests that signaling pathways in eukaryotes that are mediated by *ras* or heterotrimeric G proteins have complex feedback regulation. Adaptation in these systems may be effected via the feedback mechanisms. Methylation-independent adaptation in bacteria may be effected by a yet to be discovered feedback pathway. Such a pathway involving the CheZ phosphatase has long been postulated.

D. Amplification

Amplification is an essential property of signal transduction pathways. In *E. coli*, the binding of an aspartate molecule to a single Tar receptor changes the probability of a smooth swimming response by 10% (Segall *et al.*, 1986). In visual transduction, one photon can activate a single rhodopsin molecule and inhibit entry of more than 6×10^6 sodium ions per second to the rod cell (Stryer, 1986). In these cases and others, amplification involves one or more catalytic steps in the signal transduction pathway. In the enteric bacteria, an activated CheA protein may increase the phosphorylation rate of CheY some 300-fold (Borkovich *et al.*, 1989). Attractant-induced inactivation of CheA kinase could, in theory, be amplified further by controlling CheZ activity (Bourret *et al.*, 1991). In the eukaryote systems reviewed in the second section of this book, catalytic steps in signal transduction pathways include cyclic-GMP phosphodiesterase, phospholipase C, and adenylate cyclase.

V. Conclusion

The chapters in this section summarize major progress in identifying the components and sequences of events in signal transduction pathways in bacteria. Common themes that are emerging in receptor structure and function and in the signal processing mechanism have fundamental similarities to signal transduction in higher organisms. Considerable insight has been gained into the contributions of subtle shifts in protein conformation to the transduction process. Future studies of signal transduction will place even greater emphasis on understanding the relationships between structure and function, particularly in the sensory receptors and the response regulators.

Acknowledgments

Research in the laboratory of B. L. Taylor was supported by Public Health Service Grant GM29481-10 from the National Institutes of Health. The assistance of Lenoa Edwards and Lisa Sarmiento in the preparation of this manuscript is gratefully acknowledged.

References

Adler, J. (1965). Chemotaxis in *Escherichia coli. Cold Spring Harbor Symp. Quant. Biol.* **30**, 289–292.

Adler, J. (1969). Chemoreceptors in bacteria. *Science* **166**, 1588–1596.

Adler, J. (1973). A method for measuring chemotaxis and use of the method to determine optimum conditions for chemotaxis by *Escherichia coli. J. Gen. Microbiol.* **74**, 77–91.

Armstrong, J. B., and Adler, J. (1969). Complementation of nonchemotactic mutants of *Escherichia coli. Genetics* **61**, 61–66.

Armstrong, J. B., Adler, J., and Dahl, M. M. (1967). Nonchemotactic mutants of *Escherichia coli. J. Bacteriol.* **93**, 390–398.

Aswad, D., and Koshland, D. E., Jr. (1975). Isolation, characterization and complementation of *Salmonella typhimurium* chemotaxis mutants. *J. Mol. Biol.* **97**, 225–235.

Barak, R., Welch, M., Yanovsky, A., Oosawa, K., and Eisenbach, M. (1992). Acetyladenylate or its derivative acetylates the chemotaxis protein CheY in vitro and increases its activity at the flagellar switch. *Biochem.* **31**, 10099–10107.

Berg, H. C., and Brown, D. A. (1972). Chemotaxis in *Escherichia coli* analyzed by three-dimensional tracking. *Nature (London)* **239**, 500–504.

Black, R. A., Hobson, A. C., and Adler, J. (1980). Involvement of cyclic GMP in intracellular signaling in the chemotactic response of *Escherichia coli. Proc. Natl. Acad. Sci. U.S.A.* **77**, 3879–3883.

Borkovich, K. A., Kaplan, N., Hess, J. F., and Simon, M. I. (1989). Transmembrane signal transduction in bacterial chemotaxis involves ligand-dependent activation of phosphate group transfer. *Proc. Natl. Acad. Sci. U.S.A.* **86**, 1208–1212.

Bourret, R. B., Borkovich, K. A., and Simon, M. I. (1991). Signal transduction pathways involving protein phosphorylation in prokaryotes. *Ann. Rev. Biochem.* **60**, 401–441.

Brown, D. A., and Berg, H. C. (1974). Temporal stimulation of chemotaxis in *Escherichia coli. Proc. Natl. Acad. Sci. U.S.A.* **71**, 1388–1392.

Conley, M. P., Wolfe, A. J., Blair, D. F., and Berg, H. C. (1989). Both CheA and CheW are required for reconstitution of chemotactic signaling in *Escherichia coli. J. Bacteriol.* **171**, 5190–5193.

Dunten, P., and Koshland, D. E., Jr. (1991). Tuning the responsiveness of a sensory receptor via covalent modification. *J. Biol. Chem.* **266**, 1491–1496.

Gegner, J. A., and Dahlquist, F. W. (1991). Signal transduction in bacteria: CheW forms a reversible complex with the protein kinase CheA. *Proc. Natl. Acad. Sci. U.S.A.* **88**, 750–754.

Hazelbauer, G. L., and Engstrom, P. (1981). Multiple forms of methyl-accepting chemotaxis proteins distinguished by a factor in addition to multiple methylation. *J. Bacteriol.* **145**, 35–42.

Hess, J. F., Oosawa, K., Matsumura, P., and Simon, M. I. (1987). Protein phosphorylation is involved in bacterial chemotaxis. *Proc. Natl. Acad. Sci. U.S.A.* **84**, 7609–7613.

Jancarik, J., Scott, W. G., Milligan, D. L., Koshland, D. E., Jr., and Kim, S.-H. (1991). Crystallization and preliminary x-ray diffraction study of the ligand-binding domain of the bacterial chemotaxis-mediating aspartate receptor of *Salmonella typhimurium*. *J. Mol. Biol.* **221**, 31–34.

Koshland, D. E., Jr. (1974). Chemotaxis as a model for sensory systems. *FEBS Lett.* **40**, S3–S9.

Lengeler, J. (1975). Nature and properties of hexitol transport systems in *Escherichia coli*. *J. Bacteriol.* **124**, 39–47.

McFadden, P. N., and Koshland, D. E., Jr. (1990). Habituation in the single cell: Diminished secretion of norepinephrine with repetitive depolarization of PC12 cells. *Proc. Natl. Acad. Sci. U.S.A.* **87**, 2031–2035.

Macnab, R. M. (1987). Motility and chemotaxis. *In* "*Escherichia coli* and *Salmonella typhimurium*: Cellular and Molecular Biology" (F. C. Neidhart, J. L. Ingraham, K. B. Low, B. Magasanik, M. Schaechter, and H. E. Umbarger, eds.), Vol. 1, pp. 732–759. American Society for Microbiology, Washington, D.C.

Macnab, R. M., and Koshland, D. E., Jr. (1972). The gradient-sensing mechanism in bacterial chemotaxis. *Proc. Natl. Acad. Sci. U.S.A.* **69**, 2509–2512.

Maloney, P. M. (1987). Coupling to an energized membrane: Role of ion-motive gradients in the transduction of metabolic energy. *In* "*Escherichia coli* and *Salmonella typhimurium*: Cellular and Molecular Biology" (F. C. Neidhart, J. L. Ingraham, K. B. Low, B. Magasanik, M. Schaechter, and H. E. Umbarger, eds.), Vol. 1, pp. 222–243. American Society for Microbiology, Washington, D.C.

Mattoo, R. L., and Waygood, E. B. (1983). Determination of the levels of HPr and enzyme I of the phosphoenolpyruvate-sugar phosphotransferase system in *Escherichia coli* and *Salmonella typhimurium*. *Can. J. Biochem. Cell Biol.* **61**, 29–37.

Miller, J. B., and Koshland, D. E., Jr. (1977). Sensory electrophysiology of bacteria: Relationship of the membrane potential to motility and chemotaxis in *Bacillus subtilis*. *Proc. Natl. Acad. Sci. U.S.A.* **74**, 4752–4756.

Moe, G. R., Bollag, G. E., and Koshland, D. E., Jr. (1989). Transmembrane signaling by a chimera of the *Escherichia coli* aspartate receptor and the human insulin receptor. *Proc. Natl. Acad. Sci. U.S.A.* **86**, 5683–5687.

Nixon, B. T., Ronson, C. W., and Ausubel, F. M. (1986). Two component regulatory systems responsive to environmental stimuli share strongly conserved domains with the nitrogen assimilation regulatory genes *ntrB* and *ntrC*. *Proc. Natl. Acad. Sci. U.S.A.* **83**, 7850–7854.

Parkinson, J. S. (1976). *cheA*, *cheB*, and *cheC* genes of *Escherichia coli* and their role in chemotaxis. *J. Bacteriol.* **126**, 758–770.

Parkinson, J. S. (1981). Genetics of bacterial chemotaxis. *In* "Genetics as a Tool in Microbiology: Society for General Microbiology Symposium 31" (S. W. Glover and D. A. Hopwood, eds.), pp. 265–290. Cambridge University Press, Cambridge.

Parkinson, J. S., and Kofoid, E. C. (1992). Communication modules in bacterial signaling proteins. *Ann. Rev. Genet.* **26**, 71–112.

Segall, J. E., Block, S. M., and Berg, H. C. (1986). Temporal comparisons in bacterial chemotaxis. *Proc. Natl. Acad. Sci. U.S.A.* **83**, 8987–8991.

Shichi, H., and Somers, R. L. (1978). Light-dependent phosphorylation of rhodopsin. *J. Biol. Chem.* **253**, 7040–7046.

Shioi, J., Dang, C. V., and Taylor, B. L. (1987). Oxygen as attractant and repellent in bacterial chemotaxis. *J. Bacteriol.* **169**, 3118–3123.

Springer, M. S., Goy, M. F., and Adler, J. (1979). Protein methylation in behavioural control mechanisms and in signal transduction. *Nature (London)* **280**, 279–284.

Stock, A., Koshland, D. E., Jr., and Stock, J. (1985). Homologies between the *Salmonella typhimurium* CheY protein and proteins involved in the regulation of chemotaxis, membrane protein synthesis, and sporulation. *Proc. Natl. Acad. Sci. U.S.A.* **82**, 7989–7993.

Stock, A., Chen, T., Welsh, D., and Stock, J. (1988). CheA protein, a central regulator of bacterial chemotaxis, belongs to a family of proteins that control gene expression in response to changing environmental conditions. *Proc. Natl. Acad. Sci. U.S.A.* **85**, 1403–1407.

Stock, J. B., Ninfa, A. J., and Stock, A. M. (1989). Protein phosphorylation and regulation of adaptive responses in bacteria. *Microbiol. Rev.* **53**, 450–490.

Stock, J. B., Lukat, G. S., and Stock, A. M. (1991). Bacterial chemotaxis and the molecular logic of intracellular signal transduction networks. *Ann. Rev. Biophys. Biophys. Chem.* **20**, 109–136.

Stryer, L. (1986). Cyclic GMP cascade of vision. *Ann. Rev. Neurosci.* **9**, 87–119.

Taylor, B. L. (1983a). How do bacteria find the optimal concentration of oxygen? *Trends Biochem. Sci.* **8**, 438–441.

Taylor, B. L. (1983b). Role of proton motive force in sensory transduction in bacteria. *Ann. Rev. Microbiol.* **37**, 551–573.

Tribhuwan, R. C., Johnson, M. S., and Taylor, B. L. (1986). Evidence against direct involvement of cyclic GMP or cyclic AMP in bacterial chemotactic signaling. *J. Bacteriol.* **168**, 624–630.

Tsang, N., Macnab, R. M., and Koshland, D. E., Jr. (1973). Common mechanism for repellents and attractants in bacterial chemotaxis. *Science* **181**, 60–63.

Wolfe, A. J., Conley, M. P., Kramer, T. J., and Berg, H. C. (1987). Reconstitution of signaling in bacterial chemotaxis. *J. Bacteriol.* **169**, 1878–1885.

Wolfe, A. J., Conley, M. P., and Berg, H. C. (1988). Acetyladenylate plays a role in controlling the direction of flagellar rotation. *Proc. Natl. Acad. Sci. U.S.A.* **85**, 6711–6715.

Chapter 2

Phosphorylation in Bacterial Chemotaxis

WILLIAM R. McCLEARY and JEFFRY B. STOCK

Department of Molecular Biology
Princeton University
Princeton, New Jersey 08544

I. Introduction

One of the best characterized receptor-mediated signal transduction systems controls the swimming behavior of *Escherichia coli* and *Salmonella typhimurium*. These enteric bacteria sense many aspects of their environment, integrate multiple signals, and swim toward favorable and away from unfavorable environments. The literature on bacterial chemotaxis has been reviewed extensively (Macnab, 1987; Stewart and Dahlquist, 1987; Bourret *et al.*, 1991; J. Stock *et al.*, 1991). In this chapter we focus on the structure–function relationships of the chemotaxis proteins and their interactions with each other as well as on findings regarding multiple inputs that control the motility regulator CheY. We attempt to relate this information to observed physiological responses.

II. Physiology of Bacterial Chemotaxis

Escherichia coli and *S. typhimurium* propel themselves by rotating flagella that extend up to 10 cell lengths from their surfaces. The flagella are distributed randomly over the cell envelope, where each is attached to its own rotary motor apparatus. When the motors spin counterclockwise, individual flagellar filaments coalesce to form a rotating bundle that pushes the cell and causes it to swim smoothly. Clockwise rotation disrupts the flagellar bundle and causes the bacterium to tumble.

In the absence of a chemotactic gradient, cells swim in a three-dimensional random walk consisting of 1- to 2-sec periods of smooth swimming punctuated by 0.1- 0.2-sec tumbles (Berg and Brown, 1972). During chemotaxis, cells do not navigate toward an attractant by steering themselves in the right direction but control their net migration by tumbling less frequently when they swim toward attractants or away from repellents. The sensory machinery detects changes in the concentrations of chemoeffectors over time (Macnab and Koshland, 1972). Increases in concentrations of favorable stimuli (or decreases in repellents) lower the probability of a tumble, whereas decreases in favorable stimuli (or increases in repellents) increase the probability of a tumble. By spending greater amounts of time swimming smoothly when attractant concentrations are increasing and by more frequently sampling new directions when these concentrations are not increasing, bacteria show chemotactic behavior.

As do many other organisms, bacteria adapt to the continued presence of a stimulus. The immediate changes in swimming behavior that result when cells first are exposed to chemotactic stimuli diminish over time; cells subsequently return to prestimulus behaviors although they remain in the presence of the chemoeffector. Adaptation is necessary for additional responses in the continued presence of a chemoeffector and allows for responses to a wide range of effector concentrations.

In the *E. coli* chemotaxis system, stimuli are received at the cell surface through interactions with transmembrane receptor proteins (Fig. 2.1; Bourret *et al.*, 1991). Ligand binding information is transduced across the cytoplasmic

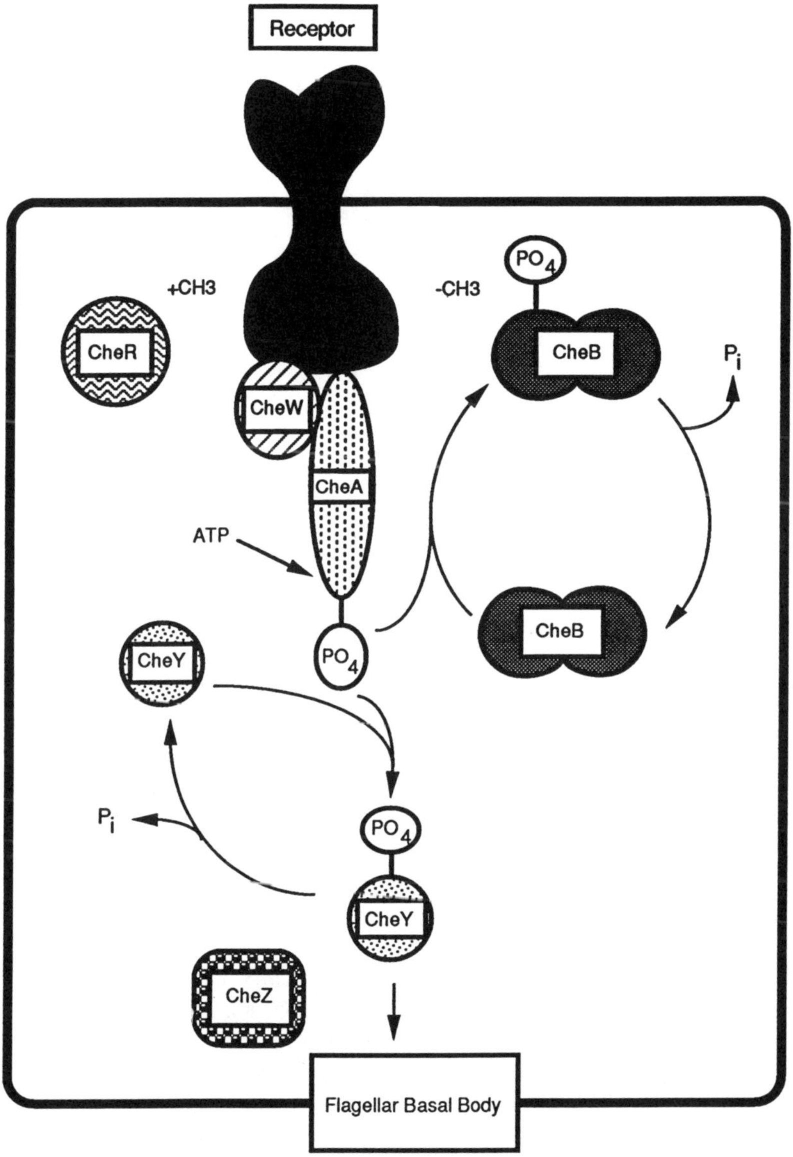

Figure 2.1 Receptor-mediated bacterial chemotaxis system.

membrane to produce conformational changes in the cytoplasmic domains of the receptors. This information is transmitted through a coupling protein, CheW, to the central processing unit of the system, CheA. CheA is an autokinase that phosphorylates itself on a histidine residue. The phosphate group of CheA can be transferred to either of two related proteins, CheY or CheB. Phospho-CheY is thought to interact at the basal structure of the rotary motor to generate a tumble. CheZ interacts with phospho-CheY to attenuate the tumbling response by increasing the rate of phospho-CheY hydrolysis. Adaptation occurs by modifying the sensitivity of the receptors; this modification has been correlated with the reversible methylation of the receptor proteins on specific glutamyl residues. CheR is an *S*-adenosylmethionine-dependent methyltransferase; CheB is a methylesterase.

III. CheY

A. Motility Regulator

Koshland originally postulated the presence of a tumble regulator in bacterial chemotaxis (Koshland, 1977). According to this idea, the swimming behavior of *E. coli* and *S. typhimurium* would be controlled by a threshold detector that is sensitive to levels of the tumble regulator. Such a system would be responsive to the rates of synthesis and decomposition of the regulator and would allow for integration of different signals. Several lines of evidence indicate that the tumble regulator of bacterial chemotaxis is phospho-CheY.

Genetic analysis points to CheY as a signaling protein that interacts with the flagellar motor. Strains that carry mutations in *cheY* swim smoothly all the time and are nonchemotactic. Parkinson and co-workers found that many of the second site suppressors of *cheY* missense mutations map in the flagellar switch genes (Parkinson *et al.*, 1983). Many of these suppressors are allele-specific, suggesting a direct interaction between CheY and the flagellar motor (Yamaguchi *et al.*, 1986). Additional support for the idea that CheY is the tumble regulator is derived from experiments on cells that are missing all chemotaxis signaling proteins. These "gutted" strains exclusively rotate their flagella counterclockwise and swim smoothly all the time. The expression of CheY in these cells is sufficient to allow clockwise flagellar rotation, provided ATP is present (Clegg and Koshland, 1984; Wolfe *et al.*, 1987; Smith *et al.*, 1988). Co-expression of CheZ antagonizes the tumble-producing effect of CheY (Kuo and Koshland, 1987; for further explanation, see Section V). In another series of experiments, flagellated cell envelopes that are supplied with an energy source are observed to rotate exclusively counterclockwise (Ravid *et al.*, 1986). However, when these envelopes are prepared in the presence of CheY, the envelopes can rotate clockwise. These observations are consistent with the hypothesis that CheY is the tumble regulator that controls the direction of flagellar rotation.

B. CheY—Structure and Function

CheY is a 14-kDa monomeric protein that is found in the cytoplasm at a concentration of approximately 8–16 μ*M* (Stock *et al.*, 1985; Kuo and Koshland, 1987). The three-dimensional structure of this protein has been determined by X-ray crystallography (A. Stock *et al.*, 1989; Volz and Matsumura, 1991). CheY is a doubly wound α/β protein that is made up of a five-stranded parallel β-sheet surrounded by five α-helices (Fig. 2.2). A prominent feature of the structure is an acidic pocket that is composed of Asp-12, Asp-13, and Asp-57. Adjacent to this pocket is Lys-109, whose ε-amino group is in bonding contact with a carboxyl oxygen of Asp-57.

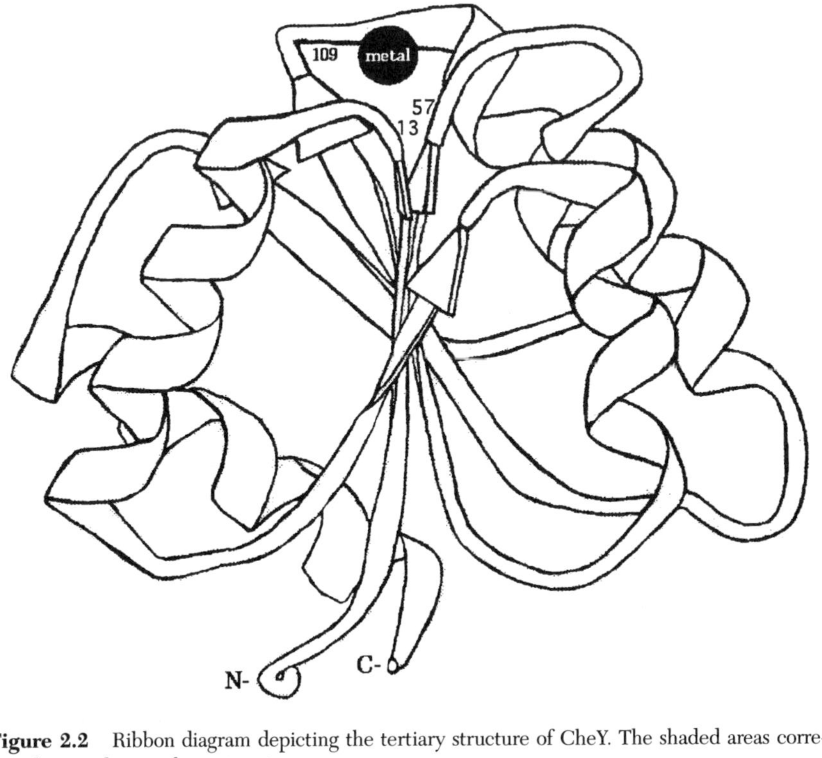

Figure 2.2 Ribbon diagram depicting the tertiary structure of CheY. The shaded areas correspond to residues at the active site.

Genetic and biochemical evidence supports the idea that CheA activates CheY. CheY has been shown to become transiently phosphorylated when it is incubated in the presence of CheA and ATP (Hess *et al.*, 1988b; Wylie *et al.*, 1988). A requirement for ATP in the CheY-dependent generation of tumbling behavior had been documented previously (Smith *et al.*, 1988). Asp-57 has been identified by tandem mass spectrometric analysis as the site of CheY phosphorylation (Sanders *et al.*, 1989a). The phosphoaspartyl group in CheY is hydrolyzed rapidly to produce P_i (Hess *et al.*, 1988b; Stock *et al.*, 1988b). At 25°C and pH 7.9, the first-order rate constant for phospho-CheY hydrolysis is approximately 2 min^{-1}. This rate is much faster than would be expected for an acyl phosphate linkage, and indicates that CheY has inherent autophosphatase activity. This autophosphatase activity has been demonstrated to require Mg^{2+} (Lukat *et al.*, 1990). In the absence of Mg^{2+} or in a denatured state, the kinetics of phospho-CheY hydrolysis are similar to the kinetics of an acyl phosphate such as acetyl phosphate. The inherent instability of the phosphate linkage in phospho-CheY is an important characteristic for a molecular switch that functions in a system that temporally responds to rapidly changing chemoeffector concentrations.

How does phosphorylation activate CheY? An attractive hypothesis for the activation of CheY is that phosphorylation at Asp-57 stabilizes an alternative protein conformation that reveals a previously masked binding site for components of the flagellar switching mechanism. One possibility is that phosphorylation disrupts the linkage between Asp-57 and Lys-109 and shifts the position of the Lys-109 side chain. This local shift would be amplified at a distal site, resulting in the activation of CheY. In support of this possibility, a conservative substitution of an arginine residue for Lys-109 has been observed to result in a protein that is completely unable to cause tumbles; however, the purified mutant protein is phosphorylated efficiently by CheA *in vitro* (Lukat *et al.*, 1991). These results are consistent with the participation of Lys-109 in an event subsequent to phosphorylation (probably the stabilization of an alternative conformation) that is involved in the activation of CheY.

If activation results from the stabilization of an alternative protein conformation, *cheY* mutations that cause tumbly swimming behavior independently of CheA should be obtainable. A *cheY* mutation in which a lysine residue is substituted for Asp-13 (CheYD13K) has this characteristic (Bourret *et al.*, 1990). Despite the fact that this mutant protein is not phosphorylated by CheA *in vitro*, cells carrying this mutation are capable of tumbling behavior.

Adjacent to Asp-57 is a tryptophan residue that has proven very useful in studying the function of the acidic pocket (Lukat *et al.*, 1990,1992). By following changes in the intrinsic fluorescence of this tryptophan, events that occur at the active site—for example, magnesium ion binding and phosphorylation—can be monitored (Lukat *et al.*, 1990,1992). Using this approach, mutations in the acidic pocket have been shown to greatly reduce or eliminate metal binding (Lukat *et al.*, 1990).

CheY is homologous to many proteins that regulate responses to changing environmental conditions (Stock *et al.*, 1985). *Escherichia coli* may encode up to 50 different proteins with domains that are homologous to CheY (J. Stock *et al.*, 1989). The members of this protein family have been termed response regulators. In most cases, these proteins are transcriptional regulators; the exceptions are proteins such as CheY and CheB that control chemotaxis (J. Stock *et al.*, 1989).

Sequence alignments between any two response regulators indicate that they show between 20 and 35% amino acid identity over approximately 120 amino acids (J. Stock *et al.*, 1989). Among all the response regulator sequences that have been determined to date, a conserved pattern of hydrophobic residues similar to the hydrophobic core of CheY is seen (Stock *et al.*, 1990). Other residues that are highly conserved correspond to Asp-12, Asp-13, Asp-57, Gly-102, Ala-103, and Lys-109. The patterns of conserved amino acids within this family strongly support the idea that these domains share a common structure. Most of the CheY homologs are larger than CheY and consist of an N-terminal regulatory domain that is homologous to the entire CheY protein and a C-terminal domain that contains the regulated activity. For example, CheB has an N-terminal regulatory domain that is homologous to the entire CheY protein and a C-terminal catalytic domain that modifies the chemotaxis receptors. In several cases, the C-terminal domains of response regulators, generated either by proteolysis or by genetic manipulation, can function independently of their N-terminal regulatory domains. The methylesterase activity of CheB is enhanced greatly on removal of its regulatory domain (Simms *et al.*, 1985; Lupas *et al.*, 1989). In addition, the transcriptional activator FixJ also functions in the absence of its N-terminal regulatory domain (Kahn and Ditta, 1991).

C. Multiple Phospho-Donors for CheY

In the presence of a stable intracellular concentration of CheY, the levels of phospho-CheY within cells and, thus, the swimming behavior of the cells can be modulated only by controlling the relative rates of phosphorylation and dephosphorylation (Fig. 2.3). To understand the chemotactic response, the physiological phospho-donors of CheY and the effectors of phospho-CheY dephosphorylation must be characterized. Phospho-CheA is the best phosphodonor for CheY; however, it is not the only phospho-donor. Other histidine kinases and certain low molecular weight phospho-donors can phosphorylate CheY *in vitro*. The following sections discuss these three pathways of CheY phosphorylation.

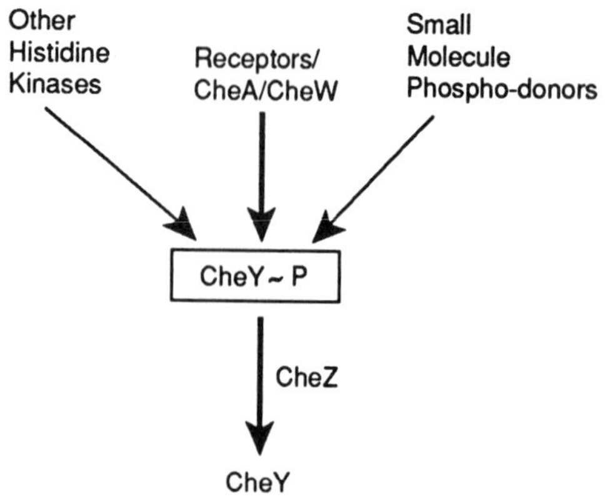

Figure 2.3 Multiple phospho-donors for CheY.

IV. Receptor-Controlled Phosphorylation of CheY

The autophosphorylation activity of CheA is controlled by the transmembrane receptor proteins in conjunction with the coupling protein CheW. The receptors, CheW, and CheA form a ternary complex that can be considered the regulated kinase. To a first approximation, the control of CheA autophosphorylation within this complex regulates chemotaxis. This section discusses findings regarding the structure–function relationships of the individual members of this ternary complex, as well as the evidence for protein–protein interactions between the individual components.

A. Receptor Structure and Function

The most effective stimuli for *E. coli* chemotaxis are certain amino acids and sugars that are sensed by a family of four membrane receptors: Tar (taxis to aspartate and repellents), Tsr (taxis to serine and repellents), Trg (taxis to ribose and galactose), and Tap (taxis to peptides) (Macnab, 1987). Mutations in a particular receptor gene eliminate responses to only the corresponding subclass of chemoeffectors. DNA sequence analysis of the four receptor genes reveals that these proteins share a common set of structural features (Krikos *et al.*, 1983). Like the eukaryotic receptors that bind epidermal growth factor or insulin, the bacterial chemotaxis receptors contain an extracellular domain that binds ligands and a cytoplasmic domain that controls signaling. Ligands are bound by an N-terminal periplasmic domain of approximately 160 amino acids that is flanked on each end by membrane-spanning segments of approximately 24 amino acids (Krikos *et al.*, 1983; Russo and Koshland, 1983). The second transmembrane segment links the ligand binding domain to the ~330 amino acid C-terminal cytoplasmic domain that effects signaling by interacting with CheA and CheW. Cross-linking studies have shown that the chemotaxis receptors, like the aforementioned eukaryotic receptors, function as homodimers (Milligan and Koshland, 1988).

The chemotaxis receptors mediate responses either by binding ligands directly or by interacting with other ligand binding proteins. For example, the Tar protein of *E. coli* binds aspartate and the liganded form of the maltose binding protein. Genetic studies have suggested that the binding sites for these two very different ligands are adjacent and partially overlapping (Gardina *et al.*, 1992). The X-ray crystal structure of the ligand binding domain of the *S. typhimurium* Tar protein has been solved (Milburn *et al.*, 1991). Each subunit within the artificially cross-linked dimer is a four-helix bundle; the dimeric structure is thus a cluster of eight helices. The aspartate binding site is situated between the two subunits of the dimer; nonidentical ligand contacts are contributed by each subunit. A comparison of the apo- and aspartate-bound forms of this domain indicates only a slight structural difference between the two forms. The two membrane-spanning regions have been proposed to extend through the membrane as alpha helices from the N- and C-terminal helices of the four-helix bundles.

Within the cytoplasmic domain of the receptors are three highly conserved regions that show greater than 90% amino acid identity among the family members (Krikos *et al.*, 1983). Two of these sequences correspond to sites of reversible carboxyl methylation and are predicted to form alpha helices. A sequence analysis algorithm identifies two regions within the cytoplasmic domain that are predicted to form coiled-coil structures or related helical bundles (Lupas *et al.*, 1991). These sites are coincident with the methylated regions. The other highly conserved region, which is flanked on each end by the methylated segments, has been identified as the signaling domain because mutations in this region can result in receptors with either a clockwise or a counterclockwise rotational bias (Oosawa *et al.*, 1988b; Parkinson and Kofoid, 1992). The signaling domain is likely to be the site of interactions with CheA and/or CheW.

1. Transmembrane Signaling Considering the general receptor structure just presented, what is the mechanism by which ligand binding information is transduced across the membrane to the cytoplasmic signaling domain? Several different mechanisms have been proposed for transmembrane signaling. One model, which calls for a change in the oligomerization state of the receptor on ligand binding, has been ruled out because disulfide cross-links between subunits of the aspartate chemoreceptor that should eliminate oligomerization changes do not interfere with signaling (Milligan and Koshland, 1988). Alternative models involve positional changes between subunits or within subunits (Milburn *et al.*, 1991; Milligan and Koshland, 1991). We previously proposed an alternating two-state model in which the periplasmic and cytoplasmic domains can exist in either a relaxed (R) or a tense (T) state (J. Stock *et al.*, 1991). According to this idea, a rigid transmembrane coupling of the domains precludes both regions from being in the T state. When the cytoplasmic signaling domain is in the T state, it activates the kinase. Attractant binding induces the sensory domain to attain a T state and, thus, precludes the signaling domain from also being in a T state to activate CheA.

The X-ray crystallographic data tend to support an intersubunit "scissors" mechanism for transmembrane signaling in which ligand binding causes a shift in the position of one subunit relative to another (Milburn *et al.*, 1991). The conformational change on aspartate binding is very slight, only a 4° rotation between subunits. However, such a small change in the periplasmic domain may be amplified a great deal across the membrane in the cytoplasmic domain. Experiments by Milligan and Koshland (1991) tend to support a signal transduction mechanism that involves only one subunit. These investigators constructed receptor dimers that contained one full-length subunit and either another full-length subunit, a subunit lacking the cytoplasmic domain, or a subunit lacking the cytoplasmic domain and the transmembrane domains. Milligan and Koshland found that, in all cases, methylation of the cytoplasmic domain was stimulated by the binding of aspartate. Interpretations of these results are complicated because the authors examined methylation and not signaling per se and because of the possibility of higher order interactions between cross-linked dimers.

2. *Adaptation: CheB and CheR* Changes in chemoeffector concentrations over time cause changes in the levels of receptor glutamyl methylation (Springer *et al.*, 1979). Highly methylated receptors send signals for clockwise flagellar rotation. Conversely, unmethylated receptors send signals for counterclockwise motion. Increasing concentrations of attractants cause an increase in the level of methylation, whereas increasing levels of repellants cause a decrease. These changes in receptor methylation levels have been correlated with behavioral adaptation. Long-term methylation changes are specific to the receptor that sensed the stimulatory ligand. The principal mechanism for this specificity appears to involve the ligand-induced stabilization of receptor conformations that have differential reactivities toward the modifying enzymes (Stock and Koshland, 1981; Springer *et al.*, 1982; Borczuk *et al.*, 1986). How-

ever, a transient global response also occurs that affects all receptors and requires CheA, CheW, CheB, and ATP (Springer and Zanolari, 1984; Russell *et al.*, 1989).

Since receptor methylation has been shown not to affect ligand binding significantly, this event must somehow exert its effects by altering the signaling capacity of the receptors (Dunten and Koshland, 1991). These results carry a caveat, however. The experiments were done in the absence of CheA and CheW and assume that these proteins have a negligible effect on the ligand binding activity of the receptors. Nevertheless, an understanding of how methylation alters receptor structure seems likely to provide clues to the mechanism of signaling.

Lupas *et al.* (1991) have proposed a model for the role of methylation in modulating the conformational changes that occur within the cytoplasmic domain of the receptor during chemotactic signaling and adaptation (see also J. Stock *et al.*, 1991). In this scheme, the oligomeric structure of the cytoplasmic domain is modulated by methylation. These researchers propose that the signaling domains of the receptor dimer exist in a dynamic equilibrium between a coiled-coil state and a four-helix bundle. The equilibrium between the two forms is influenced by ligand binding in the periplasmic domain and by the methylation state of the cytoplasmic domain. The amino-proximal methylation region generally contains three methylation sites in heptad repeat. In the fully unmethylated form of the receptor, the negative charges of the glutamyl residues would lie on one face of the helix and would, most likely, be solvent exposed. On methylation, negative charges are removed and new higher order interactions between helices would be favored. Methylation and repellant binding favor the four-helix bundle whereas demethylation and attractant binding favor the coiled-coil form. In this model, the four-helix bundle form of the receptor activates the kinase activity of CheA.

The enzymes that control the methylation state of the chemotaxis receptors are CheR, the methyltransferase, and CheB, the methylesterase (Simms *et al.*, 1985,1987). In addition to its methylesterase activity, CheB also deamidates several glutamine residues that are encoded in the receptor primary amino acid sequence (Kehry *et al.*, 1983). Deamidation allows these sites to be accessible to further methylation and demethylation. Mutations in *cheR* cause a smooth swimming phenotype whereas those in *cheB* cause a tumbly phenotype.

CheR is a 33-kDa monomer that has an apparent K_m for *S*-adenosylmethionine of approximately 8–17 μM (Simms *et al.*, 1987). CheB is a 37-kDa monomeric protein that is composed of two functional domains: a 15-kDa N-terminal regulatory domain and a 21-kDa catalytic domain (Simms *et al.*, 1985). In the absence of the N-terminal domain, the methylesterase activity of the C-terminal domain is increased at least 10-fold. DNA sequence analysis showed that the N-terminal domain of CheB is homologous to the entire CheY

(Hess *et al.*, 1988b). Phosphorylation of the regulatory domain, like proteolytic deletion, activates CheB (Lupas and Stock, 1989). Thus, in its unphosphorylated form, the regulatory domain of CheB appears to inhibit the catalytic activity of the C-terminal domain. Activation of CheB through phosphorylation provides a feedback loop to regulate the activity of the receptors. When the phosphorylation pathway is stimulated, cells tumble more frequently. The feedback loop through CheB assures a shortened tumble response in the cell.

B. CheA Structure and Function

The *cheA* gene encodes two polypeptides, $CheA_L$ (M_r 73,000) and $CheA_S$ (M_r 69,000). Both polypeptides are translated in the same reading frame but from different start sites (Kofoid and Parkinson, 1991). Equilibrium ultracentrifugation analysis shows that $CheA_L$ has a molecular weight of 146,000, the value predicted for a homodimer (Gegner and Dahlquist, 1991). However, molecular sieve chromatography under nondenaturing conditions indicates that $CheA_L$ has an apparent molecular weight of >250 kDa (Stock *et al.*, 1988a). Collectively, these results suggest that the CheA dimer has an elongated shape.

CheA is a member of a protein family that modulates the activities of response regulators (Stock *et al.*, 1988a; J. Stock *et al.*, 1989). This family is defined by a conserved region of approximately 200–300 amino acids that generally is located near the C terminus of these proteins. This conserved region probably defines sites that are important for catalysis. Members of this protein family are kinases that autophosphorylate a histidine residue in the presence of ATP. Most members of this protein family have a highly conserved histidine residue located near the N terminus of the kinase domain. Many of the histidine kinases are membrane proteins; CheA is an exception because it is cytoplasmic.

CheA catalyzes its own phosphorylation on a histidine residue in the presence of ATP and Mg^{2+} (Hess *et al.*, 1987,1988a; Wylie *et al.*, 1988). The kinetics of autophosphorylation show a first-order dependence on CheA concentration and suggest that this reaction proceeds through an intramolecular mechanism (Hess *et al.*, 1988b). Whether CheA autophosphorylation occurs within an individual monomer or between monomers in the CheA dimer is not known. However, a homologous bacterial histidine kinase, EnvZ, has been shown to autophosphorylate by an intersubunit mechanism in which the kinase domain of one subunit phosphorylates a histidine residue within another subunit (Yang and Inouye, 1992). The site of CheA phosphorylation has been identified as His-48 (Hess *et at.*, 1988a). CheA is unusual among the histidine kinases because the site of phosphorylation is located some 300 residues upstream of the kinase domain. The autokinase reaction exhibits an apparent K_m for Mg^{2+}-ATP of 0.1–0.3 m*M* (Wylie *et al.*, 1988).

CheA is composed of at least three functional regions. The N-terminal region probably interacts with CheY and CheB. This region contains the site of autophosphorylation. In addition, several mutants have been isolated in the N terminus that are defective in phosphotransfer (Oosawa *et al.*, 1988a). The central region of the protein contains the conserved kinase domain. Within this region are five short segments that show significant sequence identity with other histidine kinases (J. Stock *et al.*, 1989; Parkinson and Kofoid, 1992). The C-terminal region probably interacts with CheW and the receptors (Oosawa *et al.*, 1988a; Parkinson and Kofoid, 1992). Mutant proteins with defects in this region have been obtained that show efficient phosphorylation and dephosphorylation *in vitro* but that are, nevertheless, defective in chemotaxis.

C. CheW Structure and Function

The autokinase activity of CheA is regulated by the chemotaxis receptor proteins and CheW (Borkovich *et al.*, 1989; Ninfa *et al.*, 1991). Mutations in the *cheW* gene result in a smooth swimming phenotype, consistent with the idea that the receptors cannot effectively couple their output to the kinase if CheW is absent. The nucleotide sequence of CheW predicts a 167-amino-acid protein with a molecular mass of 18 kDa (Mutoh and Simon, 1986; Stock *et al.*, 1987). Equilibrium analytical ultracentrifugation indicates that CheW is a monomer with a molecular mass of 17.5 kDa (Gegner and Dahlquist, 1991). However, molecular sieve chromatography under nondenaturing conditions shows a molecular mass of approximately 35 kDa (Stock *et al.*, 1987). Combined, these results suggest that CheW is a monomer that has an extended prolate shape.

D. Protein Complexes in Signaling

Experiments have shown that CheW forms a reversible complex with CheA. Using the equilibrium column chromatography method of Hummel and Dreyer, Gegner and Dahlquist (1991) demonstrated an interaction between these two proteins. These researchers proposed that two CheW monomers interact with each CheA dimer. The dissociation constant for the interaction was estimated to be 17 μM. The intracellular concentrations of CheA and CheW are 2.5 μM and 5.0 μM, respectively (Gegner and Dahlquist, 1991). McNally and Matsumura (1991) were able to detect stable $CheA_L$/$CheA_S$/CheW complexes (1:1:1 stoichiometry) from extracts coexpressing CheA and CheW. These investigators found that CheW is retained on an anti-CheA column when both proteins are expressed simultaneously but not when CheW is expressed by itself. They also observed that the autophosphorylation rate of CheA is enhanced when it is in an immune complex with CheW, and that the complex is responsive to the signaling state of the receptors. These researchers saw a 50-fold decrease in the K_m for ATP when CheA is in a complex with CheW, suggesting that the role of CheW might be to increase the affinity of CheA for ATP. At present, why such a difference exists between the stabilities of the complexes reported by Gegner and Dahlquist and those of the complexes reported by McNally and Matsumura is not clear.

CheW not only interacts with CheA, but also interacts with the receptors. Like *cheW* null mutations, overexpression of CheW results in smooth swimming behavior (Liu and Parkinson, 1989; Sanders *et al.*, 1989b). The counterclockwise flagellar bias caused by overexpression of CheW can be compensated by also overexpressing the receptors (Liu and Parkinson, 1989). Stoichiometric compensation between CheW and the receptors suggests that these proteins directly interact with one another. Additional genetic evidence supports a direct interaction between CheW and one of the chemotaxis receptors, Tsr. Liu and Parkinson (1991) demonstrated that certain missense mutations in *tsr* could be suppressed by mutations in *cheW*. These investigators also isolated dominant *cheW* mutations. Sequence analysis of these mutations showed that they were either identical or very similar to the suppressible *cheW* mutations previously isolated. These results indicate that only a limited number of ways to alter CheW structure are possible to produce a dominant phenotype and that dominance is mediated most often through interactions with the receptors.

Additional evidence for the requirement of CheW to couple receptor output to CheA activity is seen in *in vitro* reconstitution experiments (Borkovich *et al.*, 1989; Ninfa *et al.*, 1991). Crude membrane extracts containing high levels of the aspartate receptor (Borkovich *et al.*, 1989) or purified aspartate receptors reconstituted in phospholipid vesicles (Ninfa *et al.*, 1991) require both CheW and CheA to yield significant rates of ATP-dependent CheY phosphorylation. Consistent with the smooth swimming "null" phenotype that is observed when CheW is overexpressed, high concentrations of CheW in the *in vitro* reconstituted assay also inhibit kinase activity (Ninfa *et al.*, 1991). Thus, high concentrations of CheW spoil receptor coupling. These results are consistent with the activation of the kinase resulting from the formation of a ternary complex between the receptors, CheW, and CheA. If, within the ternary signaling complex, CheW interacts with both CheA and the receptors, then high levels of CheW will tend to jam the system, favoring the CheA–CheW and receptor–CheW binary complexes in favor of the receptor–CheA–CheW ternary complex. In both *in vitro* systems, ligand binding inhibits the autophosphorylation of CheA. The simplest explanation for this observation is that aspartate binding somehow inhibits the formation of the active ternary complex.

V. Cross Talk: Phosphorylation of CheY by Other Histidine Kinases

Numerous histidine kinases and response regulators within a single cell are predicted to share common structural elements and to utilize common phosphotransfer mechanisms. In addition to phospho-CheA, then, other histidine kinases could donate phosphate to CheY. In fact, several studies have shown cross-specificity between noncognate kinases and response regulators using purified proteins *in vitro*. Purified nitrogen kinase, phospho-NtrB, has been demonstrated to donate phosphate to CheY (Ninfa *et al.*, 1988). However, the kinetics and efficiency of phosphotransfer are reduced drastically compared with phosphotransfer between CheA and CheY. Nevertheless, NtrB also was found to activate CheY *in vivo*. Overexpression of a constitutively active mutant form of NtrB could suppress the smooth-swimming phenotype of a *cheA* mutant and cause tumbly behavior (Ninfa *et al.*, 1988). Phospho-CheA also has been observed to donate phosphate to NtrC, OmpR, and SpoOA, the regulators of the nitrogen, osmolarity, and sporulation responses, respectively (Ninfa *et al.*, 1988; Igo *et al.*, 1989; Olmedo *et al.*, 1990). As for NtrB and CheY, the phosphotransfer between these proteins is much slower than between cognate pairs. Although these experiments have demonstrated the possibility that histidine kinases other than CheA can donate phosphate to CheY, they are unlikely to contribute significantly to the pool of phospho-CheY under most conditions.

Cross talk has been invoked in many systems to explain low level response regulator activation in the absence of cognate kinases (J. Stock *et al.*, 1989). These responses often depend on the concentration of a given regulator and generally are enhanced by overproduction of the regulator from a multicopy expression vector. Such interactions are probably nonspecific and may contribute to a background "noise" of regulator activity. However, genetic results suggest that, in some cases, cross talk may be of physiological importance (Wanner, 1992). The phosphate (PHO) regulon is controlled by a response regulator, PhoB. PhoR is the normal partner of PhoB but, in the absence of PhoR, PHO regulation still is observed in response to glucose. PHO expression, in a *phoR*$^-$ background, depends at least in part on cross talk that responds to changes in glucose metabolism.

Additionally, kinase mutants in the nitrogen (*ntrB*), osmoregulation (*envZ*), and nitrate regulation (*narL*) systems do not show null phenotypes for their various response pathways (Backman *et al.*, 1983; Egan and Stewart, 1990; Forst *et al.*, 1990). Such gene expression in the absence of a cognate histidine kinase raises the possibility of a network of histidine kinases and response regulators that is responsive to global cellular conditions. Alternatively, such regulation may be the result of another phosphorylation system that does not depend on histidine kinases (see next section).

VI. Histidine Kinase-Independent Phosphorylation of CheY

A. Low Molecular Weight Phospho-Donors

Kinase-independent phosphorylation of CheY has been demonstrated by showing that CheY can be phosphorylated by low molecular weight phospho-donors (Lukat *et al.*, 1992). As stated earlier, the intrinsic fluorescence of CheY because of Trp-58 is sensitive to events at the active-site acidic pocket. The addition of phosphoramidate, acetyl phosphate, and carbamoyl phosphate to CheY in the presence of metal causes quenching of its intrinsic fluorescence. After addition of these small molecules, the fluorescence intensity decreases over 30–60 sec. Fluorescence quenching is dependent on the presence of metal and on the concentration of phospho-donor in the reaction. In addition, the fluorescence of CheY proteins that contains mutations at the active site (CheYD13N and CheYD57N) does not change in the presence of these small molecules. Addition of catalytic amounts of CheZ reverses the fluorescence quenching. These results are consistent with a model in which CheY is phosphorylated by the low molecular weight phospho-donors. CheZ reverses the fluorescence quenching, presumably by increasing the rate of phospho-CheY dephosphorylation.

To confirm that CheY is phosphorylated by low molecular weight phospho-donors, it was reacted with acetyl [^{32}P]phosphate and examined by SDS–PAGE and autoradiography (Lukat *et al.*, 1992). In the presence of Mg^{2+}, CheY is phosphorylated rapidly. However, in the absence of Mg^{2+} or when the active-site Asp-57 is changed to Asn (CheYD57N), CheY is not labeled.

Because CheY is phosphorylated by small molecule phospho-donors and because of the short half-life of phospho-CheY, CheY could be considered a phospho-donor (R~P) phosphatase and phospho-CheY could be considered an enzyme intermediate:

$$\text{CheY} + \text{R}{\sim}\text{P} \leftrightarrow [\text{CheY/R}{\sim}\text{P}] \rightarrow \text{CheY}{\sim}\text{P} \rightarrow \text{CheY} + \text{P}_i$$

The phospho-donor phosphatase activity of CheY is measured by directly following the hydrolysis of acetyl phosphate or phosphoramidate. From the steady-state kinetics of these reactions under varying phospho-donor concentrations, determining the K_m of CheY for each compound is possible. The K_m values for acetyl phosphate, carbamoyl phosphate, and phosphoramidate are

0.7 m*M*, 1.2 m*M*, and 1.8 m*M*, respectively. Intracellular concentrations of acetyl phosphate in excess of 1 m*M* have been reported (Hunt, 1986).

To test whether the phosphorylation of CheY by low molecular weight phospho-donors is a general mechanism applicable to other response regulators, the activity of CheB was assayed in the presence of these compounds. Previously, the enzymatic activity of CheB was demonstrated to be greatly stimulated by CheA-dependent phosphorylation. CheB is activated by and can hydrolyze phosphoramidate, but is inactive toward acetyl phosphate and carbamoyl phosphate (Lukat *et al.*, 1992). In addition, when analyzed by SDS–PAGE and autoradiography, CheB is not phosphorylated by acetyl [^{32}P]phosphate.

These results show that response regulators catalyze their own phosphorylation in the absence of any auxiliary proteins. Phosphorylation by phosphoramidate probably reflects the general ability of response regulators to accept phosphoryl groups from phosphohistidines in their associated kinases. Thus, one of the roles of histidine kinases appears to be to provide an acceptable substrate for response regulator autophosphorylation. This characteristic explains how a proteolytic fragment of phospho-CheA that contains the site of phosphorylation (His-48) but that lacks the conserved histidine kinase domain can donate phosphate to CheY (Hess *et al.*, 1988a).

The specificity of response regulators for various phospho-donors suggests that these compounds may have a biological role in response regulation. For example, CheY is phosphorylated by acetyl phosphate and CheB is not. If both proteins were phosphorylated by acetyl phosphate, then any flagellar response caused by phospho-CheY would be negated by the negative feedback from the demethylated receptors created by phospho-CheB. In other systems, certain small molecules conceivably would be much better phospho-donors for a given response regulator. For example, carbamoyl phosphate is a better substrate for NtrC than is acetyl phosphate (Feng *et al.*, 1992).

B. Acetate Effect

Several earlier reports in the literature can now be understood in light of the discovery that acetyl phosphate can donate a phosphate group to CheY. As has been mentioned previously, *E. coli* strains that have been gutted for all chemotaxis genes are exclusively smooth swimming (i.e., they only show counterclockwise flagellar rotation; Wolfe *et al.*, 1987). The introduction into these "gutted" strains of a plasmid that expresses CheY is necessary for clockwise rotation. Wolfe and colleagues observed that, following a short lag, clockwise flagellar rotation is restored to "gutted" cells expressing CheY on adding 10 m*M* acetate (Wolfe *et al.*, 1987). In a subsequent study, these same investigators found that cells show a very limited response to acetate when they are starved for ATP by incubation in 25 m*M* arsenate (Wolfe *et al.*, 1988). These observations are consistent with the clockwise flagellar rotation in the "gutted" cells being caused by a metabolite that is closely related to acetate. Two pathways exist for the uptake of acetate; both require ATP (Nunn, 1987). The primary uptake pathway utilizes acetyl CoA synthetase to produce acetyl CoA from acetate and ATP; acetyl adenylate is a metabolic intermediate in this pathway. The other pathway uses acetate kinase to produce acetyl phosphate from ATP and acetate. The equilibrium constant for this latter reaction indicates that this pathway is used primarily to excrete acetate and to generate ATP. The absence of acetyl CoA synthetase, either through catabolite repression or by mutations, greatly reduces and delays the effects of acetate on the flagellar rotation of tethered cells (Wolfe *et al.*, 1988). These strains would be expected to be deficient in conversion of acetate to acetyl phosphate via acetyl CoA and phosphotransacetylase. Thus, the tumbly behavior observed by Wolfe and colleagues is likely to have been caused by the generation of acetyl phosphate.

C. Small Molecule Phospho-Donors in Signal Transduction

Some evidence suggests that small molecule phospho-donors are involved in the regulation of adaptive responses in bacteria. The PHO regulon is controlled in response to changes in the P_i concentration by the histidine kinase PhoR and the transcriptional regulator PhoB (Wanner and Wilmes-Riesenberg, 1992). Alkaline phosphatase (PhoA) is a member of the PHO regulon; its expression is used often as a marker for PHO regulon gene expression. Phosphate-independent control of the PHO regulon is observed in *phoR*$^-$ strains and is mediated by the histidine kinase, CreC, and PhoB. What CreC senses is not known. Additional P_i-independent control of the PHO regulon has been observed that requires neither PhoR nor CreC. Genetic evidence indicates that this type of control relies on the synthesis of acetyl phosphate. Expression of *phoA* is abolished in *phoR*$^-$, *creC*$^-$ double mutants that lack both acetate kinase (*ack*) and phosphotransacetylase (*pta*), but is enhanced in cells missing only acetate kinase. The *ack*$^-$, *pta*$^-$ double mutant would be expected to produce very little acetyl phosphate whereas the *ack*$^-$ strain would be expected to accumulate acetyl phosphate. Additional evidence for a possible role of acetyl phosphate in the regulation of the PHO operon was obtained when the overproduction of acetate kinase was observed to restore expression of PHO regulon genes in a *phoR*$^-$, *creC*$^-$ double mutant (Lee *et al.*, 1990). Phosphorylation of PhoB by increased levels of acetyl phosphate could explain this observation. Figure 2.4 shows the results of an experiment in which PhoB was incubated with acetyl [^{32}P]phosphate and then subjected to SDS–PAGE.

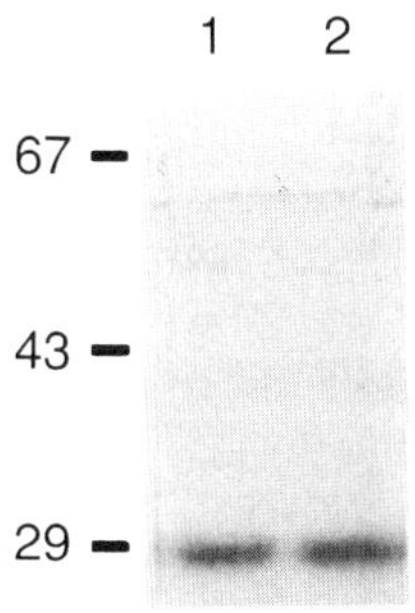

Figure 2.4 Phosphorylation of PhoB by acetyl [^{32}P]phosphate. Two aliquots of PhoB (1.5 μg) were incubated in a buffer containing 50 m*M* Tris–HCl (pH 7.0), 10 m*M* $MgCl_2$, 1 m*M* DTT, 10 m*M* acetyl [^{32}P]phosphate for 10 min at 37°C. The reactions were stopped by the addition of loading buffer and were separated by SDS–PAGE.

This experiment demonstrates that PhoB is phosphorylated in the presence of acetyl phosphate.

In the nitrogen regulation system, transcription of the glutamine synthetase gene *glnA* is regulated by the histidine kinase NtrB and the transcriptional regulator NtrC. *In vitro*, transcription of *glnA* has been shown to require phospho-NtrC (Weiss and Magasanik, 1988). However, NtrB-independent expression of *glnA* has been observed *in vivo* (Backman *et al*., 1983). This NtrB-independent activation of *glnA* has been shown to require absolutely the enzymes involved in the metabolism of acetyl phosphate: acetate kinase and phosphotransacetylase (Feng *et al*., 1992). In addition, acetyl phosphate, carbamoyl phosphate, and phosphoramidate can donate phosphate to NtrC *in vitro*; the phospho-NtrC produced by this mechanism exhibits all the activities associated with phospho-NtrC produced by NtrB.

Small-molecule-mediated phosphorylation of response regulators is an attractive hypothesis in adaptive systems that do not have an identified histidine kinase. One such system is involved in the expression of the enzymes responsible for alginate biosynthesis in *Pseudomonas aeruginosa*. The production of alginate is a major virulence factor in *P. aeruginosa* strains isolated from the lungs of cystic fibrosis patients. Two transcriptional regulators (AlgR and AlgB) have been identified that are necessary for alginate production. However, no histidine kinase has been identified (Mohr *et al*., 1991; Wozniak and Ohman, 1991). The regulation of the alginate response of *P. aeruginosa* could be due to increased intracellular levels of low molecular weight phospho-donors such as acetyl phosphate, because of growth in the specialized lung environment.

A direct link between the metabolism of acetyl phosphate and the swimming behavior of wild-type *E. coli* has not yet been demonstrated. However, the phospho-CheY generated from acetyl phosphate is likely to contribute to the cellular pool of phospho-CheY under certain growth conditions. Intracellular amounts of acetyl phosphate vary greatly, depending on the growth medium and the growth stage (W. McCleary, personal observation). Multiple independently regulated pathways leading to the production of the tumble regulator require that multiple control points exist to regulate swimming behavior.

VII. Dephosphorylation of CheY: CheZ

As stated earlier, control of the chemotaxis response can be exerted by modulating either the rate of CheY phosphorylation or the rate of phospho-CheY dephosphorylation. In addition to dephosphorylation by its inherent autophosphatase activity, phospho-CheY also can be dephosphorylated by CheZ (Kuo and Koshland, 1987; Hess *et al.*, 1988b). Cells that lack CheZ are very tumbly, but still can respond to an attractant by smooth swimming (Segall *et al.*, 1982). The response in *cheZ* mutants is much slower than the wild-type response, presumably because of the time required for the relatively high prestimulus levels of phospho-CheY to be hydrolyzed by the autophosphatase activity of CheY.

The nucleotide sequence of the *cheZ* gene predicts an extremely acidic, hydrophilic protein with a molecular weight of 23,900 (Stock and Stock, 1987). However, the native size appears to be much larger and supports higher ordered structures ranging from tetramers (115 kDa) to >20-mers (500 kDa). Genetic analysis suggests that CheZ may interact directly with the flagellar switch; however, such an interaction has yet to be observed biochemically (Sockett *et al.*, 1992).

Some as yet unidentified mechanism seems likely to modulate CheZ activity. Bacteria can change their flagellar rotational bias markedly on very slight changes in the ligand binding status of receptors (Segall *et al.*, 1986). That the smooth swimming signal (stoichiometric inhibition of CheA kinase activity) from such a small number of liganded receptors could account for the observed behavioral response does not seem plausible. An attractive hypothesis is that CheZ activity may be modulated by the receptors. Preliminary results suggest that $CheA_S$ forms a complex with CheZ (Matsumura *et al.*, 1990). This characteristic raises the possibility that a $CheA_S$–CheZ–receptor interaction could modulate the activity of CheZ.

In other systems, the existence of proteins that enhance the dephosphorylation of their respective response regulators also has been demonstrated. In the nitrogen and the porin systems, dephosphorylation is stimulated in the presence of ATP by their respective histidine kinases, NtrB and EnvZ (J. Stock *et al.*, 1989). Dephosphorylation by NtrB also requires a regulatory protein, P_{II}. In light of the multiple pathways that can lead to response regulator phosphorylation, the regulation of dephosphorylation seems likely to be the primary locus of control in adaptive responses.

VIII. Conclusions

The control of bacterial motility relies on the phosphorylation status of a molecular switch, CheY. Alternative CheY conformations probably correspond to an "on" state and an "off" state. Phosphorylation activates CheY by stabilizing its "on" state. Other bacterial response regulators probably work in a similar fashion. The phosphorylation state of CheY is controlled by its rates of phosphorylation and dephosphorylation. The primary pathway for CheY phosphorylation is mediated by the signaling state of transmembrane receptor proteins acting through the cognate histidine kinase, CheA, of CheW and CheY. However, alternative pathways exist that may, under certain physiological conditions, contribute significantly to the phosphorylation of CheY.

References

Backman, K. C., Chen, Y.-M., Ueno-Nishio, S., and Magasanik, B. (1983). The product of *glnL* is not essential for regulation of bacterial nitrogen assimilation. *J. Bacteriol.* **154**, 516–519.

Berg, H. C., and Brown, D. A. (1972). Chemotaxis in *Escherichia coli* analyzed by three-dimensional tracking. *Nature (London)* **239**, 500–504.

Borczuk, A., Staub, A., and Stock, J. (1986). Demethylation of bacterial chemoreceptors is inhibited by attractant stimuli in the complete absence of the regulatory domain of the demethylating enzyme. *Biochem. Biophys. Res. Commun.* **141**, 918–923.

Borkovich, K. A., Kaplan, N., Hess, J. F., and Simon, M. I. (1989). Transmembrane signal transduction in bacterial chemotaxis involves ligand-dependent activation of phosphate group transfer. *Proc. Natl. Acad. Sci. U.S.A.* **86**, 1208–1212.

Bourret, R. B., Hess, J. F., and Simon, M. J. (1990). Conserved aspartate residues and phosphorylation in signal transduction by the chemotaxis protein CheY. *Proc. Natl. Acad. Sci. U.S.A.* **87**, 41–45.

Bourret, R. B., Borkovich, K. A., and Simon, M. I. (1991). Signal transduction pathways involving protein phosphorylation in prokaryotes. *Ann. Rev. Biochem.* **60**, 401–441.

Clegg, D. O., and Koshland, D. E., Jr. (1984). The role of a signaling protein in bacterial sensing: Behavioral effects of increased gene expression. *Proc. Natl. Acad. Sci. U.S.A.* **81**, 5056–5060.

Dunten, P., and Koshland, D. E., Jr. (1991). Tuning the responsiveness of a sensory receptor via covalent modification. *J. Biol. Chem.* **266**, 1491–1496.

Egan, S. M., and Stewart, V. (1990). Nitrate regulation of anaerobic respiratory gene expression in *narX* deletion mutants of *Escherichia coli* K-12. *J. Bacteriol.* **172**, 5020–5029.

Feng, J., Atkinson, M. R., McCleary, W. R., Stock, J. B., Wanner, B. L., and Ninfa, A. J. (1992). Role of phosphorylated metabolic intermediates in the regulation of glutamine synthetase synthesis in *Escherichia coli*. *J. Bacteriol.*

Forst, S., Delgado, J., Rampersaud, A., and Inouye, M. (1990). *In vivo* phosphorylation of OmpR, the transcription activator of the *ompF* and *ompC* genes in *Escherichia coli*. *J. Bacteriol.* **172**, 3473–3477.

Gardina, P., Conway, C., Kossman, M., and Manson, M. (1992). Aspartate and maltose-binding protein interact with adjacent sites in the Tar chemotactic signal transducer of *Escherichia coli*. *J. Bacteriol.* **174**, 1528–1536.

Gegner, J. A., and Dahlquist, F. W. (1991). Signal transduction in bacteria: CheW forms a reversible complex with the protein kinase CheA. *Proc. Natl. Acad. Sci. U.S.A.* **88**, 750–754.

Hess, J. F., Oosawa, K., Matsumura, P., and Simon, M. I. (1987). Protein phosphorylation is involved in bacterial chemotaxis. *Proc. Natl. Acad. Sci. U.S.A.* **84**, 7609–7613.

Hess, J. F., Bourret, R. B., and Simon, M. I. (1988a). Histidine phosphorylation and phosphoryl group transfer in bacterial chemotaxis. *Nature (London)* **336**, 139–143.

Hess, J. F., Oosawa, K., Kaplan, N., and Simon, M. I. (1988b). Phosphorylation of three proteins in the signaling pathway of bacterial chemotaxis. *Cell* **53**, 79–87.

Hunt, A. G. (1986). Micromethod for the measurement of acetyl phosphate and acetyl coenzyme A. *Meth. Enzymol.* **122**, 43–50.

Igo, M. M., Ninfa, A. J., Stock, J. B., and Silhavy, T. J. (1989). Phosphorylation and dephosphorylation of a bacterial transcriptional activator by a transmembrane receptor. *Genes Dev.* **3**, 1725–1734.

Kahn, D., and Ditta, G. (1991). Modular structure of FixJ: Homology of the transcriptional activator domain with the −35 binding domain of sigma factors. *Mol. Microbiol.* **5**, 987–997.

Kehry, M. R., Bond, M. W., Hunkapiller, M. W., and Dahlquist, F. W. (1983). Enzymatic deamidation of methyl-accepting chemotaxis proteins in *Escherichia coli* catalyzed by the *cheB* gene product. *Proc. Natl. Acad. Sci. U.S.A.* **80**, 3599–3603.

Kofoid, E. C., and Parkinson, J. S. (1991). Tandem translation starts in the *cheA* locus of *Escherichia coli*. *J. Bacteriol.* **173**, 2116–2119.

Koshland, D. E. (1977). A response regulator model in a simple sensory system. *Science* **196**, 1055–1063.

Krikos, A., Mutoh, N., Boyd, A., and Simon, M. I. (1983). Sensory transducers of *E. coli* are composed of discrete structural and functional domains. *Cell* **33**, 615–522.

Kuo, S. C., and Koshland, D. E., Jr. (1987). Roles of *cheY* and *cheZ* gene products in controlling flagellar rotation in bacterial chemotaxis of *Escherichia coli*. *J. Bacteriol.* **169**, 1307–1314.

Lee, T.-Y., Makino, K., Shinagawa, H., and Nakata, A. (1990). Overproduction of acetate kinase activates the phosphate regulon in the absence of the *phoR* and *phoM* functions in *Escherichia coli*. *J. Bacteriol.* **172**, 2245–2249.

Liu, J., and Parkinson, J. S. (1989). Role of CheW protein in coupling membrane receptors to the intracellular signaling system of bacterial chemotaxis. *Proc. Natl. Acad. Sci. U.S.A.* **86**, 8703–8707.

Liu, J., and Parkinson, J. S. (1991). Genetic evidence for interaction between the CheW and Tsr proteins during chemoreceptor signaling by *Escherichia coli*. *J. Bacteriol.* **173**, 4941–4951.

Lukat, G. S., Stock, A. M., and Stock, J. B. (1990). Divalent metal ion binding to the CheY protein and its significance to phosphotransfer in bacterial chemotaxis. *Biochem.* **29**, 5436–5442.

Lukat, G. S., Lee, B. H., Mottonen, J. M., Stock, A. M., and Stock, J. B. (1991). Roles of the highly conserved aspartate and lysine residues in the response regulator of bacterial chemotaxis. *J. Biol. Chem.* **226**, 8348–8354.

Lukat, G. S., McCleary, W. R., Stock, A. M., and Stock, J. B. (1992). Phosphorylation of bacterial response regulator proteins by low molecular weight phospho-donors. *Proc. Natl. Acad. Sci. U.S.A.* **89**, 718–722.

Lupas, A., and Stock, J. (1989). Phosphorylation of an N-terminal regulatory domain activates the CheB methylesterase in bacterial chemotaxis. *J. Biol. Chem.* **264**, 17337–17342.

Lupas, A., VanDyke, M., and Stock, J. (1991). Predicting coiled coils from protein sequences. *Science* **252**, 1162–1164.

Macnab, R. M. (1987). Motility and chemotaxis. *In* "*Escherichia coli* and *Salmonella typhimurium*: Cellular and Molecular Biology" (F. C. Neidhardt, J. L. Ingraham, K. B. Low, B. Magasanik, M. U. Schaechter, and H. E. Umbarger, eds.), Vol. 1, pp. 732–759. American Society for Microbiology, Washington, D.C.

Macnab, R. M., and Koshland, D. E., Jr. (1972). The gradient-sensing mechanism in bacterial chemotaxis. *Proc. Natl. Acad. Sci. U.S.A.* **69**, 2509–2512.

McNally, D. F., and Matsumura, P. (1991). Bacterial chemotaxis signaling complexes: Formation of a CheA/CheW complex enhances autophosphorylation and affinity for CheY. *Proc. Natl. Acad. Sci. U.S.A.* **88**, 6269–6273.

Matsumura, P., Roman, S., Volz, K., and McNally, D. (1990). Signalling complexes in bacterial chemotaxis. *In* "Biology of the Chemotactic Response" (J. P. Armitage and J. M. Lackie, eds.), pp. 135–154. Cambridge University Press, Cambridge.

Milburn, M. V., Prive, G. G., Milligan, D. L., Scott, W. G., Yeh, J., Jancarik, J., Koshland, D. E. J., and Kim, S.-H. (1991). Three-dimensional structures of the ligand-binding domain of the bacterial aspartate receptor with and without a ligand. *Science* **254**, 1342–1347.

Milligan, D. L., and Koshland, D. E., Jr. (1988). Site-directed cross-linking. Establishment of the dimeric structure of the aspartate receptor of bacterial chemotaxis. *J. Biol. Chem.* **263**, 6268–6275.

Milligan, D. L., and Koshland, D. E., Jr. (1991). Intrasubunit signal transduction by the aspartate chemoreceptor. *Science* **254**, 1651–1654.

Mohr, C. D., Hibler, N. S., and Deretic, V. (1991). AlgR, a response regulator controlling mucoidy in *Pseudomonas aeruginosa*, binds to the FUS sites of the *algD* promoter located unusually far upstream from the mRNA start site. *J. Bacteriol.* **173**, 5136–5143.

Mutoh, N., and Simon, M. I. (1986). Nucleotide sequence corresponding to five chemotaxis genes in *Escherichia coli*. *J. Bacteriol.* **165**, 161–166.

Ninfa, A. J., Ninfa, E. G., Lupas, A. N., Stock, A., Magasanik, B., and Stock, J. (1988). Crosstalk between bacterial chemotaxis signal transduction proteins and the regulators of transcription of the Ntr regulon: Evidence that nitrogen assimilation and chemotaxis are controlled by a common phosphotransfer mechanism. *Proc. Natl. Acad. Sci. U.S.A.* **85**, 5492–5496.

Ninfa, E. G., Stock, A. Mowbray, S., and Stock, J. (1991). Reconstitution of the bacterial chemotaxis signal transduction system from purified components. *J. Biol. Chem.* **266**, 9764–9770.

Nunn, W. D. (1987). Two-carbon compounds and fatty acids as carbon sources. *In* "*Escherichia coli* and *Salmonella typhimurium*: cellular and Molecular Biology" (F. C. Neidhardt, J. L. Ingraham, K. B. Low, B. Magasanik, and M. U. Schaechter, eds.), pp. 732–759. American Society for Microbiology, Washington, D.C.

Olmedo, G., Ninfa, E. G., Stock, J., and Youngman, P. (1990). Novel mutations that alter the regulation of sporulation in *Bacillus subtilis*. *J. Mol. Biol.* **215**, 359–372.

Oosawa, K., Hess, F., and Simon, M. I. (1988a). Mutants defective in bacterial chemotaxis show modified protein phosphorylation. *Cell* **53**, 89–96.

Oosawa, K., Mutoh, N., and Simon, M. I. (1988b). Cloning of the C-terminal cytoplasmic fragment of the Tar protein and effects on chemotaxis of *Escherichia coli*. *J. Bacteriol.* **170**, 2521–2526.

Parkinson, J. S., and Kofoid, E. C. (1992). Communication modules in bacterial signaling proteins. *Ann. Rev. Gen.* **26**.

Parkinson, J. S., Parker, S. R., Talbert, P. B., and Houts, S. E. (1983). Interactions between chemotaxis genes and flagellar genes in *Escherichia coli*. *J. Bacteriol.* **155**, 265–274.

Ravid, S., Matsumura, P., and Eisenbach, M. (1986). Restoration of flagellar clockwise rotation in bacterial envelopes by insertion of the chemotaxis protein CheY. *Proc. Natl. Acad. Sci. U.S.A.* **83**, 7157–7161.

Russell, C. B., Stewart, R. C., and Dahlquist, F. W. (1989). Control of transducer methylation levels in *Escherichia coli*: Investigation of components essential for modulation of methylation and demethylation reactions. *J. Bacteriol.* **171**, 3609–3618.

Russo, A. F., and Koshland, D. E. (1983). Separation of signal transduction and adaptation functions of the aspartate receptor in bacterial sensing. *Science* **220**, 1016–1020.

Sanders, D. A., Gillece-Castro, B. L., Stock, A. M., Burlingame, A. L., and Koshland, D. E., Jr. (1989a). Identification of the site of phosphorylation of the chemotaxis response regulator protein, CheY. *J. Biol. Chem.* **264**, 21770–21778.

Sanders, D. A., Mendez, B., and Koshland, D. E. (1989b). Role of the CheW protein in bacterial chemotaxis: Overexpression is equivalent to absence. *J. Bacteriol.* **171**, 6271–6278.

Segall, J. E., Mansom, M. D., and Berg, H. C. (1982). Signal processing times in bacterial chemotaxis. *Nature (London)* **296**, 855–857.

Segall, J. E., Block, S. M., and Berg, H. C. (1986). Temporal comparisons in bacterial chemotaxis. *Proc. Natl. Acad. Sci. U.S.A.* **83**, 8987–8991.

Simms, S. A., Keane, M. G., and Stock, J. (1985). Multiple forms of the CheB methylesterase in bacterial chemosensing. *J. Biol. Chem.* **260**, 10161–10168.

Simms, S. A., Stock, A. M., and Stock, J. B. (1987). Purification and characterization of the S-adenosyl methionine: Glutamyl methyltransferase that modifies membrane chemoreceptor proteins in bacteria. *J. Biol. Chem.* **262**, 8537–8543.

Smith, J. M., Rowsell, E. H., Shioi, J., and Taylor, B. L. (1988). Identification of a site of ATP requirement for signal processing in bacterial chemotaxis. *J. Bacteriol.* **170**, 2698–2704.

Sockett, J., Yamaguchi, S., Kihara, M., Irikura, V. M., and Macnab, R. M. (1992). Molecular analysis of the flagellar switch protein FliM of *Salmonella typhimurium*. *J. Bacteriol.* **174**, 793–806.

Springer, M. S., and Zanolari, B. (1984). Sensory transduction in *Escherichia coli*: Regulation of the demethylation rate by the CheA protein. *Proc. Natl. Acad. Sci. U.S.A.* **81**, 5061–5065.

Springer, M. S., Goy, M. F., and Adler, J. (1979). Protein methylation in behavioral control mechanisms and in signal transduction. *Nature (London)* **280**, 279–284.

Springer, M. S., Zanolari, B., and Pierzchala, P. A. (1982). Ordered methylation of the methyl accepting chemotaxis proteins of *Escherichia coli*. *J. Biol. Chem.* **257**, 6861–6866.

Stewart, R. C., and Dahlquist, F. W. (1987). Molecular components of bacterial chemotaxis. *Chem. Rev.* **87**, 997–1025.

Stock, A. M., and Stock, J. B. (1987). Purification and characterization of the CheZ protein of bacterial chemotaxis. *J. Bacteriol.* **169**, 3301–3311.

Stock, A., Koshland, D. E., Jr., and Stock, J. (1985). Homologies between the *Salmonella typhimurium* CheY protein and proteins involved in the regulation of chemotaxis, membrane protein synthesis, and sporulation. *Proc. Natl. Acad. Sci. U.S.A.* **82**, 7989–7993.

Stock, A., Mottonen, J., Chen, T., and Stock, J. (1987). Identification of a possible nucleotide binding site in CheW, a protein required for sensory transduction in bacterial chemotaxis. *J. Biol. Chem.* **262**, 535–537.

Stock, A., Chen, T., Welsh, D., and Stock, J. (1988a). CheA protein, a central regulator of bacterial chemotaxis, belongs to a family of proteins that control gene expression in response to changing environmental conditions. *Proc. Natl. Acad. Sci. U.S.A.* **85**, 1403–1407.

Stock, A. M., Wylie, D. C., Mottonen, J. M., Lupas, A. M., Ninfa, E. G., Ninfa, A. J., Schutt, C. E., and Stock, J. B. (1988b). Phospho-proteins involved in bacterial signal transduction. *Cold Spring Harbor Symp. Quant. Biol.* **53**, 49–57.

Stock, A. M., Mottonen, J. M., Stock, J. B., and Schutt, C. E. (1989). Three-dimensional structure of CheY, the response regulator of bacterial chemotaxis. *Nature (London)* **337**, 745–749.

Stock, J. B., and Koshland, D. E., Jr. (1981). Changing reactivity of receptor carboxyl groups during bacterial sensing. *J. Biol. Chem.* **256**, 10826–10833.

Stock, J. B., Ninfa, A. J., and Stock, A. M. (1989). Protein phosphorylation and regulation of adaptive responses in bacteria. *Microbiol. Rev.* **53**, 450–490.

Stock, J. B., Stock, A. M., and Mottonen, J. M. (1990). Signal transduction in bacteria. *Nature (London)* **322**, 395–400.

Stock, J. B., Lukat, G. S., and Stock, A. M. (1991). Bacterial chemotaxis and the molecular logic of intracellular signal transduction networks. *Ann. Rev. Biophys. Biophys. Chem.* **20**, 109–136.

Volz, K., and Matsumura, P. (1991). Crystal structure of *Escherichia coli* CheY refined at 1.7-Å resolution. *J. Biol. Chem.* **266**, 15511–15519.

Wanner, B. L. (1992). Is cross regulation by phosphorylation of two-component response regulator proteins important in bacteria? *J. Bacteriol.* **174**, 2053–2058.

Wanner, B. L., and Wilmes-Riesenberg, M. R. (1992). Involvement of phosphotransacetylase, acetate kinase, and acetyl phosphate synthesis in control of the phosphate regulon in *Escherichia coli*. *J. Bacteriol.* **174**, 2124–2130.

Weiss, V., and Magasanik, B. (1988). Phosphorylation of nitrogen regulator I (NRI) of *Escherichia coli*. *Proc. Natl. Acad. Sci. U.S.A.* **85**, 8919–8923.

Wolfe, A. J., Conley, M. P., Kramer, T. J., and Berg, H. C. (1987). Reconstitution of signaling in bacterial chemotaxis. *J. Bacteriol.* **169**, 1878–1885.

Wolfe, A. J., Conley, M. P., and Berg, H. C. (1988). Acetyladenylate plays a role in controlling the direction of flagellar rotation. *Proc. Natl. Acad. Sci. U.S.A.* **85**, 6711–6715.

Wozniak, D. J., and Ohman, D. E. (1991). *Pseudomonas aeruginosa* AlgB, a two-component response regulator of the NtrC family, is required for *algD* transcription. *J. Bacteriol.* **173**, 1406–1413.

Wylie, D., Stock, A., Wong, C.-Y., and Stock, J. (1988). Sensory transduction in bacterial chemotaxis involves phosphotransfer between Che proteins. *Biochem. Biophys. Res. Commun.* **151**, 891–896.

Yamaguchi, S., Aizawa, S.-I., Kihara, M., Isomura, M., Jones, C. J., and Macnab, R. M. (1986). Genetic evidence for a switching and energy-transducing complex in the flagellar motor of *Salmonella typhimurium*. *J. Bacteriol.* **168**, 1172–1179.

Yang, Y., and Inouye, M. (1992). Intermolecular complementation between two defective mutants of *E. coli* signal-transducing receptors. *Proc. Natl. Acad. Sci. U.S.A.* **88**, 11057–11061.

Chapter 3

Methylation-Independent Behavioral Responses in Bacteria

JUDITH P. ARMITAGE

Microbiology Unit
Department of Biochemistry
University of Oxford
Oxford OX1 3QU, United Kingdom

I. Introduction

One specific pathway involved in chemotactically controlling bacterial swimming behavior is very well understood, as evidenced by the discussion in Chapter 2. In this system, dedicated membrane-spanning sensory transducing proteins (MCPs) bind a limited range of chemoeffectors and consequently stimulate an intracellular phosphorylation (Che) pathway to control the switching frequency of the flagellar motors. However, this mechanism is only part of a much more complex pathway. Enteric bacteria can sense and respond to changes in a range of different environmental effectors that do not interact with this family of dedicated sensory proteins. Signals from these other pathways must integrate with the signals from the methylation-dependent pathway to form a complete environmental sensing system that will result in the bacterium reaching, and remaining in, the optimum environment for growth. That environment will vary from species to species. At least two species of bacteria,

Rhodobacter sphaeroides and *Azospirillum brasilense*, have been shown to exhibit good behavioral responses, although both species lack the methylation-dependent transducers (Sockett *et al.*, 1987).

The stimuli that cause positive or negative behavioral changes without needing dedicated membrane-spanning receptor proteins are varied and include electron acceptors such as oxygen and nitrate, phosphotransferase sugars, light, and for the species *R. sphaeroides* the complete range of its metabolites. The very basic nature of the stimuli sensed via the MCP-independent pathways suggests that these pathways are not just a minor exception to the methylation pathway; being better understood by human scientists does not necessarily mean that a system is more important to a bacterium *in vivo*. The MCP-independent pathways may represent a simpler, possibly more ancient, mechanism of sensing changes in the energy state of the cell. In the natural environment these systems—which are involved in sensing such fundamental metabolites as oxygen, carbohydrates, and light—may be the dominant systems, maintaining cells in environments that can sustain their growth rate. The dedicated sensory transducers may become more important to cells in a "well-fed" state, guiding them to environments in which the growth rate can be increased. The MCP-dependent pathway, therefore, may be a highly specialized chemotaxis system used in addition to, and integrated with, the other systems. This chapter brings together current information about sensory pathways that are independent of at least part of the MCP-dependent system and develops an integrated model for chemosensing in general. Although far fewer data are available on MCP-independent sensing, the mechanisms involved in these pathways will probably prove to be common across the bacterial world, as will the mechanisms of interacting with signals from the dedicated transducers. The importance of each system may change with the growth conditions of each particular species.

II. Chemotaxis toward Phosphotransferase Sugars

In early experiments, Adler and Epstein (Adler *et al.*, 1974) and Lengeler (1975) showed that, although the response of *Escherichia coli* to some sugars (e.g., galactose and maltose) was dependent on the nontransporting sensory transduction pathway, a larger number of carbohydrates did not use these receptors but required transport through the phosphoenol pyruvate(PEP)-dependent phosphotransferase pathway to elicit a behavioral response. Some elegant experimental work, primarily from the Lengeler laboratory, helped elucidate some of the key steps in sugar transport that are involved in initiating a sensory signal to the motor (reviewed by Lengeler and Vogler, 1989; Taylor and Lengeler, 1990; Postma *et al.*, 1993). This work suggests a mechanism of sensory integration with the phosphorylation pathway from the nontransporting receptors. In fact, the connection between phosphorylation-dependent transport and sugar chemotaxis led to the first suggestion by Lengeler that phosphorylation may be involved in prokaryotic sensing (Lengeler *et al.*, 1981).

In many bacterial species, the PEP-dependent phosphotransferase system (PTS) is the major transport system for carbohydrates (Postma *et al.*, 1985b; Saier, 1989). (In other species, ATP or proton motive force-dependent uptake systems are more important; these do not seem to be involved directly in sensory transduction.) In enteric bacteria, each carbohydrate is translocated into the cell through a membrane-spanning protein, enzyme II (EII), that is specific for that sugar. Twelve different EII proteins have been identified to date. These membrane-spanning proteins operate in combination with a second specific protein, EIII, which in some cases may make up an integral part of a larger EII protein. The two proteins phosphorylate the carbohydrate while transporting it into the cell. Phosphorylation of the carbohydrate during transport depends on the phosphorylation of EII–EIII by a nonspecific pathway. Phosphates are transferred from the phosphoryl group of PEP via a soluble nonspecific protein EI to a histidine protein (HPr), which then phosphorylates the EII–EIII protein (Fig. 3.1).

Three groups of mutants unable to show a behavioral response to phosphotransferase carbohydrates have been identified. These mutants (1) show no response to single carbohydrates, (2) show no response to any PTS carbohydrate but respond to other groups of chemoeffectors, and (3) show no response to any group of chemoeffector. These three groups suggest the sequence of signaling required to cause a behavioral response, from specific primary receptor to shared carbohydrate transducer to a common chemotactic mechanism for controlling flagellar switching.

Many different mutants in specific PTS carbohydrate sensing have been analyzed. In all cases, a correlation exists between the transport and phosphorylation of that specific carbohydrate and its ability to elicit a chemotactic response (Adler *et al.*, 1974; Lengeler *et al.*, 1981; Pecher *et al.*, 1983). Nontransportable analogs of MCP-dependent chemoeffectors often will still elicit a chemotactic response, since binding to the periplasmic side of the MCP is sufficient to cause a chemotactic signal. Nontransportable analogs of PTS sugars do not cause chemotaxis. However, analogs that can be transported and phosphorylated but not metabolized further can cause chemotaxis although phosphorylation acts as an energy drain on the cell (Lengeler *et al.*, 1981). Some phosphorylated derivatives of PTS sugars can be transported into the bacteria in their phosphorylated form; however, transport of the already phosphorylated derivative does not induce a chemotactic response (Pecher *et al.*, 1983). These results all suggest that the process of phosphorylation during transport is essential for a chemotactic response. This idea is supported by

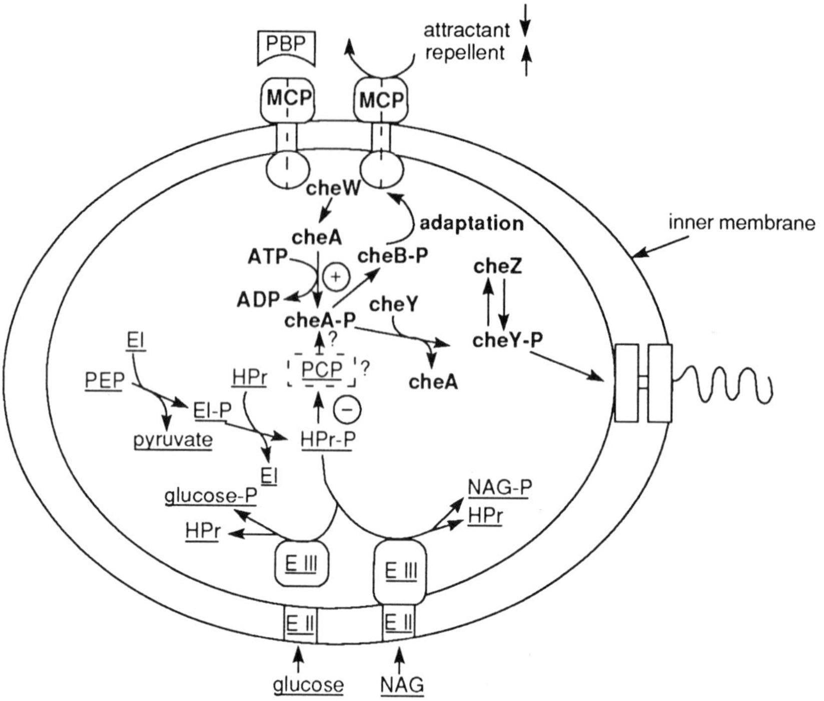

Figure 3.1 Possible pathway for the integration of receptor-dependent chemotaxis (bold letters) and PTS transport dependent chemotaxis (underlined letters). Removal of an attractant from the sensory receptors (MCPs) results in increased CheA autophosphorylation and subsequent CheY phosphorylation (+). Increased CheY–P results in increased tumbling. An increase in PTS transport results in a decrease in CheA/CheY phosphorylation, possibly as a result of increased HPr–P levels (−). This change results in increased smooth swimming. PBP, Periplasmic binding protein; PCP, hypothetical phosphotransferase chemotaxis protein; HPr, histidine protein; EII and EIII, two representative PTS transport protein systems, one for glucose and one for *N*-acetyl glucosamine (NAG). Reprinted with permission from Armitage *et al.*, 1990a.

evidence that a maximum response is seen only to a specific carbohydrate if its specific EII is fully induced. The measurements of the K_m values for PTS substrates show good correlation with the chemotactic response curves (Lengeler *et al.*, 1981). The requirement for transport and phosphorylation also has been supported by the evidence from hybrid EII–EIII complexes. The EIII component of the *N*-acetyl glucosamine PTS EII–EIII complex was deleted and replaced with the soluble EIII from the glucose PTS in a mutant lacking glucose EII, constructing an *N*-acetyl glucosamine EII-glucose–EIII hybrid. This mutant showed normal transport of and chemotaxis to *N*-acetyl glucosamine but not glucose (Lengeler and Vogler, 1989). All this evidence together indicates the requirement for transport and phosphorylation in PTS chemotactic signaling.

How can transport and phosphorylation be signaled through to the flagellar motor? Although the membrane-spanning structure of EII shows some similarity to that of the MCPs, this system is not involved in PTS chemotaxis (Niwano and Taylor, 1982a). Mutants that are unable to respond to chemoeffectors sensed through the MCPs because of *S*-adenosyl methionine depletion still respond normally to PTS sugars. Similarly, mutants in the methyltransferase and -esterase enzymes responsible for adaptation to MCP-dependent chemoeffectors do not respond to any MCP-dependent chemoattractants but again respond normally to PTS sugars (Taylor *et al.*, 1988). The chemotactic response is, therefore, independent of the primary sensing and adaptation pathway of the dedicated transducers. Indeed, transport is required for a chemotactic signal, not simply binding as for MCPs. Mutant EIIs that cannot be phosphorylated but bind the carbohydrate with normal efficiency do not elicit a chemotactic response (Weng *et al.*, 1992; Postma *et al.*, 1993).

The likelihood that the initial signal is given by the process of phosphorylation as the carbohydrate is transported into the cell is supported by the observation that mutants lacking the EI kinase responsible for HPr phosphorylation, but retaining the transport protein EII (Lengeler, 1975; Lengeler *et al.*, 1981), allow limited diffusion of the substrate into the cell but no chemotactic response to that compound. This evidence, in conjunction with the absence of a response to substrates that can be transported when already phosphorylated, suggests that the process of phosphorylation plays a role in eliciting the chemotactic signal. In addition, the sorbose EII protein from the naturally nonmotile bacterium *Klebsiella pneumoniae* allows a chemotactic response to sorbose when expressed in *E. coli* (Sprenger and Lengeler, 1984). If the chemotactic response is the result of a structural change in the transport protein, this ability would have to be lost in a nonmotile species. The intracellular signal resulting from the transport and phosphorylation of the sugar is still unknown. Early suggestions that it involves cyclic nucleotides (Black *et al.*, 1983) were found to be the indirect result of the effect of the mutations on the synthesis of the motor apparatus and the EII proteins, which in *E. coli* is under catabolite repression (Postma and Lengeler, 1985a; Tribhuwan *et al.*, 1986; Vogler and Lengeler, 1987). However, some mutants strains carrying a *cya* mutation regain PTS transport after addition of extracellular cAMP but do not regain chemotaxis, whereas other strains with similar mutations regain both activities simultaneously. The mutant allele maps close to the *crp* locus and may be involved in controlling synthesis of the phosphoryl-chemotaxis (PCP) protein that couples PTS transport to the chemotaxis pathway (Vogler and Lengeler, 1987; see subsequent discussion).

The mutants that have lost responses to all PTS sugars while retaining the other tactic responses have been found to have mutations in either the EI or HPr proteins. In *E. coli*, fructose is transported into the cell through a pathway very similar to the PTS pathway, but the phosphorylation of the sugar depends not on HPr but on a different protein, FPr (Lengeler and Vogler, 1989). When this pathway is the only one by which fructose can be transported into the cell, *E. coli* does not show a chemotactic response to fructose (mannose PTS can transport fructose and allow a chemotactic response). However, if FPr is overproduced or a "pseudo-HPr" is produced from the HPr-domain of FPr, chemotaxis to fructose occurs. Transport of carbohydrates into mutants lacking HPr has been restored by the addition of FPr, but the chemotactic responses to these PTS substrates is restored only after overexpression of FPr. This evidence suggests that, although chemotaxis and transport are normally tightly coupled, chemotaxis requires higher rates or levels of phosphorylation than does transport (Lengeler and Vogler, 1989; Grübl *et al.*, 1990). Therefore, a connection appears to exist between the extent or rate of substrate phosphorylation and the ability to cause an intracellular chemotactic signal, leading to the suggestion (Fig. 1) that a protein, designated PCP, responds to the phosphorylation state of a PTS intermediate and signals this state to either the CheA or the CheY protein involved in intracellular chemotactic signaling (Lengeler *et al.*, 1981; Grübl *et al.*, 1990). These two proteins are both (de)phosphorylated during chemotactic signaling from the MCP system and therefore are possible sites of interaction with another phosphorylation-dependent pathway. The balance of HPr and HPr-P may indirectly alter the balance of the CheA–CheY pathway to increase the level of dephosphorylated CheY. This change would result in increased smooth swimming in response to an increase in transported PTS carbohydrate, increasing the concentration of dephosphorylated HPr. The dephosphorylated HPr could cause reduced levels of CheY-P indirectly, either by inhibiting CheA autophosphorylation or by increasing CheY-P dephosphorylation. No direct evidence for this pathway is available, but in mutants lacking all chemotaxis proteins, CheA, CheW, and CheY are required for PTS chemotaxis and MCPs and CheZ are not required to restore the response (E. H. Rowsell, J. M. Smith, and B. L. Taylor, unpublished results). CheZ dephosphorylates CheY and this role may be carried out by the dephosphorylated PTS. The PTS and Che pathways are unlikely to interact directly because the kinetics and concentrations of the substrates of the two pathways are different. Mutants deleted for CheA and CheW show abnormal responses to PTS sugars (Postma *et al.*, 1993). However, CheA phosphorylation may be required for normal swimming in *E. coli*, so the altered responses may be caused by an inability of the mutants to reset the system for "normal" swimming. (For a complete description of the Che-dependent signal transduction pathway, see Chapter 2.)

Adaptation and memory are essential parts of the chemotactic response, and must be on a longer time scale than the initial excitation response. A possible mechanism for adaptation in PTS chemotaxis may be the restoration of an equilibrium of the phosphorylation enzymes responsible for PTS transport and the resetting of the Che system equilibrium (Pecher *et al.*, 1983). If these systems were to operate on different time scales, a memory of the previous phosphorylation state would be possible.

Rhodobacter sphaeroides shows an accumulation response to a wide range of sugars and polyols but has only one PTS system, the FPr-based system for fructose transport (Wu *et al.*, 1990). *Escherichia coli* only shows a chemotactic response to sugars transported through the PTS system, or interacting with the sensory transducing MCPs. Since *R. sphaeroides* lacks MCPs, the responses seen in this organism to the many non-PTS carbohydrates probably depend on an alternative sensory signaling system (W. A. Jeziorska and J. P. Armitage, unpublished data).

III Metabolism-Dependent Chemotaxis

Several chemotactic systems have been identified that require at least limited metabolism for a chemotactic response. In addition, some metabolites have been identified that appear to control the switching frequency of the flagellar motor directly, possibly indicating a site of integration of these metabolism-based chemotaxis pathways and the receptor-dependent pathways.

Chemotaxis toward proline and fumarate in *E. coli* depends on at least limited metabolism (Clancy *et al.*, 1981). The mechanisms involved are unknown, but experiments suggest that the response may be due to the role of these compounds in electron transport; proline acts as an electron donor and fumarate as an electron acceptor. Earlier results suggested that responses seen toward citrate by *S. typhimurium* could involve metabolism. However, a fifth MCP (Tcp) has been identified that signals changes in both citrate and phenol (Yamamoto and Imae, 1993).

MCP-independent chemotaxis is perhaps best understood in the purple photosynthetic bacterium *R. sphaeroides*. This organism was the first one identified to lack the dedicated sensory transducing proteins but still show good behavioral responses (Sockett *et al.*, 1987). Some data suggest that certain strains of *Rhizobium* and the rhizosphere bacterium *A. brasilense* may also lack or have very low levels of MCPs (Zhulin and Armitage, 1993; R. Schmitt, personal communication). Methionine auxotrophs of *R. sphaeroides* respond well to a wide range of compounds, and MCPs have not been detected by any biochemical or hybridization method. All the chemoeffectors identified in this organism are metabolites and no repellents (other than a reduction in light intensity or chemoeffector concentration) have been identified, all of which suggests a role for metabolism in the responses of this species. An analog of the chemoattractant alanine, 2-aminoisobutyrate, is transported into *R. sphaeroides* by the same transport mechanism as alanine at a similar rate. However, this analog is not metabolized and does not cause a chemotactic response. Chemotaxis toward alanine can be blocked by unrelated metabolites such as fructose or succinate if the cells have been induced for these metabolites. Collectively, these data again suggest that a common metabolic intermediate is involved in signaling (Poole *et al.*, 1993).

The swimming behavior of *R. sphaeroides* is different from that of many other species. This organism has a single subpolar flagellum that only rotates in one direction. Under different growth conditions this direction might be clockwise or counterclockwise, but the flagellum does not normally switch the direction of rotation (Armitage and Macnab, 1987; H. A. Packer and J. P. Armitage, unpublished results). Therefore, swimming direction change is not brought about by switching rotational direction but by briefly stopping flagellar rotation, during which time the functional flagellar helix relaxes from the distal end to form a large amplitude, short wavelength, nonfunctional helix. The stopped cells are reoriented by Brownian motion; the next period of smooth swimming may therefore be in a different direction. Experiments with tethered cells as well as with free-swimming cells, analyzed by computerized motion analysis, have shown that both the length of stops and the rate of flagellar rotation are highly variable, between cells in a population and within individuals, even under conditions of saturating light intensity, meaning the electrochemical proton gradient is well above that required for saturation of motor rotation (Poole and Armitage, 1988a). This altered flagellar behavior may be a direct consequence of the lack of feedback on the flagellar switch from the methylation and phosphorylation cycles that fine-tune the switching of enteric bacteria.

When stimulated with a metabolite, *R. sphaeroides* may show (1) a chemoattractant response, accumulating in an area of increased metabolite concentration; (2) a dispersal response, spreading from regions of bacterial concentration; or (3) both simultaneously. The response depends on the growth state of the cell and the type of metabolite.

When presented with a temporal increase (step-up) in concentration of a currently limiting metabolite the population exhibits an overall increase in mean swimming speed (Poole and Armitage, *et al.*, 1988a,b). The speed can increase by up to 25% of the unstimulated speed. Different metabolites have different chemokinetic effects on the bacteria, although all can produce strong chemotactic responses. Organic acid metabolites (the major carbon source for these organisms), or a change in potassium concentration, cause a large chemokinetic response (Armitage *et al.*, 1990a; Poole *et al.*, 1990b), but sugars such as fructose which are strong chemoattractants cause only a weak chemokinetic response (W. A. Jeziorska, H. L. Packer and J. P. Armitage, unpublished observation). Amino acids (with the exception of aspartate) cause no response, even when they constitute the sole carbon and nitrogen source for the cells. The increase in speed is sustained as long as the metabolite is present at the new elevated concentration. Therefore no measurable short-term adaptation to the stimulus occurs. This type of response would not result in a chemotactic accumulation of the population but in the chemokinetic spreading of the population. This response may be used in nutrient-limited environments (where these bacteria grow) to spread the population in response to increases in nutrient levels and, hence, the metabolic state of the cell, allowing the cells to spread and possibly encounter a metabolite gradient that will allow an additional chemotactic response (Poole *et al.*, 1991).

The cause of the chemokinetic response is still unclear. The increase in average population swimming speed is the result of an increase in the average flagellar rotation rate, as measured in both tethered and free-swimming cells, rather than the result of a reduction in stopping frequency or duration (Brown *et al.*, 1993). Since the response is seen primarily to metabolites, and most strongly to those affecting electron transport, an increased metabolic rate might lead to an increased proton motive force and, thus, to an increased rate of flagellar rotation and, subsequently, speed. However, chemokinesis is seen under conditions that caused the proton motive force and/or the electron transport rate to be either unchanged or reduced (Poole *et al.*, 1990a). The change in flagellar rotation rate is, therefore, independent of the driving force for rotation, suggesting a mechanism analogous to a gear change. In fact, metabolism does not seem to be required for chemokinesis to occur. Betaine, a transportable nonmetabolite, causes chemokinesis as does the succinate analog malonate, although neither of these compounds is a chemoattractant. Incubation with arsenate inhibits chemotaxis in general in *R. sphaeroides*, but not chemokinesis (H. L. Packer and J. P. Armitage, unpublished observation). Work in this laboratory has demonstrated that the increase in swimming speed correlates with an increase in cell size (S. Brown, P. S. Poole, W. A. Jeziorska, H. L. Packer, and J. P. Armitage, unpublished observations). Sugar-grown cells seem generally larger than organic acid-grown cells and may, exhibit less swelling and, therefore, a reduced chemokinetic response. Whether the parallel increase in cell size is a coincidence or is directly linked to swimming speed is still not known. The time courses for both cell swelling and speed increase are exactly parallel. Several minutes are required for both characteristics to reach a maximum level; the eventual return to control size is the same, and probably correlates with the time required to remove the chemoeffector metabolically. The swelling of the cells in response to an increased metabolite transport, and subsequent increased metabolic rate, may cause a physical change in the interaction of the motor (Mot) proteins in the cytoplasmic membrane and the rotor in the flagellar basal body, resulting in more efficient interaction between the motor components without a change in the size of the proton motive force. Alternatively, some component of the outer membrane such as the recently identified stretch receptors may result in a signal to the motor, altering the rotation rate (Martinac *et al.*, 1987; Li *et al.*, 1988; Saimi *et al.*, 1988). Stretch receptors have been implicated in osmotaxis in enteric bacteria and, although the response described here is independent of osmotic changes as evidenced by the lack of response to sucrose, a related phenomenon could be involved in *R. sphaeroides*.

As described earlier, *R. sphaeroides* also shows a true chemotactic accumulation response to a range of chemoeffectors if they are presented as a gradient (Armitage *et al.*, 1990a,b). This response occurs in the absence of dedicated sensory proteins and requires transport and at least limited metabolism of the effector. The intracellular signaling pathway is unclear, but no components of a pathway homologous to those of the Che sensory pathway found in enteric bacteria have been identified by antibody by DNA hybridization, or by biochemical methods (W. A. Havelka, W. A. Jeziorska, R. E. Sockett, and J. P. Armitage, unpublished data). The response is, however, independent of the chemokinetic response since (1) unlike the chemokinetic response, chemotaxis in *R. sphaeroides* requires both transport and metabolism; (2) sugars cause a strong chemotactic response but very weak chemokinesis; and (3) two independent mutants have been isolated that have lost the chemotactic response to organic acids, but retain the chemokinetic response.

The two responses can also be separated by computerized motion analysis. The addition of an organic acid to tethered *R. sphaeroides* causes an increase in swimming speed (as measured by rotation rate) and a decrease in stopping frequency. However, the stopping frequency returns to normal after a few minutes whereas the rotation rate remains high for many hours. A reduction in organic acid concentration results in an increase in stopping and a return to the prestimulus speed. The stopping frequency returns to normal within a minute or so. A change in amino acid concentration causes the change in stopping frequency without the long-term change in speed. Such a response would allow movement toward a chemoeffector presented as a gradient, whereas chemokinesis would cause dispersion of the population. Computer modeling has shown that the two responses are able to operate simultaneously. The chemokinetic response allowing a general increase in swimming speed in response to an increase in the metabolic state of growth-limited cells; the chemoattractant response alters the stopping frequency of the cells to bias their swimming up an additional chemoattractant gradient.

A gene involved in the loss of the chemoattractant response to weak acids has been cloned and partially sequenced. A transposon mutation in this gene alters not only the swimming behavior of cells grown on organic acids but also their growth characteristics, suggesting a mutation in a control gene. The sequence has some homology to a two-component regulatory system, which leads to the intriguing possibility that, although no apparent homologies exist between *R. sphaeroides* chemotaxis and the MCP-based system identified in other species, the systems still may have a similar evolutionary origin. The Che pathway is a two-component system without transcriptional activation (CheY is a truncated response regulator), but CheA can cross talk with other systems involved in regulating transcription (Ninfa *et al.*, 1988). The Che pathway may have evolved from one of these systems, but *R. sphaeroides* may have a basic level of chemosensing and motor control that uses the components of the regulatory system directly.

IV. Intracellular Metabolites That Affect Flagellar Rotation

Several investigations have identified a range of unrelated intracellular metabolites that can alter the switching frequency of the flagellar motor in the absence of a sensory stimulus in a range of bacterial species that also contain sensory receptors. For example, the addition of fumarate to a "switch" mutant of the archaebacterium *Halobacterium halobium* allows switching. Fumarate concentration in the cytoplasm of cells given a photophobic stimulus has been shown to increase (Marwan *et al.*, 1990; Oesterhelt and Marwan, 1990). The photophobic response is also dependent both on methylatable membrane receptors and on retinal-containing sensory receptors. The mechanism involved in releasing fumarate from the membrane where it is usually bound as well as its relationship to the primary photoreceptor and the sensory transducing proteins is currently unknown, as is the mechanism of interaction with the flagellar motor. Flagella on cell envelope preparations from *E. coli* and *S. typhimurium* are able to rotate but not to switch. Fumarate or a fumarate metabolite has been shown to restore motor switching in the presence of CheY in these preparations (Barak and Eisenbach, 1992a). The role of fumarate in switching is unknown.

Other metabolites have also been shown to alter switching. *In vitro* preparations of CheY can be acetylated in the presence of acetyl CoA synthetase; although the role of acetylation *in vivo* is not known, acetylated CheY can cause switching in flagellate envelopes (Barak *et al.*, 1992b). Strains of *E. coli* that lack cytoplasmic Che proteins and the sensory transducing proteins show enhanced clockwise rotation when presented with acetate but not with the enteric repellent benzoate. This effect does, however, require the switch protein CheY. Analysis of the biochemistry of these mutants suggests that acetyl phosphate, a direct product of acetate metabolism, may be needed for this response. In this case, acetyl CoA does not seem to be involved; instead, CheY is directly phosphorylated by acetyl phosphate. Acetyl phosphate is an important intracellular metabolite whose concentration changes under different growth conditions. Perhaps the direct control of CheY phosphorylation by metabolic intermediates is the basis for responses to changes in the energy status of the cell (Wolfe *et al.*, 1988), providing a connection between the metabolic state of the cell and the chemotaxis machinery at the level of either CheY or the interaction of CheY with the switch. Therefore, many different metabolites may connect the metabolic state of the cell to motor switching; fumarate, acetyl phosphate, and acetyl CoA being only three examples.

V. Electron Transport-Dependent Taxis

Most bacterial species show behavioral responses to electron acceptors of aerobic and anaerobic respiration. These responses depend on the process of electron transport rather than on specific receptors. The aerotactic response was one of the earliest identified behavioral responses, several early accounts report populations of bacteria segregating at different distances from air bubbles on microscope slides. The location of different species depended on their growth requirements for oxygen. Engelmann (1883) showed that under illumination some species of colored bacteria that normally swam close to an air bubble would move away. Although not recognized at the time, this behavior is the result of competition between respiratory and photosynthetic electron transport in photosynthetic species and provides some idea of the mechanism of signal generation in response to changes in electron transport rates.

The responses to electron acceptors and to stimuli affecting electron transport rate are likely to be among the strongest and most important for most bacteria in the natural environment since these processes provide the basic means for maintaining the size of the electrochemical proton gradient, which is essential not only for motility but also for transport and general cellular integrity. However, although one of the earliest responses to be investigated, this is probably one of the least understood sensory mechanisms. Analysis is made difficult because any change in the rate of electron transport may cause, at least transiently, a change in the electrochemical proton gradient, which is itself the driving force for flagellar rotation. Therefore transient chemokinetic effects could occur to confuse measurements of the chemotactic response. In addition, inhibition of electron transport pathways can cause major changes in other metabolic pathways in the cell. The effects measured under different growth and inhibition conditions may not be the direct result of changes in electron transport but a “knock-on” effect on cellular metabolism. Much additional work on mutants of electron transport pathways must be carried out before we can be certain of the signaling pathway. No mutants have been isolated to date that show both normal electron transport and normal chemotaxis but loss of the aerotactic signal, suggesting that the signal must be a direct response to active electron transport, and that the connection to the sensory transduction pathway is either directly from an electron transport component or through an essential metabolic intermediate.

Results from studies of *R. sphaeroides*, *E. coli*, *S. typhimurium*, and *Bacillus cereus* all show that active electron transport is an essential part of the generation of a behavioral signal (Laszlo and Taylor, 1981, Laszlo *et al.*, 1984a,b,c; Armitage *et al.*, 1985). Binding to the terminal acceptor is not sufficient to stimulate a response, showing that this is another "receptor" independent and metabolism dependent sensory pathway. Many different species can induce different terminal electron acceptors under different growth conditions, to maintain maximum electron transport rates under high and low oxygen levels or under anaerobic conditions. *Escherichia coli* can grow either aerobically with oxygen as the terminal electron acceptor or anaerobically with nitrate as a terminal acceptor. Bacteria grown with nitrate as the terminal acceptor show a positive response to nitrate but, if grown aerobically or if nitrate reductase is inhibited by the presence of oxygen, no response to nitrate is seen (Taylor *et al.*, 1979). Therefore, the terminal acceptor must be induced for a response to occur. Inhibition of electron transport by the addition of, for example, antimycin A causes the loss of a response to any electron acceptor, unless the site of electron transport inhibition is bypassed by a more electropositive electron donor (Laszlo and Taylor, 1981). The site of integration with the other sensory transduction pathways has not been identified. Signaling and adaptation are, however, independent of the membrane-spanning sensory transducing proteins or their methylation state (Niwano and Taylor, 1982b; Dang *et al.*, 1986b). Mutants that show a reversed response to chemoattractants, that is, are repelled by attractants, also show a reversed response to electron acceptors (Dang *et al.*, 1986a). The reversal of the response is the result of the loss of the methylesterase that is responsible for balancing the methylation-dependent adaptation response on the sensory receptors. These receptors are not required for the aerotactic response, so the reversal may simply be the result of a general imbalance in the phosphorylation pathway, indirectly altering all the swimming behavior of the mutants.

Rhodobacter sphaeroides has the advantage of a branched electron transport pathway that can induce two different terminal oxidases, as well as a range of different terminal reductases, depending on the extent of aeration and the acceptors available (Fig. 3.2). The terminal oxidase branching from the quinone is usually a type "*o*" cytochrome and, therefore, is not directly coupled to any change in the electrochemical proton gradient, whereas the terminal oxidase

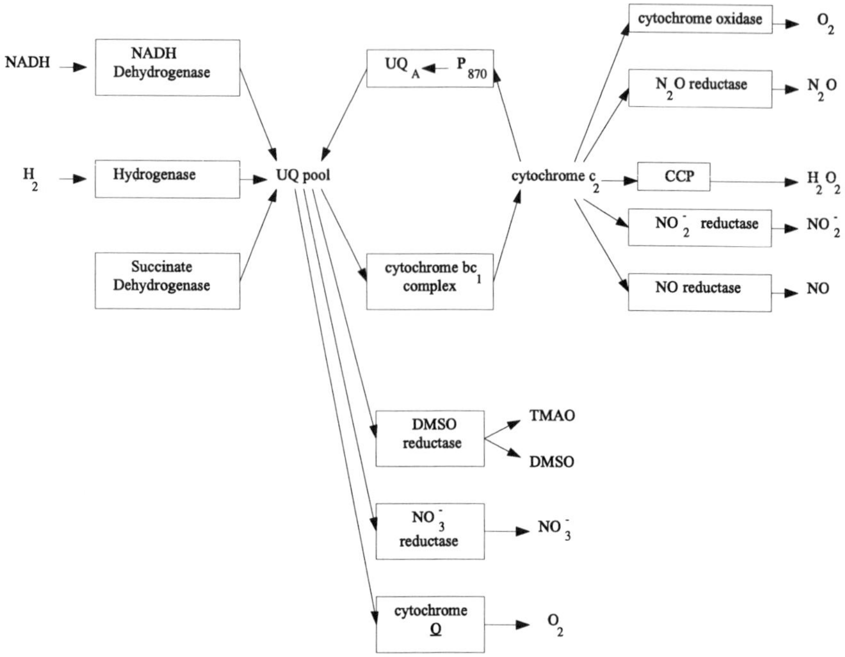

Figure 3.2 Electron transport pathways found in *Rhodobacter* species including cyclic photosynthetic electron transport. UQ, Ubiquinone; DMSO, dimethyl sulfoxide; TMAO, trimethylamine *N*-oxide; CCP, cytochrome c peroxidase.

from the cytochrome bc_1 and soluble cytochrome *c* branch may be a "coupled" cytochrome aa_3 type. The use of electron transport inhibitors and electron donors with redox potentials that allow the bypassing of the site of inhibition permits the relative roles of electron transport and changes in the electrochemical proton gradient resulting from that electron transport to be investigated, using the absorbance change of the pigment carotenoids to measure changes in the electrochemical proton gradient. Not only was electron transport found to be essential for an aerotactic response but in addition, an increase in the proton motive force seemed to be necessary (Armitage *et al.*, 1985).

Several observations suggest, however, that the signal may not always be as simple as a change in proton motive force resulting from a change in electron transport rate. *Rhodobacter sphaeroides* can be grown anaerobically in the dark with dimethyl sulfoxide (DMSO) as the terminal electron acceptor on the branched pathway (Ferguson *et al.*, 1987). Under these conditions, *R. sphaer-*

oides is attracted toward an increasing concentration of DMSO. This response is inhibited by oxygen, which is also an inhibitor of DMSO reductase and therefore inhibits electron transport to that acceptor. Unexpectedly, however, light does not inhibit the DMSO response (J. P. Armitage, unpublished observation), although photosynthetic electron transport has been shown to inhibit DMSO reductase activity (Jones *et al.*, 1990). In addition, *A. brasilense* has been demonstrated to show an attractant response to an artificial redox gradient of ferricyanide in a uniform TMPD background. Under these conditions, the organism forms a band at a specific redox potential in the absence of any natural electron donor or acceptor, suggesting that the bacteria may be able to sense the redox state of an electron transport component and that this sensing can cause a behavioral signal (Grishanin *et al.*, 1991). Therefore, perhaps the change in the redox state of one of the electron transport intermediates, not the change in electrochemical proton gradient, sends a signal to the flagellar motor. Adaptation could result from the stabilization of the redox potential of that component. Additional work is necessary before any conclusive statement can be made about electron transport signaling. The likelihood that the state of electron transport is involved in sensory signaling is supported by the response of nonphotosynthetic bacteria to high intensity blue light. The repellent response seen may be the result of inhibition of the flavoproteins in the respiratory electron transport chain by the blue light and of a reduction in the rate of electron transport. Therefore the redox state of a component of the chain may act as a signal to the flagellar motor (Taylor *et al.*, 1979).

Obligate anaerobes and aerobes subjected to hyperbaric oxygen levels show a negative response to oxygen. Pure oxygen induced a tumbling, repellent, response in *S. typhimurium* and *E. coli*. The repellent response has different concentration characteristics to the attractant response, suggesting (as might be expected, at least in the case of obligate anaerobes) that the response is the result of activation of a different receptor (Shioi *et al.*, 1987). This receptor, which might respond to toxic by-products of oxidation such as superoxides or hydrogen peroxide, has not been identified.

VI. Eubacterial Phototaxis

Much information is available about the processes involved in the positive and negative photoresponses shown by halobacteria, since these are dependent on specific photosensory retinal-containing proteins that are dedicated sensory proteins with no effect on metabolic activity (Spudich and Bogomolni, 1988; Spudich *et al.*, 1989; Oesterhelt and Marwan, 1990). In the photosynthetic eubacteria, however, the photoresponses are connected to photosynthetic electron transport and, therefore, to the electrochemical proton gradient and cellular metabolism (Armitage, 1988). As with aerotaxis, this connection has made the analysis of the mechanisms involved in signaling more difficult.

The interaction of the respiratory and photosynthetic electron transport pathways is apparent in comparisons of the responses of *R. sphaeroides* to an oxygen gradient in the dark or in high or low light. In the dark, *R. sphaeroides* swims up an oxygen gradient, but the bacterium is repelled by the same gradient if then exposed to photosynthetic light (Armitage *et al.*, 1985). Under these conditions, respiratory electron transport is inhibited by photosynthetic electron transport, probably changing the redox state of a shared electron transport component (Cotton *et al.*, 1983). In addition, measurement of the electrochemical proton gradient shows that the addition of a pulse of oxygen to bacteria incubated in low light causes a transient fall in the proton motive force, probably because oxygen interacts with the noncoupled terminal oxidase, causing a brief reduction in photosynthetic electron transport (J. P. Armitage, unpublished observation). This response, the result of the intimate connection between electron transport pathways in bacteria, would inhibit the movement of a photosynthetic bacterium in nonsaturating light into an aerobic environment, which would be inhibitory for photosynthetically growing cells.

The most obvious photoresponse shown by photosynthetic bacteria is a "photophobic" response in which the cells reverse their swimming direction when swimming over a light–dark boundary (Clayton, 1957; Pfennig, 1968; Hader, 1987; Armitage, 1988). This behavior has been seen in a wide variety of photosynthetic bacteria and results in bacteria becoming trapped in regions of light. No apparent change in swimming behavior occurs when the cells go from dark to light, although a transient change in speed may occur as the rate of photosynthetic electron transport changes. This change in speed is seen transiently when *R. sphaeroides* is exposed to an increase in light intensity. Most of the photosynthetic species used in these investigations have polar tufts of flagella and change direction by reversing their swimming direction. This swimming behavior would allow a photophobic response to cause the accumulation of cells in regions of actinic light. The mechanism must be somewhat different for species such as *R. sphaeroides* in which direction changing is brought about by stopping and reorientation through Brownian motion. Obviously if *R. sphaeroides* simply stopped when faced with a decrease in light intensity, the organism would be trapped in the dark, not the light (Armitage, 1991). Therefore, a "memory" must exist within these species that would tend to alter the direction-changing frequency for a period of time after a decrease in light intensity is sensed, increasing the chances of a new swimming direction that taking the cells back into the light. In fact, when *Rhodospirillum rubrum* is subjected to a sustained decrease in light, the cells show a 30-sec period of rapid reversals rather than the single reversal seen when swimming over a light–dark boundary, suggesting that both memory and adaptation exist (Clayton, 1953a).

The isolation of phototactically incompetent cells has always been a reliable method of isolating photosynthetic mutants of different motile species of bacteria, suggesting a close connection between phototaxis and photosynthesis. The most detailed analysis of photoresponses in photosynthetic eubacteria was carried out many years ago on *R. rubrum* by Clayton (1953a,b,c). This investigator showed that this spiral-shaped species with polar tufts of flagella responded to a 1% reduction in light intensity, and that the phototactic action spectrum coincided closely with the photosynthetic action spectrum. The bacteria responded to the transient change in photosynthetic activity rather than to an overall change in the steady state and responded only to a step down, not a step up in light intensity. Experiments since then, with inhibitors of pho tosynthetic electron transport (Harayama and Iimo, 1976; Harayama, 1977) and mutants lacking reactions centers but containing the full pigment complement (Armitage and Evans, 1981), have confirmed that a change in photosynthetic electron transport is essential for the phototactic signal. Clayton also showed that the response was additive. If the time between two subthreshold step down stimuli was less than 3 sec and more than 250 msec, the bacteria showed a photophobic response. The strength of the response was dependent on the size of the step down and the duration. For example, a 1.5-sec step down close to the response threshold caused the same response as a larger 0.5-sec step down. The time course shown in the additive experiments is much longer than the time required for photosynthetic turnover, and suggests that the signal that controls flagellar reversal is a metabolic response rather than a change in the rate of photosynthetic turnover. The period of increased rever sal seen in *R. rubrum* after a step down does suggest that, although not evident from the light spot experiments, the bacteria show both memory (since 30 sec is much longer than the time required for adaptation of the photosynthetic electron transport rate) and adaptation of the behavioral response. If memory and adaptation occur, possibly as a result of restoring the equilibrium of photosynthetically dependent metabolic pathways rather than of electron transport itself, then species such as *R. sphaeroides* may be able to return to illuminated areas rather than be trapped in the dark.

Although the experiments on *R. rubrum* and *R. sphaeroides* suggest a direct link between photosynthetic electron transport and the positive phototactic response, a eubacterial species has been identified with a specific photosensory protein that seems to be involved in negative phototactic responses, possibly via a mechanism similar to that identified in halobacteria. *Ectothiorhodospira halophila* is a halophilic eubacterium that shows both a positive and a negative phototactic response. The negative phototactic response is specifically to an increase in blue light of a wavelength similar to the absorption spectrum of a small photoactive yellow protein identified in this organism (Sprenger *et al.*, 1993).

Again, as for the behavioral response to other transducer-independent pathways, the sensory signal itself and the site of integration of the phototactic signal with the other sensory pathways are unknown. The signal could, again, be a change in electrochemical proton gradient or in the redox state of a component of the electron transport chain, although the time required for adaptation suggests that at least adaptation may again be dependent on some metabolic intermediate.

VII. Summary

Motile bacteria respond to a changes in a wide variety of different environmental stimuli that interact with a range of very different metabolic pathways. The signals produced from these different pathways, or specific sensory receptors, must be integrated to produce an overall response that either maintains the bacterium in its currently optimum environment or causes it to bias its swimming direction toward a "better" environment (Fig. 3.3).

Perhaps the strength of the behavioral response to different stimuli is dependent on the current growth state of the cell. The signals from the dedicated sensory transducers must be integrated with those from the metabolically based sensing systems, but the latter may provide stronger signals in severely growth-limited cells compared with well-fed populations. The strength of the integrated response may therefore depend on current growth conditions. Whether or not CheA or CheY will be identified as universal sites of sensory integration, perhaps responding to the phosphorylation state of the cell and therefore controling the switching frequency, remains to be seen. Much more research on the mechanisms involved in the metabolically based components of the chemotaxis system is required before a complete model for bacterial behavior can be presented.

Figure 3.3 Possible mechanisms involved in integrating receptor-dependent and receptor-independent pathways to produce a balanced response. Drawn as in Fig. 3.1 to show effect of decreased receptor-dependent chemoeffector concentration and increased PTS transport and increased metabolite concentrations. Abbreviations as for Fig. 3.1.

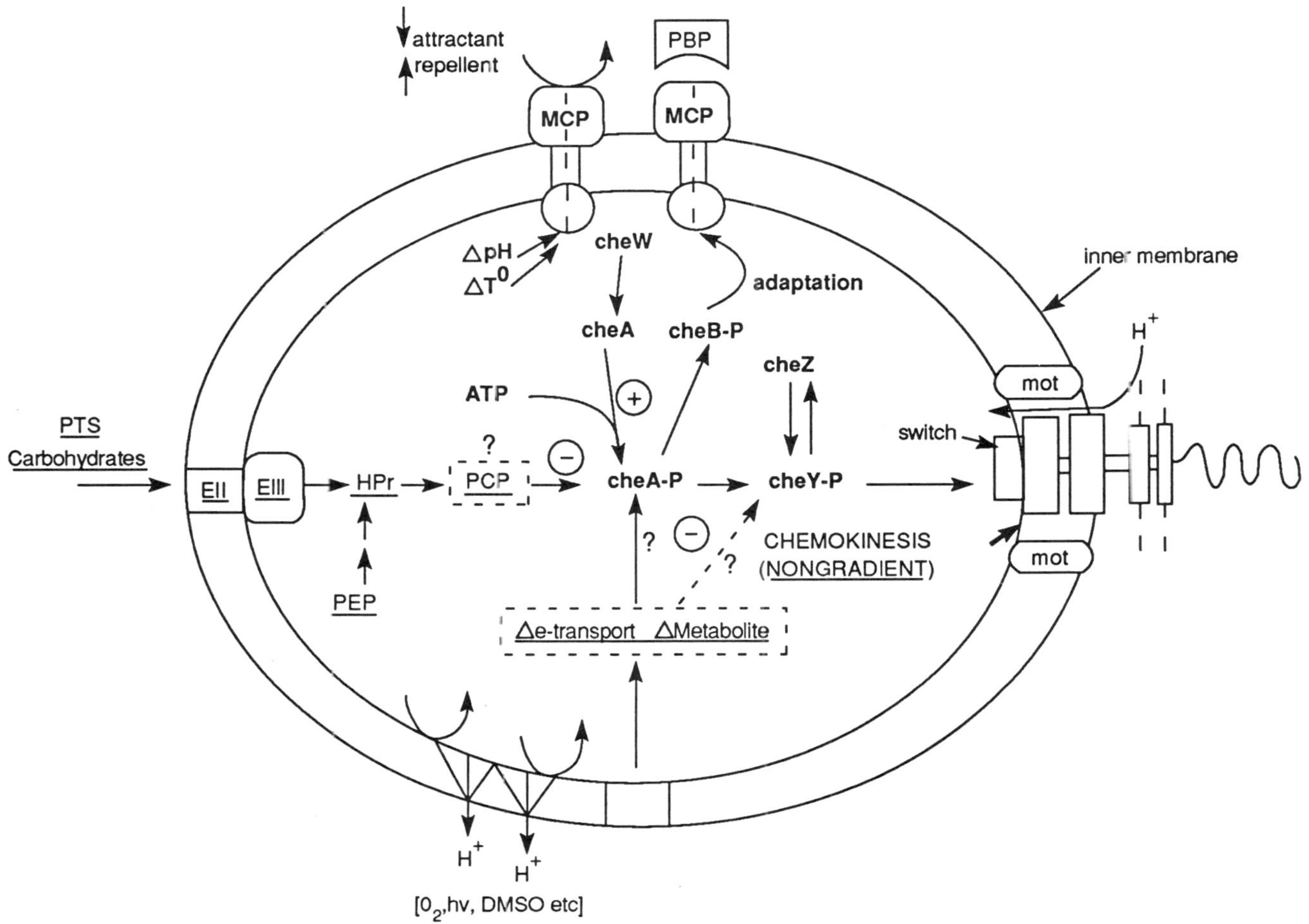
attractant
repellent
PBP
MCP
MCP
ΔpH
ΔT0
cheW
adaptation
inner membrane
cheA
cheB-P
cheZ
ATP
H+
mot
PTS
Carbohydrates
switch
?
EII
EIII
HPr
PCP
cheA-P
cheY-P
?
CHEMOKINESIS
(NONGRADIENT)
?
mot
PEP
Δe-transport
ΔMetabolite
H+
H+
[O2,hv, DMSO etc]

Acknowledgments

The work on *R. sphaeroides* described in this chapter was supported by the SERC, Wellcome Trust, and Leverhulme Foundation. I thank J. Lengeler for useful feedback on the PTS system.

References

Adler, J., and Epstein, W. (1974). Phosphotransferase-system enzymes as chemoreceptors for certain sugars in *Escherichia coli* chemotaxis. *Proc. Natl. Acad. Sci. U.S.A.* **71**, 2895–2899.

Armitage, J. P. (1988). Tactic responses in photosynthetic bacteria. *Can. J. Microbiol.* **34**, 475–481.

Armitage, J. P. (1991). Photoresponses in eubacteria. *In* "Biophysics of Photoreceptors and Photomovements in Microorganisms" (F. Lenci, F. Ghetti, G. Colombetti, D.-P. Hader, and P.-S. Song, eds.), pp. 43–52. Plenum Press, New York.

Armitage, J. P., and Evans, M. C. W. (1981). The reaction centre in the phototactic and chemotactic responses of *Rhodopseudomonas sphaeroides*. *FEMS Microbiol. Lett.* **11**, 89–92.

Armitage, J. P., and Macnab, R. M. (1987). Unidirectional intermittent rotation of the flagellum of *Rhodobacter sphaeroides*. *J. Bacteriol.* **169**, 514–518.

Armitage, J. P., Ingham, C., and Evans, M. C. W. (1985). Role of the proton motive force in phototactic and aerotactic responses of *Rhodopseudomonas sphaeroides*. *J. Bacteriol.* **163**, 967–972.

Armitage, J. P., Havelka, W. A., and Sockett, R. E. (1990a). Methylation-independent taxis in bacteria. *In* "Biology of the Chemotactic Response" (J. P. Armitage and J. M. Lackie, eds.), pp. 177–197. Cambridge University Press, Cambridge.

Armitage, J. P., Poole, P. S., and Brown, S. (1990b). Sensory signalling in *Rhodobacter sphaeroides*. *In* "Molecular Biology of Membrane Bound Complexes in Phototrophic Bacteria" (G. Drews, ed.). Plenum Press, New York.

Barak, R., and Eisenbach, M. (1992a). Fumarate or a fumarate metabolite restores switching ability to rotating flagella of bacterial envelopes. *J. Bacteriol.* **174**, 643–645.

Barak, R., Welch, M., Yanovsky, A., Oosawa, K., and Eisenbach, M. (1992b). Acetyladenylate or its derivative acetylates the chemotaxis protein CheY in vitro and increases its activity at the flagellar switch. *Biochemistry* **31**, 10099–10107.

Black, R. A., Hobson, A. C., and Adler, J. (1983). Adenlylate cyclase is required for chemotaxis to phosphotransferase system sugars by *Escherichia coli*. *J. Bacteriol.* **153**, 1187–1195.

Brown, S., Poole, P. S., Jeziorska, W., and Armitage, J. P. (1993). Chemokinesis in *Rhodobacter sphaeroides* is the result of a long term increase in the rate of flagellar rotation. *Biochim. Biophys. Acta* **1141**, 309–312.

Clancy, M., Madill, K. A., and Wood, J. M. (1981). Genetic and biochemical requirements for chemotaxis to L-proline in *Escherichia coli*. *J. Bacteriol.* **146**, 902–906.

Clayton, R. K. (1953a). Studies in the phototaxis of *Rhodospirillum rubrum*. I. Action spectrum, growth in green light, and Weber-law adherence. *Arch. Microbiol.* **19**, 107–124.

Clayton, R. K. (1953b). Studies in the phototaxis of *Rhodospirillum rubrum*. II. The relation between phototaxis and photosynthesis. *Arch. Microbiol.* **19**, 125–140.

Clayton, R. K. (1953c). Studies in the phototaxis of *Rhodospirillum rubrum*. III. Quantitative relationship between stimulus and response. *Arch. Microbiol.* **19**, 141–165.

Clayton, R. K. (1957). Patterns of accumulation resulting from taxes and changes in motility of microorganisms. *Arch. Microbiol.* **27**, 311–319.

Cotton, N. P. J., Clark, A. J., and Jackson, J. B. (1983). Interaction between the respiratory and photosynthetic electron transport chains of intact cells of *Rhodopseudomonas capsulata* mediated by membrane potential. *Eur. J. Biochem.* **130**, 581–587.

Dang, C. V., Niwano, M., Ryu, J-I., and Taylor, B. L. (1986a). Inversion of aerotactic response in *Escherichia coli* deficient in *cheB* protein methylesterase. *J. Bacteriol.* **166**, 275–280.

Dang, C. V., Niwano, M., Ryu, J.-I., and Taylor, B. L. (1986b). Interaction between methylation-dependent and methylation-independent pathways for chemotaxis and aerotaxis. *J. Bacteriol.* **166**, 275–280.

Engelmann, T. W. (1883). *Bakterium photometricum.* Ein Beitrag zur vergleichenden Physiologie des Licht- und Farbensinnes. *Pfluegers Arch. Ges. Physiol. Menschen Tiere* **42**, 183–186.

Ferguson, S. J., Jackson, J. B., and McEwan, A. G. (1987). Anaerobic respiration in the Rhodospirillaceae: Characterisation of pathways and evaluation of roles in redox balancing during photosynthesis. *FEMS Microbiol. Rev.* **46**, 117–143.

Grishanin, R. N., Chalmina, I. I., and Zhulin, I. B. (1991). Behaviour of *Azospirillum Brasiliense* in a spatial gradient of oxygen and in a "redox" gradient of an artificial electron acceptor. *J. Gen. Microbiol.* **137**, 2781–2785.

Grübl, G., Vogler, A. P., and Lengeler, J. W. (1990). Involvement of the histidine protein (HPr) of the phosphotransferase system in chemotactic signaling of *Escherichia coli* K-12. *J. Bacteriol.* **172**, 5871–5876.

Hader, D.-P. (1987). Photosensory behavior in prokaryotes. *Microbiol. Rev.* **51**, 1–21.

Harayama, S. (1977). Phototaxis and membrane potential in the photosynthetic bacterium *Rhodospirillum rubrum. J. Bacteriol.* **131**, 34–41.

Harayama, S., and Iino, T. (1976). Phototactic responses of aerobically cultivated *Rhodospirillum rubrum. J. Gen. Microbiol.* **94**, 173–179.

Jones, M. R., Richardson, D. J., McEwan, A. G., Ferguson, S. J., and Jackson, J. B. (1990). In vivo redox poising of the cyclic electron transport system of *Rhodobacter capsulatus* and the effects of the auxiliary oxidants, nitrate, nitrous oxide, and trimethylamine *N*-oxide, as revealed by multiple short flash excitation. *Biochim. Biophys. Acta Bio-Energetics* **1017**, 209–216.

Laszlo, D. J., and Taylor, B. L. (1981). Aerotaxis in *Salmonella typhimurium*: Role of electron transport. *J. Bacteriol.* **145**, 990–1001.

Laszlo, D. J., Fandrich, B. L., Sivaram, A., Chance, B., and Taylor, B. L. (1984). Cytochrome *o* as a terminal oxidase and receptor for aerotaxis in *Salmonella typhimurium. J. Bacteriol.* **159**, 663–667.

Laszlo, D. J., Niwano, M., Goral, W. W., and Taylor, B. L. (1984b). *Bacillus cereus* electron transport and proton motive force during aerotaxis. *J. Bacteriol.* **159**, 820–824.

Lengeler, J. (1975). Mutations affecting transport of the hexitols D-mannitol, D-glucitol, and galactitol in *Escherichia coli* K12: Isolation and mapping. *J. Bacteriol.* **124**, 26–38.

Lengeler, J. W., and Vogler, A. P. (1989). Molecular mechanisms of bacterial chemotaxis towards PTS-carbohydrates. *FEMS Microbiol. Rev.* **63**, 81–92.

Lengeler, J., Auburger, A.-M., Mayer, R., and Pecher, A. (1981). The phosphoenolpyruvate-dependent carbohydrate:phosphotransferase system enzyme II as chemoreceptors in chemotaxis of *Escherichia coli* K-12. *Mol. Gen. Genet.* **183**, 163–170.

Li, C., Boileau, A. J., Kung, C., and Adler, J. (1988). Osmotaxis in *Escherichia coli. Proc. Natl. Acad. Sci. U.S.A.* **85**, 9451–9455.

Martinac, B., Buechner, M., Delcour, A. H., Adler, J., and Kung, C. (1987). Pressure-sensitive ion channel in *Escherichia coli. Proc. Natl. Acad. Sci. U.S.A.* **84**, 2297–2301.

Marwan, W., Schäfer, W., and Oesterhelt, D. (1990). Signal transduction in *Halobacterium* depends on fumarate. *EMBO J.* **9**, 355–362.

Ninfa, A. J., Ninfa, E. G., Lupas, A. N., Stock, A., Magasanik, B., and Stock, J. (1988). Crosstalk between bacterial chemotaxis signal transduction proteins and regulators of transcription of the Ntr regulon: Evidence that nitrogen assimilation and chemotaxis are controlled by a common phosphorotransfer system. *Proc. Natl. Acad. Sci. U.S.A.* **85**, 5492–5496.

Niwano, M., and Taylor, B. L. (1982). Novel sensory adaptation mechanism in bacterial chemotaxis to oxygen and phosphotransferase substrates. *Proc. Natl. Acad. Sci. U.S.A.* **79**, 11–15.

Oesterhelt, D., and Marwan, W. (1990). Signal transduction in *Halobacterium halobium*. *In* "Biology of the Chemotactic Response" (J. P. Armitage and J. M. Lackie, eds.), pp. 219–240. Cambridge University Press, Cambridge.

Pecher, A., Renner, I., and Lengeler, J. (1983). The phosphoenolpyruvate-dependent carbohydrate:phosphotransferase system enzymes II, a new class of chemosensors in bacterial chemotaxis. *In* "Mobility and Recognition in Cell Biology" (H. Sund and C. Veeger, eds.), pp. 517–531. de Gruyter, Berlin.

Pfennig, N. (1968). *Chromatium akenii* (Thiorhodaceae). *In* "Encyclopedia Cinematographica" (G. Wolf, ed.), pp. 3–9. Institut für den Wissenschaftlichen Film, Göttingen, Germany.

Poole, P. S., and Armitage, J. P. (1988). Motility response of *Rhodobacter sphaeroides* to chemotactic stimulation. *J. Bacteriol.* **170**, 5673–5679.

Poole, P. S., Sinclair, D. R., and Armitage, J. P. (1988). Real time computer tracking of free-swimming and tethered rotating cells. *Anal. Biochem.* **175**, 52–58.

Poole, P. S., Brown, S., and Armitage, J. P. (1990a). Swimming changes and chemotactic responses in *Rhodobacter sphaeroides* do not involve changes in the steady state membrane potential or respiratory electron transport. *Arch. Microbiol.* **153**, 614–618.

Poole, P. S., Williams, R. L., and Armitage, J. P. (1990b). Chemotactic responses of *Rhodobacter sphaeroides* in the absence of apparent adaptation. *Arch. Microbiol.* **153**, 368–372.

Poole, P. S., Brown, S., and Armitage, J. P. (1991). Chemotaxis and chemokinesis in *Rhodobacter sphaeroides*: Modelling of the two effects. *Binary* **3**, 183–190.

Poole, P. S., Smith, M. J., and Armitage, J. P. (1993). Chemotactic signalling in *Rhodobacter sphaeroides* requires metabolism of attractants. *J. Bacteriol.* **175**, 291–294.

Postma, P. W., and Lengeler, J. (1985a). Phosphoenolpyruvate:carbohydrate phosphotransferase system of bacteria. *Microbiol. Rev.* **49**, 232–269.

Postma, P. W., and Lengeler, J. W. (1985b). Phosphoenolpyruvate:carbohydrate phosphotransferase system of bacteria. *Microbiol. Rev.* **49**, 232–269.

Postma, P. W., Lengeler, J. W., and Jacobson, G. R. (1993). The PTS as a signal transduction system in chemotaxis. *Microbiol. Rev.*, in press.

Saier, M. H., Jr. (1989). Protein phosphorylation and allosteric control of inducer exclusion and catabolite repression by the bacterial phosphoenolpyruvate:sugar phosphotransferase system. *Microbiol. Rev.* **53**, 109–120.

Saimi, Y., Marinac, B., Gustin, M. C., Culbertson, M. R., Adler, J., and Kung, C. (1988). Ion channels in *Paramecium*, yeast and *Escherichia coli*. *Trends Biochem. Sci.* **13**, 304–309.

Shioi, J., Dang, C. V., and Taylor, B. L. (1987). Oxygen as attractant and repellent in bacterial chemotaxis. *J. Bacteriol.* **169**, 3118–3123.

Sockett, R. E., Armitage, J. P., and Evans, M. C. W. (1987). Methylation-independent and methylation-dependent chemotaxis in *Rhodobacter sphaeroides* and *Rhodospirillum rubrum*. *J. Bacteriol.* **169**, 5808–5814.

Sprenger, G. A., and Lengeler, J. (1984). L-Sorbose metabolism in *Klebsiella pneumoniae* and Sor^+ derivatives of *Escherichia coli* K-12 and chemotaxis towards sorbose. *J. Bacteriol.* **157**, 39–45.

Sprenger, W. W., Hoff, W. D., Armitage, J. P., and Hellingwerf, K. J. (1993). The eubacterium *Ectothiorhodospira halophila* is negatively phototactic with a wavelength dependence that fits the absorption spectrum of the photoactive yellow protein. *J. Bacteriol.* **175**, 3096–3104.

Spudich, E. N., Takahashi, T., and Spudich, J. L. (1989). Sensory rhodopsin I and II modulate a methylation/demethylation system in *Halobacterium halobium*. *Proc. Natl. Acad. Sci. U.S.A.* **86**, 7746–7750.

Spudich, J. L., and Bogomolni, R. A. (1988). Sensory rhodopsins of halobacteria. *Ann. Rev. Biophys. Biophys. Chem.* **17**, 193–215.

Taylor, B. L., and Lengeler, J. W. (1990). Transductive coupling by methylated transducing proteins and permeases of the phosphotransferase system in bacterial chemotaxis. *In* "Membrane Transport and Information Storage" (R. C. Aloia, C. C. Curtain, and L. M. Gordon, eds.), pp. 69–90. Liss, New York.

Taylor, B. L., Miller, J. B., Warrick, H. M., and Koshland, D. E., Jr. (1979). Electron acceptor taxis and blue light effect on bacterial chemotaxis. *J. Bacteriol.* **140**, 567–573.

Taylor, B. L., Johnson, M. S., and Smith, J. M. (1988). Signaling pathways in bacterial chemotaxis. *Bot. Acta* **101**, 101–104.

Tribhuwan, R. C., Johnson, M. S., and Taylor, B. L. (1986). Evidence against the direct involvement of cyclic GMP or cyclic AMP in bacterial chemotactic signaling. *J. Bacteriol.* **168**, 624–630.

Vogler, A. P., and Lengeler, J. W. (1987). Indirect role of adenylate cyclase and cyclic AMP in chemotaxis to phosphotransferase system carbohydrates in *Escherichia coli*. *J. Bacteriol.* **169**, 593–599.

Weng, Q.-P., Elder, J., and Jacobson, G. R. (1992). Site-specific mutagenesis of residues in the *Escherichia coli* mannitol permease that have been suggested to be important for its phosphorylation and chemoreceptor functions. *J. Biol. Chem.* **267**, 19529–19535.

Wolfe, A. J., Conley, M. P., and Berg, H. C. (1988). Acetyladenylate plays a role in controlling the direction of flagellar rotation. *Proc. Natl. Acad. Sci. U.S.A.* **85**, 6711–6715.

Wu, L.-F., Tomich, J. M., and Saier, M. H., Jr. (1990). Structure and evolution of a multidomain multiphosphoryl transfer protein. Nucleotide sequence of the *fruB(HI)* gene in *Rhodobacter capsulatus* and comparisons with homologous genes from other organisms. *J. Mol. Biol.* **213**, 687–703.

Yamamoto, K., and Imae, Y. (1993). Cloning and characterization of the *Salmonella typhimurium*-specific chemoreceptor Tcp for taxis to citrate and from phenol. *Proc. Natl. Acad. Sci. U.S.A.* **90**, 217–221.

Zhulin, I. B., and Armitage, J. P. (1993). Motility, chemokinesis, and methylation-independent chemotaxis in *Azospirillum brasilense*. *J. Bacteriol.* **175**, 952–958.

Chapter 4

The *Agrobacterium* Ti Plasmid and Crown Gall Tumorigenesis: A Model for Signal Transduction in Host–Pathogen Interactions

ROBERT G. ANKENBAUER

Laboratory of Microbial Structure and Function
Rocky Mountain Laboratories
National Institutes of Health
Hamilton, Montana 59840

EUGENE W. NESTER

Department of Microbiology
University of Washington
Seattle, Washington 98195

I. Crown Gall Tumorigenesis by *Agrobacterium tumefaciens*

A. Introduction

The gram-negative bacterium *Agrobacterium tumefaciens* is the etiologic agent of crown gall, a neoplastic disease of plants. Crown gall is an agronomically important malady affecting a large number of different plant species. Crown gall tumors are the unique pathologic manifestation of the interaction between pathogen (*A. tumefaciens*) and host (plant).

Agrobacterium tumefaciens has been known to be the agent of crown gall for over 80 years (see Lippincott and Lippincott, 1975, for historical review). However, the biological basis for tumorigenesis was not elucidated until the late 1970s. The discovery of the Ti (tumor-inducing) plasmid (Van Larebeke *et al.*, 1974; Watson *et al.*, 1975) added to the growing awareness of the importance of plasmids in biologic and pathogenic phenomena. The subsequent discovery of the T-DNA (transfer-DNA) (Chilton *et al.*, 1977; Thomashow *et al.*, 1980) and its role in tumorigenesis can be deemed a cornerstone in the development of plant molecular biology as a field. The *Agrobacterium*–plant interaction is a unique pathogenic process and is one of the few examples of interkingdom DNA transfer (Heinemann and Sprague, 1989).

The molecular basis of crown gall tumorigenesis is the transfer and integration of a discrete segment of bacterial DNA, the T-DNA, from the Ti plasmid into the plant cell. Crown gall tumorigenesis is a complex process involving at least 30 *A. tumefaciens* gene products. The initial step of the tumorigenic process is the localization and attachment of the bacteria to specific sites on a susceptible host, namely, plant wounds. A large number of bacterial operons [the *vir* (*vir*ulence) regulon] subsequently are induced in response to certain plant metabolites generated by plant wounding. The products of the *vir* regulon are responsible for (1) detection of these wound-specific metabolites; (2) the intracellular processing of a segment of the Ti plasmid; and (3) transfer of this plasmid fragment (the T-DNA) into the plant cell. The T-DNA, which encodes enzymes involved in the biosynthesis of the plant growth hormones cytokinin and auxin, integrates into the plant nuclear genome. The expression of the T-DNA genes leads to the overproduction of plant hormones and the uncontrolled proliferation of plant cells, thus yielding a crown gall tumor. The T-DNA also encodes enzymes for the biosynthesis of opines, structurally unique compounds that can be catabolized by the tumorigenic *A. tumefaciens* strain. Thus, by causing crown galls, *A. tumefaciens* is believed to create a unique environmental niche for itself.

With respect to its molecular biology, the *Agrobacterium* crown gall tumorigenesis phenomenon has been recognized as the best understood interaction between pathogen and host (DiRita and Mekalanos, 1989; Miller *et al.*,

1989). The regulation of *Agrobacterium* virulence determinants in response to its specific pathogenic niche, the plant wound, has served as a prototype in the understanding of other host–pathogen interactions (Fig. 4.1). The identification of the signal transduction system responsible for the virulence of a given pathogen has assisted in the discovery of a multitude of virulence determinants regulated by such systems. Conversely, the identification of the host molecule detected by the signal transduction system of a pathogen has taught us much about the cellular and molecular biology of the host during infection. The progress achieved in these areas owes much of its success to the pioneering work on the *Agrobacterium* crown gall system.

Several excellent reviews have been written on the crown gall phenomenon. Several cover the phenomenon as a whole (Nester *et al.*, 1984; Binns

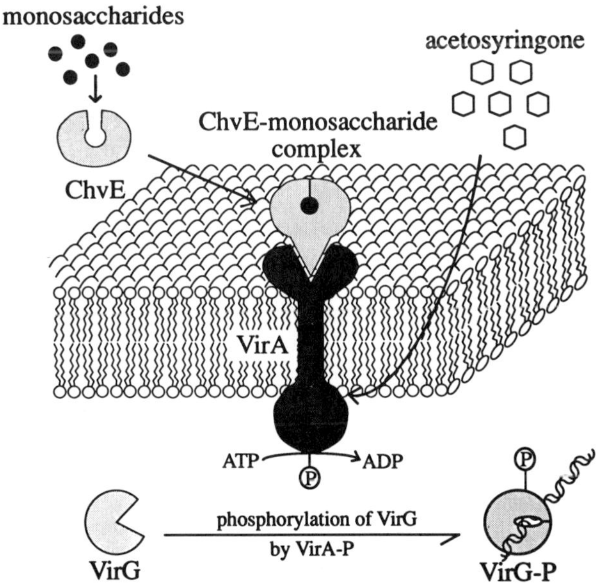

Figure 4.1 The VirA/VirG signal transduction system. The cross section shown is of the *A. tumefaciens* cytoplasmic membrane with the periplasmic space at the top and the cytoplasm at the bottom. The plant wound-specific metabolites monosaccharide and acetosyringone are detected at the cytoplasmic membrane by the sensor protein VirA. Acetosyringone penetrates the membrane by virtue of its hydrophobicity and interacts with protein sequences just inside the cytoplasmic membrane. The monosaccharides are bound by the glucose–galactose binding protein ChvE. This monosaccharide–ChvE complex interacts with the periplasmic domain of VirA. When these two chemical stimuli are detected by VirA–P, VirA–P transfers its phosphate group onto the regulator protein VirG. When phosphorylated, VirG binds to specific promoter sequences in *vir* regulon operons, thereby directing RNA polymerase to start transcription of the subject promoters. The role of other *A. tumefaciens* loci with effects on signal transduction are not included in this scheme since their role in *vir* regulon induction is not entirely clear.

and Thomashow, 1988; Winans, 1992) whereas others have emphasized the T-DNA transfer process (Zambryski, 1988,1989; Ream, 1989; Heinemann, 1991) and the early history of *Agrobacterium* research (Lippincott and Lippincott, 1975). The *Agrobacterium*–plant signaling pathways also have received attention in several reviews (Halverson and Stacey, 1986; DeCleene, 1988; Peters and Verma, 1990; Winans, 1992). This chapter concentrates on the signal transduction system responsible for crown gall tumorigenesis.

B. Chromosomal Loci

Several *A. tumefaciens* chromosomal loci are required for crown gall tumorigenesis. The *chvA*, *chvB* (Douglas *et al.*, 1982), *exoC* (*pscA*) (Cangelosi *et al.*, 1987; Thomashow *et al.*, 1987), *att* (Matthysse, 1987), and *cel* (Matthysse, 1983) loci all are involved in the attachment of *A. tumefaciens* to plant cells. No Ti plasmid-encoded loci mediate attachment. Since attachment to plant cells occurs in the normal commensal life-style of *A. tumefaciens* and no apparent involvement of the Ti plasmid is required in attachment, attachment cannot be considered in itself a pathogenic event since it occurs in the absence of any possibility of tumorigenesis. For thorough discussions of bacterial attachment in crown gall tumorigenesis, the reader is referred to any of the reviews cited earlier.

Other chromosomal loci are involved in *vir* regulon expression and have no role in attachment (Table I). The functions of the *chvD* (Winans *et al.*, 1988), *miaA* (Gray *et al.*, 1992), *ivr-211*, *ivr-223*, and *ivr-225* loci (Metts *et al.*, 1991) in *vir* regulon expression are unknown. The functions of the *chvE* and *ros* loci are better understood. The role of chromosomal loci in *vir* regulon expression is discussed in Sections II,C and II,D.

Indirect evidence suggests that chromosomal chemotaxis loci are involved in crown gall tumorigenesis (Hawes *et al.*, 1988; Hawes and Smith, 1989). Several *A. tumefaciens* mutants defective in chemotaxis are avirulent when assayed for tumorigenesis on pea plants (Hawes and Smith, 1989). Logically, *A. tumefaciens* must have some mechanism to recognize host plants in its natural environment, the soil. However, the explicit role of chemotaxis in virulence is obscure since tumorigenesis may be incident with the normal ecology of *A. tumefaciens*. *Agrobacterium tumefaciens* can be considered a plant commensal organism, a microbe that normally colonizes plants without eliciting disease. This bacterium is only capable of causing disease when the normal host defenses of the plant are breached, in this case a plant wound. Evidence indicates that *A. tumefaciens* can respond chemotactically to both wounded and unwounded plants, and that these responses are separate. Hawes *et al.* (1988) isolated a mutant of *A. tumefaciens* that was chemotactic toward plant wound exudates but not toward unwounded plant exudates. Chemotaxis toward specific phenolic plant-wound metabolites that induce the *vir* regulon (e.g., acetosyringone) is controversial. Chemotaxis is discussed in Sections II,A and II,C.

TABLE I
Loci Involved in *vir* Regulon Expression

Locus	Role of gene product in *vir* regulon expression
Ti plasmid	
virA	Sensor protein; histidine protein kinase; responsible for detecting plant wound-specific metabolites
virG	Regulator protein; positive transcriptional activator; binds *vir* box sequences and promotes transcription on phosphorylation
Chromosomal	
chvE	Periplasmic monosaccharide-binding protein; interacts with periplasmic domain of VirA to enhance signal transduction
ros	Negative regulator of *virC* and *virD* expression; acts independently of VirA/VirG signaling; precise role in *vir* regulon induction is unknown
chvD	Enhances *virG* expression in response to low pH and phosphate starvation; precise role in *vir* regulon induction is unknown
miaA	tRNA:isopentenyltransferase; affects efficiency and fidelity of overall translation within the cell; precise role in *vir* regulon induction is unknown
ivr loci	Allow *vir* regulon expression; mutations in these loci are totally defective in *vir* regulon induction; precise role in *vir* regulon induction is unknown
p10 and p21 proteins	Possibly involved in the detection of plant phenolic metabolites; shown to bind plant phenolics; precise role in *vir* regulon induction is unknown

C. Ti Plasmid

The vast majority of *A. tumefaciens* strains isolated from sources other than crown galls is nonpathogenic (Anderson and Moore, 1979; Binns and Thomashow, 1988). The difference between pathogenic and nonpathogenic isolates is the presence of a large (>200 kb) plasmid in the pathogenic strains. These large plasmids are responsible for the tumorigenic capacity of *A. tumefaciens* and have been designated tumor-inducing (Ti) plasmids (Van Larebeke *et al.*, 1974; Watson *et al.*, 1975). Of the ~200 kb that compose the Ti plasmid, only ~100 kb have known functions. These plasmid-encoded loci have roles in (1) Ti plasmid replication, conjugation, and incompatability functions and (2) tumorigenesis. Three regions of the Ti plasmid are dedicated to tumorigenesis and its aftermath. These three regions are the T-DNA, the opine catabolism genes, and the *vir* regulon.

1. T-DNA (Transfer DNA) Although the transfer of DNA from *A. tumefaciens* to plant cells long had been suspected to be the actual tumor-inducing principle, verification of this hypothesis was difficult. The discovery of the Ti plasmid was instrumental in demonstrating that DNA of this plasmid is incorporated into the plant cell DNA in crown gall tumors (Chilton *et al.*, 1977). Further work determined the specific Ti-plasmid sequences that are incorporated into the plant cell on transformation (Thomashow *et al.*, 1980); these sequences define the DNA that is transferred to the plant cell and are designated the transfer- or T-DNA.

The T-DNA encodes multiple loci that lead to uncontrolled proliferation of plant cells when expressed (Garfinkel *et al.*, 1981). The *tms1* and *tms2* (also designated *iaaM* and *iaaH*) gene products are enzymes that biosynthetic indole-3-acetic acid, an auxin. The *tmr* (or *ipt*) gene product synthesizes isopentenyl AMP, a cytokinin. Auxins and cytokinins are potent plant growth regulators that, when overproduced, cause rapid and uncontrolled cell division. The *tml* locus has an unknown function that depresses the response of the plant cell to the excess auxin and cytokinin, thereby leading to smaller tumors. The major function of these oncogenes is to alter the normal phytohormone balance in the cell to a balance that favors rapid growth.

The T-DNA also encodes loci responsible for the biosynthesis of opines, a group of amino acid and sugar derivatives. Resident on the Ti plasmid but not the T-DNA are opine catabolism genes. The transfer of the T-DNA into plant cells causes the overproduction of opines that can be catabolized by the *A. tumefaciens* strain that contains the Ti plasmid-encoded opine catabolism loci. The ability of *A. tumefaciens* selectively to utilize opines, the synthesis of which is effected directly by *A. tumefaciens*, is presumed to provide the bacterium with an exclusive ecological niche. This idea, called the opine hypothesis (Tempe and Petit, 1982), is believed to be the natural selection basis for crown gall tumorigenesis. However, in view of the fact that bacteria other than virulent *A. tumefaciens* utilize opines (Bergeron *et al.*, 1990; Nautiyal and Dion, 1990), the opine hypothesis is difficult to confirm unequivocally.

The T-DNA is a structural element only and encodes no products that mediate its own transfer. This region is flanked by *cis*-acting sequences that are necessary for transfer. These sequences, called border sequences, are imperfect 25-bp direct repeats that are recognized by *vir* regulon products and serve essentially as origins of transfer.

2. vir *Regulon* In addition to the T-DNA, another region of the Ti plasmid is required absolutely for tumorigenesis. This region initially was discovered when avirulent mutations in the Ti plasmid were localized to sequences outside of the T-DNA region (Garfinkel and Nester, 1980). Such mutations defined a region of the Ti plasmid that was required for virulence and was designated the *vir* region. Analysis of *vir* region mutants revealed that the mutants

could be complemented in *trans*, suggesting that the *vir* region encoded proteins involved in T-DNA transfer (Hille *et al.*, 1982; Klee *et al.*, 1982). Transposon mutagenesis demonstrated that the *vir* region consisted of ~35 kb and several distinct *vir* loci (Klee *et al.*, 1983). Later work established the existence of ~8 *vir* loci (the number present varies among different isolates of *A. tumefaciens*) and demonstrated that the *vir* region exists as a regulon that is controlled tightly by a common regulatory system (Stachel and Nester, 1986; Rogowsky *et al.*, 1987).

The date, the *vir* regulon in octopine strains is composed of the *virA*, *virB*, *virC*, *virD*, *virE*, *virF*, *virG*, and *virH* (formerly *pinF*) loci (Winans, 1992). Of these loci, *virA*, *virB*, *virD*, and *virG* are absolutely required for tumorigenesis. Mutations in the *virC* and *virE* loci attenuate the virulence of *A. tumefaciens* whereas lesions in *virF* and *virH* have little effect or only host-range effects on the tumorigenic potential of *A. tumefaciens*. The functions of the gene products of the various *vir* loci have been the subject of intense research. Impressive progress has been made in the understanding of how the *vir* regulon products mediate T-DNA transfer. VirA and VirG are regulatory proteins that detect a susceptible plant host and mediate the transcriptional induction of the *vir* regulon. The *virC* and *virD* loci encode two and five products, respectively, several of which recognize the T-DNA border sequences and introduce single-stranded nicks in the border sequences. The single-stranded T-DNA is displaced from the Ti plasmid by an unknown mechanism but may involve coating of the T-DNA by the two *virE* gene products, one of which (VirE2) is a single-stranded DNA binding protein. The T-DNA apparently is directed through the bacterial membrane and into the plant cell via the 11 proteins encoded by the *virB* loci. The functions of several of the *vir* regulon products (particularly the *virF* and *virH* gene products) are unknown. In addition to effecting T-DNA transfer, the *vir* regulon products also promote conjugal transfer of the Ti plasmid. Specifically, the *virA*, *virB*, *virC*, *virE*, and *virG* loci significantly enhance the frequency of conjugal transfer of the Ti plasmid between *A. tumefaciens* strains (Gelvin and Habeck, 1990; Steck and Kado, 1990).

The mechanism of T-DNA transfer across the bacterial and plant cell membranes, the fate of the T-DNA following transfer, and the role of *vir* regulon products in T-DNA integration into the plant cell genome will be major areas of *Agrobacterium* research in the next decade. Further information on the nonregulatory aspects of *vir* regulon function can be found in the reviews cited earlier.

D. Plant Hosts

Agrobacterium tumefaciens is capable of causing tumors on a large number of different plants (DeCleene and DeLey, 1976). Most dicotyledonous plants and several gymnosperms yield crown gall tumors on infection with *A. tumefaciens*.

However, very few monocotyledonous plants have been transformed successfully by *A. tumefaciens*. Despite intensive investigation into the recalcitrance of monocots to transformation, the biological basis of this difficulty remains a mystery.

Agrobacterium tumefaciens is an opportunistic pathogen that is incapable of causing disease in a structurally intact plant. Direct application or nondisruptive infiltration of *A. tumefaciens* into plants does not produce tumors. Conversely, a plant that has experienced structural damage (e.g., disruption of the waxy outer cuticle layer of the plant) is highly susceptible to infection with *A. tumefaciens* (Braun, 1978). A plant wound appears to be required for several reasons (Lippincott and Lippincott, 1975): (1) *A. tumefaciens* is not an invasive pathogen and is unable to penetrate the cuticle of the plant on its own; (2) wounding provides sites at which the bacteria must attach; (3) wound exudates are capable of serving as a growth medium for the bacterium, which must be actively growing to be induced (Culianez-Macia and Hepburn, 1988); and (4) wounding elicits a number of physiological changes in the plant cell (the wound response) that render them susceptible to transformation by *A. tumefaciens*.

On being damaged by a variety of mechanisms, plants undergo a wound response. The plant wound response is a highly complex process involving the specific induction and expression of gene products that defend the plant against infection by pathogenic microorganisms (Kahl, 1982). Several of the phenomena associated with the wound response are salient with respect to crown gall tumorigenesis. Plant cells undergo a burst of rapid cell division after a wounding event. A strong correlation exists between plant cell division and DNA synthesis and the susceptibility of such cells to transformation by *A. tumefaciens* (Binns and Thomashow, 1988). The contents of plant vacuoles are acidic (Kahl, 1982; Bolker and Weimken, 1986). Therefore disruption and cell lysis at the wound cite leads to acidic conditions. Plant cells contain a battery of glycosidases that are involved in plant defense mechanisms and expansion of the cell wall as required for cell growth (Hahn *et al.*, 1989). When the cell wall structure is damaged in wounding, these glycosidases are released and mono- and oligosaccharides are produced as a result. Specific oligosaccharides elicit the production of small antimicrobial compounds, phytoalexins, and other plant secondary metabolites (McNeil *et al.*, 1984). Further, plant wounds act as a "sink" for monosaccharides, which accumulate at high concentrations (G. An and C. Ryan, personal communication). Thus, the phytoalexins, secondary metabolites, monosaccharides, and low pH present in a plant wound can be recognized as wound-specific metabolites. *Agrobacterium tumefaciens* recognizes each of these compounds and conditions as a signal that its pathogenic niche, the plant wound, is present, so the virulence mechanisms of the bacteria are engaged. Other, yet unidentified changes in the plant that are caused by the wound response are likely to be required for tumor formation.

Analysis of the wound response as a function of transformation via *A. tumefaciens* increasingly will become an important area of research.

II. Plant–Bacteria Interaction in *vir* Regulon Induction

A. VirA and VirG

The *vir* regulon is induced when bacteria are cultured in the presence of wound-specific plant metabolites (Stachel and Zambryski, 1986; Winans *et al.*, 1986). The *virB, virC, virD, virE, virF, virG,* and *virH* (*pinF*) operons are strongly plant inducible and, except for *virG*, expressed only in plant-induced *A. tumefaciens*. In contrast, the *virA* and *virG* genes are expressed at low levels. The *virA* gene is only weakly plant inducible (Stachel and Nester, 1986; Rogowsky *et al.*, 1987; Winans *et al.*, 1988). The strongly regulated nature of the *vir* regulon and the basal expression of the *virA* and *virG* loci suggest that the *virA* and *virG* gene products served a regulatory function.

Mutational analysis of the *vir* regulon reveals that VirA and VirG are responsible for the plant-mediated induction of the *vir* regulon (Table I). Plant-inducible *vir* loci are not expressed in *virA* or *virG* mutants (Stachel and Zambryski, 1986; Winans *et al.*, 1986). Mutations in the remaining *vir* genes do not affect *vir* regulon induction (Stachel and Zambryski, 1986). Additional studies demonstrate that *virA* and *virG* are the only Ti plasmid-encoded loci required for *vir* regulon induction (Stachel and Zambryski, 1986; Rogowsky *et al.*, 1987; Das and Pazour, 1989).

Sequence analysis of the *virA* and *virG* loci indicates that the VirA–VirG protein pair is a member of a family of bacterial two-component regulatory systems responsible for sensing specific environmental stimuli (Albright *et al.*, 1989; Bourret *et al.*, 1989; Gross, *et al.*, 1989; Stock *et al.*, 1989). In most two-component regulatory systems, the sensor protein senses the environment and then activates the regulator protein which then promotes transcription of the subject regulon. The VirA protein shares homologous sequences with the sensor components and the VirG protein is homologous to the regulator components.

Several *virA* loci have been sequenced. The various genes encode a polypeptide of approximately 830 amino acids with a molecular weight of 92 kDa (Leroux *et al.*, 1987; Melchers *et al.*, 1987; Morel *et al.*, 1989). Hydropathy profiles of the predicted amino acid sequence of VirA indicate the presence of two hydrophobic stretches of amino acids at residues 20–37 (transmembrane domain 1; TM1) and 260–278 (TM2), suggesting that VirA is a transmembrane protein. VirA is associated with the *A. tumefaciens* cytoplasmic membrane after cell fractionation and immunoblot analysis (Leroux *et al.*, 1987; Melchers *et al.*, 1986). The C-terminal half of VirA is highly homologous to the C-terminal regions of EnvZ, NtrB, PhoR, CpxA, and other sensor components of the two-component regulatory systems. The N-terminal half of VirA demonstrates no homology to other sensor components, suggesting that this region may be responsible for detecting the specific environmental signals responsible for *vir* regulon induction (i.e., wound-induced plant phenolic compounds). At the extreme C terminus of VirA is a sequence that is homologous to the N-terminal domain of regulator proteins (e.g., VirG), a property shared by several sensor proteins (Chang and Winans, 1992). Investigations into the importance of this domain for VirA function have led to contradictory results. This region of VirA is proposed to be required for its biological function since deletions at the extreme C terminus of this domain render VirA inactive in induction assays (Melchers *et al.*, 1989a). In contrast, Chang and Winans (1992) have made precise deletions of this domain; the resulting VirA constructs actually yield higher *vir* regulon induction levels than wild-type VirA. Further, these truncated VirA derivatives are also proficient in tumorigenesis. These results are interpreted to indicate that this C-terminal domain of VirA is a negative modulator of VirA activity in signaling. A comparison of the constructs used in these separate studies shows that identical deletions were not used. Therefore, although the C-terminal domain is not required for induction or tumorigenesis, mutations in this region may have strong effects on the signaling capability of VirA.

Several *virG* loci also have been sequenced (Melchers *et al.*, 1986; Winans *et al.*, 1986; Powell *et al.*, 1987). These genes encode a protein of about 250 amino acids with a molecular mass of 27–28 kDa. The *virG* gene is unusual because it uses a UUG codon as its translational start site rather than AUG (Jin *et al.*, 1990c; Pazour and Das, 1990b). The use of UUG as a translational start site was hypothesized to result in lower levels of production of this regulatory protein as a result of inefficient translation. Transcriptional regulation of *virG* is complex. The *virG* gene is transcribed from two different promoters spaced approximately 50 bp apart (Stachel and Zambryski, 1986). Transcription from the 5′ promoter (P1) is promoted by acetosyringone and phosphate starvation (Stachel and Zambryski, 1986; Winans *et al.*, 1988; Winans, 1990) whereas the promoter adjacent to the *virG* coding sequence (P2) is expressed vegetatively and also induced by a variety of environmental stresses including acidic and alkaline pH, high temperature, ethanol, and heavy metals (Winans *et al.*, 1988; Winans, 1990). The *virG* P2 promoter has similarities to *Escherichia coli* heat-shock promoters and is induced in the heat-shock response to noxious agents in *A. tumefaciens*. The effects of acidic pH and phosphate starvation on *vir* regulon induction is discussed in Sections II,E,1 and II,E,2. VirG and other regulator proteins of the family of two-component regulatory systems share homology in their N-terminal halves and fewer similarities in their C termini. The similarities in the C-terminal domains of the sensors and the N-terminal domains of the regulators suggest a common mode of transcriptional activation via sensors and regulators. The unique sequences in the C-terminal regions of regulators suggest differences in the DNA sequences bound by the activated regulators (Albright *et al.*, 1989; Bourret *et al.*, 1989; Gross *et al.*, 1989; Stock *et al.*, 1989).

The hydropathy profile and cytoplasmic membrane localization of VirA has led investigators to propose the following model (Winans *et al.*, 1986; Leroux *et al.*, 1987). VirA is a transmembrane protein with its N-terminal region in the periplasmic space and its C-terminal domain in the cytoplasm. Wound-specific plant phenolic compounds (e.g., acetosyringone) that interact with the periplasmic domain of VirA elicit a conformational change through the membrane-spanning region, which causes the C-terminal domain of VirA to modify VirG covalently, presumably by phosphorylation. This modified VirG is then capable of binding *vir* regulon promoters and activating transcription.

The transmembrane topology of VirA has been confirmed genetically by Winans *et al.* (1989) and Melchers *et al.* (1989a). Both groups used alkaline phosphatase fusions to VirA to demonstrate that a stretch of 220–230 amino acids corresponding to codons 39–260 are exposed extracellularly in the periplasm. Note, however, that because of the low protein copy number of VirA in *A. tumefaciens* (~10/cell), these *virA–phoA* fusions were analyzed in *E. coli*, a heterologous host. Protease treatment of *E. coli* and *A. tumefaciens* spheroplasts expressing VirA and subsequent immunoblot analysis with anti-VirA antibodies physically demonstrates the transmembrane nature of VirA (Winans *et al.*, 1989). The native VirA protein migrates as an 88-kDa peptide on SDS–PAGE gels. Following protease treatment of the spheroplasts, a 60-kDa protein is the remaining VirA fragment, which indicates that 28 kDa of VirA is exposed extracellularly and 60 kDa is membrane bound and cytoplasmic. Although residues 260–279 of VirA TM2 are known to transverse the cytoplasmic membrane, the localization of the first hydrophobic domain TM1, residues 20–37, remains unknown. Most investigators indicate that the 19 N-terminal amino acids are cytoplasmic and that TM1 transverses the membrane leading into a periplasmic loop (Melchers *et al.*, 1989a). However, the physical data presented by Winans *et al.* (1989) suggest that approximately 40 amino acids of VirA are cleaved from the N terminus during membrane localization. Expression of native VirA in *E. coli* leads to the production of two different forms of VirA, a 92-kDa form and an 88-kDa form. Overexpression of VirA is toxic to *E. coli*, presumably because of deleterious effects associated with membrane targeting of heterologous proteins. The two different VirA forms in *E. coli* likely correspond to a full-length protein (92 kDa) and a processed form (88 kDa). Parallel analysis of the VirA expressed by *A. tumefaciens* shows that it comigrates with the 88-kDa form of VirA, indicating that the protein may exist in processed form in *A. tumefaciens*.

The extracellular periplasmic domain of VirA logically was assumed to be the binding site for the *vir* regulon-inducing plant phenolic compounds. This assumption was reasonable since the plant phenolics are of extracellular origin and sensing of an extracellular stimulus is likely to occur at a site external to the cytoplasmic membrane. Melchers *et al.* (1989a) disproved this assumption by demonstrating that in-frame deletions of the periplasmic domain of VirA still yielded high levels of *vir* regulon induction. In an additional set of experiments, these investigators constructed a series of hybrids between VirA and Tar, an *E. coli* chemosensory protein with membrane topology similar to that of VirA. These hybrids still induced the *vir* regulon and formed plant tumors, indicating that the transmembrane domain of VirA (amino acids 260–278) is absolutely required for *vir* regulon induction by plant phenolic compounds. Replacement of this transmembrane domain with the analogous transmembrane domain of Tar allowed no induction or plant tumor formation. Contradictory data have been reported by Chang and Winans (1992), who found that deleting TM2 and residues N-terminal to TM2 still allowed induction by acetosyringone, although not at the level achieved by wild-type VirA; these investigators qualified the earlier findings by determining that TM2 was required for tumorigenesis but not for *vir* regulon induction by acetosyringone. The results obtained with chimeric proteins or deletion derivates can provide indirect evidence for certain conclusions but conformational aspects of VirA that are not truly analogous in such unnatural protein constructs are likely to exist.

VirA and VirG also have been reported to mediate a chemotactic response toward acetosyringone (Shaw *et al.*, 1988). This observation was disputed by Parke *et al.* (1987), who found no chemotaxis toward acetosyringone by *A. tumefaciens*. Several plant phenolic metabolites serve as chemoattractants for *A. tumefaciens*, but acetosyringone was not among them (Parke *et al.*, 1987). Specific strain differences in chemotaxis and motility were invoked to explain these discrepancies (Shaw *et al.*, 1988). Chemotaxis toward wound-specific molecules is reasonable since this behavior would lead to the pathogenic niche of *A. tumefaciens*. Additionally, the involvement of VirA as a chemoreceptor would not be incongruous because of its homology with the *E. coli* chemotactic effector protein, CheA. However, a role for VirG in chemotaxis is more difficult to envision.

Mutants of VirA and VirG that are no longer dependent on plant phenolic metabolites (e.g., acetosyringone) to effect *vir* regulon induction have been isolated. The mutations in the *virG* gene leading to constitutive expression of the *vir* regulon are located in the N-terminal regulator domain. In both studies identifying such *virG* mutants (Han *et al.*, 1992; Pazour *et al.*, 1992), an asparagine to aspartate change at residue 54 was found to be the strongest of the mutants in expressing the *vir* regulon in the absence of acetosyringone. The *virA* mutations resulting in acetosyringone-independent signaling of the *vir* regulon were not grouped in specific sections of the VirA protein, unlike VirG

(Ankenbauer *et al.*, 1991; Pazour *et al.*, 1991). Such VirA or VirG mutants may be able to enhance the transformation of recalcitrant plant species, an idea that was supported in the case of one *virA* mutant allele, *virA112* (Ankenbauer *et al.*, 1991). *Agrobacterium tumefaciens* strains bearing the *virA112* mutation are capable of efficiently transferring DNA to the monocot maize whereas strains with the wild-type *virA* locus are not (Raineri *et al.*, 1993).

Continuous induction of the VirA–VirG signal transduction system by long-term culture of *A. tumefaciens* in the presence of acetosyringone strongly inhibits cell growth and damages the genetic integrity of the genes involved in signal transduction (Fortin *et al.*, 1992). The only cells capable of growing under these conditions are those suffering major deletions in the *vir* regulon or mutation in the *virA* or *virG* genes. All such mutants are avirulent and incapable of expressing *vir* regulon loci.

B. Acetosyringone and Plant Phenolic Metabolites

Induction of the *vir* regulon by plants was reported first in 1984 (Stachel *et al.*, 1984). Plant products were reported to be capable of inducing the expression of *lacZ* fusions to *vir* genes in *A. tumefaciens* but not in *E. coli*. In a remarkable series of studies by Stachel *et al.* (1984,1985a,b,1986), the progression from *vir* regulon induction by cocultivation with cultured plant cells to the structural elucidation of the specific plant metabolite responsible for *vir* regulon induction can be traced. Culturing *A. tumefaciens* strains in the presence of *Nicotiana* protoplasts, leaf disks, roots, and root and callus cultures yielded high expression of *vir::lacZ* fusions (Stachel *et al.*, 1985a,b,1986). The *vir* regulon-inducing compound from *Nicotiana* was found to be small (MW<1,000); resistant to boiling, freezing, lyophilization, and high and low pH; and hydrophobic (Stachel *et al.*, 1985b). After HPLC fractionation of a crude *Nicotiana* exudate, two *vir* regulon-inducing compounds were identified as acetosyringone (3′,5′-dimethoxy-4′-hydroxyacetophenone) and α-hydroxyacetosyringone (2,4′-dihydroxy-3′,5′-dimethoxyacetophenone) (Stachel *et al.*, 1985b) (Fig. 4.2). Acetosyringone appears to be a wound-specific metabolite, since wounding *Nicotiana* leaves increases acetosyringone production more than 10-fold. This hypothesis was confirmed by the observation that treatment of *Nicotiana tabacum* suspension cultures with the cell wall-degrading enzyme cellulase elicited high levels of acetosyringone production (Threlfall and Whitehead, 1988). Additionally, very little acetosyringone is present in unwounded plants, as determined by an isotope dilution method (H. Floss and E. W. Nester, unpublished observations). Only viable metabolically active plant cells are capable of producing acetosyringone; dead cells or cells inhibited by cycloheximide do not produce acetosyringone, even on wounding. Therefore, acetosyringone is likely to be a product of *de novo* synthesis and not a preexisting compound that is released on wounding.

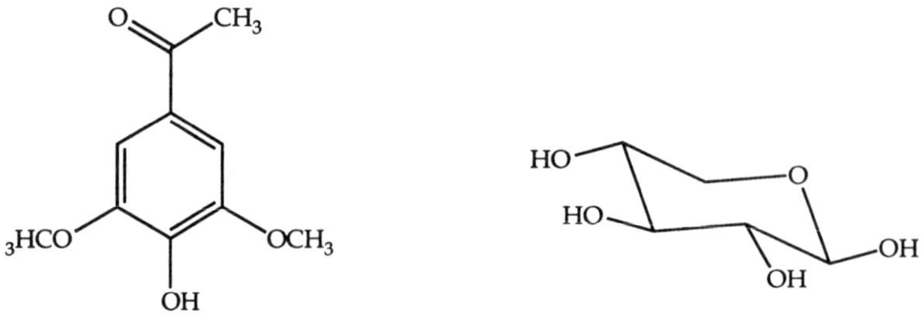

Acetosyringone

D(+)-Xylose

Figure 4.2 Structures of plant wound-specific metabolites detected by the VirA/VirG signal transduction system. Acetosyringone is shown here as a representative of the phenolic compounds detected by VirA; several other compounds also are detected among the phenolic plant metabolites (see text) and all have structural similarities to acetosyringone. D(+)-Xylose is shown here as a representative monosaccharide detected by the ChvE–VirA interaction. Several other monosaccharides also interact with ChvE and have similar three-dimensional configurations (see text).

In addition to acetosyringone and α-hydroxyacetosyringone, several other similar natural plant metabolites that induce the *vir* regulon have been identified. Messens *et al.* (1990) isolated ethyl ferulate from wheat suspension cultures and showed that it was as active as acetosyringone in *vir* regulon induction assays. A separate group of investigators also analyzing wheat metabolites identified ferulic acid and feruloyl-substituted polysaccharide fragments as *vir* regulon inducers (Usami *et al.*, 1988; Machida *et al.*, 1989). Methyl syringate from grape vines has been identified as a potent inducer of the *vir* regulon (P. A. Spencer and G. H. N. Towers, unpublished results).

Other natural products with more unusual structures also have been identified as *vir* regulon inducers. The flavonol glycosides kaempferol 3-glucosylgalactoside and quercetin 3-glucosylgalactoside were isolated from extracts of pollen and stigma of petunia (Zerback *et al.*, 1989). Although these glycosides are inducers, they are very weak in comparison to acetosyringone and have significantly different structures than do other previously identified inducers. However, these compounds possess 4′-hydroxyphenyl ring substitutions that show some structural similarity to other inducers. Their weak *vir* regulon-inducing activity has been attributed to this feature. Dehydrodiconiferyl alcohol, the aglycone of cell division promoting molecules from tobacco, has potent *vir* regulon-inducing activity (Hess *et al.*, 1991). Morris and Morris (1990) reported a unique phenomenon in which the phenolic glycoside coniferin, isolated from Douglas fir, is likely to be converted to coniferyl alcohol, a strong *vir* regulon inducer, by β-glucosidases produced by *A. tumefaciens* strains. This interpretation was supported by the observation that only those *A. tumefaciens* strains that made significant levels of β-glucosidase were induced by coniferin.

The identification of inducing compounds from natural sources and their correlation with biological function in tumorigenesis must be interpreted cautiously. All plants are likely to produce compounds that are capable of inducing the *vir* regulon since lignin precursors are inducers. However, only those compounds released by wounded plant tissue should be considered physiological effectors of *vir* regulon induction. The study by Stachel *et al.* (1985b) in which acetosyringone was identified in exudates from plant wounds should serve as the standard methodology for the identification of physiological inducers. Studies in which whole plants are extracted by vigorous mechanical disruption merely identify inducing compounds present in the plant, but not necessarily those molecules actually involved in tumorigenesis.

The structural requirements for *vir*-inducing compounds have been investigated extensively. Bolton *et al.* (1986) reported that a mixture of simple phenolic metabolites (i.e., catechol, *p*-hydroxybenzoic acid, β-resorcylic acid, protocatechuic acid, pyrogallic acid, gallic acid, and vanillin) induced a *virE::lacZ* fusion to high levels, suggesting that benzoic acid derivatives with a *p*-hydroxyl group are *vir* regulon inducers. Stachel *et al.* (1985b) investigated compounds similar to acetosyringone and determined that only those compounds with a 4-hydroxy-3-methoxyphenyl (guaiacyl) nucleus are active as inducers. Compounds with a 3,5-dimethoxy-4-hydroxyphenyl (syringyl) substituent are much more active than guaiacyl derivatives. In two separate studies analyzing a large number of phenolic compounds, Spencer and Towers (1988) and Melchers *et al.* (1989b) further determined the structural features required for strong *vir* regulon-inducing activity. Compounds with a syringyl nucleus had the highest inducing activity; guaiacyl compounds had significant but lower inducing activity. Certain compounds with *p*-hydroxyphenyl or 3,4-dihydroxyphenyl groups had barely detectable activity. With respect to the side chain on these phenolic compounds, the most active inducers had a carbonyl group or a vinyl side chain at C1. Among the inducing compounds, a correlation exists between hydrophobicity and *vir* regulon-inducing activity. In a given series of compounds with the same chemical nucleus (syringic acid, syringaldehyde, and methyl syringate), esters are more active in *vir* regulon induction than aldehydes, which in turn are more active than carboxylic acids. The most active *vir* regulon-inducing compounds identified to date are the extremely hydrophobic chalcones reported by Spencer and Towers (1988). The chemical rationale for *vir* regulon-inducing activity of compounds appears to be a direct function of the electronic properties of such compounds and the pKa of the phenolic hydroxyl (Hess *et al.*, 1991). The most active *vir* regulon inducers have electron-withdrawing substituents *para* to the phenolic hydroxyl, which lowers the pKa of this hydroxyl. Thus, a long conjugated double bond system extending from the 4′-hydroxyl groups, through the phenyl ring, and into the carbonyl group at C1 strongly enhances *vir* regulon induction. The ability of the 4′-hydroxyphenyl group to form a quinone-like structure with such a conjugated bond system suggests that a resonance form of an inducing compound may be the actual inducing species. The transfer of the phenolic proton to a basic residue in the putative acetosyringone-binding site of a protein may lead to conformational changes necessary for signal transduction.

The correlation between hydrophobicity and *vir*-inducing activity lends credence to a proposed mechanism for signal transduction pathways involving hydrophobic plant metabolites. The *nod* system of *Rhizobium leguminosarum* is induced by the hydrophobic flavonoid compound naringenin via the *nodD* gene product, a cytoplasmic membrane protein. Recourt *et al*. (1989) demonstrated the rapid partitioning of naringenin to the *R. leguminosarum* cytoplasmic membrane. Naringenin is thought to be detected by the membrane-bound region of NodD. Similar results have been observed with *A. tumefaciens* and acetosyringone. No increase occurred in association of [^{3}H]acetosyringone with *virA*$^{+}$ cells compared with *virA* mutants; acetosyringone rapidly localized to the membrane of *A. tumefaciens* (S. Lee, S. Jin, and E. W. Nester, unpublished observations; S. Turk and P. J. J. Hooykaas, personal communication). The importance of the second transmembrane domain of VirA as determined by Melchers *et al*. (1989a) provides further support for the hypothesis of intramembrane detection of acetosyringone. However, the relative shortness of the second transmembrane domain (19 residues) would seem to preclude any acetosyringone binding site in this region. Apparently, VirA deletion derivatives lacking TM2 and N-terminal residues still are able to induce the *vir* regulon in response to acetosyringone (Chang and Winans, 1992). These investigators localized the acetosyringone-responsive domain of VirA to the stretch between TM2 and the histidine protein kinase domain, which they termed the linker domain. In addition to detecting acetosyringone, the linker domain also is responsible for sensing acidic pH. (See Section II,E,1 for a further discussion of the effects of acidic pH on *vir* regulon induction.) The demonstration that VirA and VirG are the only Ti plasmid-encoded products required for *vir* regulon induction indicates that either VirA alone or a chromosomal gene product in concert with VirA actually binds and detects acetosyringone.

Several studies have provided evidence that chromosomal gene products and not VirA may be the actual binding agents of acetosyringone. Hess *et al.* (1991) synthesized a specific inhibitor of the VirA–VirG signal transduction system, α-bromoacetosyringone. In a subsequent study from the same laboratory (Lee *et al.*, 1992), α-bromoacetosyringone was modified further by replacing a methoxy group with ^{125}I. This radioiodinated derivative served as an affinity label because of the inherent reactivity of α-bromoketones. Two chromosomally encoded proteins of 10 and 21 kDa were found to be specifically labeled by the affinity substrate whereas VirA was not. Further work on the interaction of the 10- and 21-kDa proteins with acetosyringone and VirA is required to understand fully the ability of *A. tumefaciens* to sense plant phenolic compounds.

Acetosyringone does not appear to be metabolized by *A. tumefaciens* in studies using either unlabeled acetosyringone (Culianez-Macia and Hepburn, 1988; R. Kanemoto and M. P. Gordon, personal communication) or [^{3}H]acetosyringone (S. Lee, R. Ankenbauer, and E. W. Nester, unpublished observations).

C. ChvE and Plant Sugars

In addition to the *vir* regulon, several chromosomal loci also have been implicated in crown gall tumorigenesis (see Section I,B). Several—*chvD, chvE, ros,* and the *ivr* loci—have roles in *vir* regulon expression. With the exception of *chvE*, little is known of the functional mechanisms of these chromosomal gene products in signal transduction.

chvE mutants were identified first as possessing a very limited host range (i.e., only able to form tumors on a small number of plant species) compared with the wild-type strain (Garfinkel and Nester, 1980). *chvE* mutants were unable to form tumors on 4 of 7 commonly used plant hosts for *A. tumefaciens* (M. Huang *et al.*, 1990). The *chvE* mutation was localized to the *A. tumefaciens* chromosome and was among the first virulence loci identified that were not encoded by the Ti plasmid (Garfinkel and Nester, 1980; M. Huang *et al.*, 1990).

Within the tumorigenic process, the phenotype exhibited by *chvE* mutants is inefficient *vir* regulon induction. Induction of *virB::lacZ* and *virE::lacZ* fusions is reduced greatly in *chvE* mutants compared with wild-type strains in a glucose-based induction medium. Even at acetosyringone concentrations that give maximal levels of induction, the *vir* regulon induction exhibited by *chvE* mutants is 10-fold less than that of wild-type strains (M. Huang *et al.*, 1990). This defect in induction is presumably responsible for the limited host range of *chvE* mutants. The *chvE* gene and its gene product are highly homologous to the *E. coli* periplasmic receptor proteins galactose-binding protein (GBP) and ribose-binding protein (RBP). Both these proteins bind their specific monosaccharide ligands and then interact with (1) specific membrane-bound uptake complexes, resulting in the transport of the ligand, and (2) the membrane-bound chemosensory protein Trg, resulting in a positive chemotactic response (Furlong, 1987; Macnab, 1987). A glucose–galactose binding protein isolated from *Agrobacterium radiobacter*, designated GBP1, is virtually identical to ChvE; its monosaccharide-binding ability and transport activity have been confirmed by a number of biochemical and physical methods (Cornish *et al.*, 1989). The ChvE protein also is involved in *A. tumefaciens* chemotaxis toward several monosaccharides (Cangelosi *et al.*, 1990).

A direct role for ChvE in signal transduction has been proposed since indirect effects on cell growth or *virA* and *virG* expression are not observed (M. Huang *et al.*, 1990). This role was established with the discovery that a number of sugars act synergistically with acetosyringone to give high levels of *vir* regulon induction (Cangelosi *et al.*, 1990; Shimoda *et al.*, 1990). The inducing sugars are particularly effective in the presence of low concentrations of acetosyringone. Monosaccharides that promote *vir* regulon induction include D(+)glucose, D(+)galactose, L(+)arabinose, and D(+)xylose. In the absence of inducing sugars, *vir* regulon induction in wild-type strains is reduced severely and is similar to levels observed with *chvE* mutants. The induction promoted by these sugars depends on both ChvE and VirA. By constructing specific deletions in the *virA* gene, the periplasmic domain of VirA was determined to be required for sugar-mediated *vir* regulon induction but not for acetosyringone-mediated induction, thereby confirming the observations of Melchers *et al.* (1989a). Thus, these data indicate that the ChvE–sugar complex interacts with the periplasmic region of VirA to stimulate induction in synergy with acetosyringone.

The ChvE protein has at least three known functions: (1) to mediate a high-affinity monosaccharide transport system, (2) to interact with *A. tumefaciens* "Trg-like" proteins, resulting in chemotaxis toward the sugars, and (3) to interact with the periplasmic space domain of VirA to promote *vir* regulon induction (Cornish *et al.*, 1989; Cangelosi *et al.*, 1990). The overall topological similarities between the separate signal transduction systems of the *E. coli* sugar/GBP/Trg and the *A. tumefaciens* sugar/ChvE/VirA systems are remarkable.

The monosaccharide structural features capable of ChvE-mediated *vir* regulon induction also have been investigated (Ankenbauer and Nester, 1990). The basic inducing sugar structure is identical to D(+)xylose or L(+)arabinose with a number of permissible substitutions at C5 (Fig. 4.2). The most active inducing sugars identified were the acidic monosaccharide derivatives D-galacturonic acid and D-glucuronic acid. Galacturonic acid oligosaccharides (2–10 residues) were inactive in *vir* regulon induction, suggesting that the sugar-mediated induction is specific to monosaccharides (R. G. Ankenbauer, P. Albersheim, and E. W. Nester, unpublished observations). Of the 11 compounds identified as inducers, 7 (arabinose, galactose, galacturonic acid, glucose, glucuronic acid, mannose, and xylose) are known monomers of plant cell wall polysaccharides. Wounded plant tissue serves as a "sink" for free monosaccharides (G. An and C. Ryan, personal communication), an observation that correlates well with the role of sugar-mediated induction of the *vir* regulon.

The ChvE protein is likely to play a major role in the chemotaxis of *A. tumefaciens* to plant wounds (see Sections I,B and II,A). This wound chemotaxis could occur via the ChvE / "Trg-like" chemoreceptor pathway (Cangelosi *et al.*, 1990) or, alternatively, through the interaction of ChvE with the VirA-mediated chemotaxis pathway (Shaw *et al.*, 1988), possibly in concert with phenolic compounds. The bacterium initially could swim to the wound via ChvE-mediated chemotaxis toward wound-specific sugars, which are more soluble in aqueous solutions than are the phenolic compounds. The issue of solubility suggests that the bacteria would be more likely to detect sugars before phenolic compounds.

D. Chromosomal Loci Involved in *vir* Regulon Induction

Chromosomal loci other than *chvE* have effects on *vir* regulon induction. The *ros* locus has been studied in some detail. *ros* mutants are phenotypically distinct from other mutants affecting *vir* regulon induction. *ros* mutations lead to increased expression of the *virC* and *virD* operons but not of other members of the *vir* regulon (Close *et al.*, 1987). The enhanced expression of *virC* and *virD* in *ros* mutants is independent of VirA and VirG. Acetosyringone-mediated *vir* regulon induction in *ros* mutants leads to additive effects on *virC* and *virD* expression. *ros* mutants are fully virulent as expected, since *vir* regulon expression is not adversely affected. These results suggest that the *ros* gene product is a negative regulator of *virC* and *virD*. The transcriptional organization of the *virC* and *virD* genes, in which the operons possess overlapping promoters and are transcriptionally divergent, may explain the simultaneous effects of *ros* mutations on both *virC* and *virD* transcription (Tait and Kado, 1988).

The *ros* gene has been cloned and sequenced (Cooley *et al.*, 1991). The *ros* gene product has been determined to be a relatively small protein of 15.5 kDa that contains a consensus zinc finger sequence in its central region. The *ros* gene is autoregulated negatively by its gene product, suggesting that the repressing activity of Ros may be dependent on specific promoter sequences present in the *virC*, *virD*, and *ros* promoters. Sequence comparison of these promoters reveals the presence of a conserved 9-bp inverted repeat. Preliminary data are available on the binding of Ros to these sequences (Cooley *et al.*, 1991). A role for Ros in crown gall tumorigenesis is difficult to envision. Ros has been suggested to modulate *virC* and *virD* expression in response to certain conditions during plant infection. However, the need for such modulation in tumorigenesis has not been determined.

In experiments examining the transcriptional regulation of *virG*, Winans *et al.* (1988) reported on chromosomal mutations that exhibited decreased *virG* transcription. This mutation, designated *chvD*, prevents the enhanced *virG* transcription normally observed in response to low pH and phosphate starvation (see Sections II,E,1 and II,E,2). Although not totally avirulent, *chvD* mutants are attenuated in their virulence, as are *virC* and *virE* mutants. A partial sequence of the *chvD* gene indicates significant homology between the *chvD* gene product and a family of bacterial ATP binding proteins, most of which are involved in active transport (Higgins *et al.*, 1986). The mechanism of ChvD function in *virG* transcription and tumorigenesis is unknown. ChvD has been suggested not to be directly involved in tumorigenesis and, instead, to be required for productive bacterial metabolism, a defect in which leads to pleiotropic effects on the cell (S. C. Winans, personal communication).

Mutants defective in the *miaA* gene exhibit a 2- to 3-fold reduction in *vir* regulon expression (Gray *et al.*, 1992) but are not markedly different in virulence assays. The *miaA* homolog in *E. coli* encodes a tRNA:isopentenyl transferase. The effect of the *miaA* mutation on *vir* regulon induction has been speculated to be physiological, primarily because of the efficiency and fidelity of overall translation within the cell.

The *ivr* loci reported by Metts *et al.* (1991) appear to have more direct roles in *vir* regulon induction than do the *chvE*, *chvD*, or *ros* loci. The three *ivr* loci map to different regions of the *A. tumefaciens* chromosome and are therefore not allelic. Each of the *ivr* mutants is totally defective in acetosyringone-mediated *vir* regulon induction and is avirulent on the standard plants used for tumor assays. Additionally, the *ivr* mutations prevent enhanced *virG* transcription effected by low pH and limiting phosphate, as reported by Winans *et al.* (1988) (see Sections II,E,1 and II,E,2). The only physiological changes detected in the *ivr* mutants are slight differences in lipopolysaccharide substitution and outer membrane protein profiles (Metts *et al.*, 1991).

E. Other Factors Involved in *vir* Regulon Induction

1. Acidic pH When the phenomenon of plant-inducible *vir* genes first was discovered (see Section II,B), *A. tumefaciens* cells were cocultivated with cultured plant cells. The medium used for this cocultivation was MS, a medium designed for the propogation of plant cells, which has an acidic pH. Induction by plant metabolites occurred only when the pH of the induction assay was less than 6.0 (Stachel *et al.*, 1986). The fortuitous use of MS medium by these investigators was directly responsible for the discovery of plant-inducible genes in *A. tumefaciens*. Other investigations confirmed the acidic pH optimum for *vir* regulon induction (Stachel *et al.*, 1985b; Bolton *et al.*, 1986; Rogowsky *et al.*, 1987; Culianez-Macia and Hepburn, 1988; Vernade *et al.*, 1988). Although a wide range of pH values <6.0 allows strong *vir* regulon induction, in general the highest levels of induction are observed at the most acidic pH that still allows growth (~4.8–5.5). Induction is virtually nonexistent at pH 7.0.

The molecular basis for this acidic pH optimum for *vir* regulon induction has been the subject of several investigations. The transcription of *virG* was found to be inducible by acidic conditions (Veluthambi *et al.*, 1987; Winans *et al.*, 1988). This enhanced induction by acidic pH was demonstrated to occur at the *virG* promoter P2 (Winans, 1990; Mantis and Winans, 1992). In addition to acidic pH, a variety of other environmental stresses lead to transcriptional activation of the *virG* promoter P2 (Winans, 1990; Mantis and Winans, 1992). This promoter has similarities to *E. coli* heat-shock promoters and is induced in response to noxious agents such as low pH, $CdCl_2$, mitomycin C, and alkaline pH. The induction of *virG* from promoter P2 is independent of the presence of the Ti plasmid and therefore requires chromosomal genes such as *chvD* for its regulation (see Section II,D).

These observations led to the hypothesis that the acidic pH requirement for *vir* regulon induction is an effect of enhanced *virG* transcription caused by acidic conditions (Winans *et al.*, 1988; Winans, 1990). Since *vir* regulon induction necessitates the presence of VirG, induction initially would require the induction of *virG* by acidic conditions. Following the increase of VirG concentration in the cell, acetosyringone-activated VirA would phosphorylate VirG, leading to further *virG* transcription and induction of the remaining *vir* regulon. At neutral pH, the intracellular concentration of VirG would be too low to permit its VirA-mediated phosphorylation that would result in *vir* regulon induction.

Two other studies have suggested that the enhanced transcription of *virG* in response to acidic pH is not the only factor in the acidic pH optimum of *vir* regulon induction. Melchers *et al.* (1989a) demonstrated that VirA–Tar hybrids in which the periplasmic space domain of VirA was replaced by that of Tar (see Section II,A) yielded high levels of induction at pH 6.8, in contrast to the lack of significant induction observed with wild-type VirA at pH values above 6.0–6.5. This result suggests that the periplasmic domain of VirA must be able to sense acidic conditions to signal *vir* regulon induction. Additional evidence for factors other than *virG* transcription in the acidic pH optimum for *vir* regulon induction has been reported by Chen and Winans (1991). These investigators removed both *virG* promoters and placed the *virG* coding region under the control of the *E. coli lac* promoter (P*lac*), which was shown to function well in *A. tumefaciens*. Although the acidic pH-inducible P2 promoter of *virG* was absent, acidic conditions still were required for *vir* regulon induction. Although an effect through VirA was not determined in this study (Chen and Winans, 1991), the results provide indirect evidence for the detection of acidic conditions by VirA. Since the P*lac*::*virG* fusion is functional at pH 7.0 and *vir* regulon induction was not observed at this pH, VirA probably was not functioning to phosphorylate VirG at this pH. These two reports suggest that VirA is also able to detect acidic pH conditions presumably by its periplasmic domain. This hypothesis has been challenged by the observation that VirA deletion derivatives lacking residues N-terminal of and including TM2 are still able to induce the *vir* regulon in response to low pH (Chang and Winans, 1992). These investigators localized the acid-responsive domain of VirA to the stretch between TM2 and the histidine protein kinase domain, which they termed the linker domain. As mentioned earlier, this linker domain is also responsible for the detection of acetosyringone. These various, apparently contradictory data suggest that the ability to detect various stimuli may reside simultaneously in different domains of VirA and be a finely balanced system with enhancing and dampening effects on *vir* regulon induction, depending on the environment (specific plant host) and the levels and composition of wound-specific metabolites encountered by the bacteria.

An additional factor in the acidic pH optimum of *vir* regulon induction may be the protonation of acetosyringone and related phenolic compounds. The pKa of the phenolic hydroxyl of acetosyringone is 7.4; at neutral pH, a large percentage of acetosyringone would be ionized (Hess *et al.*, 1991). Perhaps only protonated phenolics are active inducers of the *vir* regulon. However, the effect of pKa on inducing activity is more complex than the mere protonation of the phenolic hydroxyl (see Section II,B for further discussion).

2. *Phosphate Starvation* Phosphate starvation also may be a factor in *vir* regulon induction. In addition to being induced by acetosyringone and acidic pH, *virG* transcription also is enhanced significantly by phosphate starvation (Winans *et al.*, 1988). Although phosphate starvation could be perceived as an environmental stress, the enhanced *virG* transcription caused by phosphate starvation occurs at *virG* promoter P1 rather than P2 (Winans, 1990). The P1 promoter has sequence homology with the *pho* box, a conserved sequence present in *E. coli* promoters that are inducible by phosphate starvation. Enhanced *virG* transcription by phosphate starvation is attenuated in *chvD* mutants, as is the effect of acidic pH (Winans *et al.*, 1988).

3. *Temperature* The effect of temperature on the formation of crown gall tumors has been known for over 60 years (Riker, 1926). Subsequent studies (Braun, 1947; Rogler, 1980) demonstrated that *A. tumefaciens* was unable to form tumors at temperatures above 30°C. These studies also determined that the thermosensitive step in tumor formation is the infective interaction between *A. tumefaciens* and the plant host.

By assaying the expression of the *virD2* gene, Alt-Moerbe *et al.* (1988) showed that the optimal temperature for *virD2* induction is 20–25°C. At temperatures of 30°C or higher, induction of *virD2* is not observed. A *virB::lacZ* fusion is expressed optimally at 29°C and similarly is inhibited by temperature >30°C (Melchers *et al.*, 1989a). Although *virB* and *virD* induction may be uniquely temperature sensitive, induction of the entire *vir* regulon is likely to be similarly inhibited by temperatures because of the coordinated regulation of these genes by VirA and VirG.

These combined results suggest that the thermosensitive step in tumor formation is the signal transduction process involving VirA and VirG. Evidence for the role of VirA in this temperature-sensitive response has been reported by Melchers *et al.* (1989a). These investigators demonstrated that induction yielded by a VirA–Tar hybrid, in which the periplasmic domain of Tar replaced that of VirA, was much less sensitive to the inhibitory effects of temperatures >30°C than wild-type VirA, suggesting that the periplasmic domain of VirA also is involved in the temperature-sensitive regulation of *vir* regulon induction. Temperature effects on VirA phosphorylation *in vitro* also have been observed. Truncated VirA derivatives undergo rapid autophosphorylation *in vitro* at temperatures below 30°C (Jin *et al.*, 1990b). Increasing the temperature to >30°C totally inhibits autophosphorylation of these VirA deletions (S. Jin and E. W. Nester, unpublished results). Since these truncated VirA derivatives are not membrane bound and do not possess intact periplasmic domains, any direct relationship between the *in vitro* phosphorylation data and the *in vivo* induction assays of Melchers *et al.* (1989a) is difficult to establish. Nevertheless, in addition to detecting phenolics, sugars, and acidic conditions, VirA is also likely to be capable of detecting changes in temperature and signaling for *vir* regulon transcription at temperatures <30°C.

This proposed thermosensing ability of VirA provides another parallel with the chemosensory proteins of *E. coli*. All four chemosensory proteins—Tar, Tsr, Trg, and Tap—are thermosensors (Nara *et al.*, 1991). Tar, Tsr, and Trg mediate positive responses to warm temperatures and Tap positively responds to cold. With respect to their thermosensing abilities, VirA and Tap appear similar in favoring positive signal transduction at cold temperatures.

4. Glycine Betaine and Osmoprotectants In studies on the optimal conditions for *vir* regulon induction, Vernade *et al.* (1988) determined that the bacterial osmoprotectant glycine betaine enhances *vir* regulon induction in a low pH (pH 5.2), high osmotic strength medium. Glycine betaine was postulated to facilitate the adaptation of the bacteria to the acidic conditions; however, the possibility that glycine betaine instead promotes bacterial adaptation to the high osmolarity of the medium has not been excluded. The mechanism responsible for this increase in induction is unknown, but may involve osmoadaptation. Alternatively, these compounds may be yet unrecognized wound-specific metabolites.

5. Opines Veluthambi *et al.* (1989) reported that the opines octopine, nopaline, leucinopine, and succinamopine enhance acetosyringone-mediated induction 2- to 10-fold whereas other opines (mannopine, acetopine, and arginine) do not enhance induction. Interestingly, all the inducing opines are iminodiacids, in contrast to the noninducing opines. The significance of this structural requirement is unknown. The roles of VirA and VirG in this opine enhancement of *vir* regulon induction is not clear since the effect appears to be synergistic with that of acetosyringone. The addition of nopaline to *A. tumefaciens* induction cultures significantly increases the transformation efficiency of *Agrobacterium* on cotton plants.

Since opines are not synthesized until the T-DNA has been transferred to and expressed in plant cells, the function of opines in *vir* regulon induction is not clear. Following the initial transformation event, opine synthesis is thought to increase induction of *A. tumefaciens* cells at the wound site and yield additional T-DNA transfer (Veluthambi *et al.*, 1989). Since certain plants are more resistant to transformation than other plants, increased T-DNA transfer to such plants may be a strategy that *A. tumefaciens* uses to enhance its tumorigenic capacity.

6. Nutritional Requirements The fact that only actively growing bacteria can be induced and that *vir* regulon induction strongly depresses the bacterial growth rate (Culianez-Macia and Hepburn, 1988; Fortin *et al.*, 1992) suggests that the nutritional state of *A. tumefaciens* is an important factor in *vir* regulon induction.

Induction of the *vir* regulon can be influenced by carbon source and other media constituents. A wide variety of sole carbon sources can support *vir* regulon induction (glucose, sucrose, mannose, galactose, glycerol, sorbitol, and succinate); carbohydrates predominate (Alt-Moerbe *et al.*, 1988; Vernade *et al.*, 1988; Melchers *et al.*, 1989b). At least two sole carbon sources are incapable of supporting *vir* regulon induction—glycine and acetate (Alt-Moerbe *et al.*, 1988). In fact, acetate was found to inhibit induction even in the presence of carbohydrates. The relationship between carbohydrates as a sole carbon source for induction and sugar-mediated induction via ChvE and VirA has not been investigated thoroughly.

Vitamins, inorganic salts, and nitrogen sources are not required but may enhance induction to a limited extent (Alt-Moerbe *et al.*, 1988). Low concentrations of $CaCl_2$, and the complex medium additives yeast extract, beef extract, casein hydrolysate, and peptone can enhance induction (Rogowsky *et al.*, 1987; Alt-Moerbe *et al.*, 1988; Vernade *et al.*, 1988). However, other investigators have reported that complex media do not increase induction (Melchers *et al.*, 1989b). Concentrations of yeast extract >0.1%, NaCl >2%, sucrose >3%, and the addition of 20 m*M* NH_4Cl or low K^+/Na^+ ratios have been reported to inhibit *vir* regulon induction (Alt-Moerbe *et al.*, 1988; Vernade *et al.*, 1988; Melchers *et al.*, 1989b). Other investigators have used relatively high concentrations of yeast extract and NH_4Cl without observing inhibition of *vir* regulon induction (Rogowsky *et al.*, 1987; Winans *et al.*, 1988; Cangelosi *et al.*, 1990).

III. Intracellular Events in *vir* Regulon Signal Transduction

A. VirA/VirG Phosphorylation and Transfer

The mechanism of signal transduction in the family of two-component regulatory systems appears to involve phosphorylation reactions. The current model proposes that, on detection of a specific stimulus, the sensor component (also termed histidine protein kinase) autophosphorylates at a conserved histidine residue and subsequently transfers this phosphate to a conserved aspartic acid residue in the N-terminal domain of the regulator protein. This phosphorylated regulator is then capable of activating transcription from the subject regulon.

As members of this family of two-component regulatory systems, VirA and VirG have been found to undergo phosphorylation. *In vitro* phosphorylation of the VirA protein has been reported by two groups (Y. Huang *et al.*, 1990; Jin *et al.*, 1990b). Difficulties associated with overproducing the native VirA protein in *E. coli* (Winans *et al.*, 1989) required working with truncated VirA derivatives (Jin *et al.*, 1990b) and TrpE′–VirA fusion proteins (Y. Huang *et al.*, 1990). Phosphorylation of these purified VirA derivatives was observed to be an ATP γ-phosphate-specific autophosphorylation since [γ-^{32}P]ATP yielded phosphorylation whereas [α-^{32}P]ATP and [γ-^{32}P]GTP did not (Y. Huang *et al.*, 1990; Jin *et al.*, 1990b). The phosphorylated site in VirA was determined to be a histidine residue. The highly conserved His-474 was implicated as the specific histidine residue by site-directed mutagenesis (Jin *et al.*, 1990b). By replacing His-474 with the conservative substitution glutamine, *in vitro* phosphorylation of VirA derivatives was abolished, as was the biological activity of the native VirA derivative (Jin *et al.*, 1990b). In contrast to these results, Y. Huang *et al.* (1990) reported that a TrpE′–VirA fusion protein lacking the 524 N-terminal residues of VirA was phosphorylated *in vitro* and that the autophosphorylation site localized to residues 524–623. This result is surprising since this region does not contain the conserved histidine residue present in all sensor proteins of the two-component regulatory systems. In light of these discrepancies, note that the precise residue at the phosphorylation site has not been identified chemically and that specific histidine residues are implicated in phosphorylation only by mutant analysis and analogy to other two-component systems.

The second step in the VirA–VirG phosphorylation cascade, transfer of a phosphoryl moiety from VirA to VirG, also has been demonstrated (Jin *et al.*, 1990a). The native VirG protein is phosphorylated by *in vitro* purified ^{32}P-phosphorylated VirA derivatives. The phosphorylation of VirG occurs by direct transfer of the phosphate group on VirA to VirG and does not require the presence of ATP. The VirG–phosphate bond is a relatively stable one in comparison to those in other regulatory proteins. VirG–phosphate has a half-life of >1 hr whereas other phosphorylated regulator proteins (e.g., CheY, CheB, NR_I) have half-lives of <5 min (Stock *et al.*, 1989; A. J. Ninfa, personal communication). The long half-life of VirG–phosphate compared with the instability of phosphorylated regulators involved in chemotaxis may be explained by the fact that adaptation in chemotaxis needs to be quite rapid as a matter of survival, whereas tumorigenesis by *A. tumefaciens* is not required for continued bacterial viability. Once *A. tumefaciens* recognizes conditions favorable for tumorigenesis, it may commit itself irreversibly to that course because of the long half-life of VirG–phosphate. The phosphorylated residue in VirG has been determined to be Asp-52 by chemical modification of the phosphorylated protein. Further confirmation of the phosphorylation of Asp-52 has been achieved by site-directed mutagenesis. Replacing Asp-52 with Asn abolishes both the transfer of phosphate to VirG and the biological activity of VirG. Further analysis of the VirG protein by site-directed mutagenesis indicates that the structure–function requirements for VirG are complex (Roitsch *et al.*, 1990). By mutating a variety of aspartic acid residues in VirG to the conservative substitution asparagine, a variety of phenotypically distinct VirG mutants have been obtained. Although all the VirG mutants are still able to bind to *vir* promoters, several are not phosphorylatable and are therefore incapable of *vir* regulon induction. One VirG mutant still able to phosphorylate and bind *vir* promoters is nonetheless inactive in *vir* regulon induction. Without a three-dimensional protein structure, the conformational changes elicited by these mutations are unpredictable.

Correlating the *in vitro* phosphorylation data with the temporal events in signal transduction has been difficult. The *in vitro* phosphorylation reactions are independent of acetosyringone and monosaccharides and, therefore, are likely to represent an unnatural artificial system in which stimulus detection is separated from phosphorylation. *In vivo* phosphorylation of VirA in the presence or absence of acetosyringone has not been observed (H. Wang and E. W. Nester, unpublished observations). However, *in vivo* phosphorylation of VirG has been demonstrated. Immunoprecipitation of VirG from *A. tumefaciens* grown in the presence of [^{32}P]phosphate shows that VirG phosphorylation occurs only in the presence of acetosyringone and that inducing monosaccharides enhance the observed phosphorylation (H. Wang and E. W. Nester, unpublished observations). These results are similar to the *in vivo* phosphorylation of OmpR (Forst *et al.*, 1990) and support the model of *vir* regulon transcriptional activation by phosphorylated VirG.

B. *vir* Promoters and *vir* Boxes

The coordinate regulation of the *vir* regulon by VirA and VirG indicates a common mechanism for transcriptional activation of each of the operons. The model for the activity of two-component regulatory systems proposes that the phosphorylated regulator protein (e.g., VirG) binds to specific promoter sequences of subject operons, thereby activating transcription. Sequence analysis of the *vir* regulon promoters revealed the presence of a conserved 12- to 14-bp sequence in each of the promoters that has been designated the *vir* box (Das *et al.*, 1986; Winans *et al.*, 1987; Das and Pazour, 1989). The consensus *vir* box sequence is 5′dRYTNCAATTGNAAY (R, purine; Y, pyrimidine; N, variable). Some *vir* promoters possess multiple *vir* boxes. *virA*, *virE*, and *virH* possess one; *virB* possesses two; *virD* and *virG* possess three; and *virC* possesses four. The *vir* box has been identified as a functional *cis*-acting regulatory sequence required for *virB* (Das and Pazour, 1989) and *virE* (Jin *et al.*, 1990c) transcription by analyzing expression from a series a 5′ deletion derivatives of the *virB* and *virE* promoters. Deletions removing all or part of the *vir* box eliminate induction from these promoters. Site-directed mutagenesis of the *vir* promoters has demonstrated that the *virB vir* box sequences cannot be moved either upstream or downstream from their original site without inhibiting *virB* induction (Das and Pazour, 1989). The proper spacing of *vir* box sequences with respect to the transcriptional start site has been proposed to be required for *vir* regulon induction. The consensus *vir* box sequence has rough dyad symmetry, and when the *virE vir* box is replaced with a synthetic *vir* box containing perfect dyad symmetry, the inducibility of the modified promoter is enhanced (Pazour and Das, 1990a). Mutagenesis of individual *vir* boxes in promoters possessing more than one such sequence (i.e., *virB*, *virC*, and *virD*) reveals that all *vir* boxes do not have equivalent biological function (Pazour and Das, 1990a). Only one *vir* box in each of the *virB* and *virD* promoters is required for induction whereas two *vir* boxes in the *virC* promoter are required for its biological function. The function of the remaining *vir* boxes in these promoters is unknown.

The DNA binding ability of VirG initially was reported as nonspecific, showing no preferential binding to *vir* regulon promoter sequences (Powell *et al.*, 1989). Later studies demonstrated that VirG bound to *vir* box sequences in gel retardation and DNase I footprinting assays (Jin *et al.*, 1990c; Pazour and Das, 1990b). The binding of VirG to DNA fragments containing *virB* and virE promoters was found to be very specific, occurring in the presence of 1000-fold excess competitor DNA (Jin *et al.*, 1990). Specific protection of *vir* box sequences by VirG in DNase I footprinting assays has been demonstrated for the *virB*, *virC–D*, *virE*, and *virG* promoters (Jin *et al.*, 1990c; Pazour and Das, 1990b). In addition to protecting the *vir* box sequences, VirG also protects a variable amount of sequence adjacent to the *vir* boxes from DNase digestion. Since the VirG protein used in these studies was isolated from *E. coli* strains bearing recombinant plasmid and therefore was unlikely to be phosphorylated, nonphosphorylated VirG appears to have the ability to bind *vir* box sequences. However, mutants of *virG* that cannot be phosphorylated do not induce the *vir* regulon (Roitsch *et al.*, 1990). Thus, the role of VirG phosphorylation in transcriptional activation of the *vir* regulon remains unknown, but may increase the binding affinity of VirG for *vir* box sequences or serve as a recognition site for RNA polymerase.

The phosphorylation-independent DNA binding activity of VirG has been confirmed by demonstrating specific *vir* box binding with VirG derivatives unable to be phosphorylated. Site-directed mutagenesis of Vir residue Asp-52 to asparagine yields a nonphosphorylatable protein that still binds *vir* box sequences specifically, as determined by gel retardation and DNase I footprinting (Roitsch *et al.*, 1990). Further, truncated VirG derivatives consisting of only the C-terminal domain responsible for DNA binding in such regulator proteins similarly are capable of specific binding to *vir* box sequences (Powell and Kado, 1990; Roitsch *et al.*, 1990).

Overproduction of VirG has been observed to promote *vir* regulon induction in the absence of acetosyringone and VirA (Rogowsky *et al.*, 1987). Perhaps the increased VirG load may provide increased opportunity for other non-VirA sensor proteins to phosphorylate VirG in cross talk interactions, as observed in other heterologous two-component regulatory systems (Stock *et al.*, 1989). Alternatively, the stoichiometrically higher levels of VirG may be able to occupy the *vir* box sites more completely than the normal levels of VirG, thereby leading to *vir* regulon transcription.

IV. Summary and Outlook

As far as can be discerned, the *Agrobacterium*–plant relationship is parasitic. The plant host gains no known benefit from crown galls whereas the bacterium benefits from a unique ecological niche, assuming the opine hypothesis to be valid. The *Agrobacterium* Ti plasmid relationship is apparently symbiotic. Because it provides the Ti plasmid with the biological means for tumorigenesis, the Ti plasmid provides the bacterium with opines designed for the growth of the tumorigenic bacterium. Because other bacteria (including certain avirulent *A. tumefaciens* isolates) can degrade opines as well as Ti plasmid-bearing *A. tumefaciens* can and because older tumor tissues usually are devoid of *A. tumefaciens*, the selective advantage of crown gall tumorigenesis to the *A. tumefaciens* bacterium is questionable.

Agrobacterium tumefaciens is in no way dependent on the Ti plasmid and the crown gall tumorigenesis phenomenon for survival. This bacterial species is a successful saprophyte distributed worldwide and colonizes a wide variety of environments and plants. Therefore, *A. tumefaciens* cannot be considered a highly adapted microorganism dependent on a pathogenic life-style or infection for its survival. Conversely, the Ti plasmid is highly adapted for and actually generates pathogenesis. The Ti plasmid should, thus, be considered the actual pathogenic agent in crown gall tumorigenesis. Disease resulting from microbial infection has been suggested to be an *inadvertent* and *unfavorable* outcome (Falkow, 1990), but this does not appear to be the case in crown gall tumorigenesis. Such a view is incompatible with the opine hypothesis since disease (tumor formation) is a prerequisite for opine production and subsequent utilization. The Ti plasmid appears to be "programmed" or "engineered" for the oncogenic transformation of plants. A necessary and physical linkage exists between the Ti plasmid "program" and tumor formation, that is, the tumorigenic determinants are encoded by the end product of *vir* regulon expression, the T-DNA. The final outcome of VirA–VirG signal transduction, *vir* regulon expression, and T-DNA transfer is plant disease. Disease (tumor formation) is an absolute requirement for fulfillment of the opine hypothesis. Therefore success, from the perspective of the Ti plasmid and the opine hypothesis, mandates disease.

Problems exist with both the opine hypothesis and the idea that crown gall tumorigenesis is an inadvertent result of plant infection by Ti plasmid-bearing *A. tumefaciens*. Our perspective is that too much emphasis has been placed on the bacteria–plant interaction instead of the Ti plasmid–plant interaction. In this view, the Ti plasmid is the etiologic agent of crown gall and *A. tumefaciens* is merely a host or vector that assists in the delivery of the T-DNA into the host cell. On detecting its ultimate host, the wounded plant, the Ti plasmid appropriates the transcriptional and translational apparatus of the bacterial cell to effect the transfer of the T-DNA. Further, the Ti plasmid, specifically the VirA–VirG regulatory system, has adopted the use of *A. tumefaciens* chromosomal gene products parasitically to enhance its signaling capacity and, therefore, the tumorigenic potential of Ti plasmid-bearing strains. This productive interaction between the Ti plasmid and the chromosome indicates a long evolutionary relationship among Ti plasmids and *A. tumefaciens*.

The literature on the phenomenon of *vir* regulon induction is vast and the multitude of gene products and environmental signals that cue this induction are potentially confusing (Fig. 4.1; Table I). However, a broad hypothesis of *vir* regulon induction can be envisioned. The initial event required for *vir* regulon induction is the wounding of a plant. Disruption of plant cells leads to rupture of vacuoles and acidification of the wound site. The disruption also releases plan cell wall glycosidases that generate mono- and oligosaccharides from the cell wall. These compounds act as elicitors of the plant wound response. At the plant wound site, a number of important events occur: (1) a burst of rapid cell division, (2) the accumulation of monosaccharides, and (3) the biosynthesis of phytoalexins and other secondary metabolites including acetosyringone and related compounds. Thus, from the perspective of *A. tumefaciens*, the wound site consists of rapidly growing plant cells bathed in an acidic solution of monosaccharides and phenolic compounds. *Agrobacterium tumefaciens*, as a plant commensal, is already in the vicinity of the wound. As the highly soluble monosaccharides diffuse from the wound exudate, the *A. tumefaciens* ChvE/"Trg-like" chemotaxis system recognizes these sugars as chemoattractants so the bacteria swim to the wound site, where they attach to plant cells. Additionally, chemotaxis to acetosyringone and related phenolics may be involved. Once bacteria arrive at the wound, *virG* transcription from promoter P2 is initiated as a result of the stress response to acidic pH and the toxic phytoalexins. Simultaneously, the VirA protein begins to sense environmental signals indicating the presence of a plant wound. Monosaccharides are detected by the ChvE/VirA signal transduction pathway while acidic pH and phenolic metabolites are sensed by the VirA periplasmic space region and cytoplasmic domains, respectively. VirA subsequently autophosphorylates at His-474 in preparation for VirG activation. As the intracellular level of VirG increases, VirA phosphorylates VirG at Asp-52, which enables VirG to bind to *vir* box sequences and activate *vir* regulon transcription. As the *vir* regulon is induced, higher *virG* transcription from promoter P1 increases the level of VirG and ultimately leads to the full induction of the *vir* regulon and transfer of the T-DNA into the plant cell. Induction of the *vir* regulon additionally promotes the conjugation of the Ti plasmid to other *A. tumefaciens* strains, thereby increasing the population of Ti plasmid-containing strains. The role of other *A. tumefaciens* loci (*ros*, *chvD*, and *ivr*) is not meant to be ignored, but is not included in this scheme since their role in *vir* regulon induction is not clear. As mentioned previously in Sections I,D, II,D, and II,E,1–6, several other as yet unidentified factors, both plant and bacterial, are likely to have important roles in *vir* regulon induction.

The sugar/ChvE/VirA-mediated signal pathway is an excellent example of evolutionary adaptation to a new host. Since only certain plants require this sugar/ChvE/VirA-mediated signaling pathway for tumorigenesis, this pathway is involved in host-specific recognition. The existence of the sugar/ChvE-mediated *vir* regulon induction pathway has a number of interesting evolutionary implications. This pathway is not required for tumorigenesis in a large number of plants, suggesting that the archetypal VirA may not have possessed the periplasmic domain of the present day VirA that interacts with ChvE. The VirA protein may have parasitically recruited the N-terminal region of the *A. tumefaciens* "Trg-like" protein, as well as its interaction with the ChvE protein, in the process of expanding the tumorigenic host range.

The *A. tumefaciens* Ti plasmid demonstrates several important characteristics of a successful parasite. Its virulence mechanisms are induced only when their deployment would be successful. The *vir* regulon signal transduction system is quiescent until it detects a susceptible host via several signals simultaneously: (1) plant phenolic metabolites (e.g., acetosyringone), (2) plant-derived monosaccharides, (3) low pH, (4) temperatures <30°C, and, possibly, (5) low phosphate concentrations and (6) osmoprotectant compounds. These signals accurately represent conditions found in the plant wound and reflect the physiology of *A. tumefaciens* itself. By this regulation of *vir* regulon induction, pathogenic *A. tumefaciens* strains demonstrate obligate tissue tropism because they attack only wounded tissue. Much value has been claimed for the economical benefits for a pathogen if its virulence genes are regulated; however, to our knowledge, the only evidence supporting such a view of regulation was reported for *A. tumefaciens* (Fortin *et al.*, 1992). Continuous induction of the *vir* regulon not only inhibits cell growth but leads to the accumulation of mutations in the signal transduction system itself. Thus, regulation of virulence factors not only allows normal cell growth and metabolism, but serves as a mechanism for the maintenance of the genetic integrity of loci necessary for the expression of virulence determinants.

The *A. tumefaciens* crown gall phenomenon has taught us much about the regulation of virulence mechanisms in host–pathogen interactions. The variety of interactions present between Ti plasmid and chromosomal products suggests that the history of host-specific pathogen evolution is a long one. The mechanisms by which a pathogen senses its host are, to human eyes, subtle and clever. Equally important is the information that such pathogens can provide us about their respective hosts. By identifying the specific environmental stimuli recognized by the signal transduction system of a pathogen, aspects of host function can be discovered. The identification of plant phenolic metabolites, monosaccharides, and acidic pH as stimuli that induce the *vir* regulon informs us about the conditions likely to be present in the plant wound. Further studies will continue to surprise us with the lessons that can be learned from microbial pathogens.

Acknowledgments

Research in the laboratory of E. W. Nester was supported by Public Health Service Grant GM 32618-18 from the National Institutes of Health. R. G. Ankenbauer was the recipient of Fellowship DRG-1005 from the Damon Runyon–Walter Winchell Cancer Research Fund.

References

Albright, L. M., Huala, E., and Ausubel, F. M. (1989). Prokaryotic signal transduction mediated by sensor and regulator protein pairs. *Ann. Rev. Genet.* **23**, 311–336.

Alt-Moerbe, J., Nedderman, P., von Lintig, J., Weiler, E. W., and Schroder, J. (1988). Temperature-sensitive step in Ti plasmid *vir*-region induction and correlation with cytokinin secretion by *Agrobacteria*. *Mol. Gen. Genet.* **213**, 1–8.

Anderson, A. R., and Moore, L. W. (1979). Host specificity in the genus *Agrobacterium*. *Phytopathology* **69**, 320–323.

Ankenbauer, R. G., and Nester, E. W. (1990). Sugar-mediated induction of *Agrobacterium tumefaciens* virulence genes: Structural specificity and activities of monosaccharides. *J. Bacteriol.* **172**, 6442–6446.

Ankenbauer, R. G., Best, E. A., Palanca, C. A., and Nester, E. W. (1991). Mutants of the *Agrobacterium tumefaciens virA* gene exhibiting acetosyringone-independent expression of the *vir* regulon. *Mol. Plant–Microbe Interact.* **4**, 400–406.

Bergeron, J., MacLeod, R. A., and Dion, P. (1990). Specificity of octopine uptake by *Rhizobium* and *Pseudomonas* strains. *Appl. Environ. Microbiol.* **56**, 1453–1458.

Binns, A. N., and Thomashow, M. F. (1988). Cell biology of *Agrobacterium* infection and transformation of plants. *Ann. Rev. Microbiol.* **42**, 575–606.

Bolker, T., and Weimken, A. (1986). Dynamics of vacuolar compartmentation. *Ann. Rev. Plant Physiol.* **37**, 137–164.

Bolton, G. W., Nester, E. W., and Gordon, M. P. (1986). Plant phenolic compounds induce expression of the *Agrobacterium tumefaciens* loci needed for virulence. *Science* **232**, 983–985.

Bourret, R. B., Hess, J. F., Borkovich, K. A., Pakula, A. A., and Simon, M. I. (1989). Protein phosphorylation in chemotaxis and two-component regulatory systems of bacteria. *J. Biol. Chem.* **264**, 7085–7088.

Braun, A. C. (1947). Thermal studies on the factors responsible for tumor initiation in crown gall. *Am. J. Botany* **34**, 234–240.

Braun, A. C. (1978). Plant tumors. *Biochim. Biophys. Acta* **516**, 167–191.

Cangelosi, G. A., Hung, L., Puvanesarajah, V., Stacey, G., Ozga, A. D., Leigh, J. A., and Nester, E. W. (1987). Common loci for *Agrobacterium tumefaciens* and *Rhizobium meliloti* exopolysaccharide synthesis and their roles in plant interactions. *J. Bacteriol.* **169**, 2086–2091.

Cangelosi, G. A., Ankenbauer, R. G., and Nester, E. W. (1990). Sugars induce the *Agrobacterium* virulence genes through a periplasmic binding protein and a transmembrane signal protein. *Proc. Natl. Acad. Sci. U.S.A.* **87**, 6708–6712.

Chang, C.-H., and Winans, S. C. (1992). Functional roles assigned to the periplasmic, linker, and receiver domains of the *Agrobacterium tumefaciens* VirA protein. *J. Bacteriol.* **174**, 7033–7039.

Chen, C.-Y., and Winans, S. C. (1991). Controlled expression of the transcriptional activator gene *virG* in *Agrobacterium tumefaciens* by using the *Escherichia coli lac* promoter. *J. Bacteriol.* **173**, 1139–1144.

Chilton, M.-D., Drummond, M. H., Merlo, D. J., Sciaky, D., Montoya, A. L., Gordon, M. P., and Nester, E. W. (1977). Stable incorporation of plasmid DNA into higher plant cells: The molecular basis of crown gall tumorigensis. *Cell* **11**, 263–271.

Close, T. J., Rogowsky, P. M., Kado, C. I., Winans, S. C., Yanofsky, M. F., and Nester, E. W. (1987). Dual control of *Agrobacterium tumefaciens* Ti plasmid virulence genes. *J. Bacteriol.* **169**, 5113–5118.

Cooley, M. B., D'Souza, M. R., and Kado, C. I. (1991). The *virC* and *virD* operons of the *Agrobacterium* Ti plasmid are regulated by the *ros* chromosomal gene: Analysis of the cloned *ros* gene. *J. Bacteriol.* **173**, 2608–2616.

Cornish, A., Greenwood, J. A., and Jones, C. W. (1989). Binding-protein-dependent sugar transport by *Agrobacterium radiobacter* and *A. tumefaciens* grown in continuous culture. *J. Gen. Microbiol.* **135**, 3001–3031.

Culianez-Macia, F. A., and Hepburn, A. G. (1988). The kinetics of T-strand production in a nopaline-type helper strain of *Agrobacterium tumefaciens*. *Mol. Plant–Microbe Interact.* **1**, 207–214.

Das, A., and Pazour, G. J. (1989). Delineation of the regulatory region sequences of *Agrobacterium tumefaciens virB* operon. *Nucleic Acids Res.* **17**, 4541–4550.

Das, A., Stachel, S., Ebert, P., Allenza, P., Montoya, A., and Nester, E. (1986). Promoters of *Agrobacterium tumefaciens* Ti-plasmid virulence genes. *Nucleic Acids Res.* **14**, 1355–1364.

DeCleene, M. (1988). The susceptibility of plants to *Agrobacterium*: A discussion of the role of phenolic compounds. *FEMS Microbiol. Rev.* **54**, 1–8.

DeCleene, M., and DeLey, J. (1976). The host range of crown gall. *Bot. Rev.* **42**, 389–466.

DiRita, V. J., and Mekalanos, J. J. (1989). Genetic control of bacterial virulence. *Ann. Rev. Genet.* **23**, 455–482.

Douglas, C. J., Halperin, W., and Nester, E. W. (1982). *Agrobacterium tumefaciens* mutants affected in attachment to plant cells. *J. Bacteriol.* **152**, 1265–1275.

Falkow, S. (1990). The "zen" of bacterial pathogenicity. *In* "Molecular Basis of Bacterial Pathogenesis" (B. H. Iglewski and V. L. Clark, eds.), pp. 3–9. Academic Press, San Diego.

Forst, S., Delgado, J., Rampersaud, A., and Inouye, M. (1990). *In vivo* phosphorylation of OmpR, the transcription activator of the *ompF* and *ompC* genes in *Escherichia coli*. *J. Bacteriol.* **172**, 3473–3477.

Fortin, C., Nester, E. W., and Dion, P. (1992). Growth inhibition and loss of virulence in cultures of *Agrobacterium tumefaciens* treated with acetosyringone. *J. Bacteriol.* **174**, 5676–5685.

Furlong, C. E. (1987). Osmotic-shock-sensitive transport systems. *In* "*Escherichia coli* and *Salmonella typhimurium*: Cellular and Molecular Biology" (F. C. Neidhardt, J. L. Ingraham, K. B. Low, B. Magasanik, M. Schaechter, and H. E. Umbarger, eds.), pp. 768–796. American Society for Microbiology, Washington, D.C.

Garfinkel, D. J., and Nester, E. W. (1980). *Agrobacterium tumefaciens* mutants affected in crown gall tumorigenesis and octopine catabolism. *J. Bacteriol.* **144**, 732–743.

Garfinkel, D. J., Simpson, R. B., Ream, L. W., White, F. F., Gordon, M. P., and Nester, E. W. (1981). Genetic analysis of crown gall: fine structure map of the T-DNA by site-directed mutagenesis. *Cell* **27**, 143–153.

Gelvin, S. B., and Habeck, L. L. (1990). *vir* genes influence conjugal transfer of the Ti plasmid of *Agrobacterium tumefaciens*. *J. Bacteriol.* **172**, 1600–1608.

Gray, J., Wang, J., and Gelvin, S. B. (1992). Mutation of the *miaA* gene of *Agrobacterium tumefaciens* results in reduced *vir* gene expression. *J. Bacteriol.* **174**, 1086–1098.

Gross, R., Arico, B., and Rappuoli, R. (1989). Families of bacterial signal-transducing proteins. *Mol. Microbiol.* **3**, 1661–1667.

Hahn, M. G., Bucheli, P., Cervone, F., Doares, S. H., O'Neill, R. A., Darvill, A., and Albersheim, P. (1989). Roles of cell wall constituents in plant-pathogen interactions. *In* "Plant–Microbe Interactions: Molecular and Genetic Perspectives" (T. Kosuge and E. W. Nester, eds.), Vol. 3, pp. 131–181. McGraw-Hill, New York.

Halverson, L. J., and Stacey, G. (1986). Signal exchange in plant-microbe interactions. *Microbiol. Rev.* **50**, 193–225.

Han, D. C., Chen, C.-Y., Chen, Y.-F., and Winans, S. C. (1992). Altered-function mutations of the transcriptional regulatory gene *virG* of *Agrobacterium tumefaciens*. *J. Bacteriol.* **174**, 7040–7043.

Hawes, M. C., and Smith, L. Y. (1989). Requirement for chemotaxis in pathogenicity of *Agrobacterium tumefaciens* on roots of soil-grown pea plants. *J. Bacteriol.* **171**, 5668–5671.

Hawes, M. C., Smith, L. Y., and Howarth, A. J. (1988). *Agrobacterium tumefaciens* mutants deficient in chemotaxis to root exudates. *Mol. Plant-Microbe Interact.* **1**, 182–186.

Heinemann, J. A. (1991). Genetics of gene transfer between species. *Trends Genet.* **7**, 181–185.

Heinemann, J. A., and Sprague, G. F. (1989). Bacterial conjugative plasmids mobilize DNA transfer between bacteria and yeast. *Nature (London)* **340**, 205–209.

Hess, K. M., Dudley, M. W., Lynn, D. G., Joerger, R. D., and Binns, A. N. (1991). Mechanism of phenolic activation of *Agrobacterium* virulence genes: Development of the first specific inhibitor of bacterial sensor/response systems. *Proc. Natl. Acad. Sci. U.S.A.* **88**, 7854–7858.

Higgins, C. F., Hiles, I. D., Salmond, G. P. C., Gill, D. R., Downie, J. A., Evans, I. J., Holland, I. B., Gray, L., Buckell, S. D., Bell, A. W., and Hermodson, M. A. (1986). A family of related ATP-binding subunits coupled to many distinct biological processes in bacteria. *Nature (London)* **323**, 448–450.

Hille, J., Klasen, I., and Schilperoort, R. (1982). Construction and application of R′ plasmids carrying different segments of an octopine Ti-plasmid from *A. tumefaciens* for complementation of *vir* genes. *Plasmid* **7**, 107–118.

Huang, M.-L. W., Cangelosi, G. A., Halperin, W., and Nester, E. W. (1990). A chromosomal *Agrobacterium tumefaciens* gene required for effective plant signal transduction. *J. Bacteriol.* **172**, 1814–1822.

Huang, Y., Morel, P., Powell, B., and Kado, C. I. (1990). VirA, a coregulator of Ti-specified virulence genes, is phosphorylated *in vitro*. *J. Bacteriol.* **172**, 1142–1144.

Jin, S., Prusti, R. K., Roitsch, T., Ankenbauer, R. G., and E. W. Nester. (1990a). Phosphorylation of the VirG protein of *Agrobacterium tumefaciens* by the autophosphorylated VirA protein: Essential role in biological activity of VirG. *J. Bacteriol.* **172**, 4945–4950.

Jin, S., Roitsch, T., Ankenbauer, R. G., Gordon, M. P., and Nester, E. W. (1990b). The VirA protein of *Agrobacterium tumefaciens* is autophosphorylated and is essential for *vir* gene regulation. *J. Bacteriol.* **172**, 525–530.

Jin, S., Roitsch, T., Christie, P. J., and Nester, E. W. (1990c). The regulatory VirG protein specifically binds to a *cis*-acting regulatory sequence involved in transcriptional activation of *Agrobacterium tumefaciens* virulence genes. *J. Bacteriol.* **172**, 531–537.

Kahl, G. (1982). Molecular biology of wound healing: The conditioning phenomenon. *In* "Molecular Biology of Plant Tumors" (G. Kahl and J. Schell, eds.), pp. 211–267. Academic Press, New York.

Klee, H. J., Gordon, M. P., and Nester, E. W. (1982). Complementation analysis of *Agrobacterium tumefaciens* Ti-plasmid mutations affecting oncogenicity. *J. Bacteriol.* **150**, 327–331.

Klee, H. J., White, F. F., Iyer, V. N., Gordon, M. P., and Nester, E. W. (1983). Mutational analysis of the virulence region of an *Agrobacterium tumefaciens* Ti-plasmid. *J. Bacteriol.* **153**, 878–883.

Lee, K., Dudley, M. W., Hess, K. M., Lynn, D. G., Joerger, R. D., and Binns, A. N. (1992). Mechanism of activation of *Agrobacterium* virulence genes: identification of phenol-binding proteins. *Proc. Natl. Acad. Sci. U.S.A.* **89**, 8666–8670.

Leroux, B., Yanofsky, M. F., Winans, S. C., Ward, J. E., Ziegler, S. F., and Nester, E. W. (1987). Characterization of the *virA* locus of *Agrobacterium tumefaciens*: A transcriptional regulator and host range determinant. *EMBO J.* **6**, 849–856.

Lippincott, J. A., and Lippincott, B. B. (1975). The genus *Agrobacterium* and plant tumorigenesis. *Ann. Rev. Microbiol.* **29**, 377–405.

Machida, Y., Okamoto, S., Matsumoto, S., Usami, S., Yamamoto, A., Niwa, Y., Jeong, S. D., Nagamine, J., Shimoda, N., Machida, C., and Iwahashi, M. (1989). Mechanisms of crown gall formation: T-DNA transfer from *Agrobacterium tumefaciens* to plant cells. *Bot. Mag. Tokyo* **102**, 331–350.

Macnab, R. M. (1987). Motility and chemotaxis. *In* "*Escherichia coli* and *Salmonella typhimurium*: Cellular and Molecular Biology" (F. C. Neidhardt, J. L. Ingraham, K. B. Low, B. Magasanik, M. Schaechter, and H. E. Umbarger, eds.), Vol. 1, pp. 732–759. American Society for Microbiology, Washington, D.C.

McNeil, M., Darvill, A. G., Fry, S. C., and Albersheim, P. (1984). Structure and function of the primary cell wall in plants. *Ann. Rev. Biochem.* **53**, 625–633.

Mantis, N. J., and Winans, S. C. (1992). The *Agrobacterium tumefaciens vir* gene transcriptional activator *virG* is transcriptionally induced by acid pH and other stress stimuli. *J. Bacteriol.* **174**, 1189–1196.

Mathews, H., Bharathan, N., Litz, R. E., Narayanan, K. R., Rao, P. S., and Bhatia, C. R. (1990). The promotion of *Agrobacterium*-mediated transformation in *Atropa belladona* L. by acetosyringone. *J. Plant Physiol.* **136**, 404–409.

Matthysse, A. G. (1983). Role of bacterial cellulose fibrils in *Agrobacterium tumefaciens* infection. *J. Bacteriol.* **154**, 906–915.

Matthysse, A. G. (1987). Characterization of nonattaching mutants of *Agrobacterium tumefaciens*. *J. Bacteriol.* **169**, 313–323.

Melchers, L. S., Thompson, D. V., Idler, K. B., Schilperoort, R. A., and Hooykaas, P. J. J. (1986). Nucleotide sequence of the virulence gene *vir*G of *Agrobacterium tumefaciens* octopine Ti-plasmid: Significant homology between *vir*G and the regulatory genes *omp*R, *pho*B, and *dye* of *E. coli*. *Nucleic Acids Res.* **24**, 9933–9942.

Melchers, L. S., Thompson, D. V., Idler, K. B., Neuteboom, S. T. C., de Maagd, R. A., Schilperoort, R. A., and Hooykaas, P. J. J. (1987). Molecular characterization of the virulence gene *vir*A of the *Agrobacterium tumefaciens* octopine Ti-plasmid. *Plant Mol. Biol.* **9**, 635–645.

Melchers, L. S., Regensburg-Tuink, T. J. G., Bourret, R. B., Sedee, N. J. A., Schilperoort, R. A., and Hooykaas, P. J. J. (1989a). Membrane topology and functional analysis of the sensory protein VirA of *Agrobacterium tumefaciens*. *EMBO J.* **8**, 1919–1925.

Melchers, L. S., Regensburg-Tuink, T. J. G., Schilperoort, R. A., and Hooykaas, P. J. J. (1989b). Specificity of signal molecules in the activation of *Agrobacterium* virulence gene expression. *Mol. Microbiol.* **3**, 969–977.

Messens, E., Dekeyser, R., and Stachel, S. E. (1990). A nontransformable *Triticum monococcum* monocotyledonous culture produces the potent *Agrobacterium vir*-inducing compound ethyl ferulate. *Proc. Natl. Acad. Sci. U.S.A.* **87**, 4368–4372.

Metts, J., West, J., Doares, S. H., and Matthysse, A. G. (1991). Characterization of three *Agrobacterium tumefaciens* avirulent mutants with chromosomal mutations that affect induction of *vir* genes. *J. Bacteriol.* **173**, 1080–1087.

Miller, J. F., Mekalanos, J. J., and Falkow, S. (1989). Coordinate regulation and sensory transduction in the control of bacterial virulence. *Science* **243**, 916–922.

Morel, P., Powell, B. S., Rogowsky, P. M., and Kado, C. I. (1989). Characterization of the *virA* virulence gene of the nopaline plasmid, pTiC58, of *Agrobacterium tumefaciens*. *Mol. Microbiol.* **3**, 1237–1246.

Morris, J. W., and Morris, R. O. (1990). Identification of an *Agrobacterium tumefaciens* virulence gene inducer from the pinaceous gymnosperm *Pseudotsuga menziesii*. *Proc. Natl. Acad. Sci. U.S.A.* **87**, 3614–3618.

Nara, T., Lee, L., and Imae, Y. (1991). Thermosensing ability of Trg and Tap chemoreceptors in *Escherichia coli*. *J. Bacteriol.* **173**, 1120–1124.

Nautiyal, C. S., and Dion, P. (1990). Characterization of the opine-utilizing microflora associated with samples of soil and plants. *Appl. Environ. Microbiol.* **56**, 2576–2579.

Nester, E. W., Gordon, M. P., Amasino, R. M., and Yanofsky, M. F. (1984). Crown gall: A molecular and physiological analysis. *Ann. Rev. Plant Physiol.* **35**, 387–413.

Parke, D., Ornston, L. N., and Nester, E. W. (1987). Chemotaxis to plant phenolic inducers of virulence genes is constitutively expressed in the absence of the Ti plasmid in *Agrobacterium tumefaciens*. *J. Bacteriol.* **169**, 5336–5338.

Pazour, G. J., and Das, A. (1990a). Characterization of the VirG binding site of *Agrobacterium tumefaciens*. *Nucleic Acids Res.* **18**, 6909–6913.

Pazour, G. J., and Das, A. (1990b). *virG*, an *Agrobacterium tumefaciens* transcriptional activator, initiates translation at a UUG codon and is a sequence-specific DNA-binding protein. *J. Bacteriol.* **172**, 1241–1249.

Pazour, G. J., Ta, C. N., and Das, A. (1991). Mutants of *Agrobacterium tumefaciens* with elevated *vir* gene expression. *Proc. Natl. Acad. Sci. U.S.A.* **88**, 6941–6945.

Pazour, G. J., Ta, C. N., and Das, A. (1992). Constitutive mutations of *Agrobacterium tumefaciens* transcriptional activator *virG*. *J. Bacteriol.* **174**, 4169–4174.

Peters, N. K., and Verma, D. P. S. (1990). Phenolic compounds as regulators of gene expression in plant–microbe interactions. *Mol. Plant–Microbe Interac.* **3**, 4–8.

Powell, B. S., and Kado, C. I. (1990). Specific binding of VirG to the *vir* box requires a C-terminal domain and exhibits a minimum concentration threshold. *Mol. Microbiol.* **4**, 2159–2166.

Powell, B. S., Powell, G. K., Morris, R. O., Rogowsky, P. M., and Kado, C. I. (1987). Nucleotide sequence of the *virG* locus of the *Agrobacterium tumefaciens* plasmid pTiC58. *Mol. Microbiol.* **1**, 309–316.

Powell, B. S., Rogowsky, P. M., and Kado, C. I. (1989). *virG* of *Agrobacterium tumefaciens* plasmid pTiC58 encodes a DNA-binding protein. *Mol. Microbiol.* **3**, 411–419.

Raineri, D. M., Boulton, M. I., Davies, J. W., and Nester, E. W. (1993). VirA, the plant signal receptor, is responsible for the Ti plasmid-specific transfer of DNA to maize by *Agrobacterium*. *Proc. Natl. Acad. Sci. U.S.A.* **90**, 3549–3553.

Ream, W. (1989). *Agrobacterium tumefaciens* and interkingdom genetic exchange. *Ann. Rev. Phytopathol.* **27**, 583–618.

Recourt, K., van Brussel, A. A. N., Driessen, A. J. M., and Lugtenberg, B. J. J. (1989). Accumulation of a *nod* gene inducer, the flavanoid naringenin, in the cytoplasmic membrane of *Rhizobium leguminosarum* biovar *viciae* is caused by the pH-dependent hydrophobicity of naringenin. *J. Bacteriol.* **171**, 4370–4377.

Riker, A. J. (1926). Studies on the influence of some environmental factors on the development of crown gall. *J. Agricult. Res.* **32**, 83–96.

Rogler, C. E. (1980). Plasmid-dependent temperature-sensitive phase in crown gall tumorigenesis. *Proc. Natl. Acad. Sci. U.S.A.* **77**, 2688–2692.

Rogowsky, P. M., Close, T. J., Chimera, J. A., Shaw, J. J., and Kado, C. I. (1987). Regulation of the *vir* genes of *Agrobacterium tumefaciens* plasmid pTiC58. *J. Bacteriol.* **169**, 5101–5112.

Roitsch, T., Wang, H., Jin, S., and Nester, E. W. (1990). Mutational analysis of the VirG protein, a transcriptional activator of *Agrobacterium tumefaciens* virulence genes. *J. Bacteriol.* **172**, 6054–6060.

Shaw, C. H., Ashby, A. M., Brown, A., Royal, C., Loake, G. J., and Shaw, C. H. (1988). *virA* and *virG* are the Ti-plasmid functions required for chemotaxis of *Agrobacterium tumefaciens* towards acetosyringone. *Mol. Microbiol.* **2**, 413–417.

Sheikholeslam, S. N., and Weeks, D. P. (1987). Acetosyringone promotes high efficiency transformation of *Arabidopsis thaliana* explants by *Agrobacterium tumefaciens*. *Plant Mol. Biol.* **8**, 291–298.

Shimoda, N., Toyoda-Yamamoto, A., Nagamine, J., Usami, S., Katayama, M., Sakagami, Y., and Machida, Y. (1990). Control of expression of *Agrobacterium vir* genes by synergistic actions of phenolic signal molecules and monosaccharides. *Proc. Natl. Acad. Sci. U.S.A.* **87**, 6684–6688.

Spencer, P. A., and Towers, G. H. N. (1988). Specificity of signal compounds detected by *Agrobacterium tumefaciens*. *Phytochemistry* **27**, 2781–2785.

Stachel, S. E., and Nester, E. W. (1986). The genetic and transcriptional organization of the *vir* region of the A6 Ti plasmid of *Agrobacterium tumefaciens*. *EMBO J.* **5**, 1445–1454.

Stachel, S. E., and Zambryski, P. C. (1986). *vir*A and *vir*G control the plant-induced activation of the T-DNA transfer process of *A. tumefaciens*. *Cell* **46**, 325–333.

Stachel, S. E., An, G., and Nester, E. (1984). Inducible expression of the virulence genes of the A6 *Agrobacterium tumefaciens* Ti-plasmid. *J. Cell. Biochem. Suppl.* **8B**, 64.

Stachel, S. E., An, G., Flores, C., and Nester, E. W. (1985a). A Tn*3 lacZ* transposon for the random generation of β-galactosidase gene fusions: Application to the analysis of gene expression in *Agrobacterium*. *EMBO J.* **4**, 891–898.

Stachel, S. E., Messens, E., van Montagu, M., and Zambryski, P. (1985b). Identification of the signal molecules produced by wounded plant cells that activate T-DNA transfer in *Agrobacterium tumefaciens*. *Nature (London)* **318**, 624–629.

Stachel, S. E., Nester, E. W., and Zambryski, P. C. (1986). A plant cell factor induces *Agrobacterium tumefaciens vir* gene expression. *Proc. Natl. Acad. Sci. U.S.A.* **83**, 379–383.

Steck, T. R., and Kado, C. I. (1990). Virulence genes promote conjugative transfer of the Ti plasmid between *Agrobacterium* strains. *J. Bacteriol.* **172**, 2191–2193.

Stock, J. B., Ninfa, A. J., and Stock, A. M. (1989). Protein phosphorylation and regulation of adaptive responses in bacteria. *Microbiol. Rev.* **53**, 450–490.

Tait, R. C., and Kado, C. I. (1988). Regulation of the *virC* and *virD* promoters of pTIC58 by the *ros* chromosomal mutation of *Agrobacterium tumefaciens*. *Mol. Microbiol.* **2**, 385–392.

Tempe, J., and Petit, A. (1982). Opine utilization by *Agrobacterium*. *In* "Molecular Biology of Plant Tumors" (G. Kahl and J. Schell, eds.), pp. 451–459. Academic Press, New York.

Thomashow, M. F., Nutter, R., Montoya, A. L., Gordon, M. P., and Nester, E. W. (1980). Integration and organization of Ti plasmid sequences in crown gall tumors. *Cell* **19**, 729–739.

Thomashow, M. F., Karlinsky, J. E., Marks, J. R., and Hurlburt, R. E. (1987). Identification of a new virulence locus in *Agrobacterium tumefaciens* that affects polysaccharide composition and plant cell attachment. *J. Bacteriol.* **169**, 2309–2316.

Threlfall, D. R., and Whitehead, I. M. (1988). Coordinated inhibition of squalene synthetase and induction of enzymes of sesquiterpenoid phytoalexin biosynthesis in cultures of *Nicotiana tabacum*. *Phytochemistry* **27**, 2567–2580.

Usami, S., Okamoto, S., Takebe, I., and Machida, Y. (1988). Factor inducing *Agrobacterium tumefaciens vir* gene expression is present in monocotyledonous plants. *Proc. Natl. Acad. Sci. U.S.A.* **85**, 3748–3752.

Van Larebeke, N., Engler, G., Holsters, M., Van den Elsacker, S., Zaenen, I., Schilperoort, R. A., and Schell, J. (1974). Large plasmid in *Agrobacterium tumefaciens* essential for crown gall-inducing ability. *Nature (London)* **252**, 169–170.

Veluthambi, K., Jayaswal, R. K., and Gelvin, S. B. (1987). Virulence genes *A*, *G*, and *D* mediate the double stranded border cleavage of T-DNA from the *Agrobacterium* Ti-plasmid. *Proc. Natl. Acad. Sci. U.S.A.* **84**, 1881–1885.

Veluthambi, K., Krishnan, M., Gould, J. H., Smith, R. H., and Gelvin, S. B. (1989). Opines stimulate induction of the *vir* genes of the *Agrobacterium tumefaciens* Ti-plasmid. *J. Bacteriol.* **171**, 3696–3703.

Vernade, D., Herrera-Estrella, A., Wang, K., and Van Montagu, M. (1988). Glycine betaine allows enhanced induction of the *Agrobacterium tumefaciens vir* genes by acetosyringone at low pH. *J. Bacteriol.* **170**, 5822–5829.

Watson, B., Currier, T. C., Gordon, M. P., Chilton, M.-D., and Nester, E. W. (1975). Plasmid required for virulence of *Agrobacterium tumefaciens*. *J. Bacteriol.* **123**, 255–264.

Winans, S. C. (1990). Transcriptional induction of an *Agrobacterium* regulatory gene at tandem promoters by plant-released phenolic compounds, phosphate starvation, and acidic growth media. *J. Bacteriol.* **172**, 2433–2438.

Winans, S. C. (1992). Two way chemical signalling in *Agrobacterium*-plant interactions. *Microbiol. Rev.* **56**, 12–31.

Winans, S. C., Ebert, P. R., Stachel, S. E., Gordon, M. P., and Nester, E. W. (1986). A gene essential for *Agrobacterium* virulence is homologous to a family of positive regulatory loci. *Proc. Natl. Acad. Sci. U.S.A.* **83**, 8278–8282.

Winans, S., Jin, S., Komari, T., Johnson, K., and Nester, E. (1987). The role of virulence regulatory loci in determining *Agrobacterium* host range. *In* "Plant Molecular Biology" (D. Von Wettstein and N.-H. Chua, eds.), pp. 573–582. Plenum Publishing, New York.

Winans, S. C., Kerstetter, R. A., and Nester, E. W. (1988). Transcriptional regulation of the *vir*A and *vir*G genes of *Agrobacterium tumefaciens*. *J. Bacteriol.* **170**, 4047–4054.

Winans, S. C., Kerstetter, R. A., Ward, J. E., and Nester, E. W. (1989). A protein required for transcriptional regulation of *Agrobacterium* virulence genes spans the cytoplasmic membrane. *J. Bacteriol.* **171**, 1616–1622.

Zambryski, P. (1988). Basic processes underlying *Agrobacterium*-mediated DNA transfer to plant cells. *Ann. Rev. Genet.* **22**, 1–30.

Zambryski, P. (1989). *Agrobacterium*-plant cell DNA transfer. *In* "Mobile DNA" (D. E. Berg and M. M. Howe, eds.), pp. 309–333. American Society Microbiology, Washington, D.C.

Zerback, R., Dressler, K., and Hess, D. (1989). Flavonoid compounds from pollen and stigma of *Petunia hybrida*: Inducers of the *vir* region of the *Agrobacterium tumefaciens* Ti-plasmid. *Plant Sci.* **62**, 83–91.

Part II

Eukaryotic Signaling Systems

Chapter 5

Signal Transduction Systems in Eukaryotic Microorganisms, *Drosophila*, and Nematodes

JANET KURJAN

Department of Microbiology and Molecular Genetics and
The Vermont Cancer Center
College of Medicine and
College of Life Sciences and Agriculture
University of Vermont
Burlington, Vermont 05405

I. Introduction

The explosion of information in the field of eukaryotic signal transduction in recent years has provided much greater understanding of signaling mechanisms. In mammalian systems, most of the information has come from biochemical and molecular approaches. In eukaryotic microorganisms, the approaches have varied depending on the particular system, and include genetic, molecular, biochemical, and physiological approaches. Research on signal transduction in *Drosophila* and the nematode *Caenorhabditis elegans* also has made rapid advances and has benefited greatly from the history of genetic and developmental studies in these systems.

One of the striking conclusions drawn from research in diverse organisms has been the extensive evolutionary conservation of many fundamental cellular processes, including signal transduction. However, in addition to detecting similarities of components among systems, observing the variations among systems is also important. In this introduction to the eukaryotic section of this book, I discuss several signal transduction pathways stimulated by ras and heterotrimeric G proteins in genetically tractable systems. I compare them with one another and with the mammalian systems to focus on some of the themes and variations on these themes that are observable in the comparisons of systems.[1]

II. Activation by Gα Proteins

Heterotrimeric G proteins were first identified in mammalian cells by the finding that GTP promoted adenylyl cyclase activity (reviewed by Iyengar and Birnbaumer, 1990). These proteins consist of three subunits: the β and γ subunits are basically inseparable and therefore can be thought of as a single entity, βγ.

Two systems in particular have been characterized extensively and give rise to similar models for the action of G proteins (reviewed by Iyengar and Birnbaumer, 1990). Response to photons of light acts through rhodopsin and the G protein transducin to activate the effector cGMP phosphodiesterase (Fig. 5.1A). Response to catecholamines acts through β-adrenergic receptors and G_s to activate the effector adenylyl cyclase (Fig. 5.1C). In both cases, in the absence of the signal the G protein is present as a heterotrimer with bound GDP. In the presence of the signal, the rhodopsin or B-adrenergic receptor is activated, allowing an interaction with the G protein. Guanine nucleotide exchange ensues, and the α subunit with bound GTP separates from βγ. α-GTP then activates the effector in its respective system. A GTPase activity intrinsic to the α subunit acts to restore the GDP-bound state, which then binds to βγ to return the system to the basal state.

1. The genetic nomenclature varies in the systems described in this chapter. I use the nomenclature specific to each system in my discussion of that system. In general, genes are indicated by italics and proteins by roman letters.

A. *Drosophila* Phototransduction

Phototransduction in *Drosophila* acts through rhodopsin homologs and is likely to be mediated by a G protein (Fig. 5.1B). Four genes for rhodopsin homologs have been identified that differ in their spectrum of expression in the eight photoreceptor cells R1–R8 (reviewed by Smith *et al.*, 1991). The *norpA* mutant, which is defective in response to light, encodes a phospholipase C (PLC) that is likely to be the effector in this system. Identification of G protein subunits that were candidates for mediators of this response was difficult; for several years, many G protein genes were identified that were expressed at high levels in the head but none of these genes was found to be expressed in the eye. Finally, an α_q gene (*dgq*) and a β gene (*gbe*) were identified that show significant expression in the eye and that are good candidates for the *Drosophila* transducin. The similarity of this α subunit gene to mammalian α_q genes is satisfying because the *Drosophila* pathway acts through a PLC. Therefore, although the *Drosophila* phototransduction system uses rhodopsin homologs, the G protein and effector are related to mammalian homologs that act through different receptors.

B. *Schizosaccharomyces pombe* Pheromone Response

Like the budding yeast *Saccharomyces cerevisiae*, the evolutionarily distant fission yeast *Schizosaccharomyces pombe* secretes pheromones that produce a set of responses in cells of the opposite mating type and are necessary for mating (described in Chapter 6). The two mating types are called h^+ and h^-; the pheromones secreted by these cells are called P-factor and M-factor, respectively. The M-factor and P-factor receptors have been identified as the *map3*$^+$ and *mam2*$^+$ gene products. Interestingly, these genes show sequence similarity to the *S. cerevisiae* pheromone receptor genes. A Gα gene (*gpa1*$^+$) has been identified as a likely component of this pathway. Loss-of-function *gpa1* mutations result in pheromone response and mating defects; an activating mutation results in morphological changes similar to those induced by exposure of wild-type cells to pheromone. These phenotypes indicate that this α subunit plays a positive role, suggesting that α-GTP is likely to activate an effector.

C. *Dictyostelium* cAMP Response

Aggregation in *Dictyostelium* involves a response to pulses of cAMP (described in Chapter 13). The cAMP receptor initially was identified by cAMP binding studies and later by the isolation of four genes (*cAR1–cAR4*), all of which show structural features characteristic of G protein-linked receptors. Eight Gα genes and one Gβ gene have been isolated. Exposure to cAMP results in increased

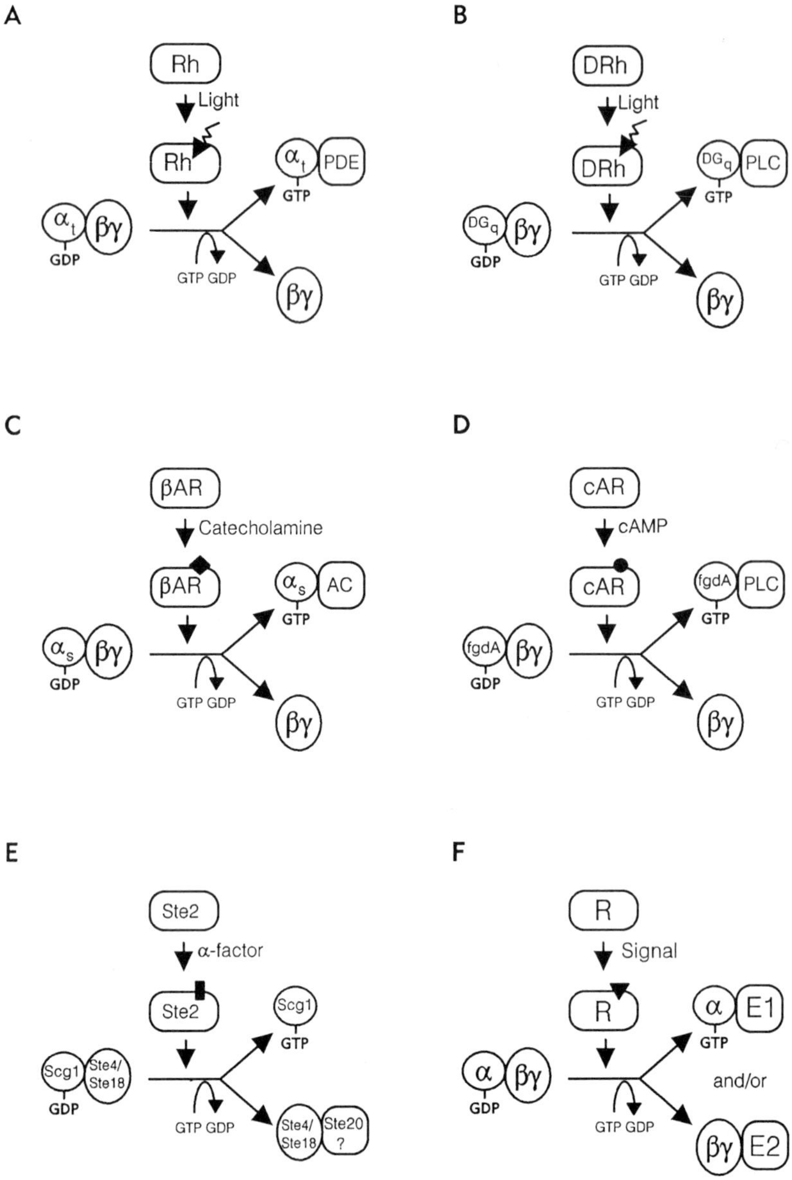

Figure 5.1 Heterotrimeric G protein-mediated systems. Several systems in which multiple components of a pathway have been identified are diagrammed. In all cases, the G protein exists as a heterotrimer composed of α, β, and γ subunits, with GDP bound to the α subunit in the absence of the stimulus. Exposure to the signal activates the receptor, causing guanine nucleotide exchange and dissociation of α-GTP from βγ. (A) In the vertebrate phototransduction system, a photon of light activates rhodopsin (Rh), which interacts with the G protein transducin. After

adenylyl cyclase and guanylyl cyclase activities. In addition, inositol 1,4,5-trisphosphate [Ins(1,4,5)P_3] levels increase, suggesting that a PLC also may be activated. The existence of several cAMP receptors and many Gα proteins complicates the determination of specific interactions between individual receptors and G proteins, and between particular G proteins and the potential effectors. In addition, some of the increased enzyme activities may occur indirectly rather than by direct activation by a G protein. The finding that the *fgdA* mutation contains a defect in the Gα2 gene has provided more detailed information on this subunit. This mutation results in a defect in stimulation of PLC by GTP analogs, suggesting that PLC is the effector for Gα2 and that this α subunit plays a positive role (Fig. 5.1D).

III. Activation by βγ Subunits

Early results in a mammalian system indicating that βγ could activate an effector directly were highly disputed (Bourne, 1989). The discovery that the *S. cerevisiae* βγ subunit of the pheromone response pathway acts downstream of α, and therefore must activate the effector, added to the controversy. To date, no other genetically tractable system has provided an additional example of βγ activating an effector. However, some results in mammalian systems have indicated cases in which βγ can play a role in activation (Birnbaumer, 1992).

A. *Saccharomyces cerevisiae* Pheromone Response

In *S. cerevisiae*, the response to the pheromones **a**- and α-factor involves a G protein-mediated pathway (described in Chapter 8). The *STE2* and *STE3* genes encode the α- and **a**-factor receptors, respectively. The Gα subunit is

dissociation from βγ, the α subunit of transducin (α_t) activates a cGMP phosphodiesterase (PDE). (B) In the *Drosophila* phototransduction system, a photon of light activates *Drosophila* rhodopsin (DRh). The α subunit (DGq) activates a phospholipase C (PLC). (C) In the vertebrate β-adrenergic system, catecholamines act through the β-adrenergic receptor (βAR) and α_s to activate adenylyl cyclase (AC). (D) In *Dictyostelium*, response to pulses of cAMP act through a cAMP receptor (cAR) to activate several enzymes. The α subunit fgdA is thought to correspond to the Gα subunit that activates the effector PLC. (E) In the *S. cerevisiae* pheromone response pathway, peptide pheromones act through receptors to elicit responses to pheromone; in **a** cells, α-factor acts through the α-factor receptor Ste2; in **α** cells, **a**-factor acts through the **a**-factor receptor Ste3 (not shown). *SCGI*, *STE4*, and *STE18* encode the α, β, and γ subunits (Scg1, Ste4, Ste18) of the G protein, respectively. βγ acts downstream of α to activate an effector. The identity of the effector has not been determined, but the *STE20* gene, which encodes a kinase homolog, is a candidate for the effector. (F) A generic model for the action of G proteins is shown. Dissociation of α-GTP from βγ can result in activation of an effector by either α-GTP or βγ. In some cases, a single signal acting through a single receptor (R) could result in the activation of two effectors (E1 and E2).

the product of a gene called, alternatively, *SCGI* or *GPA1*. The finding that *scg1* disruption mutations result in constitutive activation of the pheromone response pathway indicates that Scg1 plays a negative role, unlike the positive role played by the α subunits in the mammalian β-adrenergic and phototransduction systems. Disruption mutations in the β gene (*STE4*) or the γ gene (*STE18*) result in defects in pheromone response and mating, indicating that βγ plays a positive role. Genetic results indicate that βγ acts downstream of α, consistent with a model in which the positive role of βγ is to activate the effector and the negative role of α is to keep βγ inactive in the absence of pheromone (Fig. 5.1E). To date, this model has held up to further analysis, but the ultimate proof will require identification of the effector as well as biochemical experiments to determine whether βγ is capable of activating this effector. The identification of a candidate for the effector (*STE20*), which shows sequence similarity to protein kinases, should facilitate such experiments (Leberer *et al*., 1992; R. Akada and J. Kurjan, unpublished results).

B. Activation by Mammalian βγ Subunits

For many years, the specificity in mammalian systems was thought to lie solely in the α subunits. The suggestion that mammalian βγ could act to open K^+ channels therefore was subject to debate; this issue was made more controversial by contradictory results from different groups (reviewed by Bourne, 1989). Later results indicated that the opening of K^+ channels through muscarinic receptors acts through the α_i subunit; however, βγ also could open K^+ channels by activating phospholipase A2 (PLA2), but that this effect did not occur through muscarinic receptors.

Several experiments have indicated an important role for βγ in different systems (reviewed by Birnbaumer, 1992). Multiple adenylyl cyclase types have been identified, all of which are activated by α_s, but differ in their regulation by βγ. In some cases, βγ inhibits activation by α_s; in other cases, βγ augments activation by α_s; still other types are not affected by βγ. Although the augmentation of Type II and Type IV adenylyl cyclase by βγ indicates that βγ can play a positive role, this role is as an accessory to α_s and not as a direct activator of the effector. However, some results suggest that βγ may activate the β2 form of PLC directly.

The models for the action of G proteins in different systems are mechanistically very similar (Fig. 5.1). In all cases, the heterotrimeric G protein interacts with the receptor, guanine nucleotide exchange occurs, and α-GTP and βγ dissociate from one another. The systems differ only in which subunit then activates an effector; in some cases the α subunit is the activating component whereas in others βγ is the activating component. The ability of both α and βγ to activate effectors allows the possibility of a single agonist acting through a receptor to activate two effectors, one by the α subunit and the second by βγ (Fig. 5.1F).

IV. Ras

ras was first identified in mammalian systems as an oncogene present on retroviruses and by the identification of *ras* mutations in many types of tumors (reviewed by Barbacid, 1987). Ras exists in an equilibrium between an inactive form with bound GDP and an active form with bound GTP (Fig. 5.2A). Oncogenic mutations shift the equilibrium toward the active form. This equilibrium is controlled by guanine nucleotide exchange factors (GEFs) and GTPase activating proteins (GAPs) (reviewed by Bollag and McCormick, 1992).

The discovery of *RAS* genes in the yeast *S. cerevisiae* suggested that Ras proteins play a fundamental cellular role and led to the hope that the identification of the effector of Ras in yeast would allow the identification of the mammalian effector. This hope was dashed by the discovery that the *S. cerevisiae* Ras effector is adenylyl cyclase, whereas adenylyl cyclase in mammalian cells is the effector for heterotrimeric G proteins, not for Ras. Therefore, although the important role of Ras in growth control and its basic mechanism of activity are conserved, the downstream component(s) is not, providing an important example of a conserved mechanism with a critical variation on its theme of its action.

Ras has been identified as a component of developmental pathways in several organisms. *Drosophila* mutants affecting photoreceptor development and nematode mutants affecting vulval development have identified Ras homologs that are activated by receptor tyrosine kinases (reviewed by Greenwald and Rubin, 1992). In both systems, null mutations of some positively acting components result in lethal phenotypes. Arrest occurs at early developmental stages, indicating that these components play roles in early developmental decisions as well as in the later decisions leading to photoreceptor or vulval cell development. Luckily, the latter processes can be studied using mutations that do not inactivate such genes completely and using mosaic analysis.

A. *Saccharomyces cerevisiae* Ras

S. cerevisiae contains two *ras* homologs, *RAS1* and *RAS2* (described in Chapter 7). Disruption mutations of the individual *RAS* genes do not have significant effects; *RAS1* mutants do not show a phenotype and *RAS2* mutants show only a subtle phenotype. Double mutants, however, are lethal, indicating that yeast Ras is essential and that the two genes are redundant. The double mutants arrest in the G_1 phase of the cell cycle as small unbudded cells, similar to the arrest point of wild-type cells under conditions of nutritional starvation.

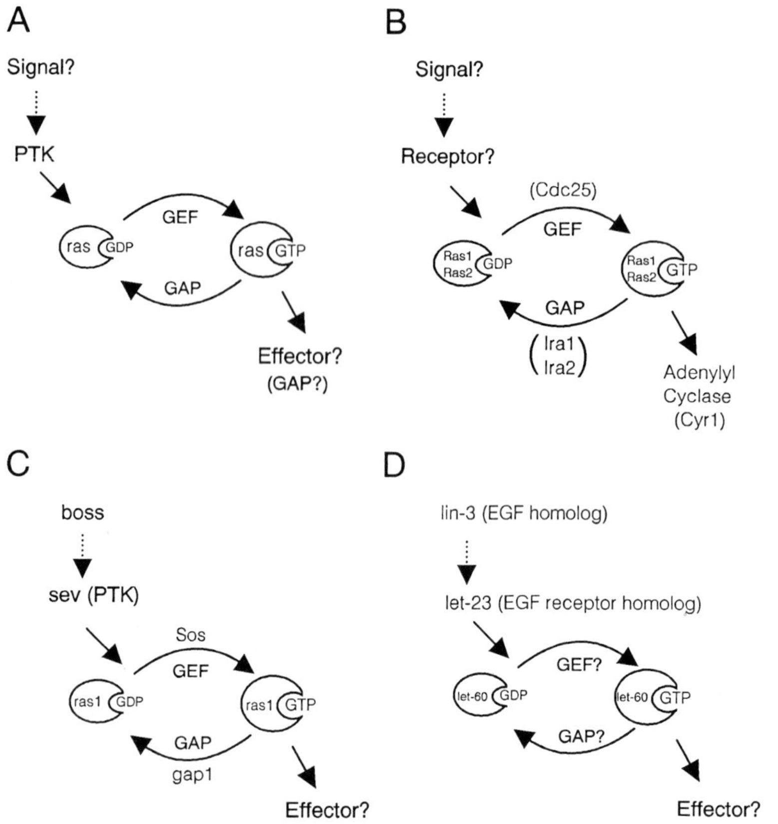

Figure 5.2 Ras-mediated systems. The components that have been identified in four systems that act through the ras pathway are shown. (A) In mammalian systems, ras is encoded by three genes: H-*ras*, K-*ras*, and N-*ras*. The GDP-bound form of ras is inactive, but the GTP-bound form is active and thought to act through a downstream effector. The equilibrium between the two forms is affected by two activities—a guanine nucleotide exchange factor (GEF) promotes replacement of GDP by GTP and a GTPase activating protein (GAP) increases the activity of the GTPase activity intrinsic to ras. Activation of protein tyrosine kinase receptors (PTK) has been shown to be involved in activation of ras. GAP has been proposed to act as the effector in this system. (B) In *S. cerevisiae*, Ras is encoded by the *RAS1* and *RAS2* genes. *CDC25* encodes a GEF and *IRA1* and *IRA2* encode GAPs. Activation of Ras acts through adenylyl cyclase (Cyrl/Cdc35) to increase cAMP levels. Although the Ras pathway is implicated in response to nutritional conditions, the actual signal and any receptor are unknown. (C) In *Drosophila*, ras1 is involved in the development of the R7 photoreceptor cell. *sevenless*(*sev*) encodes a PTK involved in this process and *bride of sevenless* (*boss*) encodes a transmembrane protein thought to correspond to the signaling molecule expressed by the adjacent R8 cell. *Son of sevenless* (*sos*) and *gap1* encode proteins with sequence similarity to GEFs and GAPs, respectively. The effector is unknown. (D) In nematodes, vulva formation requires the *let-60* product, which shows sequence similarity to Ras. *let-23*, which acts upstream of *let-60* Ras, encodes a product with similarity to the epidermal growth factor (EGF) receptor. *lin-3* is a candidate for encoding the signaling molecule and shows sequence similarity to EGF. The GEF, GAP, and effectors have not been identified in this system.

Mutations analogous to transforming mutations in mammalian *ras* result in several phenotypes that reflect an inability of cells to arrest in G_1 under conditions that induce arrest in wild-type cells.

The phenotypes resulting from both disrupting and activating mutations of *RAS* were reminiscent of phenotypes of mutants that had altered the cAMP pathway. For example, *CYR1* mutants, which are defective in adenylate cyclase, arrest at the same point as the *RAS1*, *RAS2* double mutants; *BCY1* mutants, which are defective in the regulatory subunit of cAMP-dependent protein kinase, show the same phenotype as the activating *RAS* mutants. This observation led to the determination that adenylate cyclase is the effector for RAS in *S. cerevisiae*.

Regulators of *S. cerevisiae* Ras have been identified using a variety of approaches. In the original screens for cell cycle mutants, the two mutants with the same arrest point as the *RAS1 RAS2* double mutant were *CDC25* and *CDC35*. *CDC35* was found to correspond to *CYR1*, which suggested that *CDC25* also might encode a component of the Ras pathway. Genetic studies confirmed this role and *CDC25* was found to encode a guanine nucleotide exchange factor for Ras (Fig. 5.2B). The GAP gene *IRA1* was identified among mutants affecting the cAMP pathway, although the protein initially was thought to play a different role. A second GAP gene (*IRA2*) was identified that has high sequence similarity to *IRA1*. The catalytic domains of Ira1 and Ira2 show sequence similarity to the catalytic domains of mammalian Ras-GAP and NF1, with more extensive similarity to NF1. Loss-of-function mutations in either *IRA1* or *IRA2* lead to a phenotype similar to the activated $RAS^{\mathrm{Val}\ 19}$ mutation, indicating that the yeast Ras-GAPs play negative roles.

Phenotypes of *RAS* mutants have led to the suggestion that *S. cerevisiae* Ras is involved in response to the presence of nutrients. How an extracellular signal leads to activation of Ras is not known; no components that act upstream of the yeast GEF and GAPs are known. The identification of such components is critical to the further understanding of the action of Ras in *S. cerevisiae*.

B. Vulva Formation in Nematodes

The developmental pathway involved in vulva formation in nematodes has been characterized extensively (described in Chapter 16). Mutants in this pathway can have either of two opposite phenotypes; Vul mutants are defective in vulva formation and Muv mutants produce several vulvas rather than the single vulva formed in the wild-type. For loss-of-function mutations, the Vul phenotype indicates that the component plays a positive role and the Muv phenotype indicates that the component plays a negative role in vulva formation.

The *let-23* and *let-60* loss-of-function mutants show the Vul phenotype. Cloning and sequencing of these genes indicated that *let-23* encodes a protein tyrosine kinase (PTK) that is a member of the epidermal growth factor (EGF) receptor subfamily and that *let-60* encodes a Ras homolog (Fig. 5.2D). For both genes, increased expression and/or dominant activating mutations result in a Muv phenotype, that is, the opposite as the loss-of-function phenotype. Activating mutations in *let-60* correspond to a change in codon 13, which also generates activating mutations in mammalian *ras*. Overexpression of *let-60* suppresses the Vul phenotype of *let-23* loss-of-function mutations, indicating that *let-60* (Ras) acts downstream of *let-23* (PTK), a result that correlates well with the results in mammalian systems that Ras acts downstream of PTK receptors.

The development of the vulva involves the response of several vulval precursor cells to a signal from a cell called the anchor cell. Ablation of the anchor cell results in the Vul phenotype. The signal from the anchor cell that acts on the *let-23* PTK to elicit vulva formation is clearly of great interest. The *lin-3* gene has been shown to encode an EGF homolog that is likely to correspond to the signaling molecule, which is quite striking because *let-23* shows the greatest similarity to the EGF receptor subfamily of PTK receptors. The sequence of *lin-3* suggests that it is a membrane-spanning protein with a single extracellular EGF repeat. In some mutants, the anchor cell is not in contact with the vulval precursors cells, yet a vulva is formed, indicating that the signaling molecule must be able to act at a distance. Therefore, a proteolytic processing step may result in release of the EGF repeat. Whether a membrane-spanning form also exists that allows direct contact between the anchor cell and at least one vulval precursor cell is of interest, particularly because of the indication of a direct interaction between membrane proteins in the *Drosophila* system described next.

C. R7 Photoreceptor Cell Development in *Drosophila*

The first mutant to be isolated that is defective in development of the R7 photoreceptor cell, which is one of eight clustered photoreceptor cells in each ommatidium of the *Drosophila* eye, was the *sevenless* (*sev*) mutant (reviewed by Greenwald and Rubin, 1992). The isolation of additional mutants with the same phenotype and a number of other elegant mutant isolation schemes allowed the identification of many other components of this pathway. Whereas loss-of-function mutations in *sev* and in some other positively acting genes lead to absence of the R7 cell, increased expression or activating mutations in some of these genes lead to supernumerary R7 cells.

Cloning and sequencing of the *sev* gene indicated that it encodes a transmembrane PTK receptor (Fig. 5.2C) whose closest relative among PTKs is mammalian c-*ros* (Hafen *et al.*, 1987). Additional components were identified, starting with a temperature-sensitive *sev* mutation and subsequently isolating mutants with more severe phenotypes (Simon *et al.*, 1991). One locus identified in this screen was found to correspond to the *ras1* locus, which had been identified by virtue of its homology to mammalian *ras*. Another was found to correspond to the *Son of sevenless* (*Sos*) gene, which originally was identified as a dominant mutation that suppressed *sev* (Rogge *et al.*, 1991). *Sos* encodes a protein with sequence similarity to the product of the *S. cerevisiae CDC25* gene (Bonfini *et al.*, 1992), suggesting that *Sos* encodes a GEF. Loss-of-function mutations in *ras1* or *Sos* result in decreased signaling or a complete absence of R7, depending on the severity of the mutation (Simon *et al.*, 1991), whereas dominant mutations increase signaling, suppress *sev* mutations, and induce supernumerary R7 cells even in *sev* mutants (Rogge *et al.*, 1991; Fortini *et al.*, 1992). These results indicate that these components act downstream of the *sev*-encoded PTK receptor, as would be predicted based on the roles of similar components in mammalian cells. Interestingly, both *ras1* and *Sos* also affect the pathway induced by the *Ellipse* product, which is a PTK with similarity to the EGF receptor (Rogge *et al.*, 1991; Simon *et al.*, 1991). Thus, these components may mediate responses through multiple PTK receptors.

The *gap1* gene was identified in an enhancer trap screen that gave rise to eye disk-specific expression and a rough-eye phenotype (Gaul *et al.*, 1992). This mutation also has supernumerary R7 cells and therefore resembles activated alleles of components of the *sev* pathway. *gap1* loss-of-function mutations are able to suppress *sev* mutations. These effects indicate that the enhancer trap had identified a component involved in attenuation of the signal that originates with the product of *sev*. Sequencing showed that this gene encodes a GAP homolog.

The signal for R7 development involves an interaction with the R8 photoreceptor cell. Mosaic analysis with mutations in components in the response pathway just described indicated that the action of these components is limited to the R7 cell itself. However, the *bride of sevenless* (*boss*) locus, which was identified by loss-of-function mutations that resulted in a phenotype identical to the sevenless phenotype, was shown to act in the R8 cell (Reinke and Zipursky, 1988). This result suggested that the *boss* gene product could identify the ligand for the *sev*-encoded PTK. The sequence of *boss* indicates that the gene product is a transmembrane protein (Hart *et al.*, 1990). Expression of *sev* and *boss* in separate cell lines results in aggregation of the cells and internalization of the *boss* product by the *sev*-expressing cells; these results are consistent with the idea of *boss* as the *sev* ligand (Kramer *et al.*, 1991). The identification of a presumptive transmembrane protein as a ligand for a transmembrane receptor provides a novel mechanism for stimulation of a signal transduction pathway, but is an appealing mechanism for stimulation of a developmental pathway that is mediated by direct cell–cell interaction.

D. Is GAP the Ras Effector?

When GAP first was identified, *ras* mutations that eliminated GAP binding were found to be in the domain that had been shown to be necessary for activity, which should correspond to the domain that interacts with the effector (reviewed by Bollag and McCormick, 1992). Therefore, GAP was suggested to act as the effector of Ras in addition to its role in attenuation of Ras activity. Since that time, some experimental results have been interpreted as favoring this hypothesis, whereas others have argued against this hypothesis.

Although mechanisms in other systems may be different from those in mammals, defining the role of GAP in other systems is clearly of great interest. In *S. cerevisiae*, the *IRA1* and *IRA2* GAPs play a negative role; loss-of-function mutations result in a phenotype similar to that of activated *RAS* mutations. In addition, the effector is clearly a separate protein, adenylyl cyclase. The results in *Drosophila* suggest a similar picture; loss-of-function *gap1* mutants show a phenotype similar to that of activating mutants or overexpression of positive components. To date, no evidence has been found to indicate that the *Drosophila* GAP acts as an effector. In nematodes, neither GEF- nor GAP-encoding genes have been identified. Assuming that a GAP exists in this system, the determination of the phenotype of GAP mutants will be critical. For loss-of-function mutants a Muv phenotype would indicate a negative role, consistent with attenuation of Ras activity, whereas a Vul phenotype would indicate a positive role, consistent with an additional role as an effector.

To my knowledge, GAP has not been found to act as a Ras effector in any genetically tractable system. Considering the variations that are possible in different systems, the finding that GAP does not act as an effector in other systems would not eliminate the possibility that it is the mammalian effector. Mammalian GAP contains SH2 and SH3 (*src* homology) domains, whereas the *S. cerevisiae* and *Drosophila* GAPs do not. The role of these domains in interactions with PTKs may indicate an important functional difference between GAPs in mammals and those in other systems. Whether such a functional difference is relevant to the proposed effector role of GAP in mammalian systems remains to be determined. Given the extensive genetics possible in both the *Drosophila* and the nematode system, the direct effectors of Ras are likely to be identified in the near future. Many mutants are available whose genes have not yet been characterized; these genes are likely to include those that encode the effectors. Once the effectors have been identified in these systems, determining whether similar components could correspond to the mammalian effector obviously will be of interest.

E. How Is Ras Activated?

In mammals, nematodes, and *Drosophila*, Ras clearly acts downstream of transmembrane PTKs. How activation of the PTK leads to activation of Ras is less clear. Since both GEFs and GAPs regulate the balance of the GDP- and GTP-bound forms of Ras, activation of Ras could occur either by activation of the GEF or by inhibition of the GAP. In mammalian systems, the binding of GAP to PTKs through its SH2 domain provides a possible direct link (reviewed by Bollag and McCormick, 1992). Some evidence from nematodes indicates that activation of Ras by PTK could be indirect. The *sem5* gene consists mostly of SH2 and SH3 domains and acts in the vulval developmental pathway between the *let-23* PTK and *let-60* Ras, and therefore could act as an intermediary (Clark *et al.*, 1992). A *sem5* homolog, as well as other proteins that may act between PTKs and Ras, has been found in mammals (Lowenstein *et al.*, 1992). Further analysis of interactions between such components in multiple systems should help elucidate this early step in activation.

V. Other Signaling Systems

In the examples discussed here, I concentrated on a few systems and on the elements of the pathways from the signal to the effector. Many of these systems are described in more detail in chapters in this section, with information on how these systems have been approached as well as on other components of the pathways, including downstream components and components involved in desensitization.

The *S. cerevisiae* pheromone response and Ras pathways (Chapter 7; Chapter 8) have been studied extensively for many years. The *S. pombe* Ras pathway lagged behind, but extensive information has become available (Chapter 6). In the latter system, both Ras and G proteins are necessary to transmit the pheromone signal. The *Dictyostelium* system is more complicated, with multiple receptors and G proteins, and therefore should provide the opportunity to investigate interactions or cross talk between these components in a system that is less complex than mammalian systems (Chapter 13; Chapter 14).

In *Drosophila* and nematodes, the Ras pathways are likely to play roles in multiple processes, because mutations in components of these pathways lead to lethality as a result of developmental arrest at early stages, as well as to the effects on R7 photoreceptor cell (discussed earlier) and vulva development (Chapter 16). The ability to dissect how signal transduction pathways induce particular development steps is quite powerful in these systems. Many G protein subunit genes have been identified, but most have not yet been correlated with particular processes (Chapter 15). However, the genetics possible with these systems should allow dissection of these processes as well in the foreseeable future.

Many readers are likely to be less familiar with some other systems discussed in this book. In *Chlamydomonas*, two interesting signaling systems are discussed—the induction of mating responses by interactions between flagellar surface agglutinins (Chapter 10), which activate adenylyl cyclase, and the phototaxis response, which acts through rhodopsin homologs (Chapter 11). In *Paramecium*, the system of chemotaxis is discussed (Chapter 12). In these cases, the components of the pathways are much less well defined. Some provocative evidence suggestions that G proteins may not be involved. Following progress in these systems in the years to come will be interesting.

Finally, Chapter 9 does not discuss a signal response system but addresses the production of a novel type of signaling molecule, prenylated peptide pheromones. One of these pheromones, the *S. cerevisiae* **a**-factor, initially proved quite difficult to characterize. The discovery that pheromones in the basidiomycetes *Rhodosporidium toruloides*, *Tremella mesenterica*, and *Tremella brasiliensis* were prenylated peptides helped determine the actual structure of **a**-factor. The *S. pombe* M-factor also has been shown to be a lipopeptide (Davey, 1992). Sequencing of a portion of the α mating type locus from *Cryptococcus neoformans* (Moore and Edman, 1993) and of the *a* mating type genes from *Ustilago maydis* (Bolker *et al.*, 1992) suggests that prenylated pheromones are encoded by these loci; the *U. maydis a* mating type loci also are proposed to encode G protein-linked receptors (Bolker *et al.*, 1992). Putative lipopeptide pheromone genes also have been found just outside the mating type locus in *Neurospora crassa*; disruption of one of these genes results in phenotypes consistent with the gene encoding a pheromone required for *A* mating (C. Staben, personal communication; R. Metzenberg, personal communication). Lipopeptide signaling molecules thus seem to be quite ubiquitous in fungal systems. Are they an oddity of this branch of the evolutionary tree or will they be found to be conserved more extensively?

The systems discussed in this book are diverse both with respect to the approaches of study and with respect to detail of the mechanisms involved. Hopefully aspects of each of the systems will help researchers think about their own systems in a new light.

Acknowledgments

I would like to thank Doug Johnson, Susan DeSimone, and Ian Macara for comments on the manuscript.

References

Barbacid, M. (1987). *ras* genes. *Ann. Rev. Biochem.* **56**, 779–827.

Birnbaumer, L. (1992). Receptor-to-effector signaling through G proteins: Roles for βγ dimers as well as α subunits. *Cell* **71**, 1069–1072.

Bolker, M., Urban, M., and Kahmann, R. (1992). The *a* mating type locus of *U. maydis* specifies cell signaling components. *Cell* **68**, 441–450.

Bollag, G., and McCormick, F. (1992). Regulators and effectors of *ras* proteins. *Ann. Rev. Cell Biol.* **7**, 601–632.

Bonfini, L., Karlovich, C. A., Dasgupta, C., and Banerjee, U. (1992). The *Son of sevenless* gene product: A putative activator of ras. *Science* **255**, 603–606.

Bourne, H. R. (1989). G-protein subunits: Who carries what message? *Nature (London)* **337**, 504–505.

Clark, S. G., Stern, M. J., and Horvitz, H. R. (1992). *C. elegans* cell-signaling gene *sem-5* encodes a protein with SH2 and SH3 domains. *Nature (London)* **356**, 340–344.

Davey, J. (1992). Mating pheromones of the fission yeast *Schizosaccharomyces pombe*: Purification and structural characterization of M-factor and isolation and analysis of two genes encoding the pheromone. *EMBO J* **11**, 951–960.

Fortini, M. E., Simon, M. A., and Rubin, G. M. (1992). Signaling by the *sevenless* protein tyrosine kinase is mimicked by Ras1 activation. *Nature (London)* **355**, 559–561.

Gaul, U., Mardon, G., and Rubin, G. M. (1992). A putative Ras GTPase activating protein acts as a negative regulator of signaling by the sevenless receptor tyrosine kinase. *Cell* **68**, 1007–1019.

Greenwald, I., and Rubin, G. M. (1992). Making a difference: The role of cell-cell interactions in establishing separate identities for equivalent cells. *Cell* **68**, 271–281.

Hafen, E., Basler, K., Edstroem, J.-E., and Rubin, G. M. (1987). *Sevenless*, a cell-specific homeotic gene of *Drosophila*, encodes a putative transmembrane receptor with a tyrosine kinase domain. *Science* **236**, 55–63.

Hart, A. C., Kramer, H., Van Vactor, D. L., Jr., Paidhungat, M., and Zipursky, S. L. (1990). Induction of cell fate in the *Drosophila* retina: The bride of sevenless protein is predicted to contain a large extracellular domain and seven transmembrane segments. *Genes Dev.* **4**, 1835–1847.

Iyengar, R., and Birnbaumer, L. (1990). Overview. *In* "G Proteins" (R. Iyengar and L. Birnbaumer, eds.), pp. 1–14. Academic Press, New York.

Kramer, H., Cagan, R. L., and Zipursky, S. L. (1991). Interaction of *bride of sevenless* membrane-bound ligand and the *sevenless* tyrosine-kinase receptor. *Nature (London)* **352**, 207–212.

Leberer, E., Dignard, D., Harcus, D., Thomas, D. Y., and Whiteway, M. (1992). The protein kinase homologue Ste20p is required to link the yeast pheromone response G-protein βγ subunits to downstream signaling components. *EMBO J.* **11**, 4815–4824.

Lowenstein, E. J., Daly, R. J., Batzer, A. G., Li, W., Margolis, B., Lammers, R., Ullrich, A., Skolnik, E. Y., Bar-Sagi, D., and Schlessinger, J. (1992). The SH2 and SH3 domain-containing protein GRB2 links receptor tyrosine kinases to ras signaling. *Cell* **70**, 431–442.

Moore, T. D. E., and Edman, J. C. (1993). The α-mating type locus of *Cryptococcus neoformans* contains a peptide pheromone gene. *Mol. Cell. Biol.* **13**, 1962–1970.

Reinke, R., and Zipursky, S. L. (1988). Cell–cell interaction in the *Drosophila* retina: The *bride of sevenless* gene is required to photoreceptor cell R8 for R7 cell development. *Cell* **55**, 321–330.

Rogge, R. D., Karlovich, C. A., and Banerjee, U. (1991). Genetic dissection of a neurodevelopmental pathway. *Son of sevenless* functions downstream of the *sevenless* and EGF receptor tyrosine kinases. *Cell* **64**, 39–48.

Simon, M. A., Bowtell, D. D. L., Dodson, G. S., Laverty, T. R., and Rubin, G. M. (1991). Ras1 and a putative guanine nucleotide exchange factor perform crucial steps in signaling by the sevenless protein tyrosine kinase. *Cell* **67**, 701–716.
Smith, D. P., Stamnes, M. A., and Zuker, C. S. (1991). Signal transduction in the visual system of *Drosophila*. *Ann. Rev. Cell Biol.* **7**, 161–190.

Chapter 6

ras and Signal Transduction during Sexual Differentiation in the Fission Yeast *Schizosaccharomyces pombe*

DAVID A. HUGHES

Chester Beatty Laboratories
Institute of Cancer Research
London SW3 6JB, United Kingdom

MASAYUKI YAMAMOTO

Department of Biophysics and Biochemistry
Faculty of Science
University of Tokyo
Tokyo 113, Japan

I. Introduction

A. *Schizosaccharomyces pombe*

The fission yeast *Schizosaccharomyces pombe* is a simple unicellular eukaryote with a life cycle that is amenable to genetic analysis. The application of classical and molecular genetic approaches has made *S. pombe* a favorite organism for the study of life cycle controls. Although the organism is unicellular, it has three cell types, each of which has a unique developmental potential and a different pattern of gene expression. The sexual cycle of this organism is analogous to cell differentiation in more complex organisms and involves cell–cell interactions mediated by diffusible pheromones, extensive morphological changes, and a new program of gene expression. Of particular interest is the involvement of a *ras* proto-oncogene homolog in sexual differentiation. The ease of genetic manipulation of this organism should help in the identification of elements of the Ras pathway. Homologs of some of these components may be involved in Ras-mediated signal transduction in more complex organisms.

Some aspects of sexual development and *ras* function in *S. pombe* have been reviewed in literature devoted to this organism (Egel, 1989; Nadin-Davis *et al.*, 1989). In this chapter, after an introduction to sexual differentiation and *ras* function in *S. pombe*, we focus on the regulation of the Ras pathway and on the interaction of Ras with the pheromone-response and nutrient-sensing pathways.

B. Sexual Differentiation

The sexual cycle of *S. pombe* is illustrated in Fig. 6.1. Wild-type *S. pombe* is a homothallic organism. Cells switch between two haploid cell types, designated h^+ and h^-, every few divisions by a mechanism that involves DNA transposition (Beach, 1983; for a review, see Egel, 1989). Therefore, a colony derived from a single *S. pombe* cell is a mixture of h^+ and h^- mating types. The cells continue to proliferate if nutrients are plentiful but if starved, particularly for a nitrogen source (Egel, 1971), cells of opposite mating type mate (or conjugate) to produce diploid zygotes. Interaction between cells is mediated by diffusible mating pheromones: M-factor secreted by h^- cells and P-factor secreted by h^+ cells (Fukui *et al.*, 1986b; Leupold, 1987). After nuclear fusion (karyogamy) is completed, the zygotes soon begin meiosis and ascospore

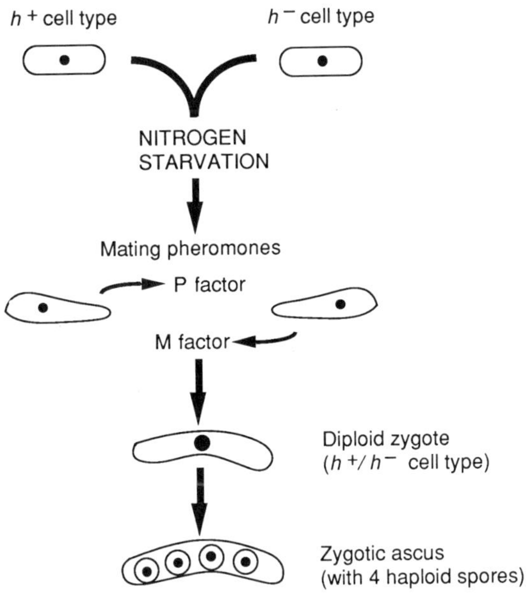

Figure 6.1 Sexual differentiation in *S. pombe*. This life cycle is described in the text (Section I,B).

formation to give four haploid spores enclosed in an ascus. Unlike in its distant relative *Saccharomyces cerevisiae*, the diploid state in *S. pombe* is usually transient, although proliferating diploid cultures can be obtained. These constitute the third cell type (h^+/h^-). Such diploids initiate meiosis directly when nitrogen starved (Egel and Egel-Mitani, 1974). Thus, mating and meiosis in *S. pombe* can be viewed as parts of a single process. As described in this chapter, a link between mating and meiosis is reflected in the genetic control of the sexual cycle.

C. Genes Involved in Sexual Differentiation

A large collection of mutants defective in the sexual process has been isolated (reviewed by Egel, 1989). Many of the mutants are defective in mating, that is, they are sterile; most of these mutants are also defective in meiosis. The sterile mutants can be divided into two classes: (1) mutants that show mating defects in both mating types (general sterility) and (2) mutants that show mating defects in only one of the mating types (mating type-specific sterility) (see Table I). Many mutants that can mate but are defective in meiosis or ascospore formation are known also, but will not be discussed in this chapter (Egel, 1989).

TABLE I
Some Genes Involved in Sexual Differentiation in *S. pombe*

Gene	Mutant phenotype (loss of function)	Protein product	Reference
General sterility genes			
ras1 (*ste5*)	Mating defect, meiosis inhibition	GTPase	Fukui and Kaziro (1985); Nadin-Davis *et al.* (1986a)
ste6	Mating defect	*ras* GNRP	Hughes *et al.* (1990)
gap1 (*sar1*)	Pheromone-supersensitive, mating inhibition	*ras* GAP	Imai *et al.* (1991); Wang *et al.* (1991a)
ral2	Mating defect, meiosis inhibition	*ras* activator?	Fukui *et al.* (1989)
gpa1	Mating and meiosis defect	α Subunit of G protein	Obara *et al.* (1991)
byr1 (*ste1*)	Mating and meiosis defect	Protein kinase	Nadin-Davis and Nasim (1988)
byr2 (*ste8*)	Mating and meiosis defect	Protein kinase	Wang *et al.* (1991b)
spk1	Mating and meiosis defect	Protein kinase	Toda *et al.* (1991)
ste11 (*aff1*)	Mating and meiosis defect	HMG-box protein, transcription factor	Sugimoto *et al.* (1991)
Mating-type-specific genes			
mat1-Pc	h^+ Mating defect, meiosis defect	Transcription factor?	Kelly *et al.* (1988)
mat1-Pi	Meiosis defect	Homeo-box protein, transcription factor?	Kelly *et al.* (1988)
mat1-Mc	h^- Mating defect, meiosis defect	HMG-box protein, transcription factor?	Kelly *et al.* (1988)
mat1-Mi	Meiosis defect	Transcription factor?	Kelly *et al.* (1988)
map3	h^+ Mating defect	M-Factor receptor	Tanaka *et al.* (1993)
mam2	h^- Mating defect	P-Factor receptor	Kitamura and Shimoda (1991)
mfm1	Not known	M-Factor pheromone	Davey (1992)
mfm2	Not known	M-Factor pheromone	Davey (1992)
map2	h^+ Mating defect	P-Factor pheromone	Y. Imai and M. Yamamoto (unpublished observations)
sxa1	h^+ Mating inhibition	Serine protease	Imai and Yamamoto (1992)
sxa2	h^- Mating inhibition	Carboxypeptidase	Imai and Yamamoto (1992)

[a]The references for each gene contain its DNA and deduced amino-acid sequences.

Some progress has been made in assigning biochemical roles to the products of the genes that are defective in these mutants (Table I). The finding that a homolog of mammalian *ras* proto-oncogenes is involved in sexual differentiation in *S. pombe* (Fukui *et al.*, 1986a; Nadin-Davis *et al.*, 1986b) suggests the possibility that some of these genes might interact with Ras and encode components of the Ras pathway.

II. Genetic Analysis of *ras* Function

A. Structure of the *ras1* Gene

Schizosaccharomyces pombe appears to contain a single *ras* gene, designated *ras1*$^{+}$, that has many of the features conserved in the mammalian *ras* genes (Fukui and Kaziro, 1985; Nadin-Davis *et al.*, 1986a; Barbacid, 1987). The predicted Ras1 protein (p25^{ras1}) is similar in size (219 amino acids) to the mammalian p21^{ras} proteins (189 amino acids) and does not carry the large inserts found in the C-terminal regions of the *S. cerevisiae* Ras proteins. The C terminus of the Ras1 protein (CCVIC) is similar to that of mammalian and *S. cerevisiae* Ras proteins. This domain is likely to be the site of posttranslational modification of cysteine residues by palmitoylation and farnesylation. The addition of these lipid modifications promotes the association of Ras proteins with the plasma membrane (Hancock *et al.*, 1989). Another region that is conserved strictly in all Ras proteins is the "effector" region (residues 32–40 in H-Ras; residues 37–45 in Ras1), which is implicated in interaction with the Ras effector (reviewed by Bourne *et al.*, 1991). The evolutionary conservation of mammalian and *S. pombe ras* genes is exemplified by the finding that human H-Ras can substitute for *S. pombe* Ras1 *in vivo* (Nadin-Davis *et al.*, 1986b), although this functional conservation does not necessarily imply that the target of mammalian p21^{ras} is the same as that of *S. pombe* Ras1.

B. Phenotypes of *ras1* Mutants

Schizosaccharomyces pombe cells carrying a null mutation in the *ras1* gene are viable and grow at a rate similar to that of wild-type cells, but show complete sterility in both mating types (Fukui *et al.*, 1986a; Nadin-Davis *et al.*, 1986b). The *ras1*$^{+}$ gene is allelic with the *ste5*$^{+}$ gene, which had been identified in a collection of sterile mutants (Girgsdies, 1982; Lund *et al.*, 1987). Homozygous diploid *ras1* mutants sporulate poorly, at a small percentage of the wild-type frequency. A curious feature of *ras1* mutant cells is that they are short and fat (dumpy) compared with the elongated shape of wild-type cells. The *ras1* mutants show normal secretion of mating pheromones but fail to respond to the pheromones (Fukui *et al.*, 1986b; Leupold *et al.*, 1991). This pheromone response defect may cause the sterility of the *ras1* mutant.

The effects of point mutations analogous to some of those in mammalian *ras* genes that confer a transforming phenotype have been examined also. These mutations are expected to activate the Ras protein by inhibiting its GTPase activity (Barbacid, 1987). The *ras1*$^{\text{Val-17}}$ mutation, which is analogous to the *ras*$^{\text{Val-12}}$ oncogene in mammals, causes increased response of cells to pheromone (Fukui *et al.*, 1986a,b; Nadin-Davis *et al.*, 1986b; Y. Imai and M. Yamamoto, unpublished results). *ras1*$^{\text{Val-17}}$ mutants of both mating types mate poorly but can complete meiosis and sporulation normally as diploids. Similar phenotypes are conferred by the *ras1*$^{\text{Leu-66}}$ allele, which is analogous to the mammalian *ras*$^{\text{Leu-61}}$ oncogene (Y. Fukui and M. Yamamoto, unpublished results). Whether the activated *ras1* mutants are supersensitive to the pheromones or respond normally but are defective in some late stage of mating is unclear. We describe this phenotype as pheromone hyperresponsive.

The phenotypes conferred by loss or activation of *ras1*$^{+}$ suggest that the Ras pathway in *S. pombe* is involved in modulating the response of nitrogen-starved cells to mating pheromones. Additional evidence for a close interaction between the Ras and pheromone-response pathways is discussed in Section IV,B.

Considering the suspected role of mammalian *ras* genes in cell proliferation (Barbacid, 1987) and the requirement of *S. cerevisiae RAS* genes for cell growth (Chapter 7), the result that the growth and cell cycle of *S. pombe* were unaffected by loss of *ras1*$^{+}$ function was perhaps surprising. Further, the amount of intracellular cAMP was unaffected by *ras1* mutations (Fukui *et al.*, 1986a), suggesting that, like mammalian cells but unlike *S. cerevisiae*, the target of Ras is not adenylyl cyclase. These observations beg the question of whether the effector of Ras proteins is conserved among this yeast and mammals. The ability of mammalian *ras* genes to function in yeast could reflect promiscuous interactions of Ras proteins with different effector molecules, or the effector could be conserved among this yeast and mammals but have different physiological functions in distantly related organisms.

C. Effect of *ras1* Mutations on Mating-Type Gene Expression

The mating type of *S. pombe* cells is determined by the expression at the mating-type locus of either *mat1-P*, which confers the h^{+} mating type, or *mat1-M*, which confers the h^{-} mating type (Egel, 1989). Each of these loci contains two genes: *mat1-P* carries the *mat1-Pc* and *mat1-Pi* genes whereas *mat1-M* carries the *mat1-Mc* and *mat1-Mi* genes (Kelly *et al.*, 1988) (see Table I). The *mat1-Pc* and *mat1-Mc* genes are essential for both mating and meiosis. They are transcribed at a low level in growing cells and at a higher level after nitrogen starvation. The *mat1-Pi* and *mat1-Mi* genes, however, are required only for meiosis and are transcribed only after nitrogen starvation. Loss of *ras1* function has little effect on the expression of the Pc and Mc transcripts, but the induction of Pi and Mi transcripts by nitrogen starvation is much lower in *ras1* mutants than in wild-type cells (Nadin-Davis and Nasim, 1990). Cells carrying the *ras1*$^{\text{Val-17}}$ allele have higher levels of all four transcripts than do wild-type cells. Expression of the Pi transcript requires both nitrogen starvation and exposure to P-factor pheromone (Nielsen *et al.*, 1992). Pheromone induction of *mat1-Pi* expression is blocked in a loss-of-function *ras1* mutant but enhanced in an activated *ras1*$^{\text{Val-17}}$ mutant. *ras1* activation does not, however, bypass the requirement for nitrogen starvation and the presence of P-factor. These observations are consistent with the proposal (Section II,B) that Ras1 controls the sensitivity of cells to the mating pheromones.

The effect of *ras1* mutations on *mat* gene expression provides an explanation for the poor sporulation of *ras1* mutant diploids, because Pi and Mi expression is reduced greatly in these cells. This explanation is supported by the observation that expression of either *mat1-P* or *mat1-M* on a multicopy plasmid can restore sporulation to diploid *ras1* mutant cells (Nadin-Davis and Nasim, 1988,1990). Because Pc and Mc transcription is not affected in *ras1* mutants, an effect on *mat* gene expression cannot explain the mating defect of *ras1* mutants.

D. Protein Kinases Downstream of *ras*

The *byr1*$^+$ gene (bypass of *ras*) was identified by its ability to suppress the sporulation defect of *ras1* mutant diploids when expressed on a multicopy plasmid (Nadin-Davis and Nasim, 1988). This gene could not, however, efficiently suppress the mating defect of homothallic *ras1* mutant haploids, although a very low frequency of mating could be observed when *ras1* mutant cells carrying multicopy *byr1*$^+$ were crossed with wild-type tester strains. The deduced amino acid sequence of this gene reveals sequence similarity to protein kinases. Disruption of the *byr1*$^+$ gene blocks both mating of haploid cells and meiosis and sporulation of diploids. Subsequent genetic analysis showed that *byr1* is allelic to *ste1* (Thuriaux *et al.*, 1980; Girgsides, 1982; Nadin-Davis and Nasim, 1990).

Another protein kinase gene (*byr2*$^+$) involved in mediating *ras1* function was isolated as a multicopy suppressor of the inhibition of mating caused by expression of the *S. cerevisiae RAS2*$^{\text{Ala-22}}$ dominant negative mutation in wild-type *S. pombe* cells (Wang *et al.*, 1991b). Disruption of *byr2*$^+$ blocks both mating and meiosis. Genetic analysis indicates that *byr2* is allelic to *ste8* (Michael and Gutz, 1987). As for *byr1*$^+$, overexpression of *byr2*$^+$ can suppress the sporulation defect of *ras1* mutant diploids but not the mating defect of *ras1* mutant haploids. Wang and colleagues suggest that both protein kinases act downstream of Ras1, and that Byr1 kinase acts downstream of Byr2 kinase because multicopy *byr1* can suppress the sporulation defect (but not the mating defect) of *byr2* mutants.

A third protein kinase gene, *spk1*$^+$, is essential for mating in *S. pombe*. Overexpression of this gene confers resistance to the protein kinase inhibitor staurosporine (Toda *et al.*, 1991). The *pap1*$^+$ gene, which encodes a transcription factor related to mammalian c-Jun, was isolated in the same screen. Although a *pap1* mutant has no apparent phenotype, *pap1*$^+$ function is required for multicopy *spk1*$^+$ to confer staurosporine resistance; thus the Pap1 transcription factor may mediate some of the effects of the Spk1 protein kinase.

These three protein kinases show sequence similarity to components of the pheromone-signaling pathway of *S. cerevisiae* (Chapter 8). The *byr1*$^+$ and *byr2*$^+$ gene products show extensive sequence similarity to the products of the *S. cerevisiae STE7* and *STE11* gene products, respectively (Teague *et al.*, 1986; Rhodes *et al.*, 1990). The *S. pombe* genes can complement partially the mating defect of the corresponding *S. cerevisiae* mutant (cited in Marsh *et al.*, 1991). The *spk1*$^+$ gene product is homologous to the *S. cerevisiae FUS3* and *KSS1* gene products (Courchesne *et al.*, 1989; Elion *et al.*, 1990) and to the vertebrate MAP kinase (mitogen-activated protein kinase) family of protein kinases (reviewed by Leevers and Marshall, 1992). In each case, the sequence similarity extends beyond the catalytic domain of the kinase. These homologies suggest that the *S. pombe* kinases act in the pheromone-signaling pathway and not directly downstream of Ras1, as discussed further in Section IV.

E. Control of Cell Morphology by *ras*

The characteristic dumpy morphology of *ras1* mutants cannot be explained by a defect in pheromone-signaling because this phenotype is not seen in mutants defective in other components of this pathway, suggesting that *ras1*$^+$ has at least two different functions. Whether the same Ras effector is involved in these different functions is unresolved. Missense mutations in the *ras1* gene give the same dumpy morphology, showing that the phenotype is not caused by the absence of Ras1 protein (D. A. Hughes and M. Yamamoto, unpublished results). A plausible explanation for the dumpy shape is that the addition of new wall material to the tips of the cells is disorganized, possibly because of changes in the cytoskeleton. In *S. cerevisiae*, the localization of the site where budding occurs during the cell cycle is regulated by *RSR1*, a *ras*-like gene that has the same effector domain sequence as *ras* genes and is related most closely to human *rap1A* (Bender and Pringle, 1989; Ruggieri *et al.*, 1992). In mammalian cells transformed by v-*ras*$^{\text{Val-12}}$, overexpression of *rap1A*, also known as K-*rev-1*, can suppress the transformed phenotype (Kitayama *et al.*, 1989).

Remarkably, overexpression of human *rap1A* and *rap1B*, a closely related gene, can suppress both the dumpy morphology and the sporulation defect of *ras1* mutant diploids (Xu *et al.*, 1990). Therefore, whereas in mammalian cells *rap1A* antagonizes mammalian *ras* function, in *S. pombe rap1A* can substitute for at least some functions of *ras1*$^+$. Thus *ras1*$^+$, in addition to its function in mating, may play a role similar to that of *S. cerevisiae RSR1* in regulating the site of cell wall growth.

III. Regulation of the Activity of Ras Protein

A. GTPase Cycle

The activity of Ras proteins and of other guanine nucleotide binding proteins is regulated by binding and hydrolysis of GTP (Bourne *et al.*, 1990,1991; Fig. 6.2A). The GTP-bound form of Ras protein is in the active conformation that interacts with effector molecules. GTP hydrolysis converts this active conformation to the inactive GDP-bound form. Two kinds of proteins that control this cycle have been identified: (1) GTPase-activating proteins (GAPs) that stimulate the intrinsic GTPase activity of Ras proteins and thus inactivate the Ras protein and (2) guanine nucleotide releasing proteins (GNRPs) that activate Ras protein by stimulating the release of bound GDP and its replacement by GTP.

B. Gap1: A Negative Regulator of Ras1

The *gap1*$^+$ gene of *S. pombe* encodes a homolog of the GTPase activating proteins GAP and NF1 of mammalian cells and Ira1 and Ira2 of *S. cerevisiae*

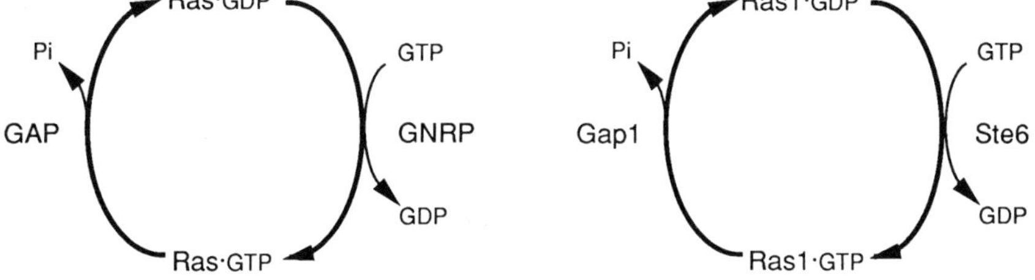

Figure 6.2 The *ras* GTPase cycle. (A) Generalized GTPase cycle for Ras proteins. GAP is the GTPase activating protein and GNRP is the guanine nucleotide release protein. (B) Proposed GTPase cycle for the *S. pombe* Ras1 protein. Gap1 is a putative GAP encoded by the *gap1* gene (also called *sar1*). Ste6 is a putative GNRP encoded by the *ste6* gene.

(Fig. 6.3A; Imai *et al.*, 1991; Wang *et al.*, 1991a). This gene was identified by the isolation of a mutant (*gap1-1*) that has a phenotype similar to those caused by activated alleles of *ras1* discussed earlier, that is, the mutant cells are hyperresponsive to M-factor pheromone and mate poorly (Imai *et al.*, 1991; Wang *et al.*, 1991a). Disruption of the *gap1* gene causes the same phenotypes.

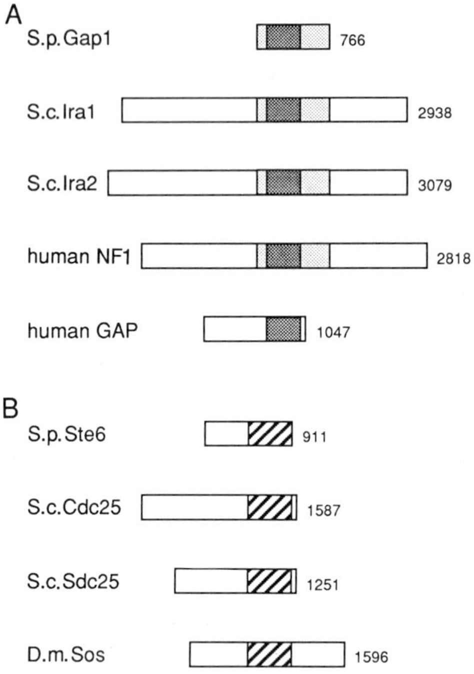

Figure 6.3 Similarities between regulators of the *ras* GTPase cycle of *S. pombe* and those of other organisms. (A) Schematic comparison of the putative *ras* GAP of *S. pombe* (Gap1) with *ras* GAPs from *S. cerevisiae* (Ira1 and Ira2) and mammals (NF1 and GAP). The core region (~370 amino acids) found in mammalian GAP and its homologs is shown by dark shading (20–25% amino acid identity). Regions found in Gap1, Ira1, Ira2, and NF1 but not in GAP are shown by light shading (10–20% amino acid identity). Ira1, Ira2, and NF1 are similar throughout most of their lengths (not shown because the comparison is to *S. pombe* Gap1). (B) Schematic comparison of the putative GNRP of *S. pombe* (Ste6) with *S. cerevisiae* GNRPs (Cdc25 and Sdc25), and a putative GNRP for the *Drosophila* Ras protein (Sos). The C terminus of the Ste6 product shows sequence similarity to the regions of Cdc25, Sdc25, and Sos shown by cross-hatching (~30% amino acid identity). The *S. cerevisiae* proteins are similar throughout their lengths (not shown). The number of amino acid residues in each protein is shown. References: Gap1, Imai *et al.* (1991), Wang *et al.* (1991a); Ira1, Tanaka *et al.* (1989); Ira2, Tanaka *et al.* (1990); NF1, G. Xu *et al.* (1990), Marchuk *et al.* (1991); GAP, Trahey *et al.* (1988), Vogel *et al.* (1988); Ste6, Hughes *et al.* (1990); Cdc25, Camonis *et al.* (1986), Broek *et al.* (1987); Sdc25, Boy-Marcotte *et al.* (1989), Damak *et al.* (1991); Sos, Simon *et al.* (1991).

The similarity of the phenotypes caused by loss of *gap1*$^+$ and activation of *ras1*$^+$ indicates that the Gap1 protein acts as an inhibitor of Ras1 activity (Fig. 6.2B), but not as an effector as has been suggested for mammalian GAP (McCormick, 1989).

The *gap1*$^+$ gene (called *sar1*$^+$ in this study) was isolated independently as a multicopy suppressor of the mating inhibition caused by the activated *ras1*$^{\text{Val-17}}$ allele (Wang *et al.*, 1991a). Mammalian GAP and NF1 are both able to rescue the mating inhibition and pheromone supersensitivity of *gap1/sar1* mutants; *S. pombe gap1*$^+$/*sar1*$^+$ rescues the heat-shock sensitivity of *S. cerevisiae IRA1* mutants. Based on the ability of GAP, NF1, and Ira2 to stimulate the GTPase activity of Ras proteins and on the phenotype of *gap1* mutants, this Gap1 protein is highly likely also to function in this way.

C. Ste6: An Activator of Ras1

Loss of *ste6*$^+$ function prevents mating of haploid cells, but sporulation of diploids is not affected (Girgsdies, 1982; Hughes *et al.*, 1990). The deduced *ste6*$^+$ gene product (Hughes *et al.*, 1990) shows similarity to the product of the *CDC25* gene of *S. cerevisiae* (Fig. 6.3B), which functions as a GNRP for the Ras1 and Ras2 proteins (Jones *et al.*, 1991). Significant similarity also is shown to the *S. cerevisiae SDC25* product (Fig. 6.3B), which was identified as a multicopy suppressor of *CDC25* mutants (Boy-Marcotte *et al.*, 1989). Sdc25 expressed in bacteria stimulates release of GDP bound to yeast and mammalian Ras proteins (Crechet *et al.*, 1990). A related protein, called Son of sevenless, is involved in the Ras-mediated signal transduction pathway in the eye of the fruit fly *Drosophila* (Simon *et al.*, 1991). All members of this gene family are likely to encode GNRPs for Ras proteins.

Evidence that Ste6 regulates Ras1 activity is seen in the observation that the *ras1*$^{\text{Val-17}}$ or *ras1*$^{\text{Leu-66}}$ allele restores mating pheromone sensitivity and mating ability to *ste6* mutants, that is, *ras1* activation is epistatic to loss of *ste6*$^+$ function (Hughes *et al.*, 1990). This result suggests that the Ste6 protein is no longer required for Ras1 function when the GTPase activity of Ras1 is inhibited, consistent with Ste6 acting as a GNRP for the Ras1 protein (Fig. 6.2B). Loss of *gap1*$^+$ function also suppresses the mating defect of *ste6* mutants (Imai *et al.*, 1991), further supporting this argument.

One problem with this interpretation of *ste6*$^+$ function is the difference in phenotype between *ras1* and *ste6* mutants. The *ste6* mutants are not inhibited for sporulation and do not have the dumpy shape characteristic of *ras1* mutants. Perhaps another GNRP for Ras1 protein exists, or the intrinsic GDP–GTP exchange rate of the Ras1 protein provides sufficient activity for sporulation and maintenance of cell shape. Investigation of GNRP activity in *S. pombe* cells will be required to settle this question.

D. Ral2: A Novel Pathway of *ras* Activation?

Mutations in the *ral2*$^+$ gene generate phenotypes similar to those of loss of *ras1*$^+$ function (Fukui and Yamamoto, 1988). Further, the activated *ras1*$^{\text{Val-17}}$ allele restores mating pheromone sensitivity and a low level of mating to *ral2* mutants (Fukui *et al.*, 1989). Thus, *ral2*$^+$ seems to function upstream of *ras1*$^+$. The deduced Ral2 protein shows no similarity to either Ste6 or *S. cerevisiae* Cdc25 (Fukui *et al.*, 1989), suggesting that this protein does not function as the catalytic subunit of a GNRP. The biochemical function of the Ral2 product remains unknown, but one possibility is that it functions as an inhibitor of Gap1 (Hughes *et al.*, 1991; Imai *et al.*, 1991).

E. Mechanism of Activation of Ras1 Protein during Mating

The phenotypes of *ste6*, *gap1*, and activated *ras1* mutants indicate that the activity of the Ras1 protein must be finely tuned to insure successful completion of the mating process; either too little (in *ste6* mutants) or too much (in *gap1* and *ras1*$^{\text{Val-17}}$ mutants) active GTP-bound Ras1 inhibits mating. The ratio of active GTP-bound to inactive GDP-bound Ras1 protein can be changed by regulating the GNRP activity of Ste6, the GAP activity of Gap1, or both.

Some idea of how the activity of Ras1 might be regulated *in vivo* comes from studies on the expression of elements of the Ras pathway. When cells are nitrogen starved, the expression of the *ste6* mRNA is induced at least sixfold (D. A. Hughes, N. Yabana, and M. Yamamoto, unpublished results), whereas the expression of *ras1*, *gap1*, and *ral2* mRNAs is unaffected (Nadin-Davis and Nasim, 1990; Y. Imai, S. Miyake, and M. Yamamoto, unpublished results). These results suggest that the Ras1 protein can be activated in response to nitrogen starvation by stimulation of GDP–GTP exchange catalyzed by Ste6 protein, although biochemical evidence to support this mechanism is lacking.

IV. Interaction of *ras* with the Pheromone-Response Pathway

A. Elements of the Pheromone-Response Pathway

1. Mating Pheromones The M-factor pheromone secreted by h^- cells has been purified and its structure analyzed by amino acid sequencing and mass spectrometry (Davey, 1991,1992). M-factor is a 9-amino-acid peptide with a C-terminal cysteine that is modified by addition of a prenyl group, probably farnesyl, and by carboxyl-methylation. These modifications appear to be identical to those on the **a**-factor pheromone of *S. cerevisiae* (Anderegg *et al.*, 1988). M-factor is produced by proteolytic maturation of precursors encoded by two genes, *mfm1*$^+$ and *mfm2*$^+$.

The P-factor pheromone has been more difficult to assay biologically; haploid h^- cells do not show elongation when exposed to h^+ cells whereas homozygous diploid h^-/h^- cells do show this appearance and show a response to P-factor (Leupold, 1987). The *map2*$^+$ gene appears to encode the precursor for P-factor, although the amino acid sequence of the mature pheromone has not yet been determined (Y. Imai and M. Yamamoto, unpublished observations). Loss of *map2*$^+$ function causes P-cell specific sterility and inability to produce active P-factor.

2. *Mating Pheromone Receptors* Genes encoding putative membrane receptors for the *S. pombe* mating pheromones have been identified (Table I). The P-factor receptor is encoded by the *mam2*$^+$ gene (Kitamura and Shimoda, 1991); mutations in *mam2*$^+$ confer M-cell specific sterility (Egel, 1973b). The M-factor receptor is encoded by P-cell specific sterility gene *map3*$^+$ (Tanaka *et al.*, 1993). A curious finding is that *mam2*$^+$ is homologous to the *STE2* gene of *S. cerevisiae*, which encodes the α-factor receptor (Kitamura and Shimoda, 1991), and *map3*$^+$ is homologous to *S. cerevisiae STE3*, which encodes the **a**-factor receptor (Tanaka *et al.*, 1993; see Chapter 8). The deduced Mam2 and Map3 proteins both contain seven potential transmembrane domains and therefore are similar to the large family of G protein-coupled membrane receptors (Lefkowitz and Caron, 1988). Haploid *mam2,map3* double mutants are sterile in both mating types; *mam2,map3* diploids are defective in meiosis and sporulation (Tanaka *et al.*, 1993). This finding suggests that mating pheromones are involved in the initiation not only of mating but also of meiosis, as had been suggested previously (Egel, 1973b; Meade and Gutz, 1976; Leupold *et al.*, 1989a; Obara *et al.*, 1991). The requirement for pheromones in both mating and meiosis presumably reflects the coupling of these two events in *S. pombe* sexual development and is quite different from the situation in *S. cerevisiae* (Chapter 8). As might be predicted expression of the receptor genes is cell type-specific and induced by nitrogen starvation (Kitamura and Shimoda, 1991; Tanaka *et al.*, 1993).

3. *G Protein* The *gpa1*$^+$ gene of *S. pombe* encodes a G protein α subunit. Loss of its function prevents both mating and sporulation (Obara *et al.*, 1991). This phenotype is the same as that conferred by mutations in both *mam2*$^+$ and *map3*$^+$, the mating pheromone receptor genes, suggesting that *gpa1*$^+$ could be involved in the pheromone-response pathway. Evidence for coupling of Gpa1 protein to the pheromone receptors is seen in the phenotypes caused by a point mutation that is proposed to inhibit GTPase activity

of the protein, thus constitutively activating the protein (Graziano and Gilman, 1989; Masters *et al.*, 1989). Introduction of a *gpa1* activated allele (*gpa1*$^{\text{Leu-224}}$; Glu to Leu at residue 224) into heterothallic cells (h^+ or h^-) causes morphological changes similar to those found in cells responding to mating pheromones (Obara *et al.*, 1991). Thus, activation of the Gα protein by mutation phenocopies the response to pheromone. However, nitrogen starvation is still necessary for this response, indicating that the expression or function of some component(s) of the pheromone-signaling system is regulated by nutrient availability (see Section IV,B). This genetic analysis indicates that the Gα protein acts as a positive element in transducing the pheromone signal required for mating. This role is similar to the role of mammalian Gα subunits in signal transduction (Neer and Clapham, 1988), but is quite different from the role of the *S. cerevisiae* Gα protein, which acts as an inhibitor of the pheromone response. Loss of function mutations in *S. cerevisiae* Gα constitutively activate the pathway (Chapter 8). The roles of the β and γ subunits of the G protein, which play a positive role in transducing the pheromone signal in *S. cerevisiae*, are not known yet in *S. pombe*.

4. Proteases Three genes were identified in a screen for mutants supersensitive to mating pheromones, that is, with a phenotype similar to that caused by activated *ras1*$^+$ (e.g., *ras1*$^{\text{Val-17}}$) (see Table I). The *gap1*$^+$ gene is described in Section III,B. The other two genes, *sxa1*$^+$ and *sxa2*$^+$ (sexually activated), encode putative proteases (Imai and Yamamoto, 1992). The *sxa1* product shows sequence similarity to the pepsin-like family of proteases, including the product of the *BAR1/SST1* gene of *S. cerevisiae*, which degrades the α-factor pheromone (MacKay *et al.*, 1988). The *sxa2* product shows sequence similarity to carboxypeptidases, including the products of the *S. cerevisiae KEX1* and *PRC1* genes (Dmochowska *et al.*, 1987; Valls *et al.*, 1987). Homothallic strains carrying mutations in either of these genes mate poorly, but crosses to heterothallic tester strains indicate that the mating deficiency is mating type specific. The *sxa1* mutants show a P-cell (h^+) specific mating defect whereas the *sxa2* mutants show an M-cell (h^-) specific defect. Why do mutations in these putative protease-encoding genes cause pheromone supersensitivity and mating type-specific sterility? The simplest explanation is that the Sxa1 and Sxa2 proteins are involved in degradation or proteolytic inactivation of the pheromones. Such a function for the *sxa1* product is likely because of its homology with Bar1, a secreted protease that is implicated in degradation of α-factor. The putative carboxypeptidase activity of the *sxa2* product also may inactivate mating pheromone, although this function is the opposite of that of the Kex1 peptidase of *S. cerevisiae*, which is involved in maturation of the α-factor precursor. The mating defect of *sxa2* mutants can be suppressed by a diffusible factor(s) produced by wild-type (*sxa2*$^+$) cells, suggesting that Sxa2 acts extracellularly.

B. Genetic Interactions

The phenotypes conferred by *ras1* mutations and by mutations in other genes, such as *gap1*$^+$ and *ste6*$^+$, that interact with *ras1*$^+$ suggest that the Ras pathway affects the response of cells to the mating pheromones. At what level does the Ras1 protein interact with the pheromone-signaling pathway and by what mechanism? The main obstacle in trying to answer this question is that the identity of the biochemical effectors of Ras1 and the G protein remains unknown.

Any model for the role of Ras1 in pheromone response must account for certain observations. First, *ras1* mutants cannot be totally insensitive to pheromones, at least in diploid cells, because they can sporulate (although at a lower frequency than wild-type cells), whereas cells lacking the pheromone receptors or the Gα protein show total sporulation defects. Second, the Ras pathway is not sufficient for transmission of the pheromone signal because activated *ras1* mutants still require exposure to pheromone to elicit the pheromone response, that is, to give the phenotype conferred by the activated Gα mutant (Section IV,A,2).

A model for the interaction between Ras and the pheromone-response pathway is illustrated in Fig. 6.4. The decision to switch from the cell cycle to sexual differentiation requires the integration of signals concerning both the availability of nutrients and the presence of mating partners. Our model proposes that Ras is activated by nitrogen starvation and Gpa1 is activated by pheromone. We propose that Ras1 and Gpa1 interact with the same effector or, more likely, regulate separate effectors. The protein kinases Byr1/Ste1, Byr2/Ste8, and Spk1 seem to function downstream of the G protein. All three kinases have close homologs in the pheromone-signaling pathway of *S. cerevisiae* (see Section II,D). The transcription factors that are responsible for pheromone-induced transcription have not yet been identified.

V. Nutritional Signaling and Gene Expression

A. Cyclic AMP Pathway

When *S. pombe* cells are nutritionally starved, they can enter stationary phase, sometimes called G_0, from either the G_1 or G_2 phase of the cell cycle (Costello *et al.*, 1986), but only cells arrested in G_1 can initiate mating or meiosis (Egel and Egel-Mitani, 1974; Nurse and Bisset, 1981). Nitrogen starvation, which arrests cells primarily in the G_1 phase (Egel and Egel-Mitani, 1974; Costello *et al.*, 1986), is usually a prerequisite for initiation of sexual differentiation. Considerable evidence now suggests that the level of the intracellular messenger cAMP and the activity of cAMP-dependent protein kinase (protein kinase

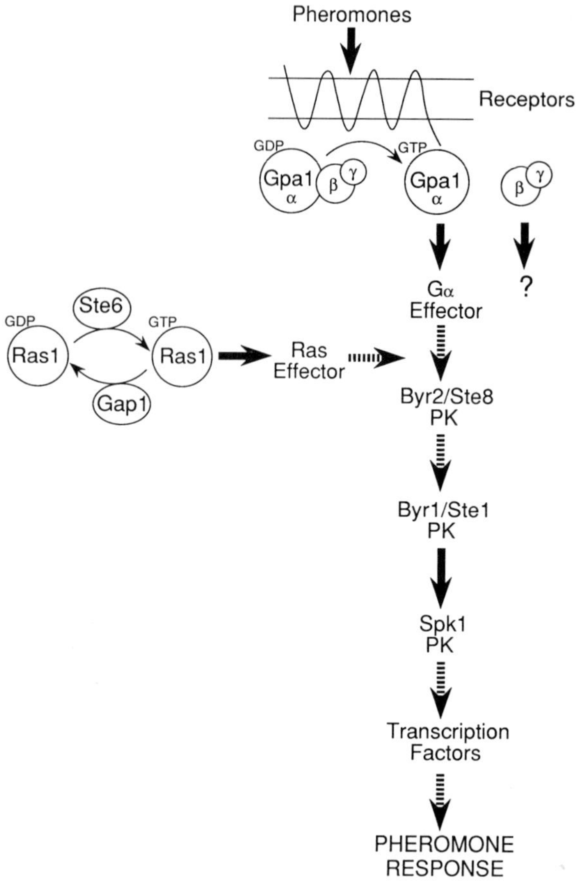

Figure 6.4 A model for interaction between the *ras* and pheromone-signaling pathways. The model is explained in the text (Section IV,B).

A) regulate the switch from the cell cycle to the sexual cycle in response to nutritional conditions. First, nutritional starvation leading to initiation of sexual differentiation causes approximately a 50% decrease in the intracellular concentration of cAMP (Fukui *et al.*, 1986a; Mochizuki and Yamamoto, 1992). Second, high concentrations of cAMP inhibit both mating and meiosis (Watanabe *et al.*, 1988). Third, cells lacking cAMP because of a mutation in the adenylyl cyclase gene $cyr1^+$ initiate sexual differentiation without nutritional starvation (Maeda *et al.*, 1990).

How are nutritional starvation and the consequent decrease in cAMP level coupled to the onset of sexual differentiation? A major mechanism seems to be the induction of the expression of many genes involved in mating and meiosis (Fig. 6.5; Sugimoto *et al.*, 1991). This new program of gene expression in response to nutritional deprivation is inhibited by high levels of cAMP (Watanabe *et al.*, 1988; Sugimoto *et al.*, 1991). The expression of all the starvation-inducible genes examined to date (*matP, matM, ste6*+, and *mei2*+) depends on the function of the *ste11*+ gene. Constitutive expression of *ste11*+ can drive cells into sexual differentiation even in rich medium (Sugimoto *et al.*, 1991). The *ste11* gene product contains an HMG-box motif found in other eukaryotic transcription factors and is a sequence-specific DNA-binding protein. The starvation-induced genes all have one or more putative binding sites for the Ste11 protein in their promoter regions (Sugimoto *et al.*, 1991; D. A. Hughes, N. Yabana, and M. Yamamoto, unpublished results).

How is *ste11*+ regulated in response to nutritional starvation? The expression of *ste11*+ is induced by nitrogen starvation, perhaps by an autoregulatory activity of the Ste11 protein that results in activation of transcription of the *ste11*+ gene. Expression of the *ste11*+ gene is inhibited by high levels of cAMP (as are the genes downstream of *ste11*+) and is constitutive in cells lacking cAMP, that is, in *cyr1* mutants (Sugimoto *et al.*, 1991). A simple hypothesis is

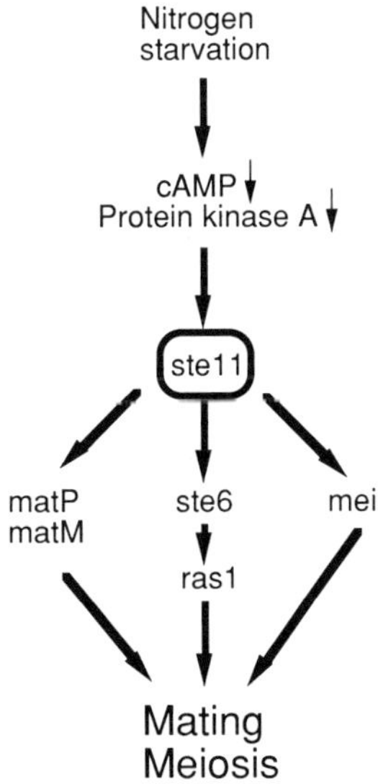

Figure 6.5 Nutritional regulation of gene expression leading to mating and meiosis. Nitrogen starvation lowers the intracellular concentration of cAMP and the activity of cAMP-dependent protein kinase (protein kinase A). Through an unidentified mechanism, this effect causes activation of the *ste11*+ transcription factor, which then induces the expression of many genes required for sexual differentiation. This process is discussed in more detail in the text (Section V,A).

that the ability of the Ste11 protein to activate transcription of its own gene and its downstream targets is inhibited by high activity of protein kinase A, and that the drop in cAMP level in response to nitrogen starvation causes activation of the Ste11 transcription activator (Fig. 6.5). The products of the target genes of Ste11 then cause cells to enter the sexual cycle.

An interesting point made by the analysis of the adenylyl cyclase gene *cyr1*$^+$ is that cAMP does not control entry into the G_0 state (Maeda *et al.*, 1990; Kawamukai *et al.*, 1991), unlike the events in *S. cerevisiae* (Chapter 7). The *cyr1* mutants are, however, highly derepressed for sexual differentiation and undergo mating and sporulation on rich medium. These results raise the issue of how nitrogen starvation causes cell cycle arrest in the G_1 phase if protein kinase A is not involved. A related issue is whether the mating pheromones have an effect on the cell cycle, as they do in *S. cerevisiae*. The availability of mutants that can differentiate sexually without nutritional deprivation could prove useful in addressing these topics.

B. Pat1 Protein Kinase

The nutritional control of sexual differentiation also can be overcome in strains carrying mutations in the *pat1*$^+$ gene (also called *ran1*$^+$) (Beach *et al.*, 1985; Iino and Yamamoto, 1985; Nurse, 1985). Temperature-sensitive (ts) mutations in this gene, which encodes a protein kinase essential for viability, cause cells to undergo mating in rich medium at the semipermissive temperature. A molecular explanation for this effect was provided by Nielsen and Egel (1990), who showed that expression of the transcripts at the *mat1-P* mating-type locus was induced by shifting cells with the *pat1-114* ts allele to the semipermissive temperature on rich medium. This partial inhibition of Pat1 kinase can bypass the requirement for *ras1*$^+$ (and *ste6*$^+$) function in pheromone-induced transcription of *mat1-Pi* (Nielsen *et al.*, 1992). The *pat1* mutant cells also have been shown to be supersensitive to the mating pheromones, even at the permissive temperature (Leupold *et al.*, 1989b). These results suggest that Pat1 protein kinase might have a general inhibitory effect on sexual differentiation, in addition to its better understood role in meiosis (reviewed by McLeod, 1989; Egel *et al.*, 1990). Does the Pat1 protein kinase act directly on the nutritional-signaling or pheromone-response pathways? *pat1*$^+$ appears to act through the *mei2*$^+$ gene because *pat1,mei2* double mutants do not show derepression of mating (Y. Iino, Y. Watanabe, and M. Yamamoto, unpublished results). However, whether the effects of *pat1*$^+$ inhibition on gene expression also are mediated through *mei2*$^+$ is not clear. The *mei2*$^+$ gene itself is not essential for mating, although *mei2* mutants mate at a lower frequency than do wild-type cells (Egel, 1973a; Y. Watanabe and M. Yamamoto, unpublished results). The effects of *pat1*$^+$ on mating and on *mat* gene expression could be explained by postulating a positive feedback loop, possibly acting through *ste11*$^+$, that helps commit cells to the sexual cycle. The fact that *pat1,mei2* double mutant strains can mate shows that such a feedback loop is not essential for mating, but could play an important role in getting to the safe end point of the sexual cycle (i.e., the resistant spore) quickly and efficiently.

VI. Similarities with Signal Transduction in Vertebrates

A. MAP Kinase Activation Cascade

An intriguing similarity has emerged between the vertebrate MAP kinase activation pathway and the pheromone-signaling pathway in *S. pombe*. The vertebrate MAP kinase is activated by phosphorylation on tyrosine and threonine residues. The protein kinase responsible for this phosphorylation has been purified. Its amino acid sequence reveals strong similarity to the *byr1*$^+$/*ste1*$^+$ gene product of *S. pombe* (Kosako *et al.*, 1992). Thus, Byr1/Ste1 is likely to phosphorylate and activate Spk1, the MAP kinase homolog in *S. pombe*. Mammalian MAP kinase is activated in response to extracellular signals; p21^{ras} is required for this activation. Likewise, in *S. pombe*, Byr1/Ste1 mediates some of the effects of the Ras1 protein. Perhaps mammalian and *S. pombe* Ras proteins activate the MAP kinase cascade by a similar mechanism. Determining whether a homolog of Byr2/Ste8 is involved in this pathway in vertebrates will be interesting.

B. Interactions between Signaling Pathways

In higher eukaryotes including mammals, signal transduction involving membrane receptors coupled to G proteins plays an important role in cellular responses to a multitude of extracellular signals, including small peptides, neurotransmitters, and growth factors (Neer and Clapham, 1988). Might some of these signal transduction pathways interact with Ras in a manner similar to the interaction of Ras and the pheromone-response pathway in *S. pombe*? Cross talk between signal transduction pathways is increasingly becoming apparent. The cellular response to a given signal is likely to depend on these interactions, which will be different in different cell types. For example, mammalian p21^{ras} regulates cell proliferation in some cell types and cell differentiation in others (Barbacid, 1987). Also, G protein-coupled receptors (Young *et al.*, 1986; Julius *et al.*, 1989) and Gα subunits (Landis *et al.*, 1989; Lyons *et al.*, 1990) have been found to cause oncogenic transformation. Unraveling the complex interactions in mammalian cells, many of which express multiple receptors and G proteins, seems a daunting task. However, this task must be tackled to understand the controls that exist for cell proliferation and differentiation.

Signal transduction during sexual differentiation in *S. pombe* provides a simpler system in which to investigate interactions between signaling pathways. Even in this unicellular organism, however, cell differentiation involves complex interactions between the Ras, G protein, and protein kinase A pathways. Understanding this system in molecular detail should provide useful information about the mechanisms of signal transduction in more complex organisms.

Acknowledgments

We thank R. Egel, C. Shimoda, and M. Wigler for providing information prior to publication. D. A. Hughes was supported by a fellowship from the Human Frontier Science Program while visiting the laboratory of M. Yamamoto.

References

Anderegg, R. J., Betz, R., Carr, S. A., Crabb, J. W., and Duntze, W. (1988). Structure of the *Saccharomyces cerevisiae* mating hormone **a** factor. *J. Biol. Chem.* **263**, 18236–18240.

Barbacid, M. (1987). *ras* genes. *Ann. Rev. Biochem.* **56**, 779–827.

Beach, D. H. (1983). Cell type switching by DNA transposition in fission yeast. *Nature (London)* **305**, 682–688.

Beach, D., Rodgers, L., and Gould, J. (1985). *ran1⁺* controls the transition from mitotic division to meiosis in fission yeast. *Curr. Genet.* **10**, 297–311.

Bender, A., and Pringle, J. R. (1989). Multicopy suppression of the *cdc24* budding defect by *CDC42* and three newly identified genes including the *ras*-related gene *RSR1*. *Proc. Natl. Acad. Sci. U.S.A.* **86**, 9976–9980.

Bourne, H. R., Sanders, D., and McCormick, F. (1990). The GTPase superfamily: A conserved switch for diverse cell functions. *Nature (London)* **348**, 125–132.

Bourne, H. R., Sanders, D., and McCormick, F. (1991). The GTPase superfamily: Conserved structure and molecular mechanism. *Nature (London)* **349**, 117–127.

Boy-Marcotte, E., Damak, F., Camonis, J., Garreau, H., and Jacquet, M. (1989). The C-terminal part of a gene partially homologous to the *CDC25* gene suppresses the *cdc25-5* mutation in *Saccharomyces cerevisiae*. *Gene* **77**, 21–30.

Broek, D., Toda, T., Michel, T., Levin, L., Birchmeier, C., Zoller, M., Powers, S., and Wigler, M. (1987). The *Saccharomyces CDC25* gene product regulates the RAS/adenylate cyclase pathway. *Cell* **48**, 789–799.

Camonis, J. H., Kalekine, M., Gondre, B., Garreau, H., Boy-Marcotte, E., and Jacquet, M. (1986). Characterization, cloning and sequence analysis of the *CDC25* gene which controls the cAMP level of *Saccharomyces cerevisiae*. *EMBO J.* **5**, 375–380.

Costello, G., Rodgers, L., and Beach, D. (1986). Fission yeast enters the stationary phase G_0 from either mitotic G_1 or G_2. *Curr. Genet.* **11**, 119–125.

Courchesne, W. E., Kunisawa, R., and Thorner, J. (1989). A putative protein kinase overcomes pheromone-induced arrest of cell cycling in *S. cerevisiae*. *Cell* **58**, 1107–1119.

Crechet, J. B., Poullet, P., Mistou, M. Y., Parmeggiani, A., Camonis, J., Boy-Marcotte, E., Damak, F., and Jacquet, M. (1990). Enhancement of the GDP–GTP exchange of RAS proteins by the carboxy-terminal domain of *SDC25*. *Science* **248**, 866–868.

Damak, F., Boy-Marcotte, E., Le-Roscouet, D., Guilbaund, R., and Jacquet, M. (1991). *SDC25*, a *CDC25*-like gene which contains a RAS-activating domain and is a dispensable gene of *Saccharomyces cerevisiae*. *Mol. Cell. Biol.* **11**, 202–212.

Davey, J. (1991). Isolation and quantitation of M-factor, a diffusible mating factor from the fission yeast *Schizosaccharomyces pombe*. *Yeast* **7**, 357–366.

Davey, J. (1992). Mating pheromones of the fission yeast *Schizosaccharomyces pombe*: Purification and structural characterization of M-factor and isolation and analysis of two genes encoding the pheromone. *EMBO J.* **11**, 951–960.

Dmochowska, A., Dignard, D., Henning, D., Thomas, D. Y., and Bussey, H. (1987). Yeast *KEX1* gene encodes a putative protease with a carboxypeptidase B-like function involved in killer toxin and α-factor precursor processing. *Cell* **50**, 573–584.

Egel, R. (1971). Physiological aspects of conjugation in fission yeast. *Planta* **98**, 89–96.

Egel, R. (1973a). Commitment to meiosis in fission yeast. *Mol. Gen. Genet.* **121**, 277–284.

Egel, R. (1973b). Genes involved in mating-type expression of fission yeast. *Mol. Gen. Genet.* **122**, 339–343.

Egel, R. (1989). Mating-type genes, meiosis and sporulation. *In* "Molecular Biology of the Fission Yeast" (A. Nasim, P. Young, and B. F. Johnston, eds.), pp. 31–73. Academic Press, New York.

Egel, R., and Egel-Mitani, M. (1974). Pre-meiotic DNA synthesis in the fission yeast. *Exp. Cell Res.* **88**, 127–134.

Egel, R., Nielsen, O., and Weilguny, D. (1990). Sexual differentiation in fission yeast. *Trends Genet.* **6**, 369–373.

Elion, E. A., Grisafi, P. L., and Fink, G. R. (1990). *FUS3* encodes a *cdc2*$^+$/*CDC28*-related kinase required for the transition from mitosis into conjugation. *Cell* **60**, 649–664.

Fukui, Y., and Kaziro, Y. (1985). Molecular cloning and sequence analysis of a *ras* gene from *Schizosaccharomyces pombe*. *EMBO J.* **4**, 687–691.

Fukui, Y., and Yamamoto, M. (1988). Isolation and characterization of *Schizosaccharomyces pombe* mutants phenotypically similar to *ras1*$^-$. *Mol. Gen. Genet.* **215**, 26–31.

Fukui, Y., Kozasa, T., Kaziro, Y., Takeda, T., and Yamamoto, M. (1986a). Role of a *ras* homolog in the life cycle of *Schizosaccharomyces pombe*. *Cell* **44**, 329–336.

Fukui, Y., Kaziro, Y., and Yamamoto, M. (1986b). Mating pheromone-like diffusible factor released by *Schizosaccharomyces pombe*. *EMBO J.* **5**, 1991–1993.

Fukui, Y., Miyake, S., and Yamamoto, M. (1989). Characterization of the *Schizosaccharomyces pombe ral2* gene implicated in activation of the *ras1* gene product. *Mol. Cell. Biol.* **9**, 5617–5622.

Girgsdies, O. (1982). Sterile mutants of *Schizosaccharomyces pombe*: Analysis by somatic hybridization. *Curr. Genet.* **6**, 223–227.

Graziano, M. P., and Gilman, A. G. (1989). Synthesis in *Escherichia coli* of GTPase-deficient mutants of $G_{s\alpha}$. *J. Biol. Chem.* **264**, 15475–15482.

Hancock, J. F., Magee, A. I., Childs, J. E., and Marshall, C. J. (1989). All *ras* proteins are polyisoprenylated but only some are palmitoylated. *Cell* **57**, 1167–1177.

Hughes, D. A., Fukui, Y., and Yamamoto, M. (1990). Homologous activators of *ras* in fission and budding yeast. *Nature (London)* **344**, 355–357.

Hughes, D. A., Imai, Y., and Yamamoto, M. (1991). Regulation of the *ras* pathway in the fission yeast *Schizosaccharomyces pombe*. *In* "The Superfamily of *ras*-Related Genes" (D. A. Spandidos, ed.), pp. 41–47. Plenum Press, New York.

Iino, Y., and Yamamoto, M. (1985). Mutants of *Schizosaccharomyces pombe* which sporulate in the haploid state. *Mol. Gen. Genet.* **198**, 416–421.

Imai, Y., and Yamamoto, M. (1992). *Schizosaccharomyces pombe sxa1*$^{+}$ and *sxa2*$^{+}$ encode putative proteases involved in the mating response. *Mol. Cell. Biol.* **12**, 1827–1842.

Imai, Y., Miyake, S., Hughes, D. A., and Yamamoto, M. (1991). Identification of a GTPase-activating protein homolog in *Schizosaccharomyces pombe*. *Mol. Cell. Biol.* **11**, 3088–3094.

Jones, S., Vignais, M.-L., and Broach, J. R. (1991). The CDC25 protein of *S. cerevisiae* promotes exchange of guanine nucleotides bound to ras. *Mol. Cell. Biol.* **11**, 2641–2646.

Julius, D., Livelli, T. J., Jessell, T. M., and Axel, R. (1989). Ectopic expression of the serotonin 1c receptor and the triggering of malignant transformation. *Science* **244**, 1057–1062.

Kawamukai, M., Ferguson, K., Wigler, M., and Young, D. (1991). Genetic and biochemical analysis of the adenylyl cyclase of *Schizosaccharomyces pombe*. *Cell Reg.* **2**, 155–164.

Kelly, M., Burke, J., Smith, M., Klar, A., and Beach, D. (1988). Four mating-type genes control sexual differentiation in the fission yeast. *EMBO J.* **7**, 1537–1547.

Kitamura, K., and Shimoda, C. (1991). The *Schizosaccharomyces pombe mam2* gene encodes a putative pheromone receptor which has significant homology with the *Saccharomyces cerevisiae* Ste2 protein. *EMBO J.* **10**, 3743–3751.

Kitayama, H., Sugimoto, Y., Matsuzaki, T., Ikawa, Y., and Noda, M. (1989). A *ras*-related gene with transformation suppressor activity. *Cell* **56**, 77–84.

Kosako, H., Gotoh, Y., Matsuda, S., Ishikawa, M., and Nishida, E. (1992). *Xenopus* MAP kinase activator is a serine/threonine/tyrosine kinase activated by threonine phosphorylation. *EMBO J.* **11**, 2903–2908.

Landis, C. A., Masters, S. B., Spada, A., Pace, A. M., Bourne, H. R., and Vallar, L. (1989). GTPase inhibiting mutations activate the α chain of Gs and stimulate adenylyl cyclase in human pituitary tumours. *Nature (London)* **340**, 692–696.

Leevers, S. J., and Marshall, C. J. (1992). MAP kinase regulation—The oncogene connection. *Trends Cell Biol.* **2**, 283–286.

Lefkowitz, R. J., and Caron, M. G. (1988). Adrenergic receptors. *J. Biol. Chem.* **263**, 4993–4996.

Leupold, U. (1987). Sex appeal in fission yeast. *Curr. Genet.* **12**, 543–545.

Leupold, U., Nielsen, O., and Egel, R. (1989a). Pheromone-induced meiosis in P-specific mutants of fission yeast. *Curr. Genet.* **15**, 403–405.

Leupold, U., Nielsen, O., and Egel, R. (1989b). Pheromone response in fission yeast: modulation by *pat1*. *Curr. Genet.* **16**, 465–468.

Leupold, U., Sipiczki, M., and Egel, R. (1991). Pheromone production and response in sterile mutants of fission yeast. *Curr. Genet.* **20**, 79–85.

Lund, P. M., Hasegawa, Y., Kitamura, K., Shimoda, C., Fukui, Y., and Yamamoto, M. (1987). Mapping of the *ras1* gene of *Schizosaccharomyces pombe*. *Mol. Gen. Genet.* **209**, 627–629.

Lyons, J., Landis, C. A., Harsh, G., Vallar, L., Grunewald, K., Feichtinger, H., Duh, Q.-Y., Clark, O. H., Kawasaki, E., Bourne, H. R., and McCormick, F. (1990). Two G protein oncogenes in human endocrine tumors. *Science* **249**, 655–659.

McCormick, F. (1989). *ras* GTPase activating protein: Signal transmitter and signal terminator. *Cell* **56**, 5–8.

MacKay, V. L., Welch, S. K., Insley, M. Y., Manney, T. R., Holly, J., Saari, G. C., and Parker, M. L. (1988). The *Saccharomyces cerevisiae BAR1* gene encodes an exported protein with homology to pepsin. *Proc. Natl. Acad. Sci. U.S.A.* **85**, 55–59.

McLeod, M. (1989). Regulation of meiosis: From DNA binding protein to protein kinase. *BioEssays* **11**, 9–14.

Maeda, T., Mochizuki, N., and Yamamoto, M. (1990). Adenylyl cyclase is dispensable for vegetative growth in fission yeast *Schizosaccharomyces pombe*. *Proc. Natl. Acad. Sci. U.S.A.* **87**, 7814–7818.

Marchuk, D. A., Saulino, A. M., Tavakkol, R., Swaroop, M., Wallace, M. R., Andersen, L. B., Mitchell, A. L., Gutmann, D. H., Boguski, M., and Collins, F. S. (1991). cDNA cloning of the type 1 neurofibromatosis gene: Complete sequence of the *NF1* gene product. *Genomics* **11**, 931–940.

Marsh, L., Neiman, A. M., and Herskowitz, I. (1991). Signal transduction during pheromone response in yeast. *Ann. Rev. Cell Biol.* **7**, 699–728.

Masters, S. B., Miller, R. T., Chi, M-H., Chang, F-H., Beiderman, B., Lopez, N. G., and Bourne, H. R. (1989). Mutations in the GTP-binding site of Gs alter stimulation of adenylyl cyclase. *J. Biol. Chem.* **264**, 15467–15474.

Meade, J. H., and Gutz, H. (1976). Mating-type mutations in *Schizosaccharomyces pombe*. Isolation of mutants and analysis of strains with an h^- or h^+ phenotype. *Genetics* **83**, 259–273.

Michael, H., and Gutz, H. (1987). Sterility (*ste*) genes of *Schizosaccharomyces pombe*. *Yeast* **3**, 5–9.

Mochizuki, N., and Yamamoto, M. (1992). Reduction in the intracellular cAMP level triggers initiation of sexual development in fission yeast. *Mol. Gen. Genet.* **233**, 17–24.

Nadin-Davis, S. A., and Nasim, A. (1988). A gene which encodes a predicted protein kinase can restore some functions of the *ras* gene in fission yeast. *EMBO J.* **7**, 985–993.

Nadin-Davis, S. A., and Nasim, A. (1990). *Schizosaccharomyces pombe ras1* and *byr1* are functionally related genes of the *ste* family that affect starvation-induced transcription of the mating-type genes. *Mol. Cell. Biol.* **10**, 549–560.

Nadin-Davis, S. A., Yang, R. C. A., Narang, S. A., and Nasim, A. (1986a). The cloning and characterization of a *RAS* gene from *Schizosaccharomyces pombe*. *J. Mol. Evol.* **23**, 41–51.

Nadin-Davis, S. A., Nasim, A., and Beach, D. (1986b). Involvement of *ras* in sexual differentiation but not in growth control in fission yeast. *EMBO J.* **5**, 2963–2971.

Nadin, Davis, S. A., Nasim, A., Fukui, Y., and Yamamoto, M. (1989). Oncogene homologs. *In* "Molecular Biology of the Fission Yeast" (A. Nasim, P. Young, and B. F. Johnston, eds.), pp. 97–126. Academic Press, New York.

Neer, E. J., and Clapham, D. E. (1988). Roles of G. protein subunits in transmembrane signaling. *Nature (London)* **333**, 129–134.

Nielsen, O., and Egel, R. (1990). The *pat1* protein kinase controls transcription of the mating-type genes in fission yeast. *EMBO J.* **9**, 1401–1406.

Nielsen, O., Davey, J., and Egel, R. (1992). The *ras1* function of *Schizosaccharomyces pombe* mediates pheromone-induced transcription. *EMBO J.* **11**, 1391–1395.

Nurse, P. (1985). Mutants of the fission yeast *Schizosaccharomyces pombe* which alter the shift between cell proliferation and sporulation. *Mol. Gen. Genet.* **198**, 497–502.

Nurse, P., and Bisset, Y. (1981). Gene required in G_1 for commitment to cell cycle and in G_2 for control of mitosis in fission yeast. *Nature (London)* **292**, 558–560.

Obara, T., Nakafuku, M., Yamamoto, M., and Kaziro, Y. (1991). Isolation and characterization of the gene encoding G protein α-subunit from *Schizosaccharomyces pombe*: Involvement in mating and sporulation pathways. *Proc. Natl. Acad. Sci. U.S.A.* **88**, 5877–5881.

Rhodes, N., Connell, L., and Errede, B. (1990). STE11 is a protein kinase required for cell-type-specific transcription and signal transduction in yeast. *Genes Dev.* **4**, 1862–1874.

Ruggieri, R., Bender, A., Matsui, Y., Powers, S., Takai, Y., Pringle, J. R., and Matsumoto, K. (1992). *RSR1*, a *ras*-like gene homologous to *Krev-1* (*smg21A/rap1A*): Role in the development of cell polarity and interactions with the *ras* pathway in *Saccharomyces cerevisiae*. *Mol. Cell. Biol.* **12**, 758–766.

Simon, M. A., Bowtell, D. D. L., Dodson, G. S., Laverty, T. R., and Rubin, G. M. (1991). Ras1 and a putative guanine nucleotide exchange factor perform crucial steps in signaling by the sevenless protein tyrosine kinase. *Cell* **67**, 701–716.

Sipiczski, M. (1988). The role of sterility genes (*ste* and *aff*) in the initiation of sexual development in *Schizosaccharomyces pombe*. *Mol. Gen. Genet.* **213**, 529–534.

Sugimoto, A., Iino, Y., Maeda, T., Watanabe, Y., and Yamamoto, M. (1991). *Schizosaccharomyces pombe ste1*$^+$ encodes a transcription factor with an HMG motif that is a critical regulator of sexual development. *Genes Dev.* **5**, 1990–1999.

Tanaka, K., Matsumoto, K., and Toh-e, A. (1989). *IRA1*, an inhibitory regulator of the RAS-cyclic AMP pathway in *Saccharomyces cerevisiae*. *Mol. Cell. Biol.* **9**, 757–768.

Tanaka, K., Nakafuku, M., Tamanoi, F., Kaziro, Y., Matsumoto, K., and Toh-e, A. (1990). *IRA2*, a second gene of *Saccharomyces cerevisiae* that encodes a protein with a domain homologous to mammalian *ras* GTPase-activating protein. *Mol. Cell. Biol.* **10**, 4303–4313.

Tanaka, K., Davey, J., Imai, Y., and Yamamoto, M. (1993). *Schizosaccharomyces pombe map3*$^{+}$ encodes the putative M-factor receptor. *Mol. Cell. Biol.* **13**, 80–88.

Teague, M. A., Chaleff, D. T., and Errede, B. (1986). Nucleotide sequence of the yeast regulatory gene *STE7* predicts a protein homologous to protein kinases. *Proc. Natl. Acad. Sci. U.S.A.* **83**, 7371–7375.

Thuriaux, P., Sipiczki, M., and Fantes, P. A. (1980). Genetical analysis of a sterile mutant by protoplast fusion in the fission yeast *Schizosaccharomyces pombe*. *J. Gen. Microbiol.* **116**, 525–528.

Toda, T., Shimanuki, M., and Yanagida, M. (1991). Fission yeast genes that confer resistance to staurosporine encode an AP-1-like transcription factor and a protein kinase related to the mammalian *ERK1/MAP2* and budding yeast *FUS3* and *KSS1* kinases. *Genes Dev.* **5**, 60–73.

Trahey, M., Wong, G., Halenbeck, R., Rubinfield, B., Martin, G., Ladner, M., Long, C., Crosier, W., Watt, K., Koths, K., and McCormick, F. (1988). Molecular cloning of two types of GAP complementary DNA from human placenta. *Science* **242**, 1697–1700.

Valls, L. A., Hunter, C. P., Rothman, J. H., and Stevens, T. H. (1987). Protein sorting in yeast: The localization determinant of yeast vacuolar carboxypeptidase Y resides in the propeptide. *Cell* **48**, 887–897.

Vogel, U., Dixson, R., Schaber, M., Diehl, R., Marshall, M., Scolnick, E., Sigal, I., and Gibbs, J. (1988). Cloning of bovine GAP and its interaction with oncogenic *ras* p21. *Nature (London)* **335**, 90–93.

Wang, Y., Boguski, M., Riggs, M., Rodgers, L., and Wigler, M. (1991a). *sar1*, a gene from *Schizosaccharomyces pombe* encoding a GAP-like protein that regulates *ras1*. *Cell Regul.* **2**, 453–465.

Wang, Y., Xu, H-P., Riggs, M., Rodgers, L., and Wigler, M. (1991b). *byr2*, a *Schizosaccharomyces pombe* gene encoding a protein kinase capable of partial phenotypic suppression of *ras1* mutant phenotype. *Mol. Cell. Biol.* **11**, 3554–3563.

Watanabe, Y., Iino, Y., Furuhata, K., Shimoda, C., and Yamamoto, M. (1988). The *S. pombe mei2* gene encoding a crucial molecule for commitment to meiosis is under the regulation of cAMP. *EMBO J.* **7**, 761–767.

Xu, G., O'Connell, P., Viskochil, D., Cawthon, R., Robertson, M., Culver, M., Dunn, D., Stevens, J., Gesteland, R., White, R., and Weiss, R. (1990). The neurofibromatosis type 1 gene encodes a protein related to GAP. *Cell* **62**, 599–608.

Xu, H-P., Wang, Y., Riggs, M., Rodgers, L., and Wigler, M. (1990). Biological activity of the mammalian RAP genes in yeast. *Cell Regul.* **1**, 763–769.

Young, D., Waitches, G., Birchmeier, C., Fasano, O., and Wigler, M. (1986). Isolation and characterization of a new cellular oncogene encoding a protein with multiple transmembrane domains. *Cell* **45**, 711–719.

Chapter 7

RAS Genes in the Budding Yeast *Saccharomyces cerevisiae*

KELLY TATCHELL

Department of Microbiology
North Carolina State University
Raleigh, North Carolina

I. Introduction

Guanine nucleotide binding proteins act in diverse sets of cellular processes, ranging from signal transduction to membrane fusion. Ras proteins constitute the most extensively studied class of guanine nucleotide binding protein, largely because of their important role in many cancers. *ras* genes clearly have a fundamental role that extends beyond neoplastic transformation. The fundamental role for Ras proteins was highlighted when *ras*-related genes were discovered in the budding yeast *Saccharomyces cerevisiae* by DeFeo-Jones and colleagues (DeFeo-Jones *et al.*, 1983) and Powers *et al.* (1984). Since that time, the yeast system has been used extensively to study Ras. Excellent articles by Gibbs and Marshall (1989) and Broach and Deschenes (1990) provide comprehensive reviews of Ras in yeast. In this Chapter, I focus on recent developments in the field. In Section II, I briefly review the role of Ras in metazoans before focusing on Ras in *S. cerevisiae*. Section III is a general overview of *RAS* genes in yeast, as well as their expression and posttranslational modification. In Section IV, I review data showing that adenylyl cyclase is an effector for Ras and briefly discuss the possibilities of other Ras effectors. Section V concerns the regulation of guanine nucleotide exchange and GTP hydrolysis, whereas Section VI reviews the evidence for feedback control in the Ras pathway. Finally, the role of Ras in nutrient response is reviewed in Section VII.

II. *ras* Genes in Metazoans

A. Members of a Class of Small GTP Binding Proteins

ras genes constitute a family that has long been implicated in cancer. Retroviral oncogenes Ha-*ras* and Ki-*ras* from Harvey and Kirsten sarcoma viruses were found to be related to cellular transforming *ras* genes from a number of tumors or tumor-derived cell lines. *ras* genes have been estimated to play a role in the etiology of 10–20% of human cancers (reviewed by Barbacid, 1987). *ras* genes have been found in widely divergent species such as yeast and *Drosophila*, and encode members of a superfamily of small guanine nucleotide binding proteins that participate in a wide range of functions ranging from vesicle trafficking to signal transduction (for review, see Downward, 1990; Bourne *et al.*, 1991). Mammalian Ras proteins, encoded by the Ha-*ras*, Ki-*ras*, and N-*ras* genes, are more than 80% identical over their 189-amino-acid length. The proteins consist of a globular N-terminal domain of 164 amino acid residues that binds guanine nucleotides and a C-terminal domain that is modified posttranslationally and is required to tether Ras to the plasma membrane. The 20 amino acids between the globular GTP-binding domain and the C terminus are not conserved and account for most of the divergence among members of the *ras* gene family.

The C terminus of all mammalian Ras proteins has the sequence CAAX, where C is cysteine, A is an aliphatic residue, and X is the C-terminal amino acid. This C terminus is modified posttranslationally, resulting in removal of the three C-terminal amino acids and carboxy-methylation and farnesylation of the cysteine (reviewed by Deschenes *et al.*, 1990; Gibbs, 1991). Mutations in the CAAX sequence result in the loss of membrane localization and biological function (Willumsen *et al.*, 1984). The biological role of modification appears to be the addition of a hydrophobic domain necessary for membrane association rather than a specific role for the farnesyl group because mutant forms of Ras that are nontransforming due to mutation of the CAAX sequence can be made transforming by engineering a site for myristoylation near the N terminus (Lacal *et al.*, 1988; Buss *et al.*, 1989).

A comparison of the properties of Ras proteins encoded by transforming genes and those encoded by normal cellular genes has revealed that most of the transforming proteins are impaired in intrinsic GTPase activity. Missense mutations at amino acid positions 12, 13, 59, 61, and 63 show increased transforming potential and decreased GTPase activity (Barbacid, 1987). The loss of GTPase activity in the oncogenic forms of Ras provides the basis for the model that Ras proteins act as molecular switches in signaling pathways that regulate cell growth. In this model, Ras proteins with bound GTP are in a conformation that allows activation of a downstream effector. Hydrolysis of bound GTP causes a conformational change in Ras that prevents activation of the effector; subsequent exchange of bound GDP for GTP reactivates the switch.

Structural analysis of Ras proteins supports the hypothesis that Ras acts as a molecular switch. The three-dimensional structure of the globular domain of Ha-Ras has been determined by X-ray crystallography in both the GDP- and GTP-bound forms (Pai *et al.*, 1989,1990; Krengel *et al.*, 1990; Milburn *et al.*, 1990; Schlichting *et al.*, 1990; Tong *et al.*, 1991). Wittinghofer and Pai (1991) describe Ras as a compact globular protein containing a nucleotide binding pocket formed by several α-helices and loops. Amino acid substitutions that alter either guanine nucleotide binding or intrinsic GTPase activity lie in these regions. A comparison of the structures of the GTP-bound and GDP-bound Ras proteins has revealed significant changes in the conformation of two loops. One of these, loop 2, has been termed the effector loop because amino acid substitutions in this region cause a loss of biological activity without loss of guanine nucleotide binding or GTPase activity (Sigal *et al.*, 1986b,1988; Adari *et al.*, 1988; Cales *et al.*, 1988).

Assuming that only Ras–GTP is able to activate downstream effector(s), the activity of Ras would be a function of the ratio of Ras with bound GTP to Ras with bound GDP. Activity could be modulated by the rate of GTP hydrolysis or by the rate of GDP–GTP exchange. The important prediction is that Ras proteins with normal GTPase activity but increased GDP–GTP exchange rate also would have transforming potential. Indeed Ha-Ras proteins in which Asp-16 is substituted with Lys (Sigal *et al.*, 1986a) or Ala-146 is substituted with Val (Feig and Cooper, 1988) show decreased affinity for guanine nucleotides, possess an increased rate of exchange *in vitro*, and have a transforming phenotype *in vivo*. Activity and transforming potential are, therefore, determined by the balance between the rate of GTPase activity and the rate of GDP–GTP exchange. Consistent with this idea, Feig and Cooper (1988) observed that two nontransforming *ras* mutations that have a small effect on GDP–GTP exchange and GTPase activity, respectively, have a transforming phenotype when combined.

A model for Ras activity based on the ratio of bound GDP–GTP lends itself to modulation by factors that regulate either GTPase activity or guanine nucleotide exchange. GTPase activating protein (GAP) is a soluble factor, first identified by Trahey and McCormick (1987), that is able to stimulate the intrinsic GTPase activity of normal Ras but not of oncogenic Ras proteins. Multiple GAPs have been identified in mammalian cells (Trahey *et al.*, 1988; Vogel *et al.*, 1988). A portion of the neurofibromatosis gene product that is similar to the catalytic domain of GAP has GAP activity *in vitro* (Martin *et al.*, 1990). GAPs that appear to act exclusively on other members of the Ras superfamily also have been isolated (Rubinfeld *et al.*, 1991, and references therein).

The observation that GAP cannot bind Ras proteins that contain amino acid substitutions in the effector domain (amino acids 32–40) (Adari *et al.*, 1988; Cales *et al.*, 1988) suggests that GAP may be an effector for Ras (McCormick, 1989). In other signal transduction pathways, one role of the downstream effector is to induce GTPase activity in the G protein. However, some studies on Ras proteins bearing amino acid substitutions near the effector domain are more consistent with the hypothesis that GAP negatively regulates the activity of Ras. A Ha-Ras protein with two amino acid substitutions (Asp-30 to Glu-30 and Glu-31 to Lys-31) retains biological function in yeast and mammalian cells, but has greatly reduced GAP binding activity *in vitro*, whereas a Ha-Ras protein with a mutation near the effector domain (Val-45 to Glu-45) binds GAP but has no biological activity (Marshall *et al.*, 1991). A mutation in the effector domain of the yeast Ras2 protein, corresponding to a change from Pro-34 to Ser in Ha-Ras, has a dominant activating phenotype *in vivo* that correlates with decreased ability of the yeast GAP homolog Ira2 to stimulate GTPase activity (Tanaka *et al.*, 1992). These results leave open the possibility that GAP negatively regulates the activity of Ras but is not a Ras effector protein.

Another potential mechanism of Ras regulation, modulation of GDP–GTP exchange, is well documented in many G protein systems but has been implicated only recently in the mammalian Ras pathway. West *et al.* (1990) and Wolfman and Macara (1990) have identified factors from mammalian cells that facilitate GDP–GTP exchange. However, the activities have not been purified to homogeneity, nor have the genes encoding these factors been cloned. However, mammalian genes similar to *S. cerevisiae CDC25*, which is thought to encode a GDP–GTP exchange factor for yeast Ras (see Section V,A), have been cloned (E. Martigani and D. Broek, personal communication). These genes could encode potential exchange factors.

B. *ras* Genes and Protein Tyrosine Kinase Receptor Pathways

The role of Ras in cellular transformation and the structural similarity between Ras and G proteins leads one to the prediction that Ras will have a role in one or more signal transduction pathways. Indeed, Ras appears to have a role in signal transduction pathways of protein tyrosine kinase receptors. Early evidence for this role was provided by studies showing that the injection of anti-Ras antibodies into cells can suppress transiently the proliferation of cells transformed by receptor tyrosine kinase-encoded oncogenes (Mulcahy *et al.*, 1985; Smith *et al.*, 1986). Injection of these antibodies has no effect on cells transformed by oncogenes that encode soluble protein kinases. Ras in the active GTP-bound form accumulates when cells are treated with either platelet derived growth factor (PDGF; Satoh *et al.*, 1990a) or epidermal growth factor (EGF; Satoh *et al.*, 1990b). Also, GAP has been shown to bind to, and be phosphorylated by, PDGF (Kaplan *et al.*, 1990; Kazlauskas *et al.*, 1990).

Genetic evidence also supports a role for Ras in protein tyrosine kinase receptor pathways (Chapter 16). The *let-60* gene, required for vulva formation in the nematode *Caenorhabditis elegans*, encodes a Ras protein (Beitel *et al.*, 1990; Han and Sternberg, 1990). Dominant mutations in *let-60* cause a multivulva phenotype (Beitel *et al.*, 1990; Han and Sternberg, 1990). These dominant *let-60* mutations suppress the vulva-less phenotype of some alleles of *let 23*, which has been shown to encode a receptor tyrosine kinase related to EGF (Aroian *et al.*, 1990). Loss-of-function alleles of *let-60* and *let-23* result in a larval lethal phenotype but increased *let-60* activity suppresses the lethality of certain strong *let-23* alleles (Han *et al.*, 1990). Thus, *let-60* Ras acts downstream of the *let-23*-encoded tyrosine kinase receptor. Ras and the receptor are required for vulva development and other essential functions.

A similar relationship between receptor tyrosine kinases and Ras has been established in *Drosophila melanogaster*, where the *Ras*1 gene was identified in a genetic screen for enhancers of the *sevenless* phenotype. A specific photoreceptor (R7) in the compound eye of *Drosophila* fails to develop in *sev* mutants. *Sevenless* (*sev*) encodes a protein tyrosine kinase receptor that is most similar to mammalian *c-ros* (Hafen *et al.*, 1987). Simon *et al.* (1991) isolated mutations in seven loci that enhance the mutant phenotype of a temperature-sensitive *sev* mutant. One of these, *E(sev)3C*, encodes the *Drosophila ras1* gene. Transgenic *Drosophila* that express a dominant Ras1 protein develop R7 photoreceptor cells in the absence of *sev* function (Fortini *et al.*, 1992). In *Drosophila*, as in *C. elegans*, Ras probably is involved in more than one pathway. In both systems, null mutations in *ras* result in lethality at an early stage of development. Also, *E(sev)3C* alleles of *ras1* also suppress the phenotype of *Ellipse*, a dominant mutation in another protein tyrosine kinase receptor (Simon *et al.*, 1991).

The *Drosophila* and *C. elegans* systems hold promise for clarifying the role of Ras in protein tyrosine kinase receptor pathways. The most difficult obstacle is that Ras has multiple functions in these organisms, since null mutations are lethal. Although genetic techniques such as mosaic analysis may circumvent this problem to some extent, a simpler genetic system such as yeast appears more promising for investigating Ras function at the cellular level. The facile transformation systems in yeast allow control of gene dosage and the creation of gene disruptions, and provide an advantage for investigating Ras. Certain advances, even in *Drosophila* and *C. elegans*, are indebted to parallel investigations in yeast. For example, the hypothesis that the *boss* gene in *Drosophila* encodes a GDP–GTP exchange factor is based on studies of its apparent homolog *CDC25* in yeast (Section VII). The yeast system is likely to continue to provide valuable information about the roles of Ras in multicellular organisms.

III. *Saccharomyces cerevisiae RAS* Genes

A. *RAS1* and *RAS2* in *Saccharomyces cerevisiae* Are Closely Related

The two *ras*-related genes in *S. cerevisiae, RAS1* and *RAS2*, were identified by their ability to hybridize with mammalian *ras* genes (DeFeo-Jones *et al.*, 1983; Powers *et al.*, 1984). The N-terminal 80 amino acid residues of the Ras1 and Ras2 products are 85% identical to mammalian Ha-Ras (Fig. 7.1). The similarity falls to 46% in the next 80 amino acids, and the remainder of the yeast and mammalian Ras proteins are completely divergent. Hereafter, the term *RAS* will be used to define collectively the *RAS1* and *RAS2* genes, as will Ras for their products, as suggested by Broach (1991). Yeast Ras proteins show some similarity to mammalian Ras at the C terminus because of the presence of conserved sequences that are required for posttranslational modification. The Ras1 and Ras2 proteins are significantly larger than metazoan Ras because of a 7-amino-acid addition at the N terminus and an ~110-amino-acid insertion near the C terminus. The large extensions in Ras1 and Ras2 are not re-

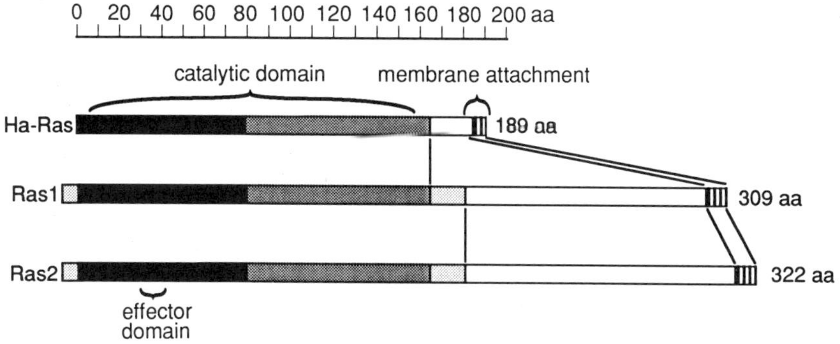

Figure 7.1 Comparison of yeast and mammalian Ras proteins. The bars represent the amino acid sequences of each protein deduced from the nucleotide sequence of *RAS1* and *RAS2* (Dhar *et al.*, 1984; Powers *et al.*, 1984). The solid region of each bar represents the amino acid sequences that are greater than 80% identical between yeast and mammalian Ras. The shaded regions are over 40% identical, whereas the open regions represent low homology. The stippled regions represent homology only between the yeast Ras proteins.

lated to each other and are found at a region that is not conserved among members of the mammalian Ras family. X-ray crystallographic studies indicate that this region of mammalian Ras is flexible and may act as a hinge or linker between the globular catalytic portion of Ras and the membrane-anchoring domain. This domain can be deleted from yeast Ras without loss of biological function (Marshall *et al.*, 1987).

In light of significant divergence between yeast and mammalian Ras and the large number of *ras*-related genes, the validity of referring to *RAS1* and *RAS2* as *ras* homologs can be questioned. Several lines of evidence suggest that the term Ras is justified. First, yeast strains that express mammalian *ras* genes grow in the absence of the yeast *RAS* genes (DeFeo-Jones *et al.*, 1985; Kataoka *et al.*, 1985b). Second, mutant forms of yeast *RAS1*, when fused to a mammalian enhancer, can transform NIH3T3 cells (DeFeo-Jones *et al.*, 1985) and give rise to lethal tumors in nude mice (Bradley *et al.*, 1986).

B. Post-translational Modification of Ras

Yeast and mammalian Ras proteins undergo similar post-translational modifications (Fujiyama and Tamanoi, 1986,1987; Deschenes *et al.*, 1989; Finegold *et al.*, 1990; Goodman *et al.*, 1990) and, as in mammalian systems, the C-terminal CAAX sequence is required for proper membrane localization and biological function. Mutations of the C terminus of Ras2 that eliminate the CAAX box severely attenuate its activity (Deschenes and Broach, 1987). However, expression of these mutants on high copy number 2μ shuttle vectors allow some growth in a *ras1*$^{-}$,*ras2*$^{-}$ background, indicating that the modification is not absolutely essential for activity.

Genetic analysis has uncovered components of the modification pathway. *ram1/dpr1* (Powers *et al.*, 1986; Fujiyama *et al.*, 1987; Goodman *et al.*, 1988) and *ram2* (Goodman *et al.*, 1990), which were identified as suppressors of the heat-shock sensitivity of the dominant *RAS2*$^{\text{Val-19}}$ mutant, are required for C-terminal processing. The farnesyl–protein transferase from rat or bovine brain (Reiss *et al.*, 1990; Moores *et al.*, 1991) is a heterodimer composed of two non-identical α and β subunits. The sequence of the α subunit is similar to that of the product of *RAM2* whereas antisera raised against Ram1 specifically react with the β subunit (Kohl *et al.*, 1991). These data are consistent with the hypothesis that *RAM1* and *RAM2* encode β and α subunits of farnesyl–protein transferase, respectively. *ram2* mutants are impaired in both farnesyl–protein transferase and geranylgeranyl–protein transferase activity, whereas *ram1* mutants are defective only in farnesyl–protein transferase activity. Ram1, Cdc43, and Bet2 are 30% identical at the amino acid level (Johnson *et al.*, 1991; Rossi *et al.*, 1991). *cdc43* mutants have impaired geranylgeranyl–protein transferase I activity whereas *bet2* mutants, which fail to localize the Ras-related Ypt2 and Sec4 proteins properly, are impaired for geranylgeranyl–protein transferase II activity (Kohl *et al.*, 1991). Collectively, these data are consistent with the hypothesis that isoprenyl transferases consist of a shared α subunit, encoded by *RAM2*, and unique β subunits, encoded by *RAM1* for the farnesyl–protein transferase and by *CDC43* and *BET2* for the geranylgeranyl–protein transferases. Ras function also can be inhibited by blocking the isoprenoid biosynthetic pathway, which is required for synthesis of the farnesine. Mutants in hydroxymethylglutaryl (HMG)-CoA reductase, the first step of the isoprenoid biosynthetic pathway, suppress the heat-shock sensitive phenotype of the *RAS2*$^{\text{Val-19}}$ mutation (Schafer *et al.*, 1989), which again implicates farnesyl modification in Ras function.

C. Ras1 and Ras2 Have an Overlapping Function

Although Ras1 and Ras2 differ dramatically in the sequence of their C-terminal amino acid extensions, no specific biological functions can be attributed to either gene product. Haploid yeast strains containing loss-of-function mutations in either *RAS1* or *RAS2* are viable under most growth conditions and are able to mate and sporulate, but *ras1*$^{-}$ *ras2*$^{-}$ double mutants are inviable, indicating that *RAS1* and *RAS2* have an essential overlapping function (Kataoka *et al.*, 1984; Tatchell *et al.*, 1984). Several differences that have been observed between *ras1*$^{-}$ and *ras2*$^{-}$ mutants can be attributed to differences in Ras expression. For example, in many genetic backgrounds, *ras2* deletion mutants fail to grow on media containing nonfermentable carbon sources such as glycerol, acetate, or ethanol (Fraenkel, 1985; Tatchell *et al.*, 1985). This difference can be attributed to the low levels of *RAS1* mRNA and Ras1 protein that accumulate in strains growing on nonfermentable carbon sources (Breviario *et al.*, 1986,1988). Consistent with this hypothesis, the growth defect can be overcome by expressing *RAS1* under the control of the constitutive *ADH1* promoter (Defeo-Jones *et al.*, 1985). Also, *sra6* mutants increase the expression of *RAS1* on nonfermentable carbon sources (Breviario *et al.*, 1986) and relieve the growth inhibition of the *ras2* mutation (Cannon *et al.*, 1986).

Differential expression of *RAS1* and *RAS2* also can explain the difference in phenotype between the Gly-19 to Val-19 mutations in Ras1 and Ras2, which are analogous to the oncogenic Ha-$ras^{\text{Val-12}}$ mutation. The $RAS2^{\text{Val-19}}$ mutant has a dominant phenotype (Toda *et al.*, 1985) that is not exhibited by the $RAS1^{\text{Val-19}}$ mutant (Marshall *et al.*, 1987). The failure to observe a dominant phenotype for $RAS1^{\text{Val-19}}$ appears to be the result of the low level of expression of *RAS1* during late log and stationary phases (Brevario *et al.*, 1986,1988) when the glycogen accumulation and heat-shock phenotypes of $RAS2^{\text{Val-19}}$ usually are assayed. In early log phase, $RAS1^{\text{Val-19}}$ strains exhibit the expected dominant phenotype (Marshall *et al.*, 1987). Further, a dominant $RAS2^{\text{Val-19}}$-like phenotype is observed when $RAS1^{\text{Val-19}}$ is expressed from the constitutive *ADH1* promoter (Marshall *et al.*, 1987).

No biological relevance can be attributed yet to the complex pattern of *RAS1* expression. Although the defects of *ras2* mutants can be explained by the expression pattern of *RAS1*, no phenotypic change has been attributed to a *ras1* mutant, at least under laboratory growth conditions. Ras1 has been reported to have a specific role in phosphatidylinositol metabolism (Kaibuchi *et al.*, 1986), but this observation has not been confirmed.

IV. Ras Effectors in *Saccharomyces cerevisiae*

A. Mutations in the cAMP-Dependent Protein Kinase Pathway Suppress *RAS* Mutations

Several observations suggest a role for Ras in promoting entry into the G_1 phase of the cell cycle. $ras1^-$ $ras2^-$ spores fail to divide (Kataoka *et al.*, 1984; Tatchell *et al.*, 1984), temperature-sensitive $ras1^-ras2^{ts}$ strains arrest at G_1 at the restrictive temperature (Fasano *et al.*, 1988; Garrett and Broach, 1989), and cells in which the only cellular *RAS* gene is expressed from a galactose-inducible promoter arrest in G_1 as unbudded nongrowing cells after a shift to glucose (Kataoka *et al.*, 1984). The arrest point under all these conditions is similar to a point in the cell cycle that is called start II (Pringle and Hartwell,

1981). This point in G_1, which is characterized by general metabolic quiescence, appears to precede the point termed start I that is typified by the arrest induced by mating pheromone or by temperature sensitive mutations in the *CDC28*-encoded protein kinase. The later arrest point specifically affects the cell cycle but has little effect on the rate of macromolecular synthesis. Although a number of different mutants and different physiological conditions cause arrest at start II (see Section VII,A), abundant evidence indicates that Ras is involved primarily in the activation of adenylyl cyclase, encoded by the *CDC35/CYR1* gene.

The first genetic evidence that adenylyl cyclase was an effector for Ras was the discovery that mutations in the BCY1 gene, which earlier had been shown to bypass the requirement for adenylyl cyclase (Matsumoto *et al.*, 1982a), also could bypass the essential requirement for Ras (Toda *et al.*, 1985). *sra1* mutants, independently identified as *ras2* suppressors (Tatchell *et al.*, 1985; Cannon *et al.*, 1986), were found to be allelic to *BCY1*. The sequence of *BCY1* (Cannon *et al.*, 1987; Kunisawa *et al.*, 1987; Toda *et al.*, 1987a) reveals that it encodes the regulatory (R) subunit of cAMP-dependent protein kinase (cAPK) (Hixon and Krebs, 1980). The *BCY1* product, expressed and purified from *Escherichia coli*, has biochemical properties that are similar to those of the mammalian type II cAPK R subunit (Johnson *et al.*, 1987). The R subunit of cAPK binds to, and represses the activity of, the catalytic subunit (Fig. 7.2). cAMP binds to the R subunit, causing the dissociation and activation of the catalytic subunit. Although other cAMP-binding proteins have been reported in yeast (Müller and Bandlow, 1991, and references therein), a biological role has been ascribed only to Bcy1.

The observation that *bcy1* mutations bypass the requirement for adenylyl cyclase is consistent with the hypothesis that the essential function of Ras is to activate adenylyl cyclase. This case is supported by the observation that all mutations that suppress loss-of-function mutations in *RAS* also suppress adenylyl cyclase mutations (Fig. 7.2). Further, most of these suppressors are also components of the cAMP-dependent protein kinase pathway. Mutants in the *TPK1/SRA3* gene, which encodes one of three cAPK catalytic subunit genes, have been isolated that suppress the temperature sensitivity of a *ras1*$^-$ *ras2*^{ts} strain (Levin *et al.*, 1988) or bypass the lethality of a *ras1*$^-$ *ras2*$^-$ mutant (Cannon and Tatchell, 1987). The C subunit encoded by one of these suppressors has a 100-fold decreased affinity for the R subunit (Levin *et al.*, 1988). Dominant mutants in adenylyl cyclase (*CYR1/SRA4*) also bypass the lethality of the *ras1*$^-$ *ras2*$^-$ genotype (Kataoka *et al.*, 1985a; Cannon *et al.*, 1986; De Vendittis *et al.*, 1986). Recessive mutations have been identified in the low K_m cAMP phosphodiesterase (*pde2/sra5*; Sass *et al.*, 1986; Wilson and Tatchell, 1988). *pde2* mutants allow *ras1*$^-$ *ras2*$^-$ double mutants to grow on minimal medium but not on rich medium, whereas mutations in both *pde2* and the

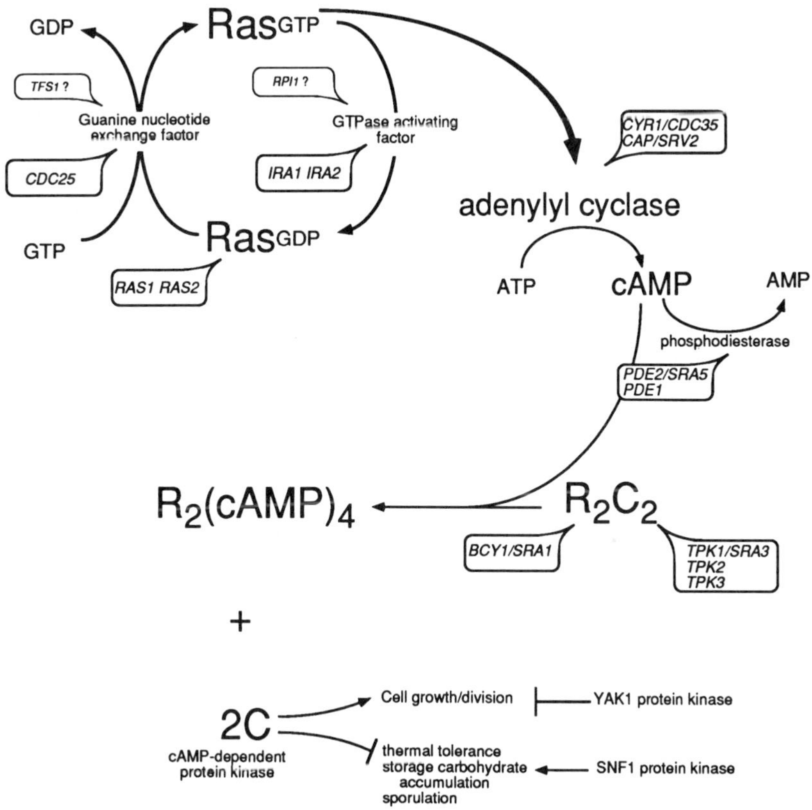

Figure 7.2 The Ras–adenylyl cyclase pathway in *Saccharomyces cerevisiae*. When relevant genes have more than one name, both are provided (e.g., *CYR1/CDC35*). A full description of each component is provided in the text.

gene encoding the low affinity cAMP phosphodiesterase (*pde1*) allow *ras1*$^-$ *ras2*$^-$ strains to grow on minimal and rich media (Nikawa *et al.*, 1987b). Thus, mutations that increase the level of cAMP in the cell can suppress the Ras requirement.

Loss-of-function mutations in *YAK1*, which encodes a novel protein kinase, bypass the entire cAPK pathway. *ras1*$^-$ *ras2*$^-$, adenylyl cyclase$^-$, and even cAPK$^-$ strains are viable in a *yak1* mutant background (Garrett and Broach, 1989). Because null mutations in *YAK1* allow cells lacking cAPK to grow, the Yak1 kinase must act as a negative regulator of cell proliferation, Consistent with this role, Garrett *et al.* (1991) have assayed Yak1 protein kinase activity in cell extracts and observed that activity is low in cells in the exponential phase of growth but high in cells arrested in G_1 at start II. Somewhat surprisingly, Yak1 kinase activity also is elevated in extracts from cells arrested by α-factor or by *cdc28*, both of which arrest cells at start I, or by *cdc8* which arrests cells in early S phase. Garrett *et al.* (1991) observed that the Yak1 kinase can be activated by at least two different mechanisms. In the case of arrest at the restrictive temperature in *ras2*^{ts}, *cdc8*^{ts}, and *cdc28*^{ts} mutants, the increase in Yak1 kinase activity correlates with an increase in the abundance of Yak1 protein determined by Western blot analysis. However, in the case of nitrogen starvation, only an increase in the specific activity of Yak1 is observed. Consistent with the hypothesis that Yak1 is an antagonist of cell proliferation, increased expression of Yak1 inhibits growth in cells that have attenuated cAPK activity. The ability of *yak1* mutations to bypass the requirement for cAPK could be explained if Yak1 were a substrate for cAPK and phosphorylation by cAPK could inhibit Yak1 kinase activity. Although Yak1 restores cell proliferation to cAPK deficient strains, it does not completely alleviate the requirement for cAPK since doubling times of *yak1 tpk1 tpk2 tpk3* strains are much longer than those in wild-type strains. Yak1 is a substrate for cAPK *in vitro* but phosphorylation of Yak1 by cAPK does not inhibit kinase activity of Yak1 assayed *in vitro* (Garrett *et al.*, 1991). Although *in vitro* kinase assays using casein as a substrate may not reflect the *in vivo* situation, the data are more consistent with Yak1 acting in a separate pathway that controls cell proliferation.

At least two other protein kinases have functions that may be related to cAPK. Increased dosage of *SCH9* suppresses both *ras* and adenylyl cyclase mutations (Toda *et al.*, 1988). *SCH9* encodes a protein kinase that is highly related to the catalytic subunit of cAPK at its C terminus but contains a large N-terminal domain that does not resemble any other known protein. Deletions of *SCH9* cause cells to grow more slowly due to a prolonged G_1 phase of the cell cycle. This defect can be suppressed by *BCY1* or *RAS2*$^{Val\text{-}19}$ mutations. Although the possibility that *SCH9* encodes an extremely divergent form of cAPK has not been excluded formally, *SCH9* is more likely to act in a separate pathway. Under normal physiological conditions, Sch9 and cAPK have different substrates, but increased activity of either kinase will suppress deficiencies in the other. In support of this model, Denis and Audino (1991) have found that Sch9 acts to regulate alcohol dehydrogenase expression in a pathway separate from cAPK.

Mutations in the Snf1 protein kinase also affect *ras* and cAPK mutants. *SNF1/CCR1/CAT1* (Ciriacy, 1977; Zimmermann *et al.*, 1977; Carlson *et al.*, 1981) encodes a protein kinase (Celenza and Carlson, 1986) well known for its role in glucose repression, but *snf1* mutants have a pleiotropic phenotype that is similar to the *bcy1* or *RAS2*$^{Val\text{-}19}$ phenotype. *snf1* and *bcy1* mutants have a defect in storage carbohydrate metabolism, sporulation, and response to nutritional stress. Mutations in *RAS2* or adenylyl cyclase moderate some *snf1*-associated traits; *bcy1* and *RAS2*$^{Val\text{-}19}$ mutations accentuate the *snf1*$^{-}$

phenotype (Thompson-Jaeger *et al.*, 1991). *SNF4/CAT3* (Entian and Zimmermann, 1982; Neigborn and Carlson, 1984) encodes a protein that physically interacts with Snfl (Celenza and Carlson, 1989; Fields and Song, 1989). Increased dosage of *PDE2* and *MSI1*, which partially alleviate defects caused by high levels of cAMP, partially suppresses the *snf4*$^-$ phenotype (Hubbard *et al.*, 1992). Collectively, these studies suggest that the cAPK and Snfl protein kinases act antagonistically on the same physiological responses. Although down-modulation of the cAPK pathway clearly suppresses some of the *snfl/snf4*-associated traits such as sporulation, glycogen accumulation, and carbon source utilization, it has no effect on glucose repression (Hubbard *et al.*, 1992). The role of the Ras pathway in glucose repression will be addressed in Section VII,A.

B. Yeast Adenylyl Cyclase Is Responsive to Ras

Yeast adenylyl cyclase is a peripheral membrane protein (Mitts *et al.*, 1990) with an estimated native molecular mass between 450 kDa (Varimo and Londesborough, 1976) and 594 kDa (Heideman *et al.*, 1987). These high molecular mass estimates, relative to the 220 kDa predicted from the *CYR1* open reading frame (Kataoka *et al.*, 1985a), indicate that the native molecule is likely to be a multimer. Casperson *et al.* (1983) first showed that adenylyl cyclase from *S. cerevisiae* is stimulated by GTP and its nonhydrolyzable analog Gpp(NH)p in the presence of Mg^{2+}-ATP, as is adenylyl cyclase from mammalian cells. Thermal inactivation of the Mg^{2+}-ATP-dependent activity implicates the presence of a distinct regulatory subunit for the cyclase. Subsequent studies from the Wigler laboratory revealed that Ras is the regulatory protein responsible for the GTP-dependent adenylyl cyclase activity. *ras1*$^-$ *ras2*$^-$ cells lack GTP-dependent adenylyl cyclase activity, whereas cells containing the dominant *RAS*$^{\text{Val-19}}$ mutation show Mg^{2+}-ATP-dependent cyclase activity in the absence of added GTP (Toda *et al.*, 1985). Subsequently, adenylyl cyclase activity in membranes prepared from Ras deficient yeast cells was shown to be activated by the addition of either mammalian or yeast Ras proteins synthesized in *E. coli* (Broek *et al.*, 1985; Field *et al.*, 1987).

The catalytic domain of the *CYR1* gene product occupies only the C-terminal 400 amino acid residues of the 2026-amino-acid protein (Kataoka *et al.*, 1985a). A truncated protein containing only the 416 C-terminal residues has basal adenylyl cyclase activity when expressed in *E. coli*. The expression of this domain in yeast results in Ras-independent activity (Kataoka *et al.*, 1985a). A point mutation in this domain has been identified that also results in Ras-independent activity (De Vendittis *et al.*, 1986).

Sequence homology between adenylyl cyclases from yeast and bovine brain (Krupinski *et al.*, 1989) is confined to the C-terminal catalytic domain of

Cyr1. The N-terminal 80% of Cyr1 does not contain membrane-spanning domains as are found in the mammalian cyclase, but residues 711–1279 of the yeast cyclase consist of roughly 26 tandem copies of a leucine-rich 23-amino-acid repeat. This repeated structure is similar to the 24-amino-acid leucine-rich repeats found in a widely divergent group of proteins, including the *Drosophila Toll* (Hashimoto *et al.*, 1988) and *Chaoptin* (Reinke *et al.*, 1988) gene products, the human serum Ib (Lopez *et al.*, 1987,1988; Titani *et al.*, 1987) and α_2 (Takahashi *et al.*, 1985) glycoproteins, and the lutropin–choriogonadotropin receptor (McFarland *et al.*, 1989). Studies on glycoprotein Ib and the lutropin–choriogonadotropin receptor suggest that the leucine-rich repeat is necessary for protein–protein interactions (Titani *et al.*, 1987; McFarland *et al.*, 1989). The leucine-rich repeats in Cyr1 are necessary for Ras responsiveness. Deletion mutations that remove all or part of the first 657 amino acids of Cyr1 retain Ras responsive activity, but deletions within the leucine-rich repeated structure or insertions of only two or three amino acids within the repeat domain eliminate Ras responsive activity (Colicelli *et al.*, 1990; Suzuki *et al.*, 1990). The C-terminal 120 amino acids of Cyr1 are required for Ras responsive cyclase activity but not for basal activity (Suzuki *et al.*, 1990). The importance of both the leucine-rich repeat region and the extreme C terminus for Ras responsiveness also is underscored by the properties of chimeric adenylyl cyclase proteins containing the catalytic domain from the fission yeast *Schizosaccharomyces pombe* and the regulatory domains from *S. cerevisiae*. Only chimeric proteins with both the N- and C-terminal domains from *S. cerevisiae* were stimulated by Ras proteins (Suzuki *et al.*, 1990), although the *S. pombe* adenylyl cyclase contains similar leucine-rich repeats (Yamawaki-Kataoka *et al.*, 1989).

Additional genetic evidence suggests that the leucine-rich repeat plays an important role in Ras activation. Field *et al.* (1990a) discovered that noncatalytically active forms of adenylyl cyclase could suppress the heat-shock sensitive phenotype of *RAS2*$^{\text{Val-19}}$. The leucine-rich repeat domain of adenylyl cyclase was the only portion that was required for suppression. The simplest interpretation of these data is that the leucine-rich repeat region of adenylyl cyclase can form a nonproductive complex with Ras. Alternatively, the leucine-rich repeats may interact nonproductively with adenylyl cyclase. This alternative hypothesis is less likely, however, since increasing the dosage of *RAS2*$^{\text{Val-19}}$ eliminates the suppression resulting from expression of the leucine-rich repeats.

Mutant forms of Ras that contain amino acid substitutions in the "effector" domain (amino acids 32–40 of Ha-Ras) have a nontransforming phenotype in mammalian systems and fail to stimulate adenylyl cyclase in yeast (Sigal *et al.*, 1986b). This result is consistent with the hypothesis that adenylyl cyclase is the effector for Ras in *S. cerevisiae*. An Asp-1547 to Tyr change in adenylyl cyclase, which lies between the catalytic domain and the leucine-rich region, results in an adenylyl cyclase that suppresses a defect in the effector region of Ras2. This mutant still requires Ras for Mg^{2+}-ATP-dependent cyclase activity, but can be stimulated by mutant Ras proteins that cannot stimulate the wild-type adenylyl cyclase (Marshall *et al.*, 1988).

Together, the studies described here demonstrate the existence of distinct regulatory domains of adenylyl cyclase that are required for Ras responsiveness. The observations that point mutations or deletions in *CYR1* can allow yeast cells to grow in the absence of Ras suggests that the regulatory domain of adenylyl cyclase is normally inhibitory in the absence of Ras stimulation. The same hypothesis has been proposed from the results of protease treatment of adenylyl cyclase, in which mild trypsin treatment results in loss of Ras-stimulated cyclase activity but increases the level of basal activity (Heideman *et al.*, 1987).

C. Is Srv2/Cap an Intermediate in the Ras–Adenylyl Cyclase Pathway?

No evidence exists that Ras activates adenylyl cyclase in any species other than *S. cerevisiae* (Beckner *et al.*, 1985; Birchmeier *et al.*, 1985), although Ras proteins from different species appear to function interchangeably. The simplest hypothesis that would resolve this paradox is that Ras does not activate adenylyl cyclase directly but does so via an intermediate(s). This intermediate could be conserved structurally among all eukaryotes but act in different biochemical pathways. All the genetic and biochemical results discussed in this chapter could be interpreted with an intermediate between Ras and adenylyl cyclase instead of a direct Ras–adenylyl cyclase interaction. Unequivocal genetic evidence for a direct protein–protein interaction, such as the isolation of allele-specific pseudorevertants, has not been obtained. Further, although Uno *et al.* (1985) reported that Ras-responsive adenylyl cyclase activity could be reconstituted from Cyr1 and Ras expressed in *E. coli*, this result has not been repeated by others (Kataoka *et al.*, 1985a) and the *CYR1* gene used for expression lacked the N-terminal portion of adenylyl cyclase that is known to be required for Ras responsiveness *in vivo*.

One potential candidate for an intermediate between adenylyl cyclase and Ras is a 70-kDa protein that copurifies with yeast adenylyl cyclase, referred to as CAP (Field *et al.*, 1990b). The *CAP* gene was found to be the same as *SUPC* (Field *et al.*, 1990b) and *SRV2* (Fedor-Chaiken *et al.*, 1990), which were identified independently in screens for suppressors of the heat-shock sensitive phenotype of *RAS2*$^{\text{Val-19}}$. *srv2-1* also suppressed the heat-shock sensitivity of a phosphodiesterase deficient strain and a strain containing a constitutively active adenylyl cyclase, but did not suppress the heat-shock sensitivity of a *bcy1* mutant (Fedor-Chaiken *et al.*, 1990). A gene disruption of *CAP/SRV2* also was able to suppress the heat-shock sensitivity of *RAS2*$^{\text{Val-19}}$ (Fedor-Chaiken *et al.*,

1990; Field *et al.*, 1990b). Ras was not able to activate adenylyl cyclase prepared from *cap*$^-$ strains (Field *et al.*, 1990b). Thus, genetic and biochemical evidence suggests that Cap/Srv2 associates with adenylyl cyclase and is required for full Ras responsiveness.

CAP/SRV2 encodes a bifunctional protein. *cap* mutants that lack the N-terminal half of the polypeptide suppress *RAS2*$^{Val\text{-}19}$, and adenylyl cyclase activity in these strains is attenuated. However, *cap* mutants that are missing the C-terminal half do not suppress *RAS2*$^{Val\text{-}19}$ but have a pleiotropic phenotype unrelated to the Ras pathway (Gerst *et al.*, 1991). C-terminal deletion mutants are unable to grow on rich media and are very sensitive to amino acid imbalances. Cell volumes are increased, actin distribution is disrupted, and the site of budding on the cell surface is altered. This phenotype is similar to that of a number of mutations in genes that encode actin binding proteins. A functional relationship between Cap and actin is confirmed by the observation that overexpression of a gene encoding profilin, an actin binding protein, partially suppress the *cap*$^-$ phenotype (Vojtek *et al.*, 1991).

Although Cap/Srv2 clearly is required for normal adenylyl cyclase activity, it is not an obligatory component of an adenylyl cyclase complex. Whereas Ras-deficient strains fail to grow because adenylyl cyclase is inactive, *cap*/srv2 deletion strains are viable (Fedor-Chaiken *et al.*, 1990; Field *et al.*, 1990b). Although *srv2* mutants are able to suppress the *RAS2*$^{Val\text{-}19}$ phenotype, they result in no phenotypic change in a wild-type Ras background (Fedor-Chaiken *et al.*, 1990). *srv2* also suppresses the heat-shock sensitivity of a *cyr1::his3* mutation that bears a deletion of part of the leucine-rich repeated region of adenylyl cyclase and has high levels of Ras-independent activity *in vivo* (Fedor-Chaiken *et al.*,1990). *srv2* would not be expected suppress *CYR1::HIS3* if the gene product only acted in cooperation with Ras. Thus Cap/Srv2 appears to have a general but not essential role in activating adenylyl cyclase. Its role as the mediator between Ras and adenylyl cyclase remains uncertain.

D. Does Ras Have Other Effectors?

Investigations in metazoans suggest that Ras may act in multiple signal transduction pathways. Is this role also possible in yeast? Although Ras clearly is required to activate adenylyl cyclase, several observations support the possibility of adenylyl cyclase-independent roles. First, in some strain backgrounds in which the *ras1 ras2* or the *cdc25* (see Section V,A) genotype is lethal and results in rapid loss of cell growth and division, *cyr1/cdc35*$^-$ strains are still able to grow, albeit very slowly (Wigler *et al.*, 1988). One interpretation of this result is that Ras and its activator Cdc25 may be absolutely essential because they are required for the activation of adenylyl cyclase and another function, whereas adenylyl cyclase mutants remain viable because this other function remains. Another example of a Ras function that is unrelated to adenylyl cyclase activity comes from the genetic analysis by Michaeli *et al.* (1989). High levels of expression of mutant forms of Ha-Ras that are unable to be processed properly because of mutations in the C-terminal processing site can suppress the heat-shock sensitivity of *RAS2*$^{\text{Val-19}}$. This suppression can be reversed by increased expression of *RAS2*$^{\text{Val-19}}$ but not by increased expression of adenylyl cyclase. cAMP levels and adenylyl cyclase activities are not appreciably lower in strains that contain these defective interfering Ha-*ras* genes, which suggests that the mechanism does not involve adenylyl cyclase. Michaeli and colleagues propose that the interfering *ras* genes are inhibiting an adenylyl cyclase-independent pathway that results in heat-shock sensitivity. However, the basis for the heat-shock sensitive phenotype for either the adenylate cyclase-dependent or -independent pathway is completely undefined. As yet, the evidence for additional pathways for Ras is suggestive but inconclusive.

V. Regulation of Ras by Controlling the Bound GDP–GTP Ratio

The transduction of a signal through a G protein-coupled pathway is accompanied by a cycle of GDP–GTP exchange followed by GTP hydrolysis. In many systems, the activation of a receptor induces GDP–GTP exchange on the coupled G protein, resulting in the accumulation of the GTP-bound form. Subsequent GTP hydrolysis returns the system to the inactive state. If Ras has a similar regulatory role in one or more signal transduction pathways, either the guanine nucleotide exchange or the GTP hydrolysis of Ras should be regulated by additional proteins (Fig. 7.3). In principle, the identification of such regulatory factors could help identify the elusive Ras-responsive signaling pathways.

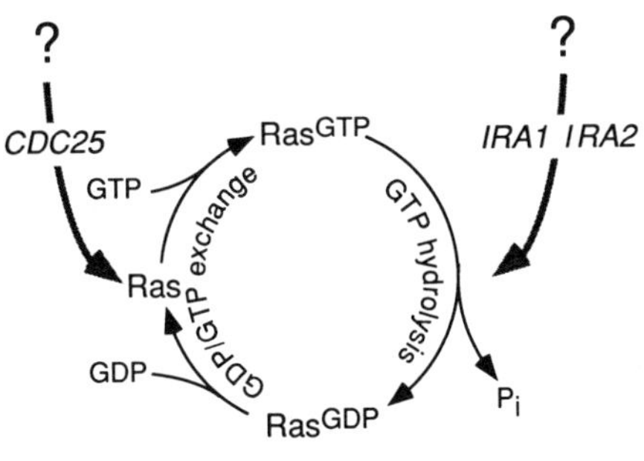

Figure 7.3 Regulation of guanine nucleotide binding and GTPase activity on Ras. The arrows represent possible input signals. A description of this process is provided in the text.

A. Cdc25 and Guanine Nucleotide Exchange

The *CDC25* locus originally was identified by temperature-sensitive mutations that arrested at the restrictive temperature at start II, at the same point as temperature-sensitive alleles of *cdc35/cyr1* (Pringle and Hartwell, 1981). This similarity prompted a number of groups to investigate the relationship between Cdc25 and the Ras–adenylyl cyclase pathway. The first indications that Cdc25 acts in the Ras pathway were observations that the temperature sensitivity of *cdc25* mutations could be suppressed by the same mutations that suppressed *ras*$^{-}$. For example, increased expression of the catalytic subunit of cAPK, TPK1, suppresses both *ras* and *cdc25* (Broek *et al.*, 1987). *bcy1* mutants suppress the lethality of *cdc25* (L. Robinson and K. Tatchell, unpublished observations) as they do for *ras1, ras2* and *cyr1*. Many *cdc25* mutants have reduced intracellular levels of cAMP (Camonis *et al.*, 1986; Broek *et al.*, 1987) and low levels of GTP-stimulated adenylyl cyclase (Broek *et al.*, 1987; Daniel *et al.*, 1987; Marshall *et al.*, 1987; Robinson *et al.*, 1987; Engelberg *et al.*, 1990; Petitjean *et al.*, 1990).

Key experiments placing Cdc25 within the Ras–adenylyl cyclase pathway were those showing that dominant *ras* mutants could eliminate the requirement for Cdc25. The temperature sensitivity of *cdc25*^{ts} alleles and the lethality of *cdc25* null alleles can be suppressed by *RAS2*$^{Val\text{-}19}$ (Broek *et al.*, 1987; Robinson *et al.*, 1987) and by *RAS2*$^{Ile\text{-}152}$ (Camonis and Jacquet, 1988), a mutation whose product is predicted to have decreased affinity for GTP (Feig *et al.*, 1986). These data are consistent with a model in which the *CDC25* product has a regulatory role in guanine nucleotide exchange. In the absence of stimulated GDP–GTP exchange, normal Ras would accumulate in the inactive GDP form, resulting in a reduction in the activity of adenylyl cyclase. However, mutant forms of Ras that lack normal GTPase activity, such as *RAS2*$^{Val\text{-}19}$, remain in the GTP-bound state and would not require Cdc25. Supporting evidence for this model is presented by measurements of the guanine nucleotide bound to Ras *in vivo*. Wild-type Ras is mostly bound to GDP whereas Ras$^{Val\text{-}19}$ is bound mostly to GTP; the increase in the amount of GTP-bound Ras in various mutants is correlated with the ability of the mutant to suppress *cdc25*, (Gibbs *et al.*, 1987).

Additional evidence indicating an interaction between Ras and Cdc25 is put forth by an investigation of the effects of dominant interfering *RAS2* mutation *RAS2*$^{Val\text{-}19Ala\text{-}22}$ (Powers *et al.*, 1989). Strains containing the *RAS2*$^{Val\text{-}19Ala\text{-}22}$ mutation are temperature sensitive for growth, even in the presence of a wild-type *RAS* gene. Increased expression of *CDC25* was found to suppress this temperature sensitivity, but only in the presence of a wild-type *RAS* gene. Powers *et al.* (1989) proposed that the dominant negative Ras2 protein binds irreversibly to Cdc25 at the restrictive temperature and that this interference can be overcome only by increasing the concentration of Cdc25. Because *RAS2*$^{\text{Val-19Ala-22}}$ does not bind guanine nucleotide and is nonfunctional at the restrictive temperature, the suppressing effects of CDC25 are only possible in the presence of wild-type Ras. As proposed by Powers *et al.* (1989), irreversible binding between Ras2 and Cdc25 may be a consequence of a high affinity of Cdc25 for nucleotide-free Ras.

The model for Cdc25 function that has arisen from these studies is that Cdc25 acts as a GDP–GTP exchange factor for Ras. This effect would be similar to the role of elongation factor Ts for the GTP-binding factor Tu (Hughes, 1983), the activity of membrane receptors on hormone-responsive G proteins (Gilman, 1984), or the activity of rhodopsin on the G protein transducin (Stryer, 1986). Biochemical data now strongly support this model. The C-terminal half of Sdc25, a protein that shares homology with the *CDC25* gene product and complements *cdc25 in vivo* (Boy-Marcotte *et al.*, 1989) was synthesized in *E. coli* and shown to have GDP–GTP exchange activity on Ras *in vitro* (Créchet *et al.*, 1990). Subsequently, Jones *et al.* (1991) showed that immunoprecipitates that contain a Cdc25–β galactosidase fusion protein stimulate GDP–GTP exchange on Ras *in vitro.* Munder and Fürst (1992) used a fusion protein system similar to the one pioneered by Fields and Song (1989) to show that Ras1 and Ras2 can interact directly with the C terminus of Cdc25 and Scd25 *in vivo.* In their assay, the DNA-binding domain of Ace1, a transcriptional activator of the metallothionein gene, was fused to Cdc25 and the transcription activation domain of Ace1 was fused to Ras. An interaction between the two fusion proteins was assayed by induction of an Ace1 responsive gene. Using this assay system, these investigators showed that the C-terminal half of Cdc25 can bind to wild-type Ras2 but not to Ras2$^{\text{Val-19}}$. An interaction was not observed under conditions in which wild-type Ras was mostly in the GTP-bound form. The authors concluded that Cdc25 can interact productively with Ras bound to GDP but not with Ras bound to GTP.

CDC25 encodes a 180-kDa protein (Camonis *et al.*, 1986; Broek *et al.*, 1987); as in the case for adenylyl cyclase, the catalytic portion of the molecule is located in the C-terminal half of the protein. Expression of only the C terminus provides Cdc25 function *in vivo* (Camonis *et al.*, 1986; Daniel, 1986; Martegani *et al.*, 1986; Broek *et al.*, 1987; Munder *et al.*, 1988) and *cdc25*^{ts} alleles are located in this region (Petitjean *et al.*, 1990). As with Ras and adenylyl cyclase, Cdc25 appears to be membrane associated. This protein has been detected immunologically in membranes (Garreau *et al.*, 1990; Vanoni *et al.*, 1990) or partitioned between cytoplasmic and membrane fractions (Jones *et al.*, 1991) from strains that overexpress Cdc25. The association of Ras, adenylyl cyclase, and Cdc25 with membrane fractions, and the physical association between Ras and Cdc25, suggests that components of the pathway may be associated as a large complex.

CDC25 is a member of an evolutionarily conserved gene family. The products of the *SCD25* (Boy-Marcotte *et al.*, 1989; Damak *et al.*, 1991), *LTE1* (Wickner *et al.*, 1987), and *BUD5* (Chant *et al.*, 1991; Powers *et al.*, 1991) genes from *S. cerevisiae* share homology with the C-terminal portion of Cdc25. The homology is most extensive between Cdc25 and the C-terminal half of Scd25, which can complement a temperature-sensitive *cdc25* mutation (Boy-Marcotte *et al.*, 1989). Surprisingly, expression of the entire *SCD25* gene fails to complement *cdc25* (Damak *et al.*, 1991), indicating that the N terminus contains a domain that is inhibitory to this function. *BUD5* cannot complement *cdc25*, but overexpression of *BUD5* or *CDC25* suppresses the dominant negative *RAS2*$^{\text{val-19Ala-22}}$ mutation (Powers *et al.*, 1991). *BUD5* is one of five genes required for the selection of the proper site of bud initiation (Chant and Herskowitz, 1991; Chant *et al.*, 1991). *BUD1* (also called *RSR1*; Bender and Pringle, 1989) and *CDC42*, which is required for bud initiation (Johnson and Pringle, 1990), both encode Ras-related proteins. An attractive hypothesis based on these relationships is that Bud5 acts as a nucleotide exchange factor for Rsr1/Bud1 or Cdc42.

Cdc25 homologs also have been identified in other organisms. The *ste6* product from *S. pombe* shows amino acid similarity to Cdc25 and may regulate the *S. pombe* Ras1 protein (Hughes *et al.*, 1990). Son of sevenless, a component of the sevenless tyrosine kinase receptor pathway in *Drosophila*, shares homology with Cdc25 (Simon *et al.*, 1991; Bonfini *et al.*, 1992). Factors from mammalian cells have been identified that enhance Ras GDP–GTP exchange *in vitro* (West *et al.*, 1990; Wolfman and Macara, 1990), and mammalian *CDC25*-related genes have been cloned (D. Broek, personal communication; E. Martigani, personal communication). A mammalian exchange factor also has been identified that acts on the Ras-related Rap1/Smg1 protein (Yamamoto *et al.*, 1990). Enhancement or modulation of guanine nucleotide exchange therefore may be a regulatory mechanism pertinent to many Ras-related proteins.

B. Ira1 and Ira2, GTPase Activating Proteins for Ras

Trahey and McCormick (1987) first identified a GTPase activating protein (GAP) in mammalian cells but initial attempts to identify GAP activity in yeast were unsuccessful (Adari *et al.*, 1988). Subsequent investigations revealed that yeast does contain two proteins (Ira1 and Ira2) that may regulate the GTPase activity of Ras. *IRA1* originally was identified as *ppd1*, a mutant with decreased cell viability that was said to be defective in phosphoprotein phosphatase activity (Matsumoto *et al.*, 1985); further analysis revealed that the mutation actually had no effect on phosphoprotein phosphatase activity but did influence the Ras–adenylyl cyclase pathway. The mnemonic for *PPD1* was changed to *IRA1* for inhibitory regulator of Ras (Tanaka *et al.*, 1989).

Gene disruption of *IRA1* results in a heat-shock sensitive phenotype and increases intracellular cAMP levels. This phenotype is similar to the *RAS2*$^{\text{Val-19}}$ phenotype (Tanaka *et al.*, 1989). *IRA2* was identified as a high-dosage suppressor of *IRA1* and its product was shown to be 45% identical to Ira1 (Tanaka *et al.*, 1990b). *IRA1* and *IRA2* appear to have an overlapping function, much like *RAS1* and *RAS2*. Single gene disruptions of *IRA1* and *IRA2* result in a heat-shock sensitive phenotype but double mutants are affected more severely (Tanaka *et al.*, 1990b). The requirement for Cdc25 also is bypassed partially by *ira1* and *ira2* mutations (Tanaka *et al.*, 1989,1990b).

Given the homology between Ira1, Ira2, and GAP, an obvious hypothesis is that the *IRA* gene products stimulate the intrinsic GTPase activity of Ras. The observation that the ratio of GDP–GTP bound to Ras *in vivo* is increased in *ira1* and *ira2* mutants is consistent with this hypothesis. (Tanaka *et al.*, 1990a). The effect of *ira1* and *ira2* mutations is additive; the GDP–GTP ratio is higher in an *ira1 ira2* double mutant than in the single mutants (Tanaka *et al.*, 1990a). *ira1* and *ira2* mutations do not alter the GDP–GTP ratio in the dominant *RAS2*$^{\text{Val-19}}$ mutant (Tanaka *et al.*, 1990a), consistent with the observations that mammalian GAP does not increase the GTPase activity of oncogenic Ras mutants.

A region encoding ~330 amino acid residues in the middle of the *IRA1* and *IRA2* open reading frames is 22% and 45% identical to the catalytic domain of mammalian GAP (Tanaka *et al.*, 1990a). More extensive homology is found between Ira1 and Ira2 and the neurofibromatosis type 1 (NF1) gene product. Whereas the similarity of Ira1 and Ira2 to GAP is confined to 330 amino acids in the central catalytic domain, the homology between Ira and NF1 is more extensive. Blocks of homology extend over 1700 amino acid residues (Ballester *et al.*, 1990; Xu *et al.*, 1990). The catalytic domains of both GAP and NF1 have been expressed in yeast; each can complement *ira1* and *ira2* mutations (Ballester *et al.*, 1989,1990; Martin *et al.*, 1990; Tanaka *et al.*, 1990a; Xu *et al.*, 1990). Finally, Ira2 has been shown directly to have GAP activity on Ras *in vitro* (Tanaka *et al.*, 1991). Although GAP has been argued to be the effector for Ras in mammalian systems (McCormick, 1989), the data from analysis in yeast are more consistent with Ira proteins acting simply as negative regulators of Ras. If Ira1 and Ira2 indeed function as Ras regulators, one would predict that *ras* mutants could be isolated that are not sensitive to the inhibitory activity of Ira1 and Ira2. Tanaka *et al.* (1992) have identified a dominant activating *RAS2* mutant (Pro-41 to Ser) that exhibits normal intrinsic GTPase activity but is refractory to stimulation by Ira2. Marshall *et al.* (1991) identified Ha-Ras mutants with similar properties.

GAP, *IRA1*, *IRA2*, and *NF1* encode large polypeptides of 1044, 2938, 3079, and at least 2485 amino acid residues, respectively (Xu *et al.*, 1990, and references therein), but their catalytic domains make up only a small fraction of their total lengths. What are the roles of the additional portions of each protein? GAP contains regions that are similar to the *src* homology domains SH2 and SH3 that may be required for signaling through receptor tyrosine kinases (Koch *et al.*, 1991). In Ira1 and Ira2 the additional sequences, which are not related to those in GAP, may provide specificity for interactions with yeast Ras. Ha-Ras expressed in yeast remains largely in the GTP-bound form *in vivo*, and its GTPase activity is not stimulated by Ira2 *in vitro* (Tanaka *et al.*, 1991). This specificity between a particular GAP and Ras may explain the earlier observation that yeast Ras2 proteins that lack the unique C-terminal region or wild-type Ha-Ras proteins confer a dominant activating phenotype in yeast. Yeast strains expressing these proteins no longer require Cdc25 activity and display reduced levels of glycogen and reduced sporulation efficiencies, traits normally associated with increased cAMP levels (Marshall *et al.*, 1987). This phenotype would be expected if Ira1 and Ira2 were unable to act on Ras proteins that do not have the polypeptide regions unique to yeast Ras.

C. Rpi1, an Additional Ras Regulator

Kim and Powers (1991) identified a high-dosage suppressor (Rpi1) of the heat-shock sensitivity caused by increased expression of adenylyl cyclase. Increased dosage of Rpi1 also suppresses the heat-shock sensitive phenotype resulting from *ira1* and *ira2* mutants or overexpression of a truncated form of Cdc25, but it does not suppress the heat-shock sensitive phenotype of $RAS2^{\text{Val-19}}$. These results indicate that Rpi1 requires a normal Ras protein to exert its effects. A deletion of *rpi1* displays a heat-shock sensitive phenotype and accumulates reduced levels of glycogen. This phenotype would be predicted if Rpi1 were to act as a negative regulator of Ras, but the exact role of Rpi1 is not clear at the present time. The *RPI1* gene product is not similar in sequence to any known protein. Rpi1 could act in conjunction with Cdc25, since an *rpi1* deletion mutation will suppress a temperature-sensitive *cdc25* mutation partially. Alternatively, Rpi1 could act as a positive regulator of Ira1 and Ira2, since increased dosage of Rpi1 suppresses *ira1* or *ira2* single mutants, but not an *ira1 ira2* double mutant. Rpi1 could also act independently of Cdc25, Ira1, or Ira2 to negatively regulate Ras. Factors have been identified in mammalian systems that inhibit the dissociation of GDP from Ras-related proteins (Sasaki *et al.*, 1990; Ueda *et al.*, 1990). Rpi1 could play an analogous role in yeast.

D. Additional Components of the Ras–Adenylyl Cyclase Pathway

Two additional genes, *TFS1* and *MSI1*, have been identified as high-dosage suppressors of mutations in the Ras pathway. No clear role can be attributed to these genes at present. Multicopy plasmids that carry the *MSI1* gene suppress the heat-shock sensitivity of *RAS2*$^{Val\text{-}19}$ and *ira1* (Ruggieri *et al.*, 1989), the low glycogen levels of *SRA4*, a dominant mutant in *CYR1* (François *et al.*, 1992), and depletion of Snf1 protein kinase activity (partially) (Hubbard *et al.*, 1992). The sequence of *MSI1* predicts a protein that shares some homology with the β subunit of trimeric G proteins, but a deletion mutation of *MSI1* does not result in any observable phenotypic change.

A similar situation exists with *TFS1/NSP1*, a high-dosage suppressor of *cdc25* mutants (Tripp *et al.*, 1989; Robinson and Tatchell, 1991). Multicopy plasmids that carry *TFS1* are able to suppress partially the temperature-sensitive phenotype of *cdc25-1*, but do not suppress a *cdc25* null mutation or any mutations in *RAS* or *CYR1*. The putative Tfs1 protein is similar to two low molecular weight proteins from bovine and rat brain (Schoentgen *et al.*, 1987; Grandy *et al.*, 1990). These proteins are over 80% identical and share similarity to Tfs1 over their entire length, with the exception of three 13- to 18-amino-acid insertions in Tfs1. Excluding the insertions, Tfs1 is 29% identical to the mammalian proteins. The bovine and rat brain proteins were identified as phosphatidylethanolamine- and 14β-bromoacetamidomorphine-binding proteins, respectively, but no biological function has been attributed to these proteins. The ability of Tfs1 to suppress only *cdc25* mutants argues that Tfs1 may have a role in guanine nucleotide exchange. However a deletion of *TFS1*, as in the case of *MSI1*, results in no phenotypic change. Perhaps *TFS1* duplicates another gene since *TFS1* hybridizes to multiple bands in a Southern blot of genomic yeast DNA (Robinson and Tatchell, 1991).

VI. Feedback Regulation in the Ras Pathway

As in most signal transduction pathways, the Ras–adenylyl cyclase system in *S. cerevisiae* is controlled by a feedback regulatory circuit. Evidence for feedback control manifests itself in the intracellular levels of cAMP found in strains with mutations in cAPK (Nikawa *et al.*, 1987a). Strains with abnormally high levels of cAPK activity, resulting from a loss of the regulatory subunit of cAPK (Bcy1), have intracellular cAMP levels at least 10-fold below wild-type levels. In contrast, strains with low levels of cAPK activity, resulting from attenuating mutations in the catalytic subunits of cAPK (see subsequent text), have cAMP levels at least 1000-fold above wild-type. These results indicate that cAPK activity can inhibit cAMP accumulation or deplete the cAMP pool. cAMP phosphodiesterases may be important for limiting cAMP accumulation under some conditions. For example, *RAS2*$^{Val\text{-}19}$ strains have high levels of cAMP only in a phosphodiesterase-deficient genetic background, but cAMP levels in a *pde1 pde2* double mutant are only several-fold above wild-type, and intracellular phosphodiesterase levels do not vary between strains that have over a 10,000-fold difference in cAMP levels.

Other likely targets for feedback regulation include adenylyl cyclase, Ras, and regulatory proteins that act on Ras. Adenylyl cyclase (Kataoka *et al.*, 1985a), Cdc25 (Camonis *et al.*, 1986; Broek *et al.*, 1987), Ira1 (Tanaka *et al.*, 1989), and Ira2 (Tanaka *et al.*, 1990) have potential phosphorylation sites for cAPK. Although no direct evidence exists for *in vivo* phosphorylation of any of these components by cAPK, Ras1 and Ras2 are phosphorylated *in vitro* (Cobitz *et al.*, 1989). The *in vitro* phosphorylation of Ras2 by cAPK reduces its ability to activate adenylyl cyclase (Resnick and Racker, 1988).

Another observation that could provide a mechanism for the cAMP feedback control is the varying distribution of adenylyl cyclase between membrane and soluble fractions (Engelberg *et al.*, 1990; Mitts *et al.*, 1990). Adenylyl cyclase normally is distributed between membrane and cytoplasmic fractions but most of the adenylyl cyclase activity in *bcy1* strains is cytoplasmic. This redistribution could be partially responsible for the low levels of cAMP found in *bcy1* strains.

Another potential mechanism for regulating the intracellular levels of cAMP is direct secretion of cAMP into the medium. Smith *et al.* (1990) observed high levels of cAMP in the medium of wild-type cells. The concentration of extracellular cAMP did not vary with the cell cycle or growth phase. Extracellular cAMP does not relieve the growth defects of adenylyl cyclase mutants in most strain backgrounds and no physiological role for extracellular cAMP is known (Matsumoto *et al.*, 1982b). Whether mutants in the cAMP pathway show differences in the ability to secrete cAMP remains to be determined.

VII. Signal Transduction and Ras

A. Does Ras Have a Role in Nutrient Assessment?

Genetic analysis in *S. cerevisiae* has led to the rapid identification of regulatory components in the Ras pathway. Mutants in the Ras–adenylyl cyclase pathway are unable to respond properly to nutrient limitation. This observation, as well as the arrest point of adenylyl cyclase mutants, has led to the hypothesis that the Ras pathway is involved in signaling nutrient availability. However, most of the traits associated with *RAS* mutations are not necessarily indicative of a role in signal transduction. For example, the terminal phenotype of *cyr1*^{ts} or *ras2*^{ts} mutants that have been arrested at the restrictive temperature is characterized by general metabolic quiescence; the same phenotype results from temperature-sensitive mutations in *cdc19*, which encodes pyruvate kinase (Fraenkel, 1982), and *cdc33*, which encodes the cap-binding translation initiation factor eIF-4E (Brenner *et al.*, 1988). Although cells that have arrested because of nutrient limitation most often arrest in start II, the same phenotype is seen within cells that have been arrested with low concentrations of protein synthesis inhibitors (Pringle and Hartwell, 1981). These results indicate that although Ras and adenylyl cyclase may be essential for general metabolism, they need not be required in cell signaling.

Preliminary observations with other mutants also suggest a role in nutrient signaling for the Ras–adenylyl cyclase pathway, but later experiments have not confirmed these primary observations. For example, *bcy1* mutants fail to arrest in G_1 when starved and lose viability rapidly. This phenotype has led to the hypothesis that this loss of viability is caused by the failure of cells with constitutive cAPK to respond properly to a starvation signal. However, *bcy1* cells actually undergo fewer rounds of division than do wild-type cells when transferred to media deficient in reduced nitrogen (cited in Broach and Deschenes, 1990; J. Cannon and K. Tatchell, unpublished observations). *bcy1* cells may not arrest in G_1 because they are unable to complete a cell cycle after nutrient deprivation as a result of their extremely low levels of nutrient reserves rather than from a failure to respond to nutrient deprivation.

Cameron *et al.* (1988) have provided strong evidence that the Ras–adenylyl cyclase pathway is not essential for the normal response to the limitation of many nutrients. Three genes—*TPK1*, *TPK2*, and *TPK3*—encode the catalytic subunit of cAPK (Toda *et al.*, 1987b). The products of these genes have similar and overlapping functions since cells need only one of the three for normal growth. The pleiotropic phenotype of *bcy1* strains also is observed when cells contain only a single functional catalytic subunit gene. Cameron *et al.* (1988) observed that the heat-shock sensitive phenotype of *bcy1* strains that contain only a single active catalytic subunit gene could be reverted by an attenuating mutation in the remaining functional *TPK* gene. The attenuating mutations, termed tpk^w (*wimp* allele), are located in the active site of the protein and result in a gene product with severely reduced protein kinase activity. Remarkably, strains that carry a null allele of *bcy1*, null alleles of two catalytic subunit genes, and an attenuating tpk^w allele in the remaining catalytic subunit gene behave much like wild-type strains with respect to starvation responses. These *bcy1* tpk^w strains accumulate glycogen and become heat-shock resistant as they enter stationary phase, and are resistant to starvation. Diploid tpk^w strains sporulate normally. As would be expected if Bcy1 were the only essential cAMP binding protein, the additional elimination of adenylyl cyclase, Ras, or Cdc25 activity does not affect the nutrient response in *bcy1* tpk^w strains drastically. The important conclusion from this work is that the Ras–cAMP pathway is not an essential component of the signaling pathway involved in nutrient limitation. The absolute level of cAPK activity may influence nutrient responses, but if the Ras–adenylyl cyclase pathway is involved in nutrient signaling, it is redundant. The failure of *RAS2*$^{\text{Val-19}}$ and *bcy1* strains to mount a normal starvation response could be the result of abnormally high levels of cAPK activity that overwhelm the normal response.

The *bcy1 tpk*^{w} strains have proven to be extremely important reagents for critically assessing the functions of Ras and cAMP. A good example of their value is seen in the controversy surrounding a role for Ras in glucose repression. Glucose repression is the term given to the phenomenon of repression by glucose of genes that are required for utilization of alternative carbon sources (see Carlson, 1987, for review). Some observations suggest a positive role of cAMP in glucose repression, but others suggest that the Ras–adenylyl cyclase pathway has no role in the response (see Hubbard *et al.*, 1992, for review). Against this backdrop of conflicting data are the clearly interpretable results using *bcy1 tpk*^{w} strains, which show that expression of invertase (Hubbard *et al.*, 1992) and alcohol dehydrogenase II (Denis and Audino, 1991) is regulated normally in *bcy1 tpk*^{w} strains. These results do not rule out the possibility that cAPK influences the level of expression of glucose repressed genes, but they eliminate the possibility that a cAMP "signal" is part of the normal glucose-repression response.

B. Glucose Induces cAMP Synthesis in Stationary Phase Cells

The only known physiological treatment that has been shown to stimulate cAMP synthesis is the addition of glucose to stationary phase cells (van der Platt and van Solingen, 1974). This addition causes a 4- to 50-fold increase in cAMP levels within 1–2 min. cAMP levels rapidly decay and usually reach a new steady-state level within 5–10 min of glucose addition. Induction is observed only in stationary phase cells or in cells grown on nonfermentable carbon sources, suggesting that some essential component of the pathway is subject to glucose repression (Argüelles *et al.*, 1990). Mechanisms involving membrane hyperpolarization (Mazon *et al.*, 1982) or intracellular acidification (Purwin *et al.*, 1986; Valle *et al.*, 1986) were proposed to account for this induction, but subsequent studies have shown that cAMP induction and acidification are not strictly correlated and that cAMP induction occurs in the absence of acidification (Eraso *et al.*, 1987; Thevelein *et al.*, 1987). However, the Ras pathway is required for this response. Neither *ras1 ras2* (Mbonyi *et al.*, 1988) nor *cdc25* (Tanaka *et al.*, 1989; Van Aelst *et al.*, 1991) mutants show this induction. Munder and Küntzel (1989) observed that mutants with in-frame deletions of the N terminus of Cdc25 are viable and must, therefore, be able to activate Ras, but these mutants lack the glucose-induced cAMP response. This result argues that Cdc25 plays a key role in the signaling process. Rpi1 also is required for the induction. Strains with elevated dosage of the *RPI1* gene show an attenuated response to glucose whereas *rpi1* deletion mutants show a slightly higher induction of cAMP (Kim and Powers, 1991).

The transient cAMP signal is subject to feedback regulation. This effect is observed most dramatically in *bcy1 tpk*^{w} strains, which have low constitutive levels of cAPK activity. The addition of glucose to these cultures results in a rapid accumulation of cAMP that increases over 1000-fold above the basal level (Nikawa *et al.*, 1987a). In contrast, a *RAS2*$^{Val\text{-}19}$ mutant does not show induction of cAMP levels (Mbonyi *et al.*, 1988; Kim and Powers, 1991), indicating that the response pathway in a *RAS2*$^{Val\text{-}19}$ strain may be down-regulated and uninducible by glucose. Cdc25, Ira1, and Ira2 may have roles in the feedback response. A mutant in which the C-terminal 38 amino acids of Cdc25 have been deleted exhibits the transient cAMP increase in response to glucose but fails to show a rapid cAMP decay. *ira1* and *ira2* mutants also exhibit the cAMP induction without the normal decay (Tanaka *et al.*, 1989,1990b).

The glucose metabolite responsible for the induction is not known. One of three kinases—hexokinase 1, hexokinase 2, or glucokinase—has been reported to be required for induction of the cAMP pulse by glucose; either of the hexokinases is required for induction by fructose (Beullens *et al.*, 1988). These results suggest that a sugar–phosphate may be the induction signal. No modification beyond phosphorylation is required, since a phosphoglucose isomerase mutant exhibits the normal cAMP pulse. However, the high affinity glucose transporter encoded by *SNF3* is not required for the response (Mbonyi *et al.*, 1988). The dependence of cAMP induction on the external concentration of glucose has a K_m of about 25 m*M* (Beullens *et al.*, 1988), which is similar to the K_m for the low affinity glucose transport system. As yet, the connection between glucose transport and the Ras–adenylyl cyclase complex has not been elucidated.

The glucose-stimulated cAMP signal is observed in stationary phase cells or in cells grown on nonfermentable carbon sources. Most log phase cultures do not exhibit the response, which argues that one or more components of the signaling pathway are glucose repressed and therefore absent in log phase cultures (Argüelles *et al.*, 1990). cAPK is also more abundant in stationary phase cultures. Expression of the regulatory subunit gene *BCY1* is induced as cells enter stationary phase (Werner-Washburn *et al.*, 1991) and cAMP-dependent protein kinase activity also increases in cells entering stationary phase (J. Gaughran and K. Tatchell, unpublished observations). cAPK activity remains high for at least 4 days after cells have reached stationary phase. These observations are consistent with the hypothesis that cAPK has an important signaling role during recovery from stationary phase. The rapid rise in cAMP and, presumably, in cAMP-dependent protein kinase activity that occurs after the addition of glucose or fresh media might provide a "jump start" for a stationary phase culture.

Those cells that could resume growth rapidly would have a clear selective advantage over those that had a long lag phase.

One important prediction of the hypothesis that the RAS–adenylyl cyclase pathway is involved in outgrowth is that cells with altered components of the pathway exhibit a longer lag phase, as has been observed for *cdc25-1*. Stationary phase *cdc25-1* cells have a significantly longer lag phase at permissive temperatures than do the isogenic wild-type strain, although the doubling times in exponential phase are identical (A. Petitjean and K. Tatchell, unpublished observations). *tpk*^{w} strains also exhibit a longer lag phase than do isogenic wild-type (J. Gaughran and K. Tatchell, unpublished observations). Although these observations need to be extended, they clearly point to a role for the cAMP signaling system in outgrowth and recovery from stationary phase.

Some evidence suggests that glucose may act independently of cAPK. Granot and Snyder (1991) have observed that preincubation of stationary phase cells in glucose prior to inoculation into rich media reduces the lag phase. The same reduction is observed in both *cdc25* and *cyr1/cdc35* mutants preincubated at the restrictive temperature. These results raise the possibility that both cAMP-dependent and -independent pathways are necessary for the response, again pointing to the possibility that the regulatory pathways may be overlapping and redundant.

C. Downstream Components in the Pathway—Substrates for cAMP-Dependent Protein Kinase

If the glucose-induced signaling pathway is relevant to outgrowth, cAPK-dependent phosphorylation would be expected to stimulate storage carbohydrate utilization and glycolysis and to inhibit gluconeogenesis. Key enzymes in these metabolic pathways are regulated by the cAMP-dependent glucose signaling pathway. Trehalase, the enzyme responsible for the mobilization of the storage carbohydrate trehalose, is activated *in vivo* by conditions that stimulate cAPK (Thevelein and Beullens, 1985) and *in vitro* by cAPK itself (Dellamora-Ortiz *et al.*, 1986). Phosphofructokinase 2, which catalyzes the synthesis of the key activator of glycolysis (fructose 2,6-bisphosphate), also is activated by cAPK (Kretschmer *et al.*, 1987; François *et al.*, 1988). The activation of these enzymes and others (reviewed by Broach and Deschenes, 1990) suggests a clear physiological role for the glucose signaling pathway. However, in most cases, the existing *in vivo* data are only a correlation between the level of cAPK activity and a biochemical response. The data do not allow the distinction of a direct phosphorylation event by cAPK from an indirect effect. The cases of glycogen synthase and the transcriptional regulator Adr1, reviewed next, illustrate the difficulty in distinguishing between direct and indirect roles for cAPK.

Genetic and biochemical evidence strongly suggests that glycogen synthase is inactivated by phosphorylation by cAPK. Glycogen is synthesized in response to nutrient deprivation (Lillie and Pringle, 1980). Glycogen levels are reduced in strains with increased cAPK activity and elevated in mutants with reduced cAPK activity (Fraenkel, 1985; Tatchell *et al.*, 1985; Toda *et al.*, 1985; Cannon *et al.*, 1986; Nakajima *et al.*, 1987; Petitjean *et al.*, 1990). Glycogen synthase, the limiting enzyme required for the synthesis of glycogen, is phosphorylated and inactivated *in vitro* by cAPK (Peng *et al.*, 1990). Based on these data, a model for cAPK inactivation predicts that glycogen synthase isolated from strains with high levels of cAPK, such as *bcy1* strains, would be primarily phosphorylated and inactive. However, *bcy1* strains fail to synthesize glycogen synthase (François *et al.*, 1992), and preliminary evidence suggests that the defect lies at the transcriptional level. mRNAs for glycogen synthase, glycogen phosphorylase, and Gac1, a positive regulator of glycogen synthase, do not accumulate in *bcy1* strains (François *et al.*, 1992; T. Hardy and P. Roach, personal communication). In addition, a gene fusion between the upstream regulatory region of *GAC1* and *lacZ* is not expressed in *bcy1* strains (J. François and K. Tatchell, unpublished observations). These data suggest that the enzymes involved in glycogen metabolism are regulated coordinately at the level of transcription, and that cAPK activity modifies this transcriptional regulation. The expression of a number of genes is repressed by high levels of cAPK activity. The polyubiquitin-encoding gene *UBI4* (Tanaka *et al.*, 1988), the catalase-encoding gene *CTT1* (Bissinger *et al.*, 1989; Belazzi *et al.*, 1991), and the heat-shock genes *SSA3* (Werner-Washburne *et al.*, 1989; Boorstein and Craig, 1990) and *HSP12* (Praekelt and Meacock, 1990) all are repressed in cells with high levels of cAPK activity. In contrast to numerous examples in mammalian cells, no examples of gene expression that are influenced positively by cAPK have been reported in yeast.

The expression of alcohol dehydrogenase II, encoded by *ADH2*, also is influenced by cAPK activity. Strains with increased cAPK activity have reduced levels of *ADH2* expression (Cherry *et al.*, 1989). Adr1, the positive transcriptional activator of *ADH2*, can be phosphorylated *in vitro* by cAPK. Dominant constitutive mutations in *ADR1* (*ADR1*^{c}) cluster around a cAPK phosphorylation site at Ser-230. These data are consistent with a model in which Adr1 activity is regulated negatively by cAPK-dependent phosphorylation at Ser-230 (Cherry *et al.*, 1989). However, data from subsequent investigations do not support this hypothesis. The ability of Adr1 to bind DNA *in vitro* is not influenced by cAMP-dependent phosphorylation (Taylor and Young, 1990) and the phosphorylation of serine residue 230 does not vary with the activity of Adr1 *in vivo* (K. Dombek and E. T. Young, personal communication). Also, additional *ADR1*^{c} alleles have been isolated that contain mutations near Ser-230 that do not inhibit the *in vitro* phosphorylation by cAPK (Denis *et al.*, 1992).

Finally, an in-frame deletion in the *ADR1* gene between codons for amino acids 220–162 also results in a mutant with the constitutive phenotype (Denis *et al.*, 1992). In summary, cAPK appears to influence the activity of Adr1 but this influence is not simply created by phosphorylation on Ser-230. The examples of glycogen synthase and Adr1 illustrate the complexity of regulatory pathways and the danger of assigning a protein as a substrate for a kinase without a thorough analysis.

VIII. Summary

Saccharomyces cerevisiae has proven to be particularly valuable to investigations dealing directly with Ras and its immediate effectors and regulators. Studies with *S. cerevisiae* will continue to provide valuable information about the role of GDP–GTP exchange in Ras regulation and the mechanism of post-translational modification. Although the structures of Ras proteins show remarkable evolutionary conservation, the pathways that utilize Ras appear to be different in yeast and metazoans. Ras activates adenylyl cyclase in yeast, but appears to act in signal transduction pathways for protein tyrosine kinase receptors in *Drosophila*, *C. elegans*, and mammals. This paradox has not been resolved. Detailed knowledge of the effectors in both systems is required before conclusions concerning the degree of divergence of the signaling pathways can be drawn. Several large gaps in our knowledge of Ras remain. First, the receptor or upstream signaling pathway for Ras is still not identified. The glucose signaling pathway is a likely possibility but the exact mechanism is far from clear. Does glucose signal through a cell-surface receptor or through an intracellular metabolite? If a metabolite is used, why is the Ras signaling pathway associated with the membrane? A second gap in our knowledge concerns the mechanism of Ras activation of adenylyl cyclase. No physical or genetic evidence supports a direct interaction between Ras and adenylyl cyclase. The third unknown concerns downstream components of the pathway or substrates for cAPK. Almost nothing is known about the cAPK substrates involved in the regulation of heat shock, sporulation, or glycogen metabolism. Since more than one signaling system acts during any given physiological response, sorting out the different input signals will be very difficult until the substrates for the various protein kinases involved are characterized.

Acknowledgments

I thank Tim Petty and Lucy Robinson for critically reading the manuscript. This work was supported by Public Health Service grants CA37702 and GM47789.

References

Adari, H., Lowy, D. R., Willumsen, B. M., Der, C. J., and McCormick, F. (1988). Guanosine triphosphatase activating protein (GAP) interacts with the p21 *ras* effector binding domain. *Science* **240**, 518–521.

Argüelles, J. C., Mbonyi, M. K., Van Aelst, L., Vanhalewyn, M., Jans, A. W. H., and Thevelein, J. M. (1990). Absence of glucose-induced cAMP signaling in the *Saccharomyces cerevisiae* mutants *cat1* and *cat3* which are deficient in derepression of glucose-repressible proteins. *Arch. Microbiol.* **154**, 199–205.

Aroian, R. V., Koga, M., Mendel, J. E., Ohshima, Y., and Sternberg, P. W. (1990). The *let-23* gene necessary for *Caenorhabditis elegans* vulval induction encodes a tyrosine kinase of the EGF receptor subfamily. *Nature (London)* **348**, 693–699.

Ballester, R., Michaeli, T., Ferguson, K., Xu, H.-P., McCormick, F., and Wigler, M. (1989). Genetic analysis of mammalian GAP expressed in yeast. *Cell* **59**, 681–686.

Ballester, R., Marchuk, D., Boguski, M., Saulino, A., Letcher, R., Wigler, M., and Collins, F. (1990). The *NF1* locus encodes a protein functionally related to mammalian GAP and yeast IRA proteins. *Cell* **63**, 851–859.

Barbacid, M. (1987). *ras* genes. *Ann. Rev. Biochem.* **56**, 779–827.

Beckner, S. K., Hattori, S., and Shih, T. (1985). The *ras* oncogene product p21 is not a regulatory component of adenylate cyclase. *Nature (London)* **317**, 71–72.

Beitel, G. J., Clark, S. G., and Horvitz, H. R. (1990). *Caenorhabditis elegans ras* gene *let-60* acts as a switch in the pathway of vulval induction. *Nature (London)* **348**, 503–509.

Belazzi, T., Wagner, A., Wieser, R., Schanz, M., Adam, G., Hartig, A., and Ruis, H. (1991). Negative regulation of transcription of the *Saccharomyces cerevisiae* catalase T (*CTT1*) gene by cAMP is mediated by a positive control element. *EMBO J.* **10**, 585–592.

Bender, A., and Pringle, J. R. (1989). Multicopy suppression of the *cdc24* budding defect in yeast by *CDC42* and three newly identified genes including the ras-related gene *RSR1*. *Proc. Natl. Acad. Sci. U.S.A.* **86**, 9976–9980.

Beullens, M., Mbonyi, K., Geerts, L., Glandines, D., Detremerie, K., Jans, A. W. H., and Thevelein, J. M. (1988). Studies on the mechanism of the glucose-induced cAMP signal in glycolysis and glucose repression mutants of the yeast Saccharomyces cerevisiae. *Eur. J. Biochem.* **172**, 227–231.

Birchmeier, C., Broek, D., and Wigler, M. (1985). RAS proteins can induce meiosis in *Xenopus* oocytes. *Cell* **43**, 615–621.

Bissinger, P. H., Wieser, R., Hamilton, B., and Ruis, H. (1989). Control of *Saccharomyces cerevisiae* catalase T gene (*CTT1*) expression by nutrient supply via the RAS-cyclic AMP pathway. *Mol. Cell. Biol.* **9**, 1309–1315.

Bonfini, L., Karlovich, C. A., Dasgupta, C., and Banerjee, U. (1992). The *Son of sevenless* gene product: A putative activator of Ras. *Science* **255**, 603–606.

Boorstein, W., and Craig, E. (1990). Regulation of a yeast *HSP70* gene by a cAMP responsive transcription control element. *EMBO J.* **9**, 2543–2553.

Bourne, H. R., Sanders, D. A., and McCormick, F. (1991). The GTPase superfamily: Conserved structure and molecular mechanism. *Nature (London)* **349**, 117–127.

Boy-Marcotte, E., Damak, F., Camonis, J., Garreau, H., and Jacquet, M. (1989). The C-terminal part of a gene partially homologous to *CDC25* gene suppresses the *cdc25-5* mutation in *Saccharomyces cerevisiae*. *Gene* **77**, 21–30.

Bradley, M. O., Kraynak, A. R., Storer, R. D., and Gibbs, J. B. (1986). Experimental metastasis in nude mice of NIH 3T3 cells containing various *ras* genes. *Proc. Natl. Acad. Sci. U.S.A.* **83**, 5277–5281.

Brenner, C., Nakayama, N., Goebl, M., Tanaka, K., Toh-E, A., and Matsumoto, K. (1988). *CDC33* encodes mRNA cap-binding protein eIF-4E of *Saccharomyces cerevisiae*. *Mol. Cell. Biol.* **8**, 3556–3559.

Breviario, D., Hinnebusch, A., Cannon, J., Tatchell, K., and Dhar, R. (1986). Carbon source regulation of *RAS1* expression in *Saccharomyces cerevisiae* and the phenotypes of *ras2*$^-$ cells. *Proc. Natl. Acad. Sci. U.S.A.* **83**, 4152–4156.

Breviario, D., Hinnebusch, A., and Dhar, R. (1988). Multiple regulatory mechanisms control the expression of the *RAS1* and *RAS2* genes of *Saccharomyces cerevisiae. EMBO J.* **7**, 1805–1813.

Broach, J. R. (1991). *RAS* genes in *Saccharomyces cerevisiae*: Signal transduction in search of a pathway. *Trends Genet.* **7**, 28–33.

Broach, J. R., and Deschenes, R. J. (1990). The function of *RAS* genes in *Saccharomyces cerevisiae. Adv. Cancer Res.* **54**, 79–139.

Broek, D., Samiy, N., Fasano, O., Fujiyama, A., Tamanoi, F., Northup, J., and Wigler, M. (1985). Differential activation of yeast adenylate cyclase by wild-type and mutant RAS proteins. *Cell* **41**, 763–769.

Broek, D., Toda, T., Michaeli, T., Levin, L., Birchmeier, C., Zoller, M., Powers, S., and Wigler, M. (1987). The *S. cerevisiae CDC25* gene product regulates the *RAS*/adenylate cyclase pathway. *Cell* **48**, 789–799.

Buss, J. E., Solski, P. A., Schaeffer, J. P., MacDonald, M. J., and Der, C. J. (1989). Activation of the cellular proto-oncogene product p21RAS by addition of a myristylation signal. *Science* **243**, 1600–1603.

Cales, C., Hancock, J. F., Marshall, C. J., and Hall, A. (1988). The cytoplasmic protein GAP is implicated as the target for regulation by the *ras* gene product. *Nature (London)* **332**, 548–551.

Cameron, S., Levin, L., Zoller, M., and Wigler, M. (1988). cAMP-independent control of sporulation, glycogen metabolism, and heat shock resistance in *S. cerevisiae. Cell* **53**, 555–566.

Camonis, J. H., and Jacquet, M. (1988). A new *RAS* mutation that suppresses the *CDC25* gene requirement for growth of *Saccharomyces cerevisiae. Mol. Cell. Biol.* **8**, 2980–2983.

Camonis, J. H., Kakeline, M., Bernard, G., Garreau, H., Boy-Marcotte, E., and Jacquet, M. (1986). Characterization, cloning and sequence analysis of the CDC25 gene which controls the cyclic AMP level of *Saccharomyces cerevisiae. EMBO J.* **5**, 375–380.

Cannon, J. F., and Tatchell, K. (1987). Characterization of *Saccharomyces cerevisiae* genes encoding subunits of cyclic AMP-dependent protein kinase. *Mol. Cell. Biol.* **7**, 2653–2663.

Cannon, J. F., Gibbs, J. G., and Tatchell, K. (1986). Suppressors of the *ras2* mutation of *Saccharomyces cerevisiae. Genetics* **113**, 247–264.

Carlson, M. (1987). Regulation of sugar utilization in *Saccharomyces* species. *J. Bacteriol.* **169**, 4873–4877.

Carlson, M., Osmond, B. C., and Botstein, D. (1981). Mutants of yeast defective in sucrose utilization. *Genetics* **98**, 25–40.

Casperson, G. F., Walker, N., Brasier, A. R., and Bourne, H. R. (1983). A guanine nucleotide-sensitive adenylate cyclase in the yeast *Saccharomyces cerevisiae. J. Biol. Chem.* **258**, 7911–7914.

Celenza, J. L., and Carlson, M. (1986). A yeast gene that is essential for release from glucose repression encodes a protein kinase. *Science* **233**, 1175–1180.

Celenza, J. L., and Carlson, M. (1989). Mutational analysis of the *Saccharomyces cerevisiae* SNF1 protein kinase and evidence for functional interaction with the SNF4 protein. *Mol. Cell. Biol.* **9**, 5034–5044.

Chant, J., and Herskowitz, I. (1991). Genetic control of bud site selection in yeast by a set of gene products that constitute a morphogenetic pathway. *Cell* **65**, 1203–1212.

Chant, J., Corrado, K., Pringle, J. R., and Herskowitz, I. (1991). Yeast *BUD5*, encoding a prutative GDP–GTP exchange factor, is necessary for bud site selection and interacts with bud formation gene *BEM1*. *Cell* **65**, 1213–1224.

Cherry, J. R., Johnson, T. R., Dollard, C., Shuster, J. R., and Denis, C. L. (1989). Cyclic AMP-dependent protein kinase phosphorylates and inactivates the yeast transcriptional activator ADR1. *Cell* **56**, 409–419.

Ciriacy, M. (1975). Genetics of alcohol dehydrogenase in *Saccharomyces cerevisiae* II. Two loci controlling synthesis of the glucose-repressible ADHII. *Mol. Gen. Genet.* **138**, 157–164.

Ciriacy, M. (1977). Isolation and characterization of yeast mutants defective in intermediary carbon metabolism and in carbon catabolite derepression. *Mol. Gen. Genet.* **154**, 213–220.

Cobitz, A. R., Yim, E. H., Brown, W. R., Perou, C. M., and Tamanoi, F. (1989). Phosphorylation of RAS1 and RAS2 proteins in *Saccharomyces cerevisiae. Proc. Natl. Acad. Sci. U.S.A.* **86**, 858–862.

Colicelli, J., Field, J., Ballester, R., Chester, N., Young, D., and Wigler, M. (1990). Mutational mapping of RAS-responsive domains of the *Saccharomyces cerevisiae* adenylyl cyclase. *Mol. Cell. Biol.* **10**, 2539–2543.

Créchet, J. B., Poullet, P., Mistou, M. Y., Parmeggiani, A., Camonis, J., Boy-Marcotte, E., Damak, F., and Jacquet, M. (1990). Enhancement of the GDP–GTP exchange of RAS proteins by the carboxy-terminal domain of *SDC25. Science* **248**, 866–868.

Dhar, R., Nieto, A., Koller, R., DeFeo-Jones, D., and Scolnick, E. M. (1984). Nucleotide sequence of two rasH-related genes isolated from the yeast *Saccharomyces cerevisiae. Nucleic Acids Res.* **12**, 3611–3618.

Damak, F., Boy-Marcotte, E., Le-Roscouet, D., Guilbaud, R., and Jacquet, M. (1991). *SDC25*, a *CDC25*-like gene which contains a RAS-activating domain and is a dispensable gene of *Saccharomyces cerevisiae. Mol. Cell. Biol.* **11**, 202–212.

Daniel, J. H. (1986). The *CDC25* "Start" gene of *Saccharomyces cerevisiae*: Sequencing of the active C-terminal fragment and regional homologies with rhodopsin and cytochrome P450. *Curr. Genet.* **10**, 879–885.

Daniel, J., Becker, J. M., Enari, E., and Levitzki, A. (1987). The activation of adenylate cyclase by guanyl nucleotides in *Saccharomyces cerevisiae* is controlled by the *CDC25* start gene product. *Mol. Cell. Biol.* **7**, 3857–3861.

DeFeo-Jones, D., Scolnick, E. M., Koller, R., and Dhar, R. (1983). *ras*-related gene sequences identified and isolated from *Saccharomyces cerevisiae. Nature (London)* **306**, 707–709.

DeFeo Jones, D., Tatchell, K., Robinson, L. C., Sigal, I., Vass, W., Lowy, D. R., and Scolnick, E. M. (1985). Mammalian and yeast ras gene products: Biological function in their heterologous systems. *Science* **228**, 179–184.

Dellamora-Ortiz, G. M., Ortiz, C. H., Maia, J. C. C., and Panek, A. D. (1986). Partial purification and characterization of the interconvertible forms of trehalase from *Saccharomyces cerevisiae. Arch. Biochem. Biophys.* **251**, 205–214.

Denis, C. L., and Audino, D. C. (1991). The CCR1(SNF1) and SCH9 protein kinases act independently of cAMP-dependent protein kinase and transcriptional activator ADR1 in controlling yeast *ADH2* expression. *Mol. Gen. Genet.* **229**, 395–399.

Denis, C. L., Fontaine, S. C., Chase, D., Kemp, B. E., and Bemis, L. T. (1992). *ADR1^c* mutations enhance the ability of ADR1 to activate transcription by a mechanism that is independent of effects on cyclic AMP-dependent protein kinase phosphorylation of Ser-230. *Mol. Cell. Biol.* **12**, 1507–1514.

Deschenes, R. J., and Broach, J. R. (1987). Fatty acylation is important but not essential for *Saccharomyces cerevisiae RAS* function. *Mol. Cell. Biol.* **7**, 2344–2351.

Deschenes, R. J., Stimmel, J. B., Clarke, S., Stock, J., and Broach, J. R. (1989). *RAS2* protein of *Saccharomyces cerevisiae* is methyl-esterfied at its carboxy terminus. *J. Biol. Chem.* **264**, 11865–11873.

Deschenes, R. J., Resh, M. D., and Broach, J. R. (1990). Acylation and prenylation of proteins. *Curr. Opin. Cell Biol.* **2**, 1108–1113.

De Vendittis, E., Vitelli, A., Zahn, R., and Fasano, O. (1986). Suppression of defective RAS1 and RAS2 functions in yeast by an adenylate cyclase activated by a single amino acid change. *EMBO J.* **5**, 3657–3663.

Downward, J. (1990). The ras superfamily of small GTP-binding proteins. *Trends Biol. Sci.* **15**, 469–472.

Drubin, D., Miller, K., and Botstein, D. (1988). Yeast actin-binding proteins: Evidence for a role in morphogenesis. *J. Cell Biol.* **107**, 2551–2561.

Engelberg, D., Simchen, G., and Levitzki, A. (1990). *In vitro* reconstitution of CDC25 regulated *S. cerevisiae* adenylyl cyclase and its kinetic properties. *EMBO J.* **9**, 641–651.

Entian, K.-D., and Zimmermann, F. K. (1982). New genes involved in carbon catabolite repression and derepression in the yeast *Saccharomyces cerevisiae. J. Bacteriol.* **151**, 1123–1128.

Eraso, P., Mazon, M. J., and Gancedo, J. M. (1987). Internal acidification and cAMP increase are not correlated in *Saccharomyces cerevisiae. Eur. J. Biochem.* **165**, 671–674.

Fasano, O., Crechet, J. B., De Vendittis, E., Zahn, R., Feger, G., Vitelli, A., and Parmeggiani, A. (1988). Yeast mutants temperature-sensitive for growth after random mutagenesis of the chromosomal *RAS2* gene and deletion of the *RAS1* gene. *EMBO J.* **7**, 3375–3383.

Fedor-Chaiken, M., Deschenes, R. J., and Broach, J. R. (1990). *SRV2*, a gene required for *RAS* activation of adenylate cyclase in yeast. *Cell* **61**, 329–340.

Feig, L. A., and Cooper, G. M. (1988). Relationship among guanine nucleotide exchange, GTP hydrolysis, and transforming potential of mutated *ras* proteins. *Mol. Cell. Biol.* **8**, 2472–2478.

Feig, L. A., Pan, B., Roberts, T. M., and Cooper, G. M. (1986). Isolation of ras GTP-binding mutants using an in situ colony-binding assay. *Proc. Natl. Acad. Sci. U.S.A.* **83**, 4607–4611.

Field, J., Broek, D., Kataoka, T., and Wigler, M. (1987). Guanine nucleotide activation of, and competition between, RAS proteins from *Saccharomyces cerevisiae. Mol. Cell. Biol.* **7**, 2128–2133.

Field, J., Nikawa, J., Broek, D., MacDonald, B., Rodgers, L., Wilson, I. A., Lerner, R. A., and Wigler, M. (1988). Purification of a RAS-responsive adenylyl cyclase complex from *Saccharomyces cerevisiae* by use of an epitope addition method. *Mol. Cell. Biol.* **8**, 2159–2165.

Field, J., Xu, H.-P., Michaeli, T., Ballester, R., Sass, P., Wigler, M., and Colicelli, J. (1990a). Mutations of the adenylyl cyclase gene that block RAS function in *Saccharomyces cerevisiae. Science* **247**, 464–467.

Field, J., Vojtek, A., Ballester, R., Bolger, G., Colicelli, J., Ferguson, K., Gerst, J., Kataoka, T., Michaeli, T., Powers, S., Riggs, M., Rodgers, L., Wieland, I., Wheland, B., and Wigler, M. (1990b). Cloning and characterization of CAP, the *S. cerevisiae* gene encoding the 70-kd adenylyl cyclase-associated protein. *Cell* **61**, 319–327.

Fields, S., and Song, O.-K. (1989). A novel genetic system to detect protein–protein interactions. *Nature (London)* **340**, 245–246.

Finegold, A. A., Schafer, W. R., Rine, J., Whiteway, M., and Tamanoi, F. (1990). Common modification of trimeric G protein and ras protein: Involvement of polyisoprenylation. *Science* **249**, 165–169.

Fortini, M. E., Simon, M. A., and Rubin, G. M. (1992). Signalling by the *sevenless* protein tyrosine kinase is mimicked by Ras1 activation. *Nature (London)* **355**, 559–561.

Fraenkel, D. (1982). Carbohydrate metabolism. *In* "The Molecular Biology of the Yeast *Saccharomyces*–Metabolism and Gene Expression" (J. Strathern, E. Jones, and J. Broach, Eds.), Cold Spring Harbor Laboratory Press, Cold Spring Harbor, New York.

Fraenkel, D. (1985). On ras gene function in yeast. *Proc. Natl. Acad. Sci. U.S.A.* **82**, 4740–4744.

François, J., Van Schaftingen, E., and Hers, H.-G. (1988). Characterization of phosphofructokinase 2 and of enzymes involved in the degradation of fructose 2,6-bisphosphate in yeast. *Eur. J. Biochem.* **171**, 599–608.

François, J. M., Thompson-Jaeger, S., Skroch, J., Zellenka, U., Spevak, W., and Tatchell, K. (1992). *GAC1* may encode a regulatory subunit for protein phosphatase type 1 in *Saccharomyces cerevisiae. EMBO J.* **11**, 87–96.

Fujiyama, A., and Tamanoi, F. (1986). Processing and fatty acylation of RAS1 and RAS2 protein in *Saccharomyces cerevisiae. Proc. Natl. Acad. Sci. U.S.A.* **83**, 1266–1270.

Fujiyama, A., and Tamanoi, F. (1987). RAS2 protein of *Saccharomyces cerevisiae* undergoes removal of methionine at N terminus and removal of three amino acids at C terminus: *J. Biol. Chem.* **265**, 3362–3368.

Fujiyama, A., Matsumoto, K., and Tamanoi, F. (1987). A novel yeast mutant defective in the processing of ras proteins: Assessment of the effect of the mutation on processing steps. *EMBO J.* **6**, 223–228.

Garrett, S., and Broach, J. (1989). Loss of Ras activity in *Saccharomyces cerevisiae* is suppressed by disruptions of a new kinase gene, *YAK1*, whose product may act downstream of the cAMP-dependent protein kinase. *Genes Dev.* **3**, 1336–1348.

Garrett, S., Menold, M. M., and Broach, J. (1991). The *Saccharomyces cerevisiae YAK1* gene encodes a protein kinase that is induced by arrest early in the cell cycle. *Mol. Cell. Biol.* **11**, 4045–4052.

Garreau, H., Camonis, J. H., Guitton, C., and Jacquet, M. (1990). The *Saccharomyces cerevisiae CDC25* gene product is a 180-kDa polypeptide and is associated with a membrane fraction. *FEBS Letts.* **269**, 53–59.

Gerst, J. E., Ferguson, K., Vojtek, A., Wigler, M., and Field, J. (1991). CAP is a bifunctional component of the *Saccharomyces cerevisiae* adenylyl cyclase complex. *Mol. Cell. Biol.* **11**, 1248–1257.

Gibbs, J. (1991). RAS C-terminal processing enzymes-new drug targets? *Cell* **65**, 1–4.

Gibbs, J. B., and Marshall, M. S. (1989). The *ras* oncogene–An important regulatory element in lower eucaryotic organisms. *Microbiol. Rev.* **53**, 171–185.

Gibbs, J. B., Schaber, M. D., Marshall, M. S., Scolnick, E. M., and Sigal, I. (1987). Identification of guanine nucleotides bound to *ras*-encoded proteins in growing yeast cells. *J. Biol. Chem.* **262**, 10426–10429.

Gilman, A. G. (1984). G proteins and dual control of adenylate cyclase. *Cell* **36**, 577–579.

Goodman, L. E., Perou, C. M., Fujiyama, A., and Tamanoi, F. (1988). Structure and expression of yeast DPR1, a gene essential for the processing and intracellular localization of ras proteins. *Yeast* **4**, 271–281.

Goodman, L. E., Judd, S. R., Farnsworth, C. C., Powers, S., Gelb, M. H., Glomset, J. A., and Tamanoi, F. (1990). Mutants of *Saccharomyces cerevisiae* defective in the farnesylation of Ras proteins. *Proc. Natl. Acad. Sci. U.S.A.* **87**, 9665–9669.

Grandy, D. K., Hanneman, E., Bunzow, J., Shih, M., Machida, C. A., Bidlack, J. M., and Civelli, O. (1990). Purification, cloning, and tissue distribution of a 23-kDa rat protein isolated by morphine affinity chromatography. *Mol. Endocrinol.* **4**, 1370–1376.

Granot, D., and Snyder, M. (1991). Glucose induces cAMP-independent growth-related changes in stationary-phase cells of *Saccharomyces cerevisiae. Proc. Natl. Acad. Sci. U.S.A.* **88**, 5724–5728.

Hafen, E., Basler, K., Edstroem, J. E., and Rubin, G. M. (1987). *Sevenless*, a cell-specific homeotic gene of *Drosophila*, encodes a putative transmembrane receptor with a tyrosine kinase domain. *Science* **236**, 55–63.

Han, M., and Sternberg, P. W. (1990). *Let-60*, a gene that specifies cell fates during *C. elegans* vulval induction, encodes a *ras* protein. *Cell* **63**, 921–931.

Han, M., Aroian, R. V., and Sternberg, P. W. (1990). The *let-60* locus controls the switch between vulval and nonvulval cell fates in *Caenorhabditis elegans. Genetics* **126**, 899–913.

Hashimoto, C., Hudson, K. L., and Anderson, K. V. (1988). The *Toll* gene of *Drosophila*, required for dorsal–ventral embryonic polarity, appears to encode a transmembrane protein. *Cell* **52**, 269–279.

Heideman, W., Casperson, G. F., and Bourne, H. R. (1987). Adenylyl cyclase in yeast. *J. Biol. Chem.* **262**, 7087–7091.

Hixson, C. S., and Krebs, E. G. (1980). Characterization of a cyclic AMP-binding protein from bakers' yeast. *J. Biol. Chem.* **255**, 2137–2145.

Hubbard, E. J. A., Yang, X., and Carlson, M. (1992). Relationship of the cAMP-dependent protein kinase pathway to the SNF1 protein kinase and invertase expression in *Saccharomyces cerevisiae*. *Genetics* **130**, 71–80.

Hughes, D. A., Fukui, Y., and Yamamoto, M. (1990). Homologous activators of ras in fission and budding yeast. *Nature (London)* **344**, 355–357.

Hughes, S. M. (1983). Are guanine nucleotide binding proteins a distinct class of regulatory proteins? *FEBS Lett.* **164**, 1–8.

Johnson, D., and Pringle, J. (1990). Molecular characterization of *CDC42*, a gene involved in the development of cell polarity. *J. Cell. Biol.* **111**, 143–152.

Johnson, D. I., O'Brien, J. M., and Jacobs, C. W. (1991). Isolation and sequence analysis of *CDC43*, a gene involved in the control of cell polarity in *Saccharomyces cerevisiae*. *Gene* **98**, 149–150.

Johnson, K. E., Cameron, S., Toda, T., Wigler, M., and Zoller, M. J. (1987). Expression in *Escherichia coli* of *BCY1*, the regulatory subunit of cyclic AMP-dependent protein kinase from *Saccharomyces cerevisiae*. *J. Biol. Chem.* **262**, 8636–8642.

Jones, S., Vignais, M.-L., and Broach, J. (1991). The CDC25 protein of *S. cerevisiae* promotes exchange of guanine nucleotide bound to Ras. *Mol. Cell. Biol.* **11**, 2641–2646.

Kaibuchi, K., Miyajima, A., Arai, K.-I., and Matsumoto, K. (1986). Possible involvement of *RAS*-encoded proteins in glucose-induced inositolphospholipid turnover in *Saccharomyces cerevisiae*. *Proc. Natl. Acad. Sci. U.S.A.* **83**, 8172–8176.

Kaplan, D. R., Morrison, D. K., Wong, G., McCormick, F., and Williams, L. T. (1990). PDGF β-receptor stimulates tyrosine phosphorylation of GAP and association of GAP with a signaling complex. *Cell* **61**, 125–133.

Kataoka, T., Powers, S., McGill, C., Fasano, O., Strathern, J., Broach, J., and Wigler, M. (1984). Genetic analysis of yeast RAS1 and RAS2 genes. *Cell* **37**, 437–445.

Kataoka, T., Broek, D., and Wigler, M. (1985a). DNA sequence and characterization of the *S. cerevisiae* gene encoding adenylate cyclase. *Cell* **43**, 493–505.

Kataoka, T., Powers, S., Cameron, S., Fasano, O., Goldfarb, M., Broach, J., and Wigler, M. (1985b). Functional homology of mammalian and yeast RAS genes. *Cell* **40**, 19–26.

Kazlauskas, A., Ellis, C., Pawson, T., and Cooper, J. A. (1990). Binding of GAP to activated PDGF receptors. *Science* **247**, 1578–1581.

Kim, J.-H., and Powers, S. (1991). Overexpression of *RPI1*, a novel inhibitor of the yeast Ras-cyclic AMP pathway, down-regulates normal but not mutationally activated Ras function. *Mol. Cell. Biol.* **11**, 3894–3904.

Koch, C. A., Anderson, D., Moran, M. F., Ellis, C., and Pawson, T. (1991). SH2 and SH3 domains: Elements that control interactions of cytoplasmic signaling proteins. *Science* **252**, 668–674.

Kohl, N. E., Diehl, R. E., Schaber, M. D., Rands, E., Soderman, D. D., He, B., Moores, S. L., Pompliano, D. L., Ferro-Novick, S., Powers, S., Thomas, K. A., and Gibbs, J. B. (1991). Structural homology among mammalian and *Saccharomyces cerevisiae* isoprenyl-protein transferases. *J. Biol. Chem.* **266**, 18884–18888.

Krengel, U., Schlichting, I., Scherer, A., Schumann, R., Frech, M., John, J., Kabsch, W., Pai, E. F., and Wittinghofer, A. (1990). Three-dimensional structures of H-*ras* p21 mutants: Molecular basis for their inability to function as switch molecules. *Cell* **62**, 539–548.

Kretschmer, M., Schellenberger, W., Otto, A., Kessler, R., and Hofmann, E. (1987). Fructose-2,6-bisphosphatase and 6-phosphofructo-2-kinase are separable in yeast. *Biochem. J.* **246**, 755–759.

Krupinski, J., Coussen, F., Bakalyar, H. A., Tang, W.-J., Feinstein, P. G., Orth, K., Slaughter, C., Reed, R. R., and Gilman, A. G. (1989). Adenylyl cyclase amino acid sequence: Possible channel- or transporter-like structure. *Science* **244**, 1558–1564.

Kunisawa, R., Davis, T. N., Urdea, M. S., and Thorner, J. (1987). Complete nucleotide sequence of the gene encoding the regulatory subunit of 3′5′-cyclic AMP-dependent protein kinase from the yeast *Saccharomyces cerevisiae*. *Nucleic Acids Res.* **15**, 368–369.

Lacal, P. M., Pennington, C. Y., and Lacal, J. C. (1988). Transforming activity of *ras* protein translocated to the plasma membrane by a myristoylation sequence from the *src* gene product. *Oncogene* **2**, 533–537.

Levin, L. R., Kuret, J., Johnson, K. E., Powers, S., Cameron, S., Michaeli, T., Wigler, M., and Zoller, M. J. (1988). A mutation in the catalytic subunit of cAMP-dependent protein kinase that disrupts regulation. *Science* **240**, 68–70.

Lillie, S. H., and Pringle, J. R. (1980). Reverse carbohydrate metabolism in *Saccharomyces*: Responses to nutrient limitation. *J. Bacteriol.* **143**, 1384–1394.

Lopez, J., Chung, D., Fujikawa, K., Hagen, F., Papayannapoulou, T., and Roth, G. (1987). Cloning of the α-chain of glycoprotein 1b: A transmembrane protein with homology to leucine-rich α_2 glycoprotein. *Proc. Natl. Acad. Sci. U.S.A.* **84**, 5615–5619.

Lopez, J., Chung, D., Fujikawa, K., Hagen, F., Davie, E. W., and Roth, G. J. (1988). The α and chain β chains of human platelet glycoprotein 1b are both transmembrane proteins containing a leucine-rich amino acid sequence. *Proc. Natl. Acad. Sci. U.S.A.* **85**, 2135–2139.

McCormick, F. (1989). *ras* GTPase activating protein: Signal transmitter and signal terminator. *Cell* **56**, 5–8.

McFarland, K. C., Sprengel, R., Phillips, H. S., Köhler, M., Rosemblit, N., Nikolics, K., Segaloff, D. L., and Seeburg, P. H. (1989). Lutropin–choriogonadotropin receptor: An unusual member of the G protein-coupled receptor family. *Science* **245**, 494–499.

Marshall, M. S., Gibbs, J. B., Scolnick, E. M., and Sigal, I. S. (1987). Regulatory function of the *Saccharomyces cerevisiae* RAS C-terminus. *Mol. Cell. Biol.* **7**, 2309–2315.

Marshall, M. S., Gibbs, J. B., Scolnick, E. M., and Sigal, I. S. (1988). An adenylate cyclase from *Saccharomyces cerevisiae* that is stimulated by RAS proteins with effector mutations. *Mol. Cell. Biol.* **8**, 52–61.

Marshall, M. S., Davis, L. J., Keys, R. D., Mosser, S. D., Hill, W. S., Scolnick, E. M., and Gibbs, J. B. (1991). Identification of amino acid residues required for Ras p21 target activation. *Mol. Cell. Biol.* **11**, 3997–4004.

Martegani, E., Baroni, M. D., Frascotti, G., and Alberghina, L. (1986). Molecular cloning and transcriptional analysis of the start gene *CDC25* of *Saccharomyces cerevisiae*. *EMBO J.* **5**, 2363–2369.

Martin, G. A., Viskochil, D., Bollag, G., McCabe, P. C., Crosier, W. J., Haubruck, H., Conroy, L., Clark, R., O'Connell, P., Cawthon, R. M., Innis, M. A., and McCormick, F. (1990). The GAP-related domain of the neurofibromatosis Type 1 gene product interacts with ras p21. *Cell* **63**, 843–849.

Matsumoto, K., Uno, I., Oshima, Y., and Ishikawa, T. (1982a). Isolation and characterization of yeast mutants deficient in adenylate cyclase and cyclic AMP dependent protein kinase. *Proc. Natl. Acad. Sci. U.S.A.* **79**, 2355–2359.

Matsumoto, K., Uno, I., Toh-e, A., Ishikawa, T., and Oshima, Y. (1982b). Cyclic AMP may not be involved in catabolite repression in *Saccharomyces cerevisiae*: Evidence from mutants capable of utilizing it as an adenine source. *J. Bacteriol.* **150**, 277–285.

Matsumoto, K., Uno, I., and Ishikawa, T. (1985). Isolation and characterization of a phosphoprotein phosphatase deficient mutant in yeast. *Yeast* **1**, 25–38.

Mazon, M. J., Gancedo, J. M., and Gancedo, C. (1982). Phosphorylation and inactivation of yeast fructose-bisphosphatase in vivo by glucose and by proton ionophores. A possible role for cAMP. *Eur. J. Biochem.* **127**, 605–608.

Mbonyi, K., and Thevelein, J. M. (1988). The high-affinity glucose uptake system is not required for induction of the RAS-mediated cAMP signal by glucose in cells of the yeast *Saccharomyces cerevisiae*. *Biochem. Biophys. Acta* **971**, 223–226.

Mbonyi, K., Buellens, M., Detremerie, K., Geerts, L., and Thevelein, J. M. (1988). Requirement of one functional RAS gene and inability of an oncogenic ras variant to mediate the glucose-induced cyclic AMP signal in the yeast *Saccharomyces cerevisiae*. *Mol. Cell. Biol.* **8**, 3051–3057.

Michaeli, T., Field, J., Ballester, R., O'Neill, K., and Wigler, M. (1989). Mutants of H-*ras* that interfere with RAS effector functions in *Saccharomyces cerevisiae. EMBO J.* **8**, 3039–3044.

Milburn, M. V., Tong, L., DeVos, A. M., Brünger, A., Yamaizumi, Z., Nishimura, S., and Kim, S.-H. (1990). Molecular switch for signal transduction: Structural differences between active and inactive forms of protooncogenic *ras* proteins. *Science* **247**, 939–945.

Mitts, M. R., Grant, D. B., and Heideman, W. (1990). Adenylate cyclase in *Saccharomyces cerevisiae* is a peripheral membrane protein. *Mol. Cell. Biol.* **10**, 3873–3883.

Moores, S. L., Schaber, M. D., Mosser, S. D., Rands, E., O'Hara, M. B., Garsky, V. M., Marshall, M. S., Pompliano, D. L., and Gibbs, J. B. (1991). Sequence dependence of protein isoprenylation. *J. Biol. Chem.* **266**, 14603–14610.

Müller, G., and Bandlow, W. (1991). Two lipid-anchored cAMP-binding proteins in the yeast *Saccharomyces cerevisiae* are unrelated to the R subunit of cytoplasmic protein kinase A. *Eur. J. Biochem.* **202**, 299–308.

Mulcahy, L. S., Smith, M. R., and Stacey, D. W. (1985). Requirement for ras proto-oncogene function during serum-stimulated growth of NIH3T3 cells. *Nature (London)* **313**, 241–243.

Munder, T., and Fürst, P. (1992). The *Saccharomyces cerevisiae CDC25* gene product bind specifically to catalytically inactive RAS proteins in vivo. *Mol. Cell. Biol.* **12**, 2091–2099.

Munder, T., and Kuntzel, H. (1989). Glucose-induced cAMP signaling in *Saccharomyces cerevisiae* is mediated by the *CDC25* protein. *FEBS Lett.* **242**, 341–345.

Munder, T., Mink, M., and Küntzel, H. (1988). Domains of the *Saccharomyces cerevisiae CDC25* gene controlling mitosis and meiosis. *Mol. Gen. Genet.* **214**, 271–277.

Nakajima, H., Oshima, I., Yashiro, M., Yoda, K., Tamasaki, M., and Tamura, G. (1987). Novel *Saccharomyces cerevisiae* temperature-sensitive *cyr1* mutant which accumulates glycogen particles. *Agric. Biol. Chem.* **10**, 2679–2689.

Neigeborn, L., and Carlson, M. (1984). Genes affecting the regulation of *SUC2* gene expression by glucose repression in *Saccharomyces cerevisiae. Genetics* **108**, 845–858.

Nikawa, J.-I., Cameron, S., Toda, T., Ferguson, K. M., and Wigler, M. (1987a). Rigorous feedback control of cAMP levels in *Saccharomyces cerevisiae. Genes Dev.* **1**, 931–937.

Nikawa, J.-I., Sass, P., and Wigler, M. (1987b). Cloning and characterization of the low-affinity cyclic AMP phosphodiesterase gene of *Saccharomyces cerevisiae. Mol. Cell. Biol.* **7**, 3629–3636.

Pai, E. F., Kabasch, W., Krengel, U., Holmes, K. C., John, J., and Wittinghofer, A. (1989). Structure of the guanine-nucleotide-binding domain of the Ha-*ras* oncogene product p21 in the triphosphate conformation. *Nature (London)* **341**, 209–214.

Pai, E. F., Krengel, U., Petsko, G., Goody, R. S., Kabasch, W., and Wittinghofer, A. (1990). Refined crystal structure of the triphosphate conformation of H-ras p21 at 1.35-Å resolution: Implications for the mechanism of GTP hydrolysis. *EMBO J.* **9**, 2351–2359.

Peng, Z. Y., Trumbly, R. J., and Reimann, E. M. (1990). Purification and characterization of glycogen synthase from a glycogen-deficient strain of *Saccharomyces cerevisiae. J. Biol. Chem.* **265**, 13871–13877.

Petitjean, A., Hilger, F., and Tatchell, K. (1990). Comparison of thermosensitive alleles of the *cdc25* gene involved in the cAMP metabolism of *Saccharomyces cerevisiae. Genetics* **124**, 797–806.

Powers, S., Kataoka, T., Fasano, O., Goldfarb, M., Broach, J., and Wigler, M. (1984). Genes in *S. cerevisiae* encoding proteins with domains homologous to the mammalian ras proteins. *Cell* **36**, 607–612.

Powers, S., Michaelis, S., Broek, D., Santa-Anna, A. S., Field, J., Herskowitz, I., and Wigler, M. (1986). *RAM*, a gene of yeast required for a functional modification of RAS proteins and for production of mating pheromone. *Cell* **47**, 413–422.

Powers, S., O'Neill, K., and Wigler, M. (1989). Dominant yeast and mammalian *RAS* mutants that interfere with the *CDC25*-dependent activation of wild-type *RAS* in *Saccharomyces cerevisiae. Mol. Cell. Biol.* **9**, 390–395.

Powers, S., Gonzales, E., Christensen, T., Cubert, J., and Broek, D. (1991). Functional cloning of *BUD5*, a *CDC25*-related gene from *S. cerevisiae* that can suppress a dominant-negative *RAS2* mutant. *Cell* **65**, 1225–1231.

Praekelt, U. M., and Meacock, P. A. (1990). *HSP12*, a new small heat shock gene of *Saccharomyces cerevisiae*: Analysis of structure, regulation and function. *Mol. Gen. Genet.* **223**, 97–106.

Pringle, J. R., and Hartwell, L. H. (1981). The *Saccharomyces cerevisiae* cell cycle. *In* "The Molecular Biology of the Yeast *Saccharomyces*—Life Cycle and Inheritance" (J. Strathern, E. Jones and J. Broach, Eds.), pp. 97–142. Cold Spring Harbor Laboratory Press, Cold Spring Harbor, New York.

Purwin, C., Nicolay, K., Scheffers, W. A., and Holzer, H. (1986). Mechanism of control of adenylate cyclase activity in yeast by fermentable sugars and carbonyl cyanide *m*-chlorophenylhydrazone, *J. Biol. Chem.* **261**, 8744–8749.

Reinke, R., Krantz, D. E., Yen, D., and Zipursky, S. L. (1988). Chaoptin, a cell surface glycoprotein required for *Drosophila* photoreceptor cell morphogenesis, contains a repeat motif found in yeast and human. *Cell* **52**, 291–301.

Reiss, Y., Seabra, M. C., Armstrong, S. A., Slaughter, C. A., Goldstein, J., and Brown, M. S. (1991). Nonidentical subunits of $p21^{H\text{-}ras}$ farnesyltransferase. *J. Biol. Chem.* **266**, 10672–10677.

Resnick, R. J., and Racker, E. (1988). Phosphorylation of the *RAS2* gene product by protein kinase A inhibits the activation of yeast adenylate cyclase. *Proc. Natl. Acad. Sci. U.S.A.* **85**, 2474–2478.

Robinson, L. C., and Tatchell, K. (1991). *TFS1*, a suppressor of *cdc25* mutations in *Saccharomyces cerevisiae*. *Mol. Gen. Genet.* **230**, 241–250.

Robinson, L. C., Gibbs, J. B., Marshall, M., Sigal, I. S., and Tatchell, K. (1987). *CDC25*: A component of the RAS-adenylate cyclase pathway in *Saccharomyces cerevisiae*. *Science* **235**, 1218–1221.

Rossi, G., Jiang, Y., Newman, A. P., and Ferro-Novick, S. (1991). Dependence of Ypt1 and Sec4 membrane attachment on Bet2. *Nature (London)* **351**, 158–161.

Rubinfeld, B., Munemitsu, S., Clark, R., Conroy, L., Watt, K., Crosier, W. J., McCormick, F., and Polakis, P. (1991). Molecular cloning of a GTPase activating protein specific for the K *rev-1* protein $p21^{rap1}$. *Cell* **65**, 1033–1042.

Ruggieri, R., Tanaka, K., Nakafuku, M., Kaziro, Y., Toh-e, A., and Matsumoto, K. (1989). MSI1, a negative regulator of the RAS-cAMP pathway in *Saccharomyces cerevisiae*. *Proc. Natl. Acad. Sci. U.S.A.* **86**, 8778–8782.

Sasaki, T., Kikuchi, A., Araki, S., Hata, Y., Isomura, M., Kuroda, S., and Takai, Y. (1990). Purification and characterization from bovine brain cytosol of a protein that inhibits the dissociation of GDP from the subsequent binding of GTP to smg p25A, a *ras* p21-like GTP-binding protein, *J. Biol. Chem.* **265**, 2333–2337.

Sass, P., Field, J., Nikawa, J., Toda, T., and Wigler, M. (1986). Cloning and characterization of the high affinity cAMP phosphodiesterase of *S. cerevisiae*. *Proc. Natl. Acad. Sci. U.S.A.* **83**, 9303–9307.

Satoh, T., Endo, M., Nakafuku, M., Nakamura, S., and Kaziro, Y. (1990a). Platelet-derived growth factor stimulates formation of active $p21^{ras}$·GTP complex in swiss mouse 3T3 cells. *Proc. Natl. Acad. Sci. U.S.A.* **87**, 5993–5997.

Satoh, T., Endo, M., Nakafuku, M., Akiyama, T., Yamamoto, T., and Kaziro, Y. (1990b). Accumulation of p21ras·GTP in response to stimulation with epidermal growth factor and oncogene products with tyrosine kinase activity. *Proc. Natl. Acad. Sci. U.S.A.* **87**, 7926–7929.

Schafer, W. R., Kim, R., Sterne, R., Thorner, Kim, S.-H., and Rine, J. (1989). Genetic and pharmacological suppression of oncogenic mutations in *RAS* genes of yeast and humans. *Science* **245**, 379–385.

Schlichting, I., Almo, S. C., Rapp, G., Wilson, K., Petratos, K., Lentfer, A., Wittinghofer, A., Kabsch, W., Pai, E. F., Petsko, G. A., and Goody, R. S. (1990). Time-resolved X-ray crystallographic study of the confromational change in Ha-Ras p21 protein on GTP hydrolysis. *Nature (London)* **345**, 309–315.

Schoentgen, F., Saccocio, F., Jolles, J., Bernier, I., and Jolles, P. (1987). Complete amino acid sequence of a basic 21-kDa protein from bovine brain cytosol. *Eur. J. Biochem.* **166**, 333–338.

Sigal, I. S., Gibbs, J. B., D'Alonzo, J. S., Temeles, G. L., Wolanski, B. S., Socher, S. H., and Scolnick, E. M. (1986a). Mutant *ras*-encoded proteins with altered nucleotide binding exert dominant biological effects. *Proc. Natl. Acad. Sci. U.S.A.* **83**, 952–956.

Sigal, I. S., Gibbs, J. B., D'Alonzo, J. S., and Scolnick, E. M. (1986b). Identification of effector residues and a neutralizing epitope of Ha-*ras*-encoded p21. *Proc. Natl. Acad. Sci. U.S.A.* **83**, 4725–4729.

Sigal, I. S., Marshall, M. S., Schaber, M. D., Vogel, U. S., Scolnick, E. M., and Gibbs, J. B. (1988). Structure/function studies of the ras protein. *Cold Spring Harbor Symp. Quant. Biol.* **53**, 863–869.

Simon, M. A., Bowtell, D. D. L., Dodson, G. S., Laverty, T. R., and Rubin, G. M. (1991). Ras1 and a putative guanine nucleotide exchange factor perform crucial steps in signaling by the Sevenless protein tyrosine kinase. *Cell* **67**, 701–716.

Smith, M. E., Dickinson, J. R., and Wheals, A. E. (1990). Intracellular and extracellular levels of cyclic AMP during the cell cycle of *Saccharomyces cerevisiae*. *Yeast* **6**, 53–60.

Smith, M. R., DeGudicibus, S. J., and Stacey, D. W. (1986). Requirement for c-*ras* protein during viral oncogene transformation. *Nature (London)* **320**, 540–543.

Stryer, L. (1986). Cyclic GMP cascade of vision. *Ann. Rev. Neurosci.* **9**, 87–119.

Suzuki, N., Choe, H.-R., Nishida, Y., Yamawaki-Kataoka, Y., Ohnishi, S., Tamaoki, T., and Kataoka, T. (1990). Leucine-rich repeats and carboxyl terminus are required for interaction of yeast adenylate cyclase with RAS proteins. *Proc. Natl. Acad. Sci. U.S.A.* **87**, 8711–8715.

Takahashi, N., Takahashi, Y., and Putnam, F. W. (1985). Periodicity of leucine and tandem repetition of a 24-amino acid segment in the primary structure of leucine-rich α_2-glycoprotein of human serum. *Proc. Natl. Acad. Sci. U.S.A.* **82**, 1906–1910.

Tanaka, K., Matsumoto, K., and Toh-e, A. (1988). Dual regulation of the expression of the polyubiquitin gene by cyclic AMP and heat shock in yeast. *EMBO J.* **7**, 495–502.

Tanaka, K., Matsumoto, K., and Toh-e, A. (1989). *IRA1*, an inhibitory regulator of the RAS-cyclic AMP pathway in *Saccharomyces cerevisiae*. *Mol. Cell. Biol.* **9**, 757–768.

Tanaka, K., Nakafuku, M., Satoh, T., Marshall, M. S., Gibbs, J. B., Matsumoto, K., Kaziro, Y., and Toh-e, A. (1990a). *S. cerevisiae* genes *IRA1* and *IRA2* encode proteins that may be functionally equivalent to mammalian ras GTPase activating protein. *Cell* **60**, 803–807.

Tanaka, K., Nakafuku, M., Tamanoi, F., Kaziro, Y., Matsumoto, K., and Toh-e, A. (1990b). *IRA2*, a second gene of *Saccharomyces cerevisiae* that encodes a protein with a domain homologous to mammalian *ras* GTPase-activating protein. *Mol. Cell. Biol.* **10**, 4303–4313.

Tanaka, K., Lin, B. K., Wood, D. R., and Tamanoi, F. (1991). IRA2, an upstream negative regulator of RAS in yeast, in a RAS GTPase-activating protein. *Proc. Natl. Acad. Sci. U.S.A.* **88**, 468–472.

Tanaka, K., Wood, D. R., Lin, B. K., Khalil, M., Tamanoi, F., and Cannon, J. F. (1992). A dominant activating mutation in the effector region of RAS abolishes IRA2 sensitivity. *Mol. Cell. Biol.* **12**, 631–637.

Tatchell, K., Chaleff, D. T., DeFeo-Jones, D., and Scolnick, E. M. (1984). Requirement of either of a pair of *ras*-related genes of *Saccharomyces cerevisiae* for spore viability. *Nature (London)* **309**, 523–527.

Tatchell, K., Robinson, L. C., and Breitenbach, M. (1985). *RAS2* of *Saccharomyces cerevisiae* is required for gluconeogenic growth and proper response to nutrient limitation. *Proc. Natl. Acad. Sci. U.S.A.* **82**, 3785–3789.

Taylor, W. E., and Young, E. T. (1990). cAMP-dependent phosphorylation and inactivation of yeast transcription factor ADR1 does not affect DNA binding. *Proc. Natl. Acad. Sci. U.S.A.* **87**, 4098–4102.

Thevelein, J. M., and Beullens, M. (1985). Cyclic AMP and the stimulation of trehalase activity in the yeast *Saccharomyces cerevisiae* by carbon sources, nitrogen sources and inhibitors of protein synthesis. *J. Gen. Microbiol.* **131**, 3199–3209.

Thevelein, J. M., Beullens, M., Honshoven, F., Hoebeeck, G., Detremerie, K., Griewel, B., Den Hollander, J. A., and Jans, A. W. H. (1987). Regulation of the cAMP level in the yeast *Saccharomyces cerevisiae*: the glucose-induced cAMP signal is not mediated by a transient drop in the intracellular pH. *J. Gen. Microbiol.* **133**, 2197–2205.

Thompson-Jaeger, S., François, J., Gaughran, J. P., and Tatchell, K. (1991). Deletion of *SNF1* affects the nutrient response of yeast and resembles mutations which activate the adenylate cyclase pathway. *Genetics* **129**, 697–706.

Titani, K., Takio, K., Handa, M., and Ruggeri, Z. M. (1987). Amino acid sequence of von Willebrand factor-binding domain of platelet membrane glycoprotein 1b. *Proc. Natl. Acad. Sci. U.S.A.* **84**, 5610–5614.

Toda, T., Uno, I., Ishikawa, T., Powers, S., Kataoka, T., Broek, D., Cameron, S., Broach, J., Matsumoto, K., and Wigler, M. (1985). In yeast, RAS proteins are controlling elements of adenylate cyclase. *Cell* **40**, 27–36.

Toda, T., Cameron, S., Sass, P., Zoller, M., Scott, J. D., McMullen, B., Hurwitz, M., Krebs, E. G., and Wigler, M. (1987a). Cloning and characterization of *BCY1*, a locus encoding a regulatory subunit of the cyclic AMP-dependent protein kinase in *Saccharomyces cerevisiae*. *Mol. Cell. Biol.* **7**, 1371–1377.

Toda, T., Cameron, S., Sass, P., Zoller, M., and Wigler, M. (1987b). Three different genes in *S. cerevisiae* encode the catalytic subunits of the cAMP-dependent protein kinase. *Cell* **50**, 277–287.

Toda, T., Cameron, S., Sass, P., and Wigler, M. (1988). *SCH9*, a gene of *Saccharomyces cerevisiae* that encodes a protein distinct from, but functionally and structurally related to cAMP-dependent protein kinase catalytic subunits. *Genes Dev.* **2**, 517–527.

Tong, L., deVos, A. M., Milburn, M. V., and Kim, S.-H. (1991). Crystal structures at 2.2-Å resolution of the catalytic domains of normal *ras* protein and an oncogenic mutant complexed with GDP. *J. Mol. Biol.* **217**, 503–516.

Trahey, M., and McCormick, F. (1987). A cytoplasmic protein stimulates normal N-*ras* p21 GTPase but does not affect oncogenic mutants. *Science* **238**, 542–545.

Trahey, M., Wong, G., Halenbeck, R., Rubinfeld, B., Martin, G. A., Ladner, M., Long, C. M., Crosier, W. J., Watt, K., Koths, K., and McCormick, F. (1988). Molecular cloning of two types of GAP complementary DNA from human placenta. *Science* **242**, 1697–1700.

Tripp, M. L., Bouchard, R. A., and Piñon, R. (1989). Cloning and characterization of *NSP1*, a locus encoding a component of a *CDC25*-dependent nutrient responsive pathway in *Saccharomyces cerevisiae*. *Mol. Microbiol.* **3**, 1319–1327.

Ueda, T., Kikuchi, A., Ohga, N., Yamamoto, J., and Takai, Y. (1990). Purification and characterization from bovine brain cytosol of a novel regulatory protein inhibiting the dissociation of GDP from and subsequent binding of GTP to rhoB, a ras p21-like GTP-binding protein. *J. Biol. Chem.* **265**, 9373–9380.

Uno, I., Mitsuzawa, H., Matsumoto, K., Tanaka, K., Oshima, T., and Ishikawa, T. (1985). Reconstitution of the GTP-dependent adenylate cyclase from products of the yeast CYR1 and RAS2 genes in *Escherichia coli*. *Proc. Natl. Acad. Sci. U.S.A.* **82**, 7855–7859.

Valle, E., Bergillos, L., Gascon, S., Parra, F., and Ramos, S. (1986). Trehalase activation in yeast is mediated by an internal acidification. *Eur. J. Biochem.* **154**, 247–251.

Van Aelst, L., Jans, A. W. H., and Thevelein, J. M. (1991). Involvement of the *CDC25* gene product in the signal transmission pathway of the glucose-induced RAS-mediated cAMP signal in the yeast *Saccharomyces cerevisiae*. *J. Gen. Micro.* **137**, 341–349.

Van der Plaat, J. B., and Van Solingen, P. (1974). Cyclic 3′,5′-adenosine monophosphate stimulates trehalose degradation in Baker's yeast. *Biochem. Biophys. Res. Commun.* **56**, 580–587.

Vanoni, M., Vavassori, M., Frascotti, G., Martegani, E., and Alberghina, L. (1990). Overexpression of the *CDC25* gene, an upstream element of the RAS-adenylyl cyclase pathway in *Saccharomyces cerevisiae*, allows immunological identification and characterization of its gene product. *Biochem. Biophys. Res. Commun.* **72**, 61–69.

Varimo, K., and Londesborough, J. (1976). Solubilization and other studies on adenylate cyclase of baker's yeast. *Biochem. J.* **159**, 363–370.

Vogel, U. S., Dixon, R. A. F., Schaber, M. D., Diehl, R. E., Marshall, M. S., Scolnick, E. M., Sigal, I. S., and Gibbs, J. B. (1988). Cloning of bovine GAP and its interaction with oncogenic ras p21. *Nature (London)* **335**, 90–93.

Vojtek, A., Haarer, B., Field, J., Gerst, J., Pollard, T. D., Brown, S., and Wigler, M. (1991). Evidence for a functional link between proflin and CAP in the yeast *S. cerevisiae*. *Cell* **66**, 497–505.

Werner-Washburne, M., Becker, J., Kosic-Smithers, J., and Craig, E. (1989). Yeast Hsp70 RNA levels vary in response to the physiological status of the cell. *J. Bacteriol.* **171**, 2680–2688.

Werner-Washburne, M., Brown, D., and Braun, E. (1991). Bcy1, the regulatory subunit of cAMP-dependent protein kinase in yeast, is differentially modified in response to the physiological status of the cell. *J. Biol. Chem.* **266**, 19704–19709.

West, M., Kung, H. F., and Kamata, T. (1990). A novel membrane factor stimulates guanine nucleotide exchange reaction of *ras* proteins. *FEBS Lett.* **259**, 245–248.

Wickner, W. J., Koh, T. J., Crowley, J. C., O'Neil, J., and Kaback, D. B. (1987). Molecular cloning of chromosome I DNA from *Saccharomyces cerevisiae*: Isolation of the *MAK16* gene and analysis of an adjacent gene essential for growth at low temperatures. *Yeast* **3**, 51–57.

Wigler, M., Field, J., Powers, S., Broek, D., Toda, T., Cameron, S., Nikawa, J., Michaeli, T., Colicelli, J., and Ferguson, K. (1988). Studies of RAS function in the yeast *Saccharomyces cerevisiae*. *Cold Spring Harbor Symp. Quant. Biol.* **53**, 649–655.

Willumsen, B. M., Norris, K., Papageorge, A. G., Hubbert, N. L., and Lowy, D. R. (1984). Harvey murine sarcoma virus p21 ras protein: Biological and biochemical significance of the cysteine nearest the carboxy terminus. *EMBO J.* **3**, 2581–2585.

Wilson, R. B., and Tatchell, K. (1988). *SRA5* encodes the low-K_m cAMP phosphodiesterase of *Saccharomyces cerevisiae*. *Mol. Cell. Biol.* **8**, 505–510.

Wittinghofer, A., and Pai, E. F. (1991). The structure of Ras protein: A model for a universal molecular switch. *Trends Biochem. Sci.* **16**, 382–387.

Wolfman, A., and Macara, I. G. (1990). A cytosolic protein catalyzes the release of GDP from p21ras. *Science* **248**, 67–69.

Xu, G., Lin, B., Tanaka, K., Dunn, D., Wood, D., Gesteland, R., White, R., Weiss, R., and Tamanoi, F. (1990). The catalytic domain of the neurofibromatosis Type 1 gene product stimulates ras GTPase and complements ira mutants of *S. cerevisiae*. *Cell* **63**, 835–841.

Yamamoto, T., Kaibuchi, K., Mizuno, T., Hiroyoshi, M., Shirataki, H., and Takai, Y. (1990). Purification and characterization from bovine brain cytosol of proteins that regulate the GDP/GTP exchange reaction of smg p21s, ras p21-like GTP-binding proteins. *J. Biol. Chem.* **265**, 16626–16634.

Yamawaki-Kataoka, Y., Tamaoki, T., Choe, H.-R., Tanaka, H., and Kataoka, T. (1989). Adenylate cyclases in yeast: A comparison of the genes from *Schizosaccharomyces pombe* and *Saccharomyces cerevisiae*. *Proc. Natl. Acad. Sci. U.S.A.* **86**, 5693–5697.

Zimmermann, F. K., Kaufmann, I., Rasenberger, H., and Haussmann, P. (1977). Genetics of carbon catabolite repression in *Saccharomyces cerevisiae*: Genes involved in the derepression process. *Mol. Gen. Genet.* **151**, 95–103.

Chapter 8

Signal Transduction Pathway for Pheromone Response in *Saccharomyces cerevisiae*

MALCOLM WHITEWAY

National Research Council Canada
Biotechnology Research Institute
Montreal, Quebec, Canada H4P 2R2

BEVERLY ERREDE

Department of Chemistry
University of North Carolina
Chapel Hill, North Carolina

I. Introduction

In the simple life cycle of the yeast *Saccharomyces cerevisiae*, two haploid cells of opposite mating types (designated **a** and α) conjugate to form a meiotically competent **a**/α diploid cell. Under starvation conditions, the diploid cell undergoes meiosis to generate four haploid spores, two of each mating type. To conjugate, haploid cells of the opposite mating type must recognize each other and then differentiate from vegetatively growing cells into mating competent cells. Recognition occurs through diffusible peptide pheromones, **a**- and α-factor, that are characteristic of the respective **a** and α cell types. The peptide pheromone from one cell type binds to its membrane receptor on the opposite cell type. This pheromone–receptor interaction activates an intracellular signal transduction pathway that initiates and coordinates the differentiation program. The primary responses to signal activation are growth arrest in the G_1 phase of the cell cycle and transcriptional induction of a variety of genes, some of which encode products required to catalyze the actual events of conjugation.

Receptor-activated signal transduction pathways play a predominant role in directing a variety of cells to differentiate (Morgan, 1989; Ullrich and Schlessinger, 1990; Cantley *et al.*, 1991; Shepherd, 1991; Gurdon, 1992). The pheromone response pathway of *S. cerevisiae* is an example of this process in an experimentally accessible organism. The defined genetic methods available with *S. cerevisiae* have allowed the identification of many genes that encode components of the pheromone-induced signal pathway (Table I). The subsequent cloning and DNA sequencing of these genes has suggested a biochemical function for some of them. The ability to manipulate components of the pathway both *in vitro* and *in vivo* has advanced our understanding of how different components are integrated into a coherent signaling pathway.

Our current model for pheromone-induced signal transmission is the framework for this chapter (Fig. 8.1). As an introduction, we briefly trace the pathway from the receptor to the targets that mediate the primary physiological responses to pheromone. The underlying experimental evidence and logic are provided throughout the chapter. According to the model, the mating response pathway is activated by the binding of pheromone to its receptor. This interaction is coupled to the intracellular transmission components through a heterotrimeric G protein. The three subunits are the *SCG1* (*GPA1*)-encoded G_α, the *STE4*-encoded G_β, and the *STE18*-encoded G_γ. By analogy with mammalian G proteins, the pheromone-bound receptor promotes the G_α subunit to exchange GDP for GTP. This exchange in turn leads to dissociation of the GTP-bound G_α subunit from $G_{\beta\gamma}$. In contrast to the situation in other systems, in yeast the free $G_{\beta\gamma}$ complex transmits the signal. The actual target of the $G_{\beta\gamma}$ signal is still not identified. However, the first component known to be required downstream of the G protein is the Ste5 protein whose function is unknown. A kinase cascade then appears to transmit the intracellular signal. The Ste11 kinase function is necessary for pheromone-induced phosphorylation, and presumably activation, of the Ste7, Fus3, and Kss1 kinases. The Fus3 and Kss1 kinases are redundant for their role in transmitting the pheromone-induced signal; the Ste7 and Fus3/Kss1 steps in signal transmission appear to be interdependent. A likely nuclear target of the signal is the *STE12*-encoded transcription factor. The Ste12 protein binds to a DNA sequence motif (PRE) that is required for pheromone-induced transcription. Among the genes dependent on Ste12 activity are those whose products encode the catalysts of conjugation and some components of the signaling system itself. Other targets of the pheromone-induced signal are products that affect progression through the cell cycle. The latter include the *FAR1* product and the *CLN1*-, *CLN2*-, and *CLN3*-encoded G_1 cyclins.

Cell type specialization underlies the ability of a cell to produce and respond to pheromone. The regulatory mechanisms that establish yeast cell type have been reviewed (Nasmyth and Shore, 1987; Herskowitz, 1989; Sprague, 1990; Dolan and Fields, 1991) and will not be covered here. Several reviews provide a clear outline of pheromone production, phermone response, and the mating process (Cross *et al.*, 1988; Kurjan and Whiteway, 1990; Blumer and Thorner, 1991; Marsh *et al.*, 1991; Sprague, 1991; Chapter 9). We concentrate on the more recent evidence that suggests functional roles for the different pathway components. We also focus on those studies that suggest the relationship between different gene products as outlined in our model (Fig. 8.1). In the process, we hope to underscore areas of uncertainty and directions for continuing research.

II. Receptors

Cells have evolved a variety of ways to sense their external environment. One major mechanism involves the use of specific proteins that span the plasma membrane of the cell and can bind specific external compounds. These receptor proteins and the ligands they bind can take many forms. Some receptors, such as the steroid hormone-binding proteins, are cytoplasmic, and after forming a complex with their ligand, translocate to the nucleus where they can regulate transcription of target genes directly (Evans, 1988). Other receptors, such as those of the receptor tyrosine kinase class, remain at the cell surface and effect intracellular changes through protein phosphorylation (Ullrich and Schlessinger, 1990). Receptors of a third class interact with a very diverse class of ligands but have a common structural motif. They characteristically contain seven hydrophobic α-helices that serve as membrane-spanning domains. These receptors do not have catalytic activity but influence internal cellular

TABLE I
Genes Involved in Peromone-Induced Signal Transmission

		Expression			Pheromone response		
Gene	Product	Cell type	Pheromone inducible	Allele	Transcription activation	G1 arrest	References
STE2	α-Factor receptor	**a**	Yes	Null	No	No	MacKay and Manney (1974); Hartwell (1980); Burkholder and Hartwell (1985); Nakayama *et al.*, (1985); Hartig *et al.* (1986).
STE3	**a**-Factor receptor	α	Yes	Null	No	No	MacKay and Manney (1974); Hagen and Sprague (1984); Nakayama *et al.* (1985); Hagen *et al.* (1986)
SCG1/GPA1	G_{α} Subunit	**a** and α	Yes	Null	Constitutive	Constitutive (lethal)	Dietzel and Kurjan (1987a); Miyajma *et al.* (1987, 1989); Jahng *et al.* (1988); Nakayama *et al.* (1988a); Fujimura (1989)
				Gain	No	No	
STE4	G_{β} Subunit	**a** and α	?	Null	No	No	Blinder *et al.* (1989); Whiteway *et al.* (1989, 1990); Cole *et al.* (1990); Nomoto *et al.* (1990)
				Gain	Constitutive	Constitutive (lethal)	
				Overexpressed	Constitutive	Constitutive (lethal)	
STE18	G_{γ} Subunit	**a** and α	?	Null	No	No	Whiteway *et al.* (1989)
STE5	Unknown	**a** and α	?	Null	No	No	MacKay and Manney (1974); Hartwell (1980); Fields and Herskowitz (1985)

STE11	Protein kinase	All	No	Null	No	No	Hartwell (1980); Chaleff and Tatchell (1985); Teague *et al.* (1986); C. Adler and B. Errede (unpublished observations)
				Gain	Constitutive	Inducible	
STE7	Protein kinase	All	No	Null	No	No	Hartwell (1980); Chaleff and Tatchell (1985); Rhodes *et al.* (1990); Stevenson *et al.* (1992)
FUS3	Protein kinase	**a** and α	Yes	Null	Yes	No	Elion *et al.* (1990, 1991); J. Brill and G. Fink (unpublished observations)
				Gain	Constitutive	Inducible	
KSS1	Protein kinase	All	No	Null	Yes	Supersensitive	Courchesne *et al.* (1989)
STE12	Transcription factor	**a** and α	No	Null	No	No	Fields and Herskowitz (1987); Dolan *et al.* (1989); Errede and Ammerer (1989); Dolan and Fields (1990); Hartwell (1990)
				Overexpressed	Yes	Constitutive (lethal)	
FAR1	Unknown	**a** and α	Yes	Null	Yes	No	Chang and Herskowitz (1990)
CDC36	Unknown	All	?	Null	Constitutive	Constitutive (lethal)	de Barros Lopes *et al.* (1990); Neiman *et al.* (1990)
CDC39	Unknown	All	?	Null	Constitutive	Constitutive (lethal)	de Barros Lopes *et al.* (1990); Neiman *et al.* (1990)
SRM1	Unknown	All	No	Null	Constitutive	Constitutive (lethal)	Clark and Sprague (1989)

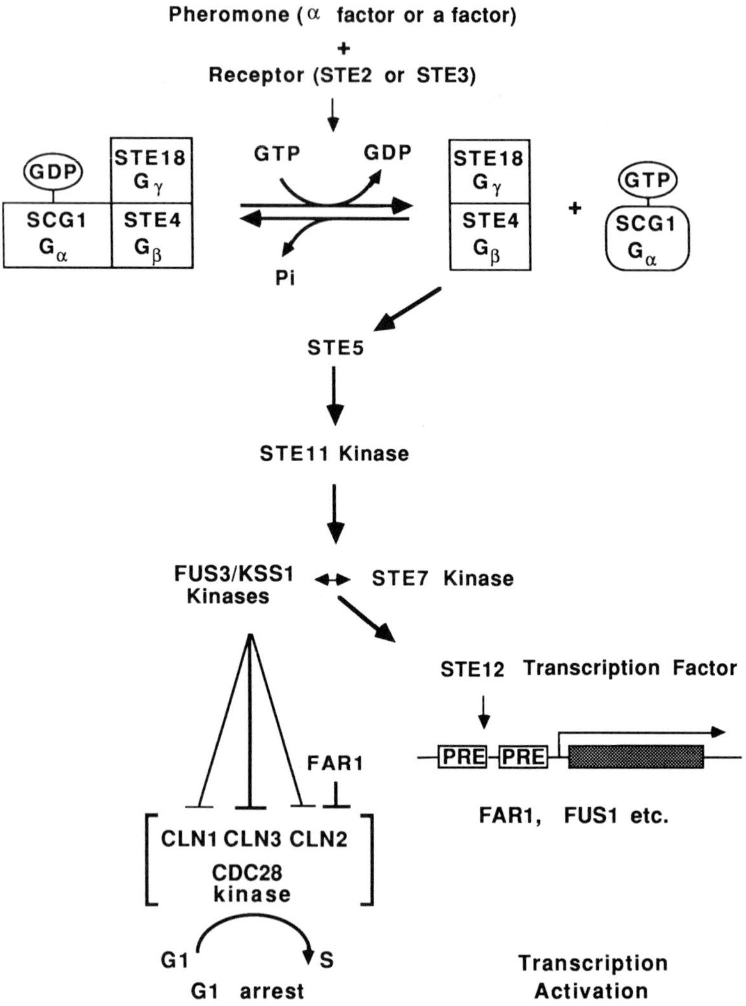

Figure 8.1 Model for the pheromone-induced signal transmission pathway, that is, the postulated relationship of signal transmission components that function after the pheromone–receptor complex is formed. Arrows indicate activation or stimulation; blunt-ended lines indicate repression or inhibition. PRE is the pheromone response element, which is a binding site for the Ste12 transcription factor and is required for pheromone-inducible transcription of target genes (shaded bar). The holoenzyme consisting of the Cdc28 kinase catalytic subunit and at least one G_1 cyclin protein (Cln) is required for progression of the cell cycle from the G_1 to S phase. The evidence that Scg1, Ste4, and Ste18 are G protein subunits (G_α, G_β, G_γ) involved in a guanine nucleotide exchange cycle is discussed in Section III in the text. The relationship between the G-protein complex and Ste5 is inferred from results of Blinder *et al.* (1989). The relationship between the Ste5 protein and the Ste11 kinase is deduced from results of Stevenson *et al.* (1992). The relationships between the Ste11, Ste7, and Fus3/Kss1 kinases are deduced from evidence discussed in Section IV,B,4. The relationship of Fus3 and Far1 to the various cyclins is discussed in Section V (see also Nakayama *et al.*, 1988b).

processes through their activation of a signal transducer molecule called a G protein (Ross, 1989).

In the mating response pathway, the **a** and α cell types produce unique receptor proteins encoded by the *STE2* and *STE3* genes, respectively. Secondary structure predictions show that the two proteins have the seven potential membrane-spanning α-helices characteristic of the G protein-linked receptor family (Fig. 8.2). However, the two receptors have essentially no primary sequence identity. The ligands that stimulate these receptors also are produced in a cell type-specific manner. *MAT*α cells produce a 13-amino-acid peptide called α-factor that is secreted and processed through the secretory pathway (Julius *et al.*, 1983, 1984a,b; Dmochowska *et al.*, 1987). *MAT***a** cells produce a farnesylated 12-amino-acid peptide called **a**-factor that is exported by the Ste6 transporter protein (McGrath and Varshavsky, 1989; Kuchler *et al.*, 1989). The characterization of **a**-factor and its secretion are discussed in Chapter 9.

A. α-Factor Receptor (Ste2)

Mutations in the *STE2* gene prevent *MAT***a** cells from mating, but have no effect on the mating of *MAT*α cells (MacKay and Manney, 1974). The cell type-specific sterility associated with *ste2* mutations was the first support for the possibility that the *STE2* gene might encode a component of the α-factor pheromone receptor. Additional strong support for this possibility was provided by the demonstration that *MAT***a** cells had an α factor binding capacity that was abolished in *ste2* mutant cells (Jenness *et al.*, 1983). Finally, cloning and sequencing studies established that the *STE2* gene product had the seven potential transmembrane hydrophobic α-helices that characterize the rhodopsin superfamily of receptor proteins (Burkholder and Hartwell, 1985; Nakayama *et al.*, 1985).

Various lines of evidence are consistent with the topology of the receptor protein containing an N terminus that is extracellular, a C terminus that is intracellular, and seven transmembrane domains that divide the protein into three extracellular and three intracellular loops (Fig. 8.2). Three potential sites for glycosylation lie in the putative N-terminal extracellular domain; two others lie in potential extracellular loops; only one lies in the potentially intracellular C terminus. Because inhibition of glycosylation significantly changes the molecular weight of the protein, more than one of these sites is likely actually to be glycosylated. Analysis of a C terminal fragment shows that this region does not contain any glycosylated residues. Thus, at least one of the potentially extracellular regions, in fact, must be extracellular (Blumer *et al.*, 1988). Similarly, the fact that the C terminus contains sites that are phosphorylated during pheromone response suggests that this region must lie within the cell (Reneke *et al.*, 1988). Together, these results imply that the receptor must traverse the membrane an odd number of times.

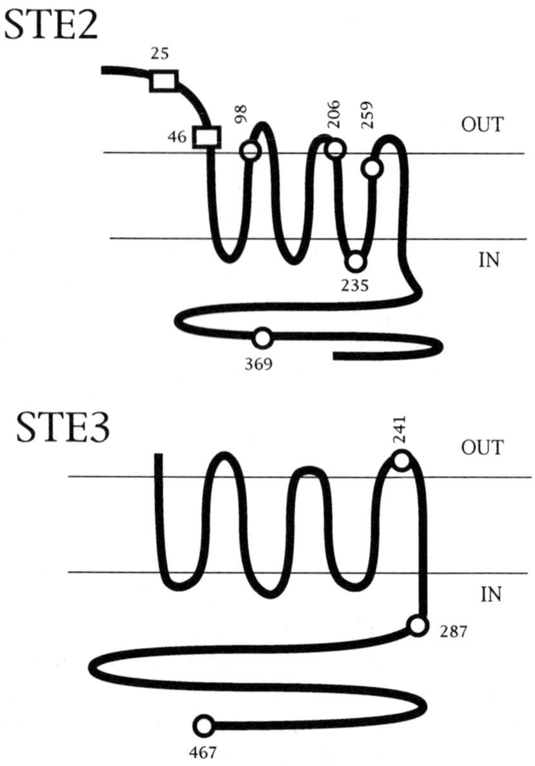

Figure 8.2 Structure of the Ste2 (*top*) and Ste3 (*bottom*) receptor proteins. The receptor are shown with their seven transmembrane domains of approximately 25 amino acids embedded in the membrane. The N termini of the proteins are extracellular (out), whereas the C termini are cytoplasmic (in). The Ste3 protein has a very short extracellular N terminus with no glycosylation sites, whereas the Ste2 protein has a longer extracellular N terminus with two potential glycosylation sites at positions 25 and 46 (squares). The circles represent the sites of fusion proteins that were constructed to assess the topology of the proteins. For Ste2, fusions of β-lactamase at positions 98, 206, and 259 are apparently extracellular, whereas those at positions 235 and 369 are apparently internal (Cartwright and Tipper, 1991). For Ste3, fusions with invertase at positions 467 and 287 are apparently internal, whereas the fusion at position 241 is apparently external (Clark *et al.*, 1988). The C-terminal tails are sites of phosphorylation in both proteins, and are predicted to be cytoplasmic. Details of protein modification are described in the text.

A fusion protein approach has been applied to investigate the topology of the Ste2 protein (Cartwright and Tipper, 1991). In this study, chimeras between the Ste2 protein and β-lactamase were constructed with a Kex2 cleavage site linking the two proteins (Figure 8.2). Liberation of extracellular β-lactamase activity was detected, indicative of Kex2 cleavage when the fusion joint was in a domain of Ste2 predicted to be extracellular, whereas the activity was intracellular when the joint was in a region predicted to be intracellular (Cartwright and Tipper, 1991). Although direct evidence for the presence of three intracellular and three extracellular loops has not been obtained, the rhodopsin structure provides a logical working model for the Ste2 protein and is consistent with all the current data.

Direct evidence for ligand-binding activity of the Ste2 protein has been presented. Affinity labeling procedures have established high affinity binding of α-factor to the Ste2 protein in yeast cells (Blumer *et al.*, 1988). Expression of the Ste2 protein in *Xenopus* oocytes established that the ligand-binding capability resided solely in the Ste2 protein (Yu *et al.*, 1989). In addition, expression of a related receptor protein from *Saccharomyces kluyveri* in *ste2* mutant cells generated a *S. cerevisiae* cell that could respond to the heterologous *S. kluyveri*-specific pheromone (Marsh and Herskowitz, 1988).

Ligand binding to the Ste2 receptor protein has two major consequences for the protein: receptor endocytosis and receptor phosphorylation. Treatment of *MAT***a** cells with α-factor causes a rapid decrease in the number of α-factor binding sites on the surface of the cell. These binding sites subsequently are replaced in a protein synthesis-dependent manner (Jenness and Spatrick, 1986). The initial loss of binding sites occurs with the internalization of the bound α factor (Jenness and Spatrick, 1986). The internalization of α-factor is dependent on the presence of the α-factor receptor. The internalized α-factor subsequently is degraded (Chvatchko *et al.*, 1986). This result suggests that the receptor–ligand complex is endocytosed; whether this endocytosis is essential for pheromone response remains to be determined.

Receptor phosphorylation is another rapid consequence of ligand binding. Within 5 min of ligand binding, enhanced phosphorylation of the Ste2 receptor C terminus can be detected (Reneke *et al.*, 1988). Similar C-terminal phosphorylation is detected for rhodopsin and the β-adrenergic receptor. In the case of the mammalian receptors, the phosphorylation has been suggested to play a role in desensitization (Bouvier *et al.*, 1988). Mutations of the Ste2 protein that remove the C terminus lead to enhanced pheromone sensitivity without a significant change in receptor binding affinity (Konopka *et al.*, 1988; Reneke *et al.*, 1988). Thus, the Ste2 C terminus appears to play a regulatory role in receptor desensitization. Whether specific phosphorylation events are involved in this desensitization process remains to be determined. Site-directed mutagenesis of potential phosphorylation sites and identification of the kinase that causes the phosphorylation should prove informative.

B. a-Factor Receptor (Ste3)

Null mutations in the *STE3* gene cause the reciprocal phenotype of *ste2* mutants; defects in *STE3* abolish mating in *MAT*α cell but not in *MAT***a** cells (MacKay and Manney, 1974). Cloning and sequencing of the *STE3* gene revealed that the *STE3* gene product, like the *STE2* gene product, was a member of the rhodopsin class of proteins (Nakayama *et at.*, 1985; Hagen *et al.*, 1986). Although the amino acid sequence predicts a secondary structure for Ste3 with strong similarities to Ste2 (see subsequent discussion), the amino acid sequences of the two proteins are very different. Interestingly, the Ste3 protein does have two regions of sequence similarity with the Ste6 protein (Kuchler *et al.*, 1989). Because the Ste6 and Ste3 proteins both must interact with **a**-factor, seeing whether the regions of Ste6 and Ste3 similarity are involved in **a**-factor binding will be interesting.

Analysis of the topology of the Ste3 protein suggests that it, like Ste2, has an extracellular N terminus and an intracellular C terminus (Fig. 8.2). The Ste3 protein is not glycosylated (Clark *et al.*, 1988) although it has five potential glycosylation sites. None of the potential glycosylation sites would be extracellular if the protein had the predicted topology with an intracellular C terminus and an extracellular N terminus. Although the absence of glycosylation is consistent with the predicted topology, this absence is not proof that all potential glycosylation sites are intracellular. Fusions between the *STE3* gene and the invertase gene were used to test the topology of the Ste3 protein (Fig. 8.2). In a construction that should have directed the invertase protein to the outside of the cell, the invertase moiety was glycosylated, whereas in two constructions that should have directed the invertase intracellularly, the invertase was not glycosylated (Clark *et al.*, 1988). Because the invertase chimeras may influence the topology of the protein, and because the absence of glycosylation could result from other effects, the support for the Ste3 protein fitting the rhodopsin structural model is circumstantial. However, no data exist that fail to fit the general model that the Ste3 receptor protein, like the Ste2 protein, has a membrane-spanning structure topologically similar to that of rhodopsin.

An interesting mutation of *STE3*, *DAF2-2*, was selected as a dominant α-factor resistant mutant (Cross, 1990). The *DAF2-2* allele also suppressed the constitutive G_1 arrest response associated with certain mutations of either the G_α or G_β subunits of the G protein (Table I). The dominant allele also caused weak but constitutive expression of a normally pheromone-inducible gene (*FUS1*). The transcriptional phenotype of *DAF2-2* was suppressed by null alleles of either the G_β or the G_γ subunit of the G protein. An obvious prediction was that the normal *DAF2* product might affect activity of the G protein directly (Cross, 1990). Another observation was that the dominant *DAF2-2* mutation had a more dramatic effect in **a** cells than in α cells. At the time, this result was surprising because no cell type-specific components were known to be involved in pheromone response after the binding of pheromones to their receptors (Bender and Sprague, 1986; Nakayama *et al.*, 1987).

Subsequent cloning and analysis of the *DAF2-2* allele revealed that it is a promoter insertion mutation of *STE3* that causes the **a**-factor receptor to be expressed in both **a** and α cells (J. Hirsch and F. Cross, unpublished results). The constitutive transcription activation caused by the *DAF2-2* mutation can be explained as an autocrine response similar to that previously described for ectopic expression of **a**-and α-receptors in the inappropriate cell type (Bender and Sprague, 1986; Nakayama *et al.*, 1987). Interestingly, the inappropriate expression of the Ste3 receptor interferes with certain of the mating responses. A better understanding of this phenomenon may reveal additional aspects of receptor interactions that are not yet fully appreciated.

C. Mammalian Receptor Function in Yeast

The observation that the yeast pheromone receptors are members of the rhodopsin-like superfamily of G protein-linked receptors and the identification of yeast G protein homologs (see subsequent discussion) have encouraged expression of mammalian receptor proteins in yeast cells with the hope of coupling them to downstream components of the pheromone response pathway. Studies of mammalian receptor structure and function in yeast could be of practical importance. For example, coupling mammalian receptor proteins that are targets for therapeutic drugs to the yeast mating pathway could provide a simple and inexpensive screening system for chemicals that affect the function of these receptors.

The first report of expression of mammalian receptor proteins in yeast cells was the expression of the M1 muscarinic receptor (Payette *et al.*, 1990). Although the pharmacological properties of the receptor were comparable to those of the native protein, the level of expression was low and no evidence suggested that the receptor could couple to the yeast pheromone response pathway. A second report of a mammalian G protein-linked receptor expressed in yeast involves the expression of the β_2-adrenergic receptor (King *et al.*, 1990). In this study, a high level of expression was obtained and evidence was provided that the receptor could couple to the phermone response pathway in cells that had a mammalian G_α subunit in place of the endogenous yeast G_α subunit. Treatment of cells expressing the receptor with β-adrenergic agonists resulted in the induction of the *FUS1* gene, one of the target genes of the yeast response pathway. In addition, the cells arrested at the G_1 phase of the cell cycle and underwent morphological changes characteristic of cells responding to mating pheromones.

Intriguingly, the cells responding to the β-adrenergic agonists were not able to mate, although their phenotypic characteristics resembled those of mating cells. In the absence of the yeast receptor and G_α subunit, the β-adrenergic receptor-triggered response may not activate the complete mating response pathway (King *et al.*, 1990). The absence of the yeast receptor or G_α subunit cannot be limiting the mating response in these cells since an absence of the G_α subunit is known to restore mating to receptor-deficient cells (Miyajima *et al.*, 1987). One possibility is that the heterologous β-adrenergic receptor and mammalian G_α subunit interfere with some aspect of signal activation. Various reports suggest that the β-adrenergic receptor can trigger responses independently of G proteins (Ganze *et al.*, 1990). Thus, another possibility is that the β-adrenergic receptor bypasses the normal G protein-mediated response pathway for some but not all yeast responses. Whether all the components of the normal yeast mating pathway are necessary for the β-adrenergic agonist-stimulated events observed in cells overexpressing the β-adrenergic receptor remains to be seen.

III. G Protein Components

Members of the seven membrane-spanning domain class of receptors are coupled to the intracellular response machinery through a G protein. The G protein complex consists of three separate subunits, designated α, β, γ. Generally the largest of the subunits is α. This subunit has the capacity to bind guanine nucleotides (hence the term G proteins). The other two subunits (β and γ) do not bind guanine nucleotides, but associate tightly together to form a heterodimer. In yeast, three genes have been identified that encode proteins with similarity to G_α, G_β, and G_γ. Superficially, this finding indicates that the signal generated by ligand binding to receptor will be transduced by a mechanism similar to that of mammalian systems that employ G proteins.

Analysis of purified and partially purified components, primarily from the rhodopsin–transducin system in mammalian retinas and from the β-adrenergic receptor–G_s system, have led to a general model of G protein function (Stryer and Bourne, 1986; Gilman, 1987; Ross, 1989; Simon *et al.*, 1991). This model involves a guanine nucleotide exchange cycle. The resting state of the signaling system is believed to involve a receptor that is associated physically with the heterotrimeric G protein containing GDP bound to the α subunit. The presence of the associated G protein results in the receptor having a high affinity for the ligand. The binding of ligand to this high affinity state receptor triggers a variety of changes in the system. The GDP bound to the α subunit is replaced by GTP, the G protein subunits dissociate from each other generating G_α and $G_{\beta\gamma}$ elements, and these subunits dissociate from the receptor, switching it to the low affinity state. This state is the active form of the G protein complex. The released G protein subunits are free to stimulate effector enzymes and thus trigger the generation of second messenger signaling molecules. This activated state persists as long as GTP remains bound to the α subunit; eventually, the endogenous GTPase activity of the α subunit triggers the hydrolysis of GTP to GDP, which allows the reassociation of the receptor and the heterotrimeric G protein, re-establishing the resting state. This model is widely accepted and provides the general framework for much of the current work in the field of G proteins. Several of the predictions of the model have been tested. Subunit dissociation in response to ligand binding has been demonstrated in a number of systems, including transducin, G_s, and G_i (Fung, 1983; Takahashi *et al.*, 1991; Lee *et al.*, 1992). In addition, nonhydrolyzable analogs of GTP and treatment with some bacterial toxins can activate G protein pathways permanently, consistent with the idea that GTP hydrolysis is necessary for deactivation (see Ross, 1989).

A. G Protein α Subunit (Scg1/Gpa1)

1. Role in Pheromone Response The gene encoding a component of the yeast pheromone response pathway with sequence similarity to the α subunit of G proteins was discovered independently by several different approaches. This gene was designated as *SCG1, GPA1, CDC70,* or *DAC1* by the various investigators. Each of the approaches that led to its discovery serves to underline some of the functions of this protein in the mating pathway. The initial identification of a G_α homolog in *S. cerevisiae* (*GPA1*) involved the cloning of yeast genes that had sequence homology to the mammalian $G_{\alpha s}$ gene through low stringency hybridization (Nakafuku *et al.*, 1987). A second approach identified the G_α homolog (*SCG1*) by the isolation of genes that, when overexpressed, reduced the hypersensitivity of a yeast strain defective in adaptation to pheromones (Dietzel and Kurjan, 1987a). A third group isolated mutations that conferred mating ability to a receptorless yeast strain. One of the receptor bypass mutations was designated *cdc70*. A clone of the G_α homolog was isolated on the basis of its ability to complement the growth arrest phenotype of the *cdc70* mutation (Jahng *et al.*, 1988). A fourth approach involved isolation of mutations that prevented growth of haploid cells but allowed normal growth in diploids (Blinder *et al.*, 1989). Finally, a fifth identification of the gene (*DAC1*) was made from a search of pheromone resistant mutants (Fujimura, 1989). For the purpose of consistency within this chapter, we use the designation of *SCG1* to refer to this gene and its various mutant alleles.

Null mutations of the *SCG1* gene were constructed by gene disruptions. Cells that lacked the *SCG1* function arrested at the G_1 point of the cell cycle in a mating-competent state. In fact, deletion mutants were able to mate even in the absence of a functional receptor (Miyajima *et al.*, 1987). The behavior of the arrested cells was reminiscent of cells responding to mating pheromones; although cell division was arrested, the cells continued to grow and change their morphology (Dietzel and Kurjan, 1987a; Miyajima *et al.*, 1987). Activation of the mating response pathway independently of pheromone in *scg1* null mutants explains why mutations in the gene restored mating competence to receptorless mutants (Jahng *et al.*, 1988). Although haploid cells containing null mutations in the *SCG1* gene arrest at the G_1 phase of the cell cycle and do not divide, diploid cells homozygous for the null mutations grow normally (Dietzel and Kurjan, 1987a; Miyajima *et al.*, 1987). Further, an absence of the *SCG1* gene causes cell cycle arrest only in cells that contain an intact pheromone response pathway (Nakayama *et al.*, 1988b). The requirement for an intact response pathway (several of the components of which are not expressed in diploid cells) explains why the *SCG1* gene was identified in screens for genes whose expression is needed only for viability in haploid cells (Blinder *et al.*, 1989).

2. Functional Domains Many point mutations in the *SCG1* gene have been generated to investigate its function in the mating process in more detail. One strategy has been to use random mutagenesis to define important residues through some phenotypic change in the cells containing the mutant protein. Another strategy has been to use directed mutagenesis to change residues assumed to be important based on the predicted structure of the protein.

Random mutagenesis of the *SCG1* gene was done by hydroxylamine treatment (Stone and Reed, 1990). The mutagenized population was screened for variants that were unable to transmit the mating response signal. These mutant alleles were designated *phun* because they created a pheromone unresponsive phenotype. The *phun* alleles, therefore, had the opposite phenotype of null mutations. Five residues were found that generated the *phun* phenotype when mutated (Fig. 8.3a). One substitution was recovered at position Gly-50 (Gly-50 → Asp), which is predicted to be part of the guanine nucleotide binding site and to be important in the GTPase activity of G protein α subunits (see subsequent discussion). Two others were at position Gly-322 (Gly-322 → Glu or Gly-322 → Arg), which is part of the switch box implicated in conformational changes to the protein during guanine nucleotide binding (Miller *et al.*, 1988). In addition, substitution mutations Glu-355 → Lys, Glu-364 → Lys, and Gly-470 → Asp created a nonresponsive phenotype, as did the double mutation Glu-364 → Lys, Gly-470 → Asp. Although all these mutations resulted in a loss of pheromone responsiveness, two distinct mechanisms appeared to be causing the pheromone resistance. The substitution derivatives at position Gly-322 and, to a lesser extent, at Gly-470 appeared to reduce signal transduction directly. This alteration in function is compatible with the predictions that the Gly-470 substitution would be defective in receptor coupling and the Gly-322 substitutions would be defective in $G_{\beta\gamma}$ release. In contrast, the substitution derivatives Gly-50 → Asp, Gly-355 → Lys, or Glu-364 → Lys caused the cells to be initially hypersensitive to pheromone but subsequently to adapt. This behavior has been suggested to indicate that an activated form

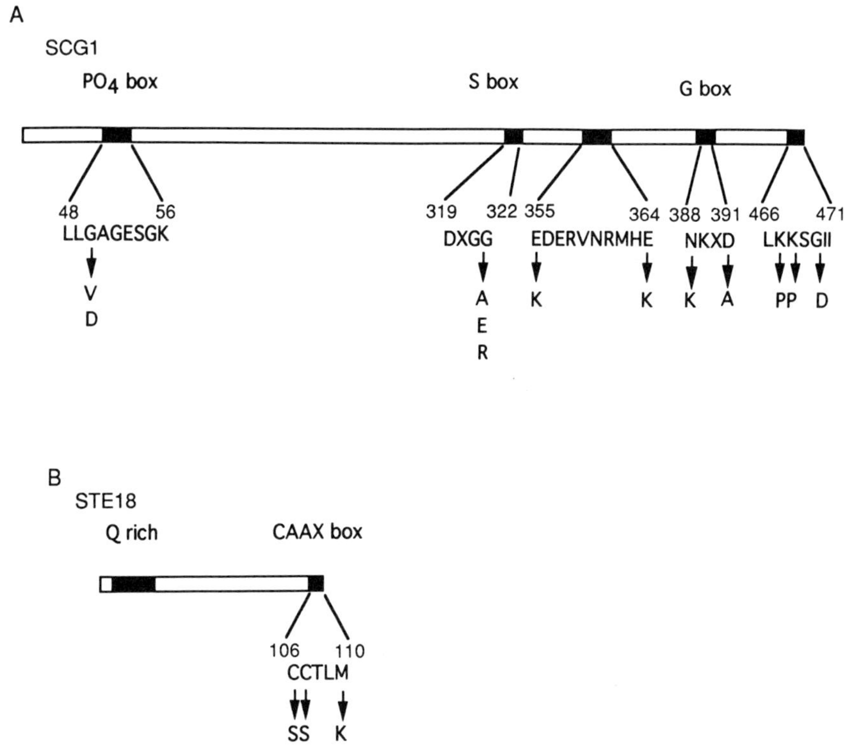

Figure 8.3 Schematic representation of G protein α and γ subunits. (A) Mutagenesis of the Scg1 G_α subunit. Mutations were created either by site-directed mutagenesis (Miyajima *et al.*, 1989; Hirsch *et al.*, 1991; Kurjan *et al.*, 1991) or by random mutagenesis (Stone and Reed, 1990). The PO_4 box is believed to contain the residues important in catalyzing the hydrolysis of GTP; mutations in this region appear to be activating. The G box contains residues that interact with the bound guanine nucleotide; mutations in this region are also activating. The S box defines residues that appear to be important in the conformational switch required to release $G_{\beta\gamma}$ and activate the pathway; mutations in this region lock the protein in an inactive state. The C-terminal residues are proposed to be important for interactions with the receptor. The phenotypes of these various mutations are described in the text. (B) Mutagenesis of the Ste18 G_γ subunit. Mutations in the CAAX box were created by site-directed mutagenesis (Finegold *et al.*, 1990). The C-106 → S mutation leads to reduced mating, the C-107 → S mutation totally inactivates Ste18 function, and the M-110 → K mutation prevents proper isoprenylation and mimics a *dpr1* defect. The phenotypes are described in the text.

of the G_α subunit has a role in signal transduction (initially sensitive) and in adaptation (subsequently pheromone resistant).

The structural relationship of the predicted Scg1 protein to biochemically characterized guanine nucleotide binding proteins such as Ras and G_s has allowed the application of site-directed mutagenesis to test the role of specific residues for Scg1 function. The first such test was the analysis of a Gly-50 → Val substitution in Scg1 (Fig. 8.3A; Miyajima *et al.*, 1989). This substitution is analogous to the oncogenic Val-12 substitution in the guanine nucleotide binding site of Ras. The Ras Val-12 substitution causes reduced GTPase activity of the protein (Gibbs *et al.*, 1984). Therefore, the Gly-50 → Val change was expected to generate a GTPase defective and, hence, "activated" form of the Scg1 protein. Consistent with this expectation, the mutant version of the protein causes hyperactivation of the response pathway. However, the cells eventually adapt or recover, even in the absence of the normal desensitization machinery (Miyajima *et al.*, 1989). As noted, the Gly-50 → Val substitution isolated as a *phun* allele causes a similar phenotype (Stone and Reed, 1990).

Another study evaluated the phenotype of the Gly-50 → Val substitution when the mutant allele replaces the gene at its chromosomal locus. In contrast to the phenotype of the plasmid-expressed allele, the chromosomally expressed Gly-50 → Val substitution protein causes partial activation of the response pathway (Kurjan *et al.*, 1991). This result suggests that the hyperactivation and adaptation responses previously reported could have been the result of aberrant expression of the Gly-50 → Val substituted protein from plasmids. Therefore, the best interpretation at present is that the presumed defect in GTP hydrolysis resulting from the Gly-50 → Val substitution in Scg1 leads to partial activation of the pheromone response pathway. Additional work will be necessary to establish whether the activated form of the protein also influences an adaptive pathway.

Site-directed mutagenesis has been used to change other residues that are predicted to affect the guanine nucleotide binding domain of Scg1 (Kurjan *et al.*, 1991). Changes at positions Asn-388 (Asn-388 → Lys) and Asp-391 (Asp-391 → Ala), analogous to the Ras transforming mutations at positions 116 and 119, resulted in activation of the response pathway. Thus, these mutant alleles behave as nulls, either because activation of the G protein turns on the response pathway or because the substitutions destabilized the protein. Another substitution was introduced by site-directed mutagenesis into a predicted hinge region (Gly-322 → Ala). The substitution was expected to interfere with a conformational change associated with guanine nucleotide exchange. The substitution was analogous to the Gly-322 substitutions identified as *phun* alleles (Stone and Reed, 1990). When the site-directed substitution allele replaced the chromosomal gene, it prevented response to pheromone and resulted in the same nonmating phenotype observed with the plasmid-borne *phun* allele. The nonresponsive phenotype of the mutation is consistent with the interpretation that the altered protein does not undergo the conformational change associated with the guanine nucleotide exchange and $G_{\beta\gamma}$ release (Kurjan *et al.*, 1991).

Mutations at the C terminus of Scg1 have been used to test whether this region is involved in receptor interaction, as has been suggested from a variety of studies on mammalian G protein–receptor interactions. When the wild-type *SCG1* gene was replaced with a mutant version lacking the codons for the C-terminal 5 amino acids, the resulting strain was sterile in both mating types (Hirsch *et al.*, 1991). This phenotype is the opposite of the phenotype of the null mutant and is similar to the effects of the *SCG1* allele isolated by Fujimura (1989) and the Gly-470 → Asp *phun* allele characterized by Stone and Reed (1990). Further evidence that the C terminus of the G_α subunit is important in receptor interactions is the observation that mutations changing adjacent lysine residues of Scg1 have differential effects on mating in *MAT***a** and *MAT*α cells. The Lys-467 → Pro substitution changes a conserved residue implicated in receptor interaction in mammalian cells whereas the Lys-468 → Pro substitution involves a nonconserved residue (Sullivan *et al.*, 1987). The Lys-467 → Pro substitution reduces mating in MAT**a** cells by three orders of magnitude and in *MAT*α cells by more than four orders of magnitude. In contrast, the Lys-468 → Pro substitution reduces mating in MAT**a** cells by four orders of magnitude but has essentially no effect on mating in *MAT*α cells (Hirsch *et al.*, 1991). These results suggest that the adjacent lysine residues may be differentially important in coupling of the G protein to the Ste2 and Ste3 receptors.

3. Arguments in Support of a Guanine Nucleotide Binding Role for Scg1 Although presently no biochemical evidence suggests that the *SCG1* gene product binds guanine nucleotides, the argument that it represents a G protein α subunit is strong. The gene shows extensive similarity to the mammalian G_α subunits and, in fact, some mammalian proteins can complement the cell cycle arrest caused by null mutations in *SCG1* (Dietzel and Kurjan, 1987a). Similar results have been obtained by the construction of chimeras between the yeast and mammalian genes (Kang *et al.*, 1989). In addition, the phenotypes of the various site-directed mutants suggest that domains of function (e.g., guanine nucleotide binding, receptor interactions) are conserved between the mammalian and yeast genes. Finally, the genetic and biochemical interactions between the *SCG1* gene product and the other components of the putative receptor–G protein complex (see subsequent discussion) are all consistent with *SCG1* encoding a G_α subunit. Biochemical evidence for guanine nucleotide binding and exchange, although necessary, will come as no surprise.

B. G Protein β Subunit (Ste4)

1. Role in Pheromone Response The *STE4* gene was identified through several independent mutational selections. *STE4* alleles first were identified in the screen for mating defective (sterile) mutations (MacKay and Manney,

1974). *STE4* alleles were also among the mutations uncovered in a screen for α-factor resistant mutations that caused temperature-sensitive sterile phenotypes (Hartwell, 1980). Finally, alleles of *STE4* that modified pheromone responsiveness without eliminating mating ability were identified in a selection of *MAT***a** cells that were resistant to endogenously produced α-factor (Whiteway *et al.*, 1988) and that gave rise to a growth defect in haploid but not diploid cells (Blinder *et al.*, 1989).

Null alleles of the *STE4* gene, either nonsense mutations or gene disruptions, cause a sterile phenotype in cells of both mating types (Hartwell, 1980; Whiteway *et al.*, 1989). Moreover, the mutant cells are defective in all aspects of the normal mating response; they are unresponsive to pheromone-mediated arrest of the cell cycle and they do not express genes such as *FUS1* that normally are induced by pheromone treatment. Thus, the *STE4* gene product is an essential component of the signal transduction machinery.

Similar to the *SCG1* gene, point mutations have been identified that have the opposite phenotype to that of null mutations. One such *STE4* allele was identified in the selection for mutations of genes whose functions are essential in haploid cells but are dispensable in diploid cells (Blinder *et al.*, 1989). The allele was designated $STE4^{Hpl}$ to indicate its haploid lethal phenotype. The $STE4^{Hpl}$ allele activates the response pathway in a manner similar to null mutations in *SCG1*. Sequence analysis of the $STE4^{Hpl}$ showed it to be a single amino acid change. Further mutagenesis in the same region of the protein identified a number of single amino acid changes that resulted in the same haploid lethal phenotype (E. Leberer, D. Dignard, and M. Whiteway, unpublished observations).

Activation of the pheromone response pathway also can be achieved through overproduction of the wild-type Ste4 protein by expression under control of the strong galactose-inducible promoter (Cole *et al.*, 1990; Nomoto *et al.*, 1990; Whiteway *et al.*, 1990). The galactose-induced overproduction of Ste4 allows mating in the absence of the pheromone receptor in either mating type. Moreover, Ste4 overproduction is sufficient to cause *FUS1* gene induction and cell cycle arrest. Activation of the pathway by Ste4 overproduction can be eliminated by overproduction of Scg1. Thus Ste4 and Scg1 appear to have antagonistic effects on the pheromone response pathway.

Evidence has shown phosphorylation of the Ste4 subunit in response to pheromone treatment of yeast cells (Cole and Reed, 1991). Removal of the region of the protein that appears to contain the bulk of the sites for phosphorylation does not prevent Ste4 from functioning in the signal transduction pathway but does enhance pheromone responsiveness to a small extent (Cole *et al.*, 1991). This effect is reminiscent of the situation with the pheromone receptors, in which removal of the phosphorylation sites at the C terminus of the Ste2 protein generates cells that are hypersensitive to pheromone (Konopka *et al.*, 1988; Reneke *et al.*, 1988). Thus, phosphorylation of various components of the signal transduction pathway may be important for attenuating the pheromone response.

2. *Arguments in support of Ste4 as the G_β subunit* Overall, the structural and genetic evidence strongly suggests that the role of the *STE4* gene product in the yeast mating response pathway is as a G protein β subunit. The sequence similarity between the Ste4 protein and the mammalian $G_{\beta1}$ and $G_{\beta2}$ proteins is strong. The only regions of nonhomology lie in two blocks that are found in Ste4 but are not present in the mammalian proteins. Deletion of these blocks from Ste4 does not severely affect its function (Cole and Reed, 1991). Therefore, these blocks do not appear to define unique structural elements involved in the positive action of the putative yeast G_β in the signaling pathway. Further, the consequences of Ste4 perturbations are consistent with its proposed role as the G_β subunit. For example, the *STE4* gene product is needed for the α-factor receptor to exhibit high affinity binding of its ligand, α-factor (Jenness *et al.*, 1987; Blumer and Thorner, 1990). Such high affinity binding is characteristic of receptors that are coupled to G proteins (Lefkowitz *et al.*, 1983). Therefore, as might be expected from the absence of a G_β subunit, a deficiency of the Ste4 protein compromises G protein function. Additional indirect evidence consistent with the role of Ste4 as the mating response pathway G_β subunit is derived from the phenotype of cells that overproduce the Ste4 protein. The activation of the response pathway caused by overproduction of Ste4 is shut off by overproduction of Scg1, consistent with the α and β subunits of the yeast G protein playing opposing roles, as they do in the mammalian systems.

C. G Protein γ Subunit (Ste18)

1. *Role in Pheromone Response* The *STE18* gene was identified through a selection of *MAT***a** cells that were resistant to endogenously produced α-factor (Whiteway *et al.*, 1988). Cloning and construction of null mutations of the gene established that the *STE18* function was necessary for mating and pheromone response in both haploid cell types. Null mutations of the *STE18* gene also suppressed cell cycle arrest caused by a deficiency in Scg1 or overproduction of Ste4 (Cole *et al.*, 1990; Nomoto *et al.*, 1990; Whiteway *et al.*, 1990).

2. *Domain Structure and Processing* The predicted Ste18 protein shares two motifs with other proteins. One, near the N terminus, is a polyglutamine stretch (Fig. 8.3B). Polyglutamine regions have been found in a number of

proteins, but any function has not been established (Courey and Tijan, 1988). A second motif is found at the C terminus of the protein. This sequence is similar to the C termini of the yeast Ras proteins, with a pair of cysteine residues three amino acids from the C-terminal amino acid, which is a methionine in Ste18 (Dhar *et al.*, 1984; Powers *et al.*, 1984; Chapter 7). Such C termini are part of a general motif that has been termed a CAAX box (Magee and Hanley, 1988), which defines a site of complex processing.

The processing of Ste18 involves the addition of an isoprenoid lipid to the cysteine residue of the CAAX box, C-terminal proteolysis leaving the isoprenylated cysteine as the C-terminal residue, and C-terminal methylation of the cysteine residue. Isoprenylation of the Ste18 protein requires the action of the *DPR1* (*RAM1*, *STE16*) gene product (Finegold *et al.*, 1990). The *DPR1* gene product also is required for C-terminal isoprenylation of the **a**-factor mating pheromone by farnesyl (Powers *et al.*, 1986; Fujiyama *et al.*, 1987). This observation suggests that the isoprenoid lipid added to Ste18 is farnesyl, but biochemical determination of the structure has not yet been achieved.

Site-directed mutations have been constructed in the CAAX box region to assess the role of various residues in the function of the Ste18 protein. In particular, the paired cysteine residues have been changed to serines (Fig. 8.3B). When Cys 107, which becomes isoprenylated, is changed to serine, all biological function of the Ste18 protein is abolished, resulting in sterility in both mating types (Finegold *et al.*, 1990). Intriguingly, loss of farnesyl transferase activity does not severely compromise mating in *MAT***a** cells (Finegold *et al.*, 1990). Therefore, the cysteine at position 107 appears to serve as more than simply the site of isoprenylation. The cysteine at position 106 is important but not essential for function of the Ste18 protein. When this amino acid is substituted with serine, cells containing the mutant protein show greatly reduced pheromone sensitivity, and the mating efficiency of the cells is reduced (Finegold *et al.*, 1990).

3. Arguments in Support of Ste18 as the G_γ Subunit Several lines of evidence suggest that the *STE18* gene product is a yeast homolog of a G protein γ subunit. The gene itself encodes a small protein (110 amino acids) and G protein γ subunits are small. The sequence similarity between Ste18 and other G_γ subunits is not strong, but all currently identified G_γ subunits share the common C-terminal CAAX box. The phenotype that initially identified a mutant allele of the *STE18* gene was a decreased responsiveness to high but not to low levels of α factor. This phenotype is the same as the one found for some mutant alleles of the *STE4* gene. This result is consistent with these proteins acting at a similar step or having similar roles in the pheromone response pathway (Whiteway *et al.*, 1989).

Additional evidence that Ste18 functions as the G protein γ subunit is derived from studies on the requirement for this protein in the formation of a high affinity receptor. As previously noted, when G protein-linked receptors are coupled to their cognate G protein, they show enhanced affinity for their ligands. The formation of a high affinity Ste2 receptor protein was shown to be dependent on all three predicted G protein subunits, Scg1, Ste4, and Ste18 (Blumer and Thorner, 1990). The role of the Ste18 protein in generating a high affinity receptor may be indirect. For example, Ste18 could be required for the proper functioning of some other G protein subunit. However, the simplest explanation is that Ste18 is a yeast G_γ subunit. In keeping with this interpretation, we used an *in vivo* protein interaction assay to show that the Ste4 and Ste18 proteins can associate physically, as would be predicted if they were functioning as G protein β and γ subunits (Clark *et al.*, 1993).

Because the G protein subunits interact with a membrane-associated receptor, and yet themselves have no hydrophobic membrane-spanning domains, their membrane attachment may be mediated through the posttranslational modifications. The farnesyl transferase (*DPR1*) gene is not essential for signal transduction in wild-type cells, but is needed for the constitutive activation of the pathway in cells lacking Scg1 (Nakayama *et al.*, 1988a). This observation suggests that although farnesylation is dispensable in the presence of Scg1, at least one component of the signal transduction machinery must be farnesylated for signal transmission in the absence of Scg1. Ste18 may be the component that requires farnesylation. In support of this notion is the observation that some CAAX substitution mutants of Ste18 are functional for signal transmission in otherwise wild-type cells but suppress the constitutive activation of the pathway caused by the absence of Scg1 (M. Whiteway, unpublished observations). Since the Scg1 protein is membrane-associated in the presence or absence of the Ste4 or Ste18 proteins (Blumer and Thorner, 1990) unfarnesylated Ste18 proteins may be localized correctly to the membrane via their association with the Scg1 protein.

IV. Post-G Protein Signal Transmission Components

A. Effectors and Second Messengers

Mammalian heterotrimeric G proteins function as signal transducers between a membrane-spanning receptor molecule and a second message-generating enzyme termed an effector (Gilman, 1987; Neer and Clapham, 1988; Simon *et al.*, 1991). Some well-studied effectors are adenylate cyclase, phospholipase C, and ion channels. The effector enzyme-generated second messenger molecules serve three general purposes. First, the second messengers are mobile signaling molecules that move the signal away from the plasma membrane to other parts of the cell. Second, they allow for signal amplification. Third, they provide a level of signal integration.

As yet, broad genetic screens for pheromone response mutants have not uncovered a second message-generating enzyme in this pathway. Mutants defective in an effector activity may have remained unidentified for a variety of reasons. For example, if the effector is encoded by more than one gene, the redundancy would make mutational inactivation of the function infrequent. An alternative strategy to identify the pheromone response pathway effector is to test small molecules for effects on the yeast pheromone response. Specifically, cAMP and Ca^{2+} have been tested because each has a second message function that is associated with an effector enzyme in mammalian signal transduction pathways. However, in yeast, neither of these molecules has effects that are consistent with a second message function in pheromone response (Casperson *et al.*, 1983; Iida *et al.*, 1990).

The failure to identify any obvious "effector-like" activity is not proof that such a function does not exist. However, the yeast pheromone response pathway may function without an effector enzyme that generates a second messenger. In general, the second messenger molecules of mammalian cells serve to stimulate a variety of protein kinases. As detailed in this section, the pheromone pathway contains several protein kinases. $G_{\beta\gamma}$ may activate a kinase cascade directly that is sufficient for signal transmission and amplification. Additionally, a signal integration function may be dispensable because of the relative simplicity of signaling pathways in *S. cerevisiae*.

B. Protein Kinases (Ste7, Ste11, Fus3, Kss1)

Signal transmission events after the ligand–receptor interaction often involve the activation of one or more protein kinases (Hunter *et al.*, 1988, Ullrich and Schlessinger, 1990). The catalytic activity of any one protein kinase has the potential to alter the phosphorylation state of many substrate proteins. Thus, a single ligand–receptor interaction can generate changes in state of many intracellular target proteins that ultimately cause a functional response to the stimulus. Continuing challenges are to identify kinases that participate in signal transmission and response, to learn what regulates these kinases in response to signal, and to identify their target proteins. However, understanding the precise role of any given protein kinase can be complicated. Some complications arise because of functional redundancies among kinase proteins. Others arise because one kinase can have multiple targets, each of which contributes to a distinct event in the overall response. Finally, activation of any given target may require phosphorylation at separate sites by more than one kinase. Several studies on kinases that are components of the yeast pheromone response pathway address some of these issues.

The *STE11*, *STE7*, *FUS3*, and *KSS1* genes that are necessary for pheromone-induced signal transmission encode predicted protein kinases. Different genetic strategies led to the initial identification of these genes. A screen for cells that become resistant to pheromone-induced growth arrest uncovered mutations in the *STE7* and *STE11* genes. These *ste7* and *ste11* mutations block pheromone-induced gene expression and cause a nonmating phenotype in both **a** and α cells (Hartwell, 1980). The nonmating phenotype of the mutants allowed direct cloning of the respective genes by complementation (Chaleff and Tatchell, 1985). The *KSS1* gene was identified because its overproduction by expression on high copy number plasmids allowed recovery of cells that are defective in recovery from the pheromone-induced signal (Courchesne *et al.*, 1989). Two distinct genetic screens uncovered mutations in the *FUS3* gene. One was a screen for mutations that caused cells to be defective in cell fusion during mating (Elion *et al.*, 1990). The other was a screen for mutant cells that were defective in the growth arrest response to pheromone but still remained capable of undergoing morphological alterations (shmoo-formation) characteristic of the pheromone-induced state (Fujimura, 1990a,b). The *FUS3* gene was cloned by complementation of the mating defect associated with the mutant alleles. The Ste7, Ste11, and Fus3 proteins function in signal transmission after the G protein because null mutations of the corresponding genes block the constitutive G_1 arrest caused by loss of G_α function (Nakayama *et al.*, 1988b; Elion *et al.*, 1990; Fujimura, 1990a).

The DNA sequence of all four genes revealed a protein coding region related to the catalytic domain of the protein kinase family of polypeptides. Certain conserved sequence motifs within the domain are highly characteristic of this family of proteins (Hanks *et al.*, 1988). An alignment of the putative catalytic domain sequences of the pheromone response pathway kinases is presented by Rhodes *et al.* (1990). These proteins have in common all the amino acids that are conserved in the kinase catalytic domain (Hanks *et al.*, 1988). In addition, these kinases have amino acid sequence motifs in subdomains VI and VIII that are indicative of the serine/threonine subfamily of protein kinases. A schematic representation of the alignments is given here to show the catalytic domain in relation to the overall structure of each kinase (Fig. 8.4). The boundaries of subdomains I–XI, with GXGXXG . . . AXK sequence motifs that are associated with the ATP binding region, and the DFG . . . PE motifs that define the acceptor region provide points of reference (Hanks *et al.*, 1988; Knighton *et al.*, 1991).

The predicted Fus3 and Kss1 proteins are closely related to each other (54% identity) (Courchesne *et al.*, 1989; Elion *et al.*, 1990). These proteins are members of an evolutionarily conserved family of protein kinases that includes

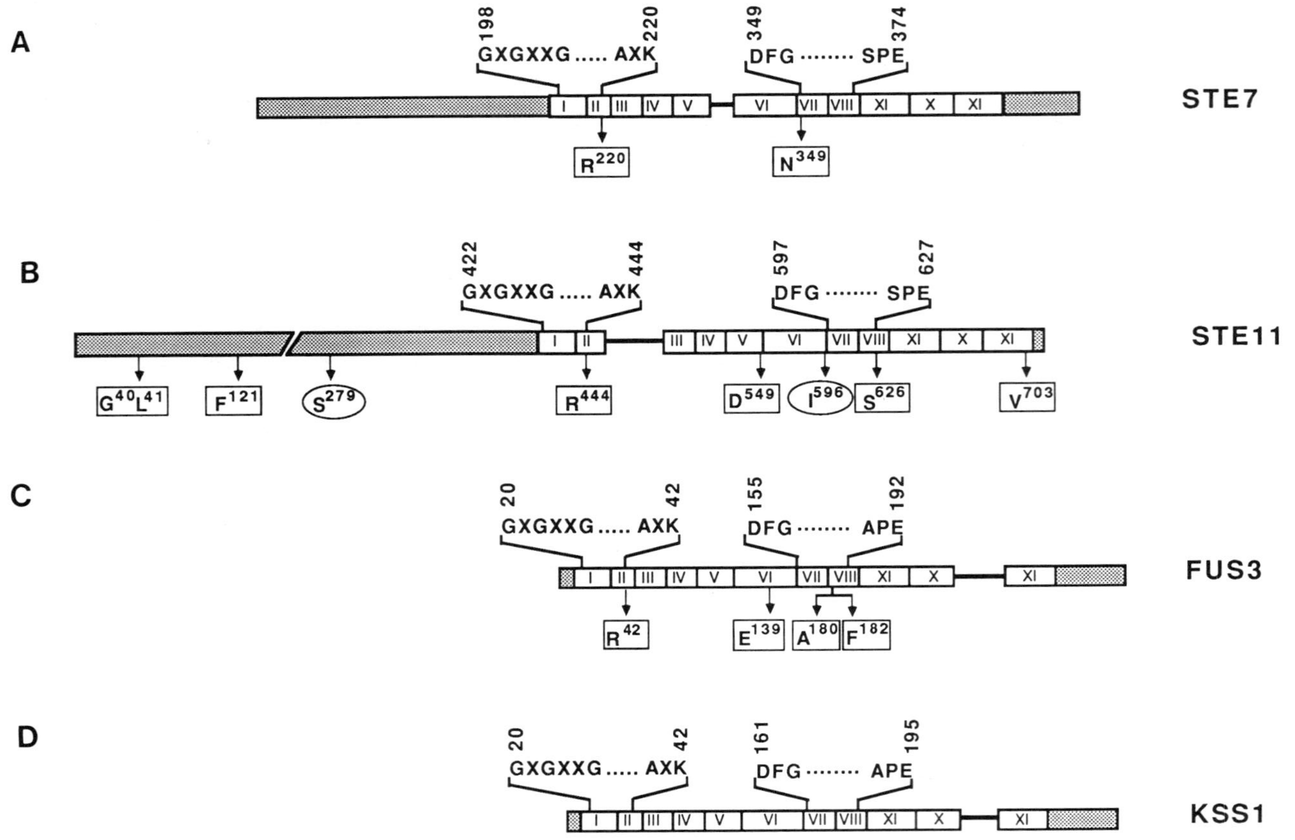

A
198
GXGXXG..... AXK
220
349
DFG .... SPE
374
I II III IV V VI VII VIII XI X XI
R220
N349
STE7
B
422
GXGXXG..... AXK
444
597
DFG .... SPE
627
I II III IV V VI VII VIII XI X XI
G40L41
F121
S279
R444
D549
I596
S626
V703
STE11
C
20
GXGXXG..... AXK
42
155
DFG .... APE
192
I II III IV V VI VII VIII XI X XI
R42
E139
A180
F182
FUS3
D
20
GXGXXG..... AXK
42
161
DFG .... APE
195
I II III IV V VI VII VIII XI X XI
KSS1

Spk1, Erk1 (Map2) and Erk2 (Boulton *et al.*, 1990, 1991; Toda *et al.*, 1991). In contrast, the Ste7 and Ste11 kinases are no more related to each other or to the Kss1 and Fus3 kinases (22–27% identity) than to the protein kinase family in general (Teague *et al.*, 1986; Rhodes *et al.*, 1990). Nonetheless, identification of Ste7 and Ste11 kinase homologs in other organisms makes it likely that each is a member of another evolutionarily conserved kinase subfamily. For example, the Byr1 and Byr2 proteins of the fission yeast *Schizosaccharomyces pombe* are homologous to the Ste7 and Ste11 proteins, respectively (Nadin-Davis and Nasim, 1988; Wang *et al.*, 1991). Additionally, the *S. pombe* proteins can complement partially mutations of their counterparts in *S. cerevisiae* (A. Neiman and I. Herskowitz, unpublished observations). Homologs of Ste7 also have been identified in *Drosophila* and *Xenopus* (A. Neiman and I. Herskowitz, unpublished observations; B. Yashar, L. Zon, and B. Errede, unpublished observations).

1. Ste7 The first *ste7* mutations to be isolated were conditional alleles (temperature-sensitive; Hartwell, 1980). At the nonpermissive temperature, *ste7*^{ts} strains of either the **a** or α mating type were defective for more than

Figure 8.4 Schematic alignment of the Ste7, Ste11, Fus3, and Kss1 protein kinases. The conserved kinase catalytic region of each kinase is indicated by the open boxes designated I–XI that specify subdomains as defined by Hanks *et al.*, (1988). Insertions between conserved subdomains are indicated by solid lines. The locations of conserved motifs associated with the ATP binding (GXGXXG . . . AXK) and acceptor (DFG . . . PE) regions are shown. The amino acid positions at the boundaries of the conserved motifs are specified above each diagram for reference. Regions of the proteins outside the catalytic domains are indicated by the shaded boxes. The amino acid substitutions corresponding to different mutant alleles of each kinase are specified below the respective diagrams. Substitutions corresponding to nonresponsive alleles are in boxes; those corresponding to gain of function alleles are in ovals. The diagrams in A, B, C, and D are derived from the predicted protein sequences of Teague *et al.* (1986), Rhodes *et al.* (1990), Elion *et al.* (1990), and Courchesne *et al.* (1989), respectively. (A) The Ste7R-220 and Ste7N-349 substitution mutations were generated by site-directed mutagenesis (M. Teague and B. Errede, unpublished observations). (B) The Ste11S-626 and Ste11V-703 substitution mutations represent the original *ste11 ts* alleles of strains 381G-53A and 381G-41A, respectively (Hartwell, 1980; Rhodes, 1991). Random mutagenesis of the cloned *STE11* gene uncovered three additional substitution mutations: Ste11D-549, Ste11G-40L-41, and Ste11F-121 that also cause a *ste11 ts* phenotype (C. Cassavant and B. Errede, unpublished observations). The Ste11I-596 and Ste11S-279 substitutions were recovered in a screen for mutations that activate the pheromone-response pathway in the absence of the G protein (Stevenson *et al.*, 1992). (C) The Fus3R-42, Fus3A-180, and Fus3F-182 substitution mutations were generated by site-directed mutagenesis of the cloned gene (Gartner *et al.*, 1992). The Fus3E-139 substitution corresponds to the *fus3A* mutation. This allele was isolated in a screen for mutations that cause cells to be defective in the division arrest response to pheromone but still capable of activating the pheromone-inducible *FUS1:lacZ* reporter gene (G. Ammerer, A. Amon, and K. Nasmyth, unpublished observations).

one function that contributes to the mating process. The mutant strains produced reduced amounts of **a**-factor, of an α-factor protease (Bar1), and of α-factor receptor (Ste2) (Hartwell, 1980; Jenness *et al.*, 1987). The mutant strains also failed to agglutinate and to arrest at the G_1 stage of the cell cycle in response to pheromone (Hartwell, 1980). Because strains with a complete deletion of *ste7* exhibit the same phenotype, these defects in the mating response are characteristic of null mutations. These pleiotropic effects of *ste7* null mutants suggest that the Ste7 product has a regulatory role in the mating process.

Some evidence supports the hypothesis that Ste7 kinase activity is important for its regulatory role in the mating response. First, *in vitro* assays showed that the Ste7 protein autophosphorylates and, by this criterion, is a protein kinase (Zhou *et al.*, 1993). Second, Ste7 substitution mutations that correspond to amino acids important for the catalytic activity of known protein kinases have a null phenotype. For example, in the yeast cAMP dependent protein kinase (Tpk1) Lys-116 and Asp-228 appear to have a role in catalysis (Gibbs and Zoller, 1991). The three-dimensional structure of the analogous mammalian enzyme shows that the location of the corresponding residues (Lys-72 and Asp-184, respectively) is consistent with their being important for catalysis (Knighton *et al.*, 1991). The corresponding positions in Ste7 are Lys-220 and Asp-349 (Fig. 8.4A). Strains expressing either the Ste7 Arg-220 or Ste7 Glu-349 substitution proteins do not mate and are completely defective for the transcriptional and G_1 arrest responses to pheromone (R. Cade, M. Teague, and B. Errede, unpublished results).

Because Ste7 kinase function is important only for the mating response, its expression could be restricted to cell types or conditions in which its activity is required. However, no evidence suggests that Ste7 expression is regulated transcriptionally. *STE7* mRNA is present in the three cell types of yeast (Chaleff and Tatchell, 1985). Also, pheromone induction does not affect the amount of *STE7* mRNA (C. Adler and B. Errede, unpublished results). In contrast, posttranscriptional events appear to be key regulators of Ste7 activity. Under pheromone-induced conditions the Ste7 protein becomes hyperphosphorylated. The timing of the increase in Ste7 hyperphosphorylation correlates with the appearance of physiological responses to pheromone. Additionally, the hyperphosphorylated Ste7 from extracts of pheromone-induced culture is more active in autophosphorylation assays than is the protein from uninduced cultures (Zhou *et al.*, 1993). These results are consistent with the model that Ste7 kinase activity is stimulated in response to pheromone.

2. *Ste11* The first *ste11* mutations to be isolated were conditional alleles (temperature-sensitive) and were recovered in the same screen that uncovered *ste7* mutations (Hartwell, 1980). At the nonpermissive temperature, $ste11^{ts}$

strains of either the **a** or α mating type have the identical phenotype as *ste7*^{ts} strains (Hartwell, 1980; Jenness *et al.*, 1987). Amino acid substitutions resulting from the original conditional alleles and other temperature-sensitive substitution mutations are located throughout the protein (Fig. 8.4B). Because strains with a deletion of 99% of the *STE11* coding sequence exhibit the same phenotype, the phenotype of the conditional mutants at the restriction temperature represents the null phenotype (Rhodes *et al.*, 1990). Finally, a substitution of the catalytically important lysine residue at Ste11 position 444 (Fig. 8.4B, Ste11 Arg-444) also has a null mutant phenotype (Rhodes *et al.*, 1990). The response defect resulting from the substitution mutations in the kinase catalytic domain suggests that Ste11 kinase activity is essential for its role in regulation of the mating response. The recovery of mutations in the N-terminal region suggests that this domain is necessary either for the function or for the stability of the protein.

Two dominant alleles of *STE11* have the interesting property that they activate expression of pheromone-responsive genes even in the absence of the pheromone-induced signal (Fig. 8.4B; Ste11 Ser-279 and Ste Ile-596). For example, the activity of a *FUS1–lacZ* reporter gene in either Ste11 Ser-279 or Ste11 Ile-576 strains in the absence of pheromone is twofold higher than that in Ste11 strains under pheromone-induced conditions (Stevenson *et al.*, 1992). Therefore, these are constitutive or gain-of-function alleles. The constitutive phenotype conferred by these mutations indicates that the pheromone-induced signal results in activation of wild-type Ste11 kinase. The location of the Ste11 Ser-279 and Ste11 Ile-596 substitutions is compatible with a model in which the N-terminal region negatively interacts with the catalytic domain. The pheromone-induced signal then would relieve the negative interaction. Precedents exist for negative regulatory domains in other protein kinases (see Taylor *et al.*, 1990).

How pheromone induction stimulates Ste11 function is not yet clear. The stimulation does not involve increased transcription because the amount of *STE11* mRNA is the same in all cell types (**a**, α, and **a**/α) and is not increased under pheromone-inducing conditions (Chaleff and Tatchell, 1985; Rhodes *et al.*, 1990). The discovery of Ste11 gain-of-function alleles suggests that pheromone induction may regulate Ste11 kinase activity. The specific phosphorylation of a 78-kDa protein (p78) that coprecipitates with Ste11 demonstrates that Ste11 is a kinase (Rhodes *et al.*, 1990). However, no difference in p78 phosphorylation occurred with Ste11 immunoprecipitated from culture extracts isolated under pheromone-induced conditions. This particular *in vitro* kinase assay may be insensitive to critical regulatory mechanisms.

The p78 protein, whose identity is unknown, may be a physiological substrate of the Ste11 kinase. p78 is present and phosphorylated by Ste11 in immune complexes from protein extracts of strains with deletions of many known genes that function in pheromone response or recovery (i.e., *ste2*Δ, *ste4*Δ, *ste5*Δ, *ste7*Δ, *ste12*Δ, *fus3*Δ *kss1*Δ, *sst2*Δ, and *cln2*Δ *cln3*Δ). These results indicate that none of these genes encode p78 or regulate the ability of Ste11 to phosphorylate p78 *in vitro* (Rhodes, 1991).

3. *Fus3 and Kss1* Fus3 has two distinct roles in the pheromone response. One role is signal transmission leading to pheromone-induced transcription. The other is regulation of cyclins that control progression of the cell cycle (discussed subsequently). However, the effects of *fus3* mutations on these responses are variable and appear to be highly dependent on factors that are not fully characterized. In the initial characterization, *fus3* null mutants were reported to have a phenotype identical with that of *ste7* or *ste11* null mutants, that is, complete defects for pheromone-induced transcriptional and division arrest responses, resulting in sterility (Elion *et al.*, 1990). However, this strain background later was found to have a *kss1* mutation, indicating that the sterile phenotype reflects the *fus3 kss1* double mutant phenotype (Elion *et al.*, 1991). Single *kss1* null mutations have little effect on the mating response in strains that otherwise have an intact signal pathway (Courchesne *et al.*, 1989). Single *fus3* null mutants also have a much less severe mating defect than the *fus3 kss1* double mutants. In some strain backgrounds, *fus3* null mutants are defective for the pheromone-induced G_1 arrest response (Fujimura, 1990a; Elion *et al.*, 1991), but in other strain backgrounds they show only subtle effects on pheromone-induced division response (B. Errede, unpublished results). The severe defect of the *fus3 kss1* double mutant in comparison to the single mutants indicates that Fus3 and Kss1 are completely redundant for a step in signal transmission leading to transcription activation. Because strain differences influence the *FUS3* role in signal transmission, finding that variations in genetic background similarly influence its role in cell division arrest would not be surprising (see subsequent text).

In addition to its redundant role in transcription activation, Kss1 is implicated in recovery from pheromone-induced G_1 arrest (Courchesne *et al.*, 1989). *Kss1* deletion mutants have been reported to be defective in recovery from pheromone-induced G_1 arrest. In some strains, *kss1*Δ alleles cause a doubling time that is about twofold longer than that of *KSS1* strains. The predominant phenotype associated with Kss1 is that its overproduction promotes recovery from pheromone-induced growth arrest. Additional observations suggested that Kss1 might positively regulate Cln3 to promote recovery from G_1 arrest. However, the phenotype associated with overproduction of a protein could be misleading and may not give an accurate insight into its normal role. For example, if the proposed model is correct, suppression of the growth defect of an *scg1* null mutation by Kss1 overproduction should require a functional Cln3 protein, yet suppression of the growth defect occurs even in the absence of Cln3 (unpublished data cited in Irie *et al.*, 1991).

No current direct biochemical evidence exists that either Kss1 or Fus3 is a protein kinase. Nevertheless, the high sequence similarity of these two yeast proteins to mammalian MAP kinases suggests that they possess this enzymatic function. Whether these kinases are regulated in response to pheromone is still unknown. Like *STE7* and *STE11*, *KSS1* is transcribed constitutively in haploids and is expressed even in **a**/α diploids (Courchesne *et al.*, 1989). In contrast, *FUS3* expression is repressed in diploid cells and enhanced by pheromone treatment (Elion *et al.*, 1990). Because high level expression of Fus3 from a constitutive promoter does not cause a transcriptional or division arrest response, merely increasing the amount of Fus3 by pheromone induction is not likely to be sufficient for its function (Gartner *et al.*, 1992).

Posttranscriptional events are critical for stimulating Fus3 activity and also are likely to be important for Kss1 (Gartner *et al.*, 1992). Pheromone induction causes a rapid phosphorylation of Fus3 and Kss1 on threonine and tyrosine residues. Similar to mammalian MAP kinases, these modifications occur at two amino acids: Fus3 Thr-180 and Fus3 Tyr-182. Substitution mutations introduced at either of these positions in Fus3 cause complete loss of function *in vivo* (Fig. 8.4C, Fus3 Ala-180 and Fus3 Phe-182). These results show that pheromone-induced phosphorylation is critical for activating Fus3 function. Similarly, a substitution mutation of the catalytic lysine residue of Fus3 (Fus3 Arg-42) causes complete loss of function. Therefore, the activity of this protein as a kinase must be essential for its *in vivo* function. Interestingly, the inactive Fus3 Arg-42 protein is phosphorylated on both Thr-180 and Tyr-182 under pheromone-induced conditions. This result excludes the possibility that the phosphorylations are autocatalytic (Gartner *et al.*, 1992). In contrast, Fus3 phosphorylation does require function of both the Ste11 and Ste7 gene products (see subsequent discussion).

4. Functional Relationship between Multiple Protein Kinases We derived the present model for the functional relationship of Ste11, Ste7, and Fus3/Kss1 kinases in signal transmission (Fig. 8.1) from the results of two types of experiment. One involves genetic epistasis analyses in which the order of function for two gene products in a common pathway is inferred from the phenotype of the double mutant when the single mutants show different phenotypes. To determine the order of action of the kinases in the pheromone response pathway, strains were constructed with a constitutive response allele of *STE11* and nonresponsive alleles of the other genes. The most rigorous establishment of functional order requires reciprocal combinations of the two types of alleles. Both constitutive and nonresponsive alleles for the gene pairs in question were not available in the reported studies, but the physiological response of protein phosphorylation provided an independent criterion to clarify deductions made from the genetic logic.

Strains with a constitutive allele encoding an activated form of Ste11 and a nonresponsive *ste7* deletion alelle have the phenotype of the nonresponsive mutant. Specifically, the double mutant strain fails to express high levels of a reporter gene under control of the *FUS1* upstream activating sequence (UAS). The same result was obtained for strains that have a constitutive *STE11* allele and deletion alleles of both *fus3* and *kss1* (Stevenson *et al.*, 1992). From these results we infer that Ste7 and Fus3/Kss1 function in signal transmission either interdependently or after Ste11. The observation that Ste11 activity is required for phosphorylations that apparently activate the Ste7 and Fus3/Kss1 kinases favors the interpretation that these kinases function after Stell (Gartner *et al.*, 1992; Zhou *et al.*, 1993) This interpretation gains additional support from the observation that the constitutive alleles of *STE11* cause Ste7, Fus3, and Kss1 phosphorylation even in the absence of pheromone (Gartner *et al.*, 1992; Stevenson *et al.*, 1992). Another implication is that Ste7 and Fus3 kinases are activated directly or indirectly by an Ste11-dependent phosphorylation. Thus, signal transmission from the G protein to downstream targets is likely to involve a protein kinase cascade.

The relationship between the Ste7 and Fus3/Kss1 kinases is not straightforward. The simplest model would be that the two kinases function after Ste11, but in some sequential order. However, predictions based on a linear pathway are inconsistent with the available data. Analysis of protein phosphorylation in different mutant backgrounds shows that Ste7 phosphorylation requires a functional Fus3 or Kss1 kinase (Zhou *et al.*, 1993) and that phosphorylation of Fus3 requires a functional Ste7 kinase (Gartner *et al.*, 1992). This outcome suggests that Ste7 and the Fus3/Kss1 kinases may be interdependent, which could be important for rapid signal amplification and subsequent attenuation.

V. Potential Targets of the Pheromone-Induced Signal

Exposure to pheromone affects the expression of genes whose products execute functions associated with multiple aspects of the mating process (Table II). An obvious prediction is that at least one transcription factor is a target of the pheromone-induced signal. The product of the Ste12 gene has properties consistent with this role. The pheromone-induced state also causes cells to arrest in the G_1 phase of the cell cycle. Therefore, regulators of cell cycle progression also are likely to be direct or indirect targets of the pheromone-induced signal.

TABLE II
Pheromone Responsive Genes

Mating response function	Gene	Change in mRNA (+Pheromone/ −Pheromone)	References
Pheromone production	*MFα1*	~ 3-fold ↑	Achstetter (1989)
	STE13	~ 3-fold ↑	Achstetter (1989)
	MFα2	~ 6-fold ↑[a]	Jarvis *et al.* (1988)
Receptor production	*STE2*	~ 4-fold ↑	Nakayama *et al.* (1985); Hartig *et al.* (1986)
	STE3	~ 4-fold ↑	Hagen and Sprague (1984); Jarvis *et al.* (1988)
Signal transmission	*GPA1*	~ 3-fold ↑[b]	Jahng *et al.* (1988)
	FUS3	~ 4-fold ↑[b]	Elion *et al.* (1990)
Growth arrest	*FAR1*	~ 5-fold ↑	Chang and Herskowitz (1990)
	CLN1	>10-fold ↓[b]	Wittenberg *et al.* (1990); Cross and Tinkelenberg (1991)
	CLN2	~ 5-fold ↓	Wittenberg *et al.* (1990); Dirick and Nasmyth (1991); Cross and Tinkelenberg (1991)
	CLN3	~ 3-fold ↑	Nash *et al.* (1988); Wittenberg *et al.* (1990)
Nuclear migration	*KAR3*	~20-fold ↑	Meluh and Rose (1990)
Morphological alteration and cell fusion	*FUS1*	>40-fold ↑	McCaffrey *et al.* (1987); Trueheart *et al.* (1987)
	FUS2	>40-fold ↑[b]	Elion *et al.* (1990)
	CHS1	~ 6-fold ↑	Appeltauer and Achstetter (1989)
Recovery	*SST2*	~50-fold ↑[c]	Dietzel and Kurjan (1987b)
	BAR1	~ 5-fold ↑[c]	Kronstad *et al.* (1987)

[a]Results of unpublished data reported by Jarvis *et al.* (1988).
[b]Our estimate of data presented in the cited work.
[c]Measurements were from reporter gene activities of promoter fusions to *lacZ* and not directly on the steady state mRNA.

A. Transcription Factor (Ste12)

1. Role in Pheromone Response The first *Ste12* mutations to be identified were conditional null alleles (temperature-sensitive) that caused a pheromone nonresponsive phenotype (Hartwell, 1980; Jenness *et al.*, 1987). The phenotype of various double mutant strains indicated the relative order of Ste12 function with respect to other pathway components. First, strains that have an *ste12* null mutation and a mutation in either the G_α subunit (*scg1*) or the G_β subunit ($STE4^{Hpl}$) have the nonresponsive phenotype of the Ste12 mutation and not the constitutive response phenotype of the G protein mutations (Nakayama *et al.*, 1988b; Blinder *et al.*, 1989). From these results, we infer that Ste12 functions after the G protein α and β subunits in signal transmission. Overexpression of Ste12 causes constitutive responses to pheromone that are similar to the effect of the *STE4*^{Hpl} mutant or overexpression of Ste4. Overexpression of Ste12 in *ste7* or *ste11* single mutants or in *kss1 fus3* double mutants produces the constitutive response phenotype (Dolan and Fields, 1990; Elion *et al.*, 1991), indicating that Ste12 functions after the Ste7, Ste11, and Fus3/Kss1 kinases. This placement in the signal pathway is consistent with its proposed role as a pheromone response pathway transcription factor.

An early indication that Ste12 regulates transcription was derived from analysis of the *CYC7-H2* gene, which has a Ty1 transposable element insertion in the promoter region of the structural gene for iso-2-cytochrome *c*. The Ty1 insertion causes expression of iso-2-cytochrome *c* to be haploid specific and to depend on a functional *STE12* product (Errede *et al.*, 1980). This pattern of regulation results from the fortuitous presence in the Ty1 element of transcriptional control sequences that respond to mating-type specific regulatory factors (Errede *et al.*, 1987; Company and Errede, 1988; Company *et al.*, 1988; Errede and Ammerer, 1989). These early observations on regulation of *CYC7-H2* were important because they established that Ste12 directly or indirectly regulates gene expression at the level of transcription. Other regulatory roles for Ste12 could be excluded since iso-2-cytochrome *c* is an electron transfer protein with no role in mating. The role of Ste12 in regulating transcription of genes involved in mating was confirmed when such genes were cloned. For example, even in the absence of pheromone, the amounts of mRNA for the pheromone receptors (*STE2* and *STE3*), the pheromones (*MF***a**2 and *MF*α2), and other genes (*FUS1* and *FAR1*) are reduced in *ste12* mutants compared with normal strains (McCaffrey *et al.*, 1987; Fields *et al.*, 1988; Chang and Herskowitz, 1990; Hagen *et al.*, 1991).

2. *DNA Binding* Ample evidence is now available that Ste12 is a DNA binding protein, supporting its role as a transcription factor. As is common for transcription factors, Ste12 interacts with other proteins when binding to DNA (Fig. 8.5A). The different associations allow flexibility in DNA recognition and underlie the qualitative and quantitative differences in the pattern of expression for various genes. For example, *STE2* is expressed at a significant level under basal conditions and is stimulated approximately four-fold under pheromone-induced conditions (Nakayama *et al.*, 1985; Hartig *et al.*, 1986) whereas *FUS1* expression is barely detectable under basal conditions and is stimulated at least 40-fold by pheromone (McCaffrey *et al.*, 1987; Trueheart *et al.*, 1987).

The *STE2*, Ty1, and *FUS1* UASs represent three distinct types of Ste12-dependent elements (Fig. 8.5B). Each of these UASs is sufficient to activate reporter gene expression in yeast. This activation is dependent on the presence of a functional Ste12 protein (Company *et al.*, 1988; Errede and Ammerer, 1989; Hagen *et al.*, 1991). The most noticeable sequence similarity between

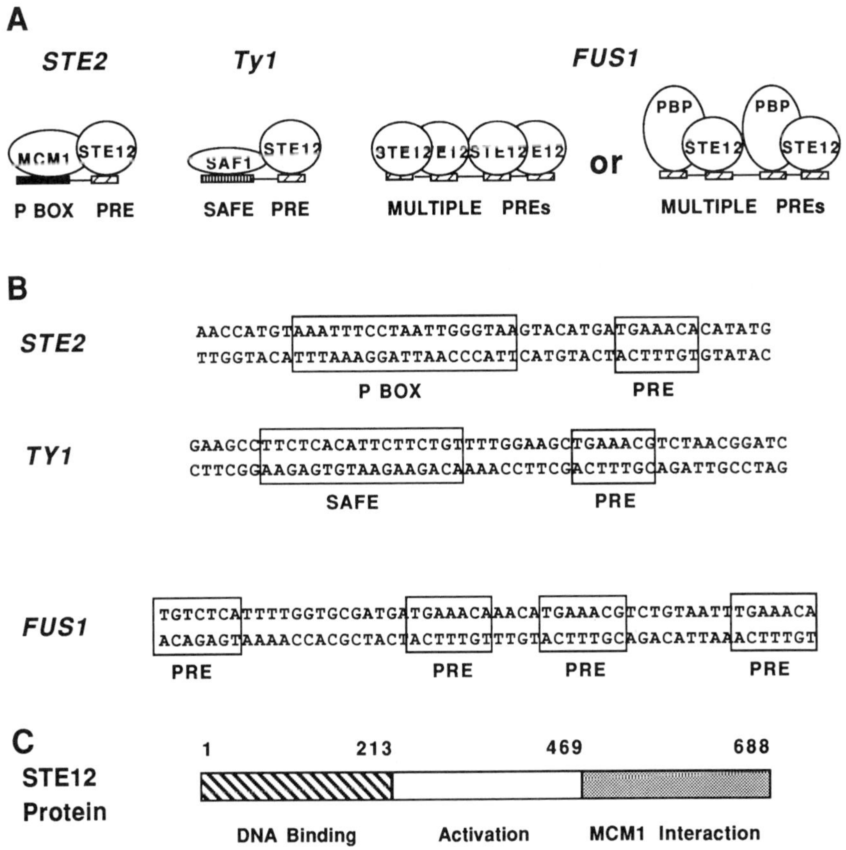

Figure 8.5 Representative Ste12-dependent upstream activating sequences (UAS). (A) Models for Ste12 binding at different UAS sites. Ste12 binds to PRE sequences present in three distinct Ste12-dependent UASs. In addition to a single PRE, the *STE2* UAS has a P box that is a binding site for the Mcm1 protein, whereas the *Ty1* UAS has a SAFE sequence that is a binding site for Saf1. The *FUS1* UAS is composed of multiple PRE sequences without a distinct binding site for any other recognized factor. Two mechanisms for pheromone induction from this element are proposed: either multiple Ste12 proteins bind without ancillary factors at the multiple sites or Ste12 binds in association with an as yet unidentified PRE binding protein (PBP) that is postulated to cooperate with Ste12 for stable complex formation. (B) The pheromone-inducible *STE2* UAS sequence is at nucleotide positions 305–351 (Nakayama *et al.*, 1985). The *Ty1* UAS sequence is at nucleotide positions 283–234 (Errede *et al.*, 1985). The *FUS1* UAS sequence is at nucleotide positions 3017–3071 (van Arsdell and Thorner, 1988). (C) The DNA binding, activation, and Mcm1 interaction domains of Ste12. Amino acid positions for the predicted protein sequence are given above the diagram for reference (Errede and Ammerer, 1989).

the three UASs is the presence of a pheromone response element (PRE). A clear distinction between the three representative Ste12-dependent UAS elements is the presence or absence of recognition sequences for ancillary proteins (see subsequent discussion).

The PRE consensus sequence (TGAAACA) is present in the promoter region of pheromone-inducible genes (Kronstad *et al.*, 1987; van Arsdell and Thorner, 1987; see Table II). At least two copies of the PRE sequence are necessary for pheromone inducibility (Kronstad *et al.*, 1987; Hagen *et al.*, 1991). However, the presence of PRE sequences in the promoter region of a gene does not insure the pheromone-inducible expression of that gene. For example, the *STE7* and *STE12* genes have multiple PREs but measurements of their mRNA do not show that their expression is pheromone inducible (C. Adler and B. Errede, unpublished observation).

Analyses of protein–DNA complexes show that Ste12 binds at the PRE(s) present in the *STE2*, *Ty1*, and *FUS1* UASs (Fig. 8.5A). DNase protection analysis of protein–DNA complexes that form with the *STE2* and Ty1 DNA probes show that protection of the PRE sequence is Ste12 dependent (Company *et al.*, 1988; Errede and Ammerer, 1989). Also, the effects of antibodies directed against Ste12 show that Ste12 is a component of protein complexes observed with both these DNA probles (Errede and Ammerer, 1989). An Ste12-dependent protein complex also forms with a synthetic DNA sequence that consists of two PRE elements (Dolan *et al.*, 1989). Finally, gel mobility shift experiments show that Ste12 is a component of the protein complexes with the *FUS1* UAS. The mobility of the complexes increases as the size of different Ste12 derivatives in the binding extracts decreases (B. Errede, unpublished observation). Evidence that Ste12 binds directly to the PRE sequence is provided by studies with *in vitro* synthesized Ste12, as described next.

The *STE2* UAS has one PRE sequence and a P-box sequence that is a binding site for Mcm1 (Fig. 8.5). Mcm1 also has been called PRTF (pheromone receptor transcription factor) and GRM (general regulator of mating) in the literature (Bender and Sprague, 1987; Keleher *et al.*, 1988; Jarvis *et al.*, 1989; Passmore *et al.*, 1989; Ammerer, 1990). The presence of Mcm1 in the protein–DNA complexes that bind to the *STE2* UAS was demonstrated with antibodies against Mcm1. Full-length Ste12 protein synthesized *in vitro* does not bind to the *STE2* UAS. However, when both *in vitro* synthesized Ste12 and Mcm1 are present in the binding reaction, a protein–DNA complex is formed of the same mobility as formed with cellular extracts (M. Primig and G. Ammerer, unpublished results). A deletion derivative of Ste12 shows that the C-terminal third of the protein (amino acids 469–688) is important for its interactions with Mcm1 in protein–DNA complex formation (Fig. 8.5C; Errede and Ammerer, 1989). Other studies using deletion and substitution derivatives of Mcm1 identified the domain and specific amino acids in Mcm1 that are critical for Ste12-dependent DNA complex formation (Mueller and Nordheim, 1991; Primig *et al.*, 1991).

The Ty1 UAS includes one PRE and a region (SAFE) that was proposed to be the binding site for another protein (Errede and Ammerer, 1989). The presence of this additional protein binding site is likely because base substitutions in SAFE abolish UAS functions; DNase footprint analyses show that proteins are bound to both the PRE and the SAFE sequence (Company *et al.*, 1988). Consistent with the hypothesis, UV-cross-linking of proteins in the complex with the Ty1 DNA probe show that two different size proteins are present; Ste12 and a 72-kDa protein Saf1 (Ste12 associate factor). DNA binding experiments using *in vitro* synthesized Ste12 and protein extracts from *ste12* deletion strains establish that neither Ste12 nor Saf1 binds to Ty1 UAS in the absence of the other, but addition of *in vitro* synthesized Ste12 to an extract that is devoid of Ste12 allows reconstitution of the Ste12–Saf1 complex (M. Baur and B. Errede, unpublished result). The cooperative associations of Ste12 with Mcm1 and Saf1 can account for the ability of Ste12 to bind at the two different UASs.

The *FUS1* UAS has four sequences that show a perfect or 6/7bp match to the PRE consensus sequence (Fig. 8.5B). A subfragment from this UAS that includes only three PREs is sufficient for pheromone-inducible expression of a reporter gene (Rhodes *et al.*, 1990). No other known recognition sequence within the fragment is sufficient for pheromone-inducible expression. However, an unidentified element upstream from this region contributes to increased expression of *FUS1* (Hagen *et al.*, 1991). Protein binding to this class of Ste12-dependent promoter element has not been studied as extensively as that with the Ty1 or *STE2* UASs. Possibilities are that multiple Ste12 molecules cooperate to bind at this class of UAS or that an unidentified ancillary factor(s) cooperates with Ste12 for stable binding (Fig. 8.5A).

The DNA binding domain of Ste12 is within the N-terminal third of the protein (Fig. 8.5C). An N-terminal Ste12 fragment consisting of amino acids 1–215 binds to a synthetic PRE sequence element (Yuan and Fields, 1991). Even in the absence of ancillary factors, an N-terminal fragment of Ste12 consisting of amino acids 1–213 can bind to the *STE2* and Ty1 UASs (M. Primig and G. Ammerer, unpublished results; M. Baur and B. Errede, unpublished results). Yuan and Fields (1991) deduced that the Ste12 DNA binding domain contains similarities to the homeodomain of other DNA binding proteins. However, the Ste12 sequence is highly diverged from other known examples of this motif and requires loops after both the putative DNA binding helices.

3. Transcription Activation Rigorous assignment of Ste12 as a transcription factor requires establishment that it has a direct role in transcription activation. This evidence was provided by a strategy in which the Gal4 DNA binding domain was fused to the complete Ste12 protein or to different Ste12 subdomains. The hybrid proteins then were assayed for their ability to activate transcription of reporter genes containing a Gal4 binding site. The Gal4 DNA domain alone is incapable of transcription activation. The hybrid that contains the complete Ste12 protein causes pheromone-inducible expression of the reporter gene. This result establishes that the complete Ste12 protein is a pheromone-dependent transcriptional activator. Fusions containing different portions of Ste12 localized the transcriptional activation domain to the central portion of Ste12. This central domain of the protein (amino acids 214–473) fused to Gal4 functions as a potent transcriptional activation domain, even in the absence of pheromone (Fig. 8.5C; Song *et al.*, 1991).

4. Target of the Pheromone-Induced Signal The evidence discussed here makes a compelling argument that activity of Ste12 is affected directly by the pheromone-induced signal. The role of protein kinases in the pathway suggests that changes in protein phosphorylation could affect Ste12 function. Evidence in support of this model is that Ste12 is a phosphoprotein (Song *et al.*, 1991; B. Errede, unpublished observations; C. Petritsch and G. Ammerer, unpublished observations) and that the Gal4–Ste12 fusion protein from pheromone-induced cells has a slower mobility on SDS–PAGE than the protein from uninduced cells. This altered mobility is reported to be caused by increase in phosphorylation of the Gal4–Ste12 fusion protein (Song *et al.*, 1991). However, no alteration in the mobility or net phosphorylation of native Ste12 has been detected (D. C. Li and B. Errede, unpublished results; C. Petritsch and G. Ammerer, unpublished results). The different behavior of the Gal4-Ste12 fusion protein and the native Ste12 highlight the necessity to map and compare specific phosphorylation sites of Ste12 under basal and pheromone-induced conditions. Mutational analysis of the target sites then could establish any role for phosphorylation in Ste12 function.

B. Regulators of Cell Cycle Progression (Far1, Fus3)

The cell cycle is a sequence of events that leads to duplication of cellular components and to their segregation to daughter cells. The Cdc28 protein kinase is an essential regulator of events that allows cells to progress from the G_1 phase to the S phase (DNA synthesis) (Pardee, 1989; Richardson *et al.*, 1989). Cdc28 activity in G_1 is thought to require its association with at least one of three cyclins—Cln1, Cln2, and Cln3 (Wittenberg *et al.*, 1990). The Cln1 and Cln2 proteins show periodic accumulation, with maximum levels observed during late G_1 in each cell cycle (Wittenberg *et al.*, 1990). Although the Cln3 protein is also unstable, it does not show striking oscillations (B. Tyers and

A. B. Futcher, unpublished results). The periodic accumulation of Cln1 and Cln2 appears to be regulated both transcriptionally and posttranscriptionally (Wittenberg *et al.*, 1990; Cross and Tinkelenberg, 1991; Dirick and Nasmyth, 1991; Nasmyth and Dirick, 1991; Ogas *et al.*, 1991). Because *CLN3* mRNA and protein do not show significant variation through the cell cycle, the current model is that its activity is regulated posttranscriptionally (Nash *et al.*, 1988; Wittenberg *et al.* 1990; B. Tyers and A. B. Futcher, unpublished results).

Progression from the G_1 to the S phase is blocked transiently in mating cells, presumably to insure that nuclear fusion occurs between nuclei containing a haploid complement of chromosomes. The pheromone-induced signal causes this arrest by inactivating the G_1 cyclins. Transcription of *CLN1* and *CLN2* is repressed under pheromone-induced conditions whereas that of *CLN3* is increased 2- to 3-fold (Nash *et al.*, 1988; Wittenberg *et al.*, 1990; Cross and Tinkelenberg, 1991; Dirick and Nasmyth, 1991; Elion *et al.*, 1991). Therefore, both transcriptional and posttranscriptional mechanisms are important for pheromone-induced division arrest. A current model is that the pheromone signal pathway induces the function of inhibitors specific for each of the G_1 cyclins (Chang and Herskowitz, 1990) *FAR1* and *FUS3* are two genes encoding such functions. (Chang and Herskowitz, 1990; Elion *et al.*, 1990, 1991).

1. Far1 The *FAR1* gene was identified in a screen for mutants that were defective for pheromone-induced division arrest but still had an intact signal pathway leading to induction of *FUS1–lacZ* reporter gene. The further characterization of *FAR1* expression and Far1 function provided the first clues to how the signal leading to transcriptional induction acts to regulate cell cycle progression (Chang and Herskowitz, 1990). *FAR1* mRNA expression is haploid specific, dependent on the presence of the Ste12 transcription factor and induced 4- to 5-fold by pheromone (Chang and Herskowitz, 1990). Far1 was found to act by specifically inhibiting Cln2. The dependence of *FAR1* expression on Ste12 provides an explanation for the resistance of *ste12* mutants to pheromone-induced division arrest, that is, in a *ste12* mutant, no Far1 would be expressed, leading to an inability to inactivate Cln2. Cells that express Cln2 but not the other two G_1 cyclins (Cln1 and Cln3) and overproduce Far1 from a galactose-inducible promoter do not have a division arrest phenotype in the absence of pheromone (unpublished data cited in Chang and Herskowitz, 1990). Therefore, the increased amount of Far1 produced in response to pheromone is not sufficient to account for its function as an inhibitor of Cln2. The pheromone-induced signal also must stimulate the inhibitory function of Far1 by a mechanism other than transcriptional induction. Pheromone induces the phosphorylation of Far1 (F. Chang and I. Herskowitz, unpublished results), suggesting that the phosphorylation state of Far1 may control its activity.

2. *Fus3* The variable mutant phenotypes associated with different *fus3* alleles show that, in addition to its role as a signal transmitter leading to transcription activation, Fus3 has a more direct role in pheromone-induced division arrest by inhibiting function of all three cyclins. Fus3 is proposed to inhibit Cln1 and Cln2 through transcriptional repression but its most critical role is to inactivate Cln3, which occurs posttranscriptionally (Elion *et al.*, 1991). As yet, no direct evidence exists for a Fus3-dependent modification of any cyclin protein, but an effect on modification is an attractive model, particularly for the inhibition of Cln3 since its transcription is not affected by Fus3 (Elion *et al.*, 1991). Also, the absence of evidence concerning direct modification leaves open the possibility that Fus3 has both a transcriptional and a more direct post-transcriptional role in the inhibition of Cln1 and Cln2.

Results in some strain backgrounds suggest that Kss1 is not redundant with Fus3 for the inhibition of cyclins (Elion *et al.*, 1991). In another strain background, a substitution allele, *fus3-A*, (Fig. 8.4C, Fus3 Glu-139) was isolated that causes cells to be more resistant to pheromone-induced division arrest than does a deletion allele. One possible interpretation is that the Fus3A protein is inactive for its role in division arrest but is able to interfere with a redundant Kss1 activity. Further examination of the proposal that Kss1 is completely redundant with Fus3 will require the determination of the role of factors that modify effects of Fus3 function in different genetic backgrounds.

VI. Enigmatic Components

A. Ste5

The *STE5* gene is the last of the initially identified genes required for mating to be understood at the molecular level (MacKay and Manney, 1974; Hartwell, 1980). Initial attempts to clone the *STE4* gene from multicopy plasmid libraries failed because of the cell cycle arrest caused by high level *STE4* expression. Interestingly, *STE5* clones were recovered in these screens because point mutants of *STE4* were suppressed by high level expression of the Ste5 protein (V. MacKay, unpublished results). This type of suppression suggests a possible physical association between Ste4 and Ste5 proteins. However, subsequent analyses in a number of laboratories have shown that overproduction of the Ste5 protein can have pleiotropic effects. For example, mutations of the G_γ subunit (Ste18), the Ste7 kinase, or the Ste11 kinase that cause attenuated function are all suppressed by Ste5 overproduction (M. Whiteway, unpublished results; D. Higgins and K. Tatchell, unpublished results; R. Cade and B. Errede, unpublished results). Similarly, dominant mutations causing increased stability of the Cln3 cyclin molecule are suppressed by Ste5 overproduction (L. Hougan and M. Whiteway, unpublished results). That Ste5 overproduction supresses all these different mutations by direct protein–protein interactions is unlikely.

A more plausible proposal is that Ste5 overproduction serves to amplify the weak signal associated with the various partially functional mutants. A prediction of this model is that Ste5 overproduction also should amplify the internal signal that is associated with the normal level of free $G_{\beta\gamma}$ present under uninduced conditions (Whiteway *et al.*, 1989). The finding that high copy plasmids containing the *STE5* gene are unstable in wild-type cells but not in cells lacking a functional pheromone response pathway is consistent with this prediction (M. Whiteway, unpublished results). The ability to amplify the signal coming from the activated G protein is one of the hallmarks expected for an effector component of a signal transduction pathway. Thus, Ste5 may be an effector protein, but further study will be necessary to clarify its role in the response pathway.

B. Srm1

The *CDC70* mutation, an allele of the *SCG1* (*GPA1*) gene, was identified in a search for mutations allowing mating of an **a** strain lacking the Ste2 receptor (Jahng *et al.*, 1988). An independent selection for suppressors that allow mating of an α strain lacking the Ste3 receptor generated another mutation that allowed mating of a receptorless mutant. This mutation defined a new gene, designated *SRM1*, with a proposed role in the pheromone response pathway (Clark and Sprague, 1989). The *srm1* mutant phenotype was superficially similar to the *scg1*^{ts} mutants. *srm1* mutants are temperature sensitive lethals; at the restrictive temperature, they exhibit cell cycle arrest in the G_1 phase, morphological changes characteristic of pheromone response, *FUS1* induction, and mating in the absence of functional pheromone receptor proteins (Clark and Sprague, 1989). However, one distinct difference is that, whereas the *SCG1* gene is completely dispensable in diploid cells, the *SRM1* gene is essential. Also, the cell cycle arrest of diploid *srm1* mutants in diploid cells is not G_1 specific (Clark and Sprague, 1989).

The sequence of the *SRM1* gene has strong similarity to a mammalian gene designated *RCC1* (Ohtsubo *et al.*, 1987). Defects in the *RCC1* gene lead to G_1 arrest in hamster cells, suggesting that Rcc1 plays a role in cell cycle decisions in mammalian cells (Nishimoto *et al.*, 1978). The *SRM1* and *RCC1* genes have functional similarities because the Rcc1 protein can complement the temperature-sensitive phenotype of an allele of the *srm1* gene (Clark *et al.*, 1991; Fleischmann *et al.*, 1991), and the *SRM1* gene can replace the *RCC1* function in the temperature-sensitive hamster cells (Ohtsubo *et al.*, 1991). However, the Rcc1 protein cannot complement disruption mutations in the

SRM1 gene, suggesting that the Rcc1 protein cannot provide all the functions of the *Srm1* protein (Clark *et al.*, 1991).

The role of the *SRM1* gene in the response pathway remains vague. Not all alleles of the gene exhibit the activation of the mating response (Aebi *et al.*, 1990). The requirements for the gene in diploid cells shows that its function is not restricted to the mating response pathway. The *SRM1* function may be involved in cell cycle regulation, and only under specific conditions (e.g., partial activity) does the mutant protein create a mating competent arrest. Because cyclins appears to be the ultimate target of the response pathway, gene products involved in cell cycle control are likely to affect the mating behavior of cells.

C. Cdc36 and Cdc39

Mutations in the *CDC36* and *CDC39* genes cause cells to arrest in G_1 in a mating competent state (Reed, 1980). The phenotype of temperature-sensitive mutations at the restrictive temperature is identical to that of *srm1* mutants: haploid cells arrest at G_1 whereas diploid cells arrest randomly throughout the cell cycle. At the restrictive temperature, these mutants also activate *FUS1* expression and allow mating in the absence of pheromone receptors (deBarros Lopes *et al.*, 1990; Neiman *et al.*, 1990). The activation of the pheromone response pathway requires the *STE4* and *STE18* gene products suggesting that the *CDC36* and *CDC39* gene products function early in the pathway.

Overproduction of the G_α subunit suppresses the response pathway activation of the *cdc36* and *cdc39* mutations. This result suggests that Cdc36 and Cdc39 may modulate the function of G_α and are, perhaps, necessary to stimulate GTPase activity or to stabilize the GDP-bound form of the protein. In either case, loss-of-function mutants of *CDC36* and *CDC39* would result in an increase in free $G_{\beta\gamma}$ subunits by reducing sequestration by the G_α subunit. Overproduction of G_α would counter this release of $G_{\beta\gamma}$ (deBarros Lopes *et al.*, 1990).

Mutations in the *SCG1/GPA1* gene product also were tested for an effect on *cdc36* and *cdc39* mutations (Neiman *et al.*, 1990). The *cdc39* mutation is suppressed by a 5-amino-acid truncation that is proposed to prevent receptor-stimulated G protein activation (Hirsch *et al.*, 1991). In contrast, the truncation does not suppress the *cdc36* mutation. These results are consistent with a role for Cdc39 in stabilizing the GDP-bound form of G_α and a role for Cdc36 in the production of G_α (Neiman *et al.*, 1990). However, a requirement for production of G_α seems inconsistent with the suppression of *cdc36* mutations by high copy plasmids carrying *SCG1/GPA1*.

The observation that the *CDC36* and *CDC39* gene products are necesary for viability of diploid cells shows that their role in the pheromone response pathway is not their only role. They could have roles in modulating the function of GTP binding proteins other than the *SCG1*-encoded subunit. However, the *cdc36* and *cdc39* mutations may amplify the weak basal signal of the response pathway in the absence of pheromone (Whiteway *et al.*, 1989), in which case the mutations could influence any step of the pathway including cell cycle regulatory steps. The suppression of the *cdc36* and *cdc39* mutations by overproduction of the G_α subunit would be explained only if the overproduction reduced this basal signal, which was not observed in the current study (de Barros Lopes *et al.*, 1990). Additional work will be necessary to define whether the genetic predictions that genes such as *CDC36* and *CDC39* function early in the response pathway are correct.

VII. Conclusions

The mating response pathway in yeast can be traced from the pheromone-bound receptor to targets that control transcriptional activation and cell division arrest. Isolation of mutants in the mating response and cloning of the corresponding genes provides a molecular handle for signal pathway components and their functions. The biochemical analysis of some components has begun. The prospects for defining molecular details of signal transduction are good. Although much remains to be learned, we know that pheromone binding to receptor activates a G protein complex. Stimulation of the G protein initiates a cascade involving four protein kinases. The resulting protein phosphorylations appear to regulate the kinases and the proteins that directly mediate the physiological responses to pheromone. Signal transduction components such as the receptor, G protein, and kinases are conserved between yeast and vertebrates, suggesting that the mechanism for signal transduction uncovered in yeast will be a useful model for other organisms.

Acknowledgments

We wish to thank J. Hirsch, F. Cross, K. Clark, G. Ammerer, G. Sprague, A. Neiman, I. Herskowitz, A. B. Futcher, V. MacKay, and K. Tatchell for providing results prior to publication. Also we wish to thank K. Clark, E. Leberer, B. Yashar, T. Gartner, and A. Amon for comments on the manuscript. Previously unpublished data from B. Errede was supported by U.S. Public Health Service grants GM-30619 and GM-39852. This is NRC Publication No. 33658 of M. Whiteway.

References

Achstetter, T. (1989). Regulation of α-factor in *Saccharomyces cerevisiae:* **a**-Factor pheromone-induced expression of the *MF*α1 and *STE13* genes. *Mol. Cell. Biol.* **9**, 4507–4514.

Aebi, M., Clark, M. W., Vijayraghavan, U., Abelson, J. (1990). A yeast mutant, PRP20, altered in mRNA metabolism and maintenance of the nuclear structure, is defective in a gene homologous to the human gene *RCC1* which is involved in the control of chromosome condensation. *Mol. Gen. Genet.* **224**, 72–80.

Ammerer, G. (1990). Identification, purification and cloning of a polypeptide (PRTF/GRM) that binds to mating-specific promoter elements in yeast. *Genes Dev.* **4**, 299–312.

Appeltaurer, U., and Achstetter, T. (1989). Hormone-induced expression of the *CHS1* gene from *Saccharomyces cerevisiae*. *Eur. J. Biochem.* **181**, 243-247.

Bender, A., and Sprague, G. F., Jr. (1986). Yeast peptide pheromones, **a**-factor and α-factor, activate a common response mechanism in their target cells. *Cell* **47**, 929–937.

Bender, A., and Sprague, G. F., Jr. (1987). MATα1 Protein, a yeast transcription activator, binds synergistically with a second protein to a set of cell-type-specific genes. *Cell* **50**, 681–691.

Blinder, D., Bouvier, S., and Jenness, D. D. (1989). Constitutive mutants in the yeast pheromone response: Ordered function of the gene products. *Cell* **56**, 479–486.

Blumer, K. J., and Thorner, J. (1990). β and γ subunits of a yeast guanine nucleotide-binding protein are not essential for membrane association of the α subunit but are required for receptor coupling. *Proc. Natl. Acad. Sci. U.S.A.* **87**, 4363–4367.

Blumer, K. J., and Thorner, J. (1991). Receptor–G protein signaling in yeast. *Ann. Rev. Physiol.* **53**, 37–57.

Blumer, K. J., Reneke, J. E., and Thorner, J. (1988). The *STE2* gene product is the ligand binding component of the α-factor receptor of *Saccharomyces cerevisiae*. *J. Biol. Chem.* **263**, 10836–10842.

Boulton, T. G., Yancopoulos, G. D., Gregory, J. S., Slaughter, C., Moomaw, C., Hsu, J., and Cobb, M. H. (1990). An insulin-stimulated protein kinase similar to yeast kinases involved in cell cycle control. *Science* **249**, 64–67.

Boulton, T. G., Nye, S. H., Robbins, D. J., Ip, N. Y., Radziejewska, E., Morgenbesser, S. D., DePinho, R. A., Panayotatos, N., Cobb, M. H., and Yancopoulos, G. D. (1991). ERKs: A family of protein serine/threonine kinases that are activated and tyrosine phosphorylated in response to insulin and NGF. *Cell* **65**, 663–675.

Bouvier, M., Hausdorff, W. P., de Blasi, A. O'Dowd, B. F., Kobilka, B. K., Caron, M. G., and Lefkowitz, R. J. (1988). Removal of phosphorylation sites from the β-adrenergic receptor delays onset of agonist-promoted desensitization. *Nature (London)* **333**, 370–373.

Burkholder, A. C., and Hartwell, L. (1985). The yeast α-factor receptor: Structural properties deduced from the sequence of the *STE2* gene. *Nucleic Acids Res.* **13**, 8463–8475.

Cantley, L. C., Auger, K. R., Carpenter, C., Duckworth, B., Graziani, A., Kapeller, R., and Soltoff, S. (1991). Oncogenes and signal transduction. *Cell* **64**, 281–302.

Cartwright, C. P., and Tipper, D. J. (1991). *In vivo* topological analysis of STE2, a yeast plasma membrane protein, by using β-lactamase gene fusions. *Mol. Cell. Biol.* **11**, 2620–2628.

Casperson, G. F., Walker, N., Brasier, A. R., and Bourne, H. R. (1983). A guanine nucleotide-sensitive adenylate cyclase in the yeast *Saccharomyces cerevisiae*. *J. Biol. Chem.* **256**, 7911–7914.

Chaleff, D. T., and Tatchell, K. (1985). Molecular cloning and characterization of the *STE7* and *STE11* genes of *Saccharomyces cerevisiae*. *Mol. Cell. Biol.* **5**, 1878–1886.

Chang, F., and Herskowitz, I. (1990). Identification of a gene necessary for cell cycle arrest by a negative growth factor of yeast: FAR1 is an inhibitor of a G1 cyclin, CLN2. *Cell* **63**, 999–1011.

Chvatchko, Y., Howald, I., and Riezman, H. (1986). Two yeast mutants defective in endocytosis are defective in pheromone response. *Cell* **46**, 355–364.

Clark, K. L., and Sprague, G. F., Jr. (1989). Yeast pheromone response pathway: Characterization of a suppressor that restores mating to receptorless mutants. *Mol. Cell. Biol.* **9**, 2682–2694.

Clark, K. L., Davis, N. G., Wiest, D. K., Hwang-Shun, J.-J., and Sprague, G. F., Jr. (1988). Response of yeast α cells to **a**-factor pheromone: Topology of the receptor and identification of a component of the response pathway. *Cold Spring Harbor Symp. Quant. Biol.* **53**, 611–620.

Clark, K. L., Ohtsubo, M., Nishimoto, T., Goebl, M., and Sprague, G. F., Jr. (1991). The yeast SRM1 and human RCC1 proteins share analogous functions. *Cell Reg.* **2**, 781–792.

Clark, K. L., Dignard, D., Thomas, D. Y., and Whiteway, M. (1993). Interactions among the subunits of the G protein involved in *Saccharomyces cerevisiae* mating. *Mol. Cell. Biol.* **13**, 1–8.

Cole, G. M., and Reed, S. I. (1991). Pheromone-induced phosphorylation of a G protein β subunit in *S. cerevisiae* is associated with an adaptive response to mating pheromone. *Cell* **64**, 703–716.

Cole, G. M., Stone, D. E., and Reed, S. I. (1990). Stoichiometry of G protein subunits affects the *Saccharomyces cerevisiae* mating pheromone signal transduction pathway. *Mol. Cell. Biol.* **10**, 510–517.

Company, M., and Errede, B. (1988). A Ty1 cell-type-specific regulatory sequence is a recognition element for a constitutive binding factor. *Mol. Cell. Biol.* **8**, 5299–5309.

Company, M., Adler, C., and Errede, B. (1988). Identification of a Ty1 regulatory sequence responsive to *STE7* and *STE12*. *Mol. Cell. Biol.* **8**, 2545-2554.

Courchesne, W. E., Kunisawa, R., and Thorner, J. (1989). A putative protein kinase overcomes pheromone-induced arrest of cell cycling in *S. cerevisiae*. *Cell* **58**, 1107–1119.

Courey, A. J., and Tijian, R. (1988). Analysis of Sp1 in vivo reveals multiple transcriptional domains, including a novel glutamine rich activation motif. *Cell* **55**, 887–898.

Cross, F. R. (1990). The *DAF2-2* mutation, a dominant inhibitor of the *STE4* step in the α-factor signaling pathway of *Saccharomyces cerevisiae MAT*α cells. *Genetics* **126**, 301–308.

Cross, F. R., and Tinkelenberg, A. H. (1991). A potential positive feedback loop controlling *CLN1* and *CLN2* gene expression at the start of the yeast cell cycle. *Cell* **65**, 875–883.

Cross, F., Hartwell, L. H., Jackson, C., and Konopka, J. B. (1988). Conjugation in *Saccharomyces cerevisiae*. *Ann. Rev. Cell Biol.* **4**, 429–457.

deBarros Lopes, M., Ho, J-Y., and Reed, S. I. (1990). Mutations in cell division cycle genes *CDC36* and *CDC39* activate the *Saccharomyces cerevisiae* mating pheromone response pathway. *Mol. Cell. Biol.* **10**, 2966–2972.

Dhar, R., Neito, A., Koller, R., DeFeo-Jones, D., and Scolnick, E. (1984). Nucleotide sequence of two *rasH* related genes isolated from the yeast *S. cerevisiae*. *Nucleic Acids Res.* **12**, 3611–3618.

Dietzel, C., and Kurjan, J. (1987a). The yeast *SCG1* gene: A G_{α}-like protein implicated in the **a**- and α-factor response pathway. *Cell* **50**, 1001–1010.

Dietzel, C., and Kurjan, J. (1987b). Pheromonal regulation and sequence of the *Saccharomyces cerevisiae* STT2 gene: A model for desensitization to pheromone. *Mol. Cell. Biol.* **7**, 4169–4177.

Dirick, L., and Nasmyth, K. (1991). Positive feedback in the activation of G1 cyclins in yeast. *Nature (London)* **351**, 754–757.

Dmochowska, A., Dignard, D., Henning, D., Thomas, D. Y., and Bussey, H. (1987). Yeast *KEX1* gene encodes a putative protease with carboxypeptidase B-like function involved in killer toxin and alpha-factor processing. *Cell* **50**, 573–584.

Dolan, J. W., and Fields, S. (1990). Overproduction of the yeast STE12 protein leads to constitutive transcriptional induction. *Genes Dev.* **4**, 492–502.

Dolan, J. W., and Fields, S. (1991). Cell-type-specific transcription in yeast. *Biochim. Biophys. Acta* **1088**, 155–169.

Dolan, J. W., Kirkman, C., and Fields, S. (1989). The yeast STE12 protein binds to the DNA sequence mediating pheromone induction. *Proc. Natl. Acad. Sci. U.S.A.* **86**, 5703–5707.

Elion, E. A., Grisafi, P. L., and Fink, G. R. (1990). *FUS3* encodes a cdc2^{+}/CDC28-related kinase required for the transition from mitosis into conjugation. *Cell* **60**, 649–664.

Elion, E. A., Brill, J. A., and Fink, G. R. (1991). FUS3 represses CLN1 and CLN2 and in concert with KSS1 promotes signal transduction. *Proc. Natl. Acad. Sci. U.S.A.* **88**, 9392–9396.

Errede, B., and Ammerer, G. (1989). STE12, a protein involved in cell-type-specific transcription and signal transduction in yeast, is part of protein-DNA complexes. *Genes Dev.* **3**, 1349–1361.

Errede, B., Cardillo, T. S., Sherman, F., Dubois, E., Deschamps, J., and Wiame, J. M. (1980). Mating signals control expression of mutations resulting from insertion of a transposable repetitive element adjacent to diverse yeast genes. *Cell* **22**, 427–436.

Errede, B., Company, M., Ferchak, J. D., Hutchison, C. A., III, and Yarnell, W. S. (1985). Activation regions in a yeast transposon have homology to mating type control sequences and to mammalian enhancers. *Proc. Natl. Acad. Sci. U.S.A.* **82**, 5423–5427.

Errede, B., Company, M., and Hutchison, C. A., III (1987). A Ty1 sequence with enhancer and mating-type-dependent regulatory activities. *Mol. Cell. Biol.* **7**, 258–265.

Evans, R. M. (1988). The steroid and thyroid hormone receptor superfamily. *Science* **240**, 889–895.

Fields, S., and Herskowitz, I. (1985). The yeast *STE12* product is required for expression of two sets of cell-type-specific genes. *Cell* **42**, 923–930.

Fields, S., and Herskowitz, I. (1987). Regulation by the yeast mating-type locus of *STE12*, a gene required for cell-type-specific expression. *Mol. Cell. Biol.* **7**, 3818–3821.

Fields, S., Chaleff, D. T., and Sprague, G. F., Jr. (1988). Yeast *STE7*, *STE11*, and *STE12* genes are required for expression of cell-type-specific genes. *Mol. Cell. Biol.* **8**, 551–556.

Finegold, A. A., Schafer, W. R., Rine, J., Whiteway, M., and Tamanoi, F. (1990). Common modifications of trimeric G proteins and ras protein: Involvement of polyisoprenylation. *Science* **249**, 165–169.

Fleishmann, M., Clark, M. W., Forrester, W., Wickens, M., Nishimoto, T., and Aebi, M. (1991). Analysis of yeast *prp20* mutations and functional complementation by the human homologue *RCC1*, a protein involved in the control of chromosome condensation. *Mol. Gen. Genet.* **227**, 417–423.

Fujimura, H. (1989). The yeast G-protein homolog is involved in the mating pheromone signal transduction system. *Mol. Cell. Biol.* **9**, 152–158.

Fujimura, H. (1990a). Identification and characterization of a mutation affecting division arrest signaling of the pheromone response pathway in *Saccharomyces cerevisiae*. *Genetics* **124**, 275–282.

Fujimura, H. (1990b). Molecular cloning of the *DAC2/FUS3* gene essential for pheromone-induced G_1-arrest of the cell cycle in *Saccharomyces cerevisiae*. *Curr. Genet.* **18**, 395–400.

Fujiyama, A., Matsumoto, K., and Tamanoi, F. (1987). A novel yeast mutant defective in the processing of ras proteins, assessment of the effect of the mutation on processing steps. *EMBO J.* **6**, 223–228.

Fung, B. K. K. (1983). Characterization of tranducin from bovine retinal rod outer segments. I. Separation and reconstitution of the subunits. *J. Biol. Chem.* **258**, 10495–10502.

Ganz, M. B., Pachter, J. A., and Barber, D. L. (1990). Multiple receptors coupled to adenylate cyclase regulate Na-H exchange independent of cAMP. *J. Biol. Chem.* **265**, 8989–8992.

Gartner, A., Nasmyth, K., and Ammerer, G. (1992). Signal transduction in *Saccharomyces cerevisiae* requires tyrosine and threonine phosphorylation of FUS3 and KSS1. *Genes Dev.* **6**, 1280–1292.

Gibbs, C. S., and Zoller, M. J. (1991). Rational scanning mutagenesis of a protein kinase identifies functional regions involved in catalysis and substrate interactions. *J. Biol. Chem.* **266**, 8923–8931.

Gibbs, I., Sigal, S., Poe, M., and Scolnick, E. M. (1984). Intrinsic GTPase activity distinguishes normal and oncogenic ras p21 molecules. *Proc. Natl. Acad. Sci. U.S.A.* **81**, 5704–5708.

Gilman, A. G. (1987). G proteins: Transducers of receptor-generated signals. *Ann. Rev. Biochem.* **56**, 615–649.

Gurdon, J. B. (1992). The generation of diversity and pattern in animal development. *Cell* **68**, 185–199.

Hagen, D. C., and Sprague, G. F., Jr. (1984). Induction of the yeast **a**-specific *STE3* gene by the peptide pheromone **a**-factor. *J. Mol. Biol.* **178**, 835–852.

Hagen, D. C., McCaffrey, G., and Sprague, G. F., Jr. (1986). Evidence the yeast *STE3* gene encodes a receptor for the peptide pheromone **a**-factor: Gene sequence and implications for the structure of the presumed receptor. *Proc. Natl. Acad. Sci. U.S.A.* **83**, 1418–1422.

Hagen, D. C., McCaffrey, G., and Sprague, G. F., Jr. (1991). Pheromone response elements are necessary and sufficient for basal and pheromone-induced transcription of the *FUS1* gene of *Saccharomyces cerevisiae*. *Mol. Cell. Biol.* **11**, 2952–2961.

Hanks, S. K. Quinn, A. M., and Hunter, T. (1988). The protein kinase family: Conserved features and deduced phylogeny of the catalytic domains. *Science* **241**, 42–52.

Hartig, A., Holly, J., Saari, G., and MacKay, V. L. (1986). Multiple regulation of *STE2*, a mating-type-specific gene of *Saccharomyces cerevisiae*. *Mol. Cell. Biol.* **6**, 2106–2114.

Hartwell, L. (1980). Mutants of *Saccharomyces cerevisiae* unresponsive to cell division control by polypeptide mating hormones. *J. Cell Biol.* **85**, 811–822.

Herskowitz, I. (1989). A regulatory hierarchy for cell specialization in yeast. *Nature (London)* **342**, 749–757.

Hirsch, J. P., Dietzel, G., and Kurjan, J. (1991). The carboxy terminus of Scg1, the Gα subunit involved in yeast mating, is implicated in interactions with the pheromone receptors. *Genes Dev.* **5**, 467–474.

Hunter, T., Angel, P., Boyle, W. J., Chiu, R., Freed, E., Gould, K. L., Isacke, C. M., Karin, M., Lindberg, R. A., and van der Geer, P. (1988). Targets for signal-transducing protein kinases. *Cold Spring Harbor Symp. Quant. Biol.* **53**, 131–142.

Iida, H., Yagawa, Y., and Anraku, Y. (1990). Essential role for induced Ca^{2+} influx followed by $[Ca^{2+}]_i$ rise in maintaining viability of yeast cells late in the mating pheromone response pathway. *J. Biol. Chem.* **264**, 13391–13399.

Irie, K., Nomoto, S., Miyajima, I., and Matsumoto, K. (1991). *SGV1* encodes a CDC28/cdc2-related kinase required for a Gα subunit-mediated adaptive response to pheromone in *S. cerevisiae*. *Cell* **65**, 785–795.

Jahng, K-Y., Ferguson, J., and Reed, S. I. (1988). Mutations in a gene encoding the α subunit of a *Saccharomyces cerevisiae* G protein indicate a role in mating pheromone signaling. *Mol. Cell Biol.* **8**, 2484–2493.

Jarvis, E. E., Hagen, D. C., and Sprague, G. F., Jr. (1988). Identification of a DNA segment that is necessary and sufficient for α-specific gene control in *Saccharomyces cerevisiae*: Implications for regulation of α-specific and **a**-specific gene. *Mol. Cell. Biol.* **8**, 309–320.

Jarvis, E. E., Clark, K. L., and Sprague, G. F., Jr. (1989). The yeast transcription activator PRTF, a homolog of the mammalian serum response factor, is encoded by the *MCM1* gene. *Genes Dev.* **3**, 936–945.

Jenness, D., and Spatrick, P. (1986). Down-regulation of the α-factor pheromone receptor in *S. cerevisiae*. *Cell* **46**, 345–353.

Jenness, D. D., Burkholder, A. C., and Hartwell, L. H. (1983). Binding of α-factor pheromone to yeast **a** cells: Chemical and genetic evidence for an α-factor receptor. *Cell* **35**, 521–529.

Jenness, D. D., Goldman, B. S., and Hartwell, L. H. (1987). *Saccharomyces cerevisiae* mutants unresponsive to α-factor pheromone: α-Factor binding and extragenic suppression. *Mol. Cell. Biol.* **7**, 1311–1319.

Julius, D., Blair, L., Brake, A., Sprague, G., and Thorner, J. (1983). Yeast alpha-factor is processed from a larger precursor polypeptide, the essential role of a membrane bound dipeptidyl aminopeptidase. *Cell* **32**, 839–852.

Julius, D., Brake, A., Blair, L., Kunisawa, R., and Thorner, J. (1984a). Isolation of the putative structural gene for the lysine–arginine-cleaving endopeptidase required for processing of yeast prepro-alpha-factor. *Cell* **37**, 1075–1089.

Julius, D., Schekman, R., and Thorner, J. (1984b). Glycosylation and processing of prepro-alpha-factor through the yeast secretory pathway. *Cell* **36**, 309–318.

Kang, Y.-S., Kane, J., Kurjan, J., Stadel, J. M., and Tipper, D. J. (1989). Effects of expression of mammalian Gα and hybrid mammalian–yeast Gα proteins on the yeast pheromone response signal transduction pathway. *Mol. Cell. Biol.* **10**, 2582–2590.

Keleher, C. A., Goutte, C., and Johnson, A. D. (1988). The yeast cell-type-specific repressor $\alpha 2$ acts cooperatively with a non-cell-type-specific protein. *Cell* **53**, 927–936.

King, K., Dohlman, H. G., Thorner, J., Caron, M. G., and Lefkowitz, R. J. (1990). Control of yeast mating signal transduction by a mammalian β_2-adrenergic receptor and G_s α-subunit. *Science* **250**, 121–123.

Knighton, D. R., Zheng, J., Ten Eyck, L. F., Ashford, V. A., Xuong, N., Taylor, S. S., and Sowadski, J. M. (1991). Crystal structure of the catalytic subunit of cyclic adenosine monophosphate-dependent protein kinase. *Science* **253**, 407–414.

Konopka, J. B., Jenness, D. D., and Hartwell, L. H. (1988). The C terminus of the *Saccharomyces cerevisiae* a pheromone receptor mediates an adaptive response to pheromone. *Cell* **54**, 609–620.

Kronstad, J. W., Holly, J. A., and MacKay, V. L. (1987). A yeast operator overlaps an upstream activation site. *Cell* **50**, 369–377.

Kuchler, K., Sterne, R. E., and Thorner, J. (1989). *Saccharomyces cerevisiae STE6* gene product, a novel pathway for protein export in eukaryotic cells. *EMBO J.* **8**, 3973–3984.

Kurjan, J., and Whiteway, M. (1990). The *Saccharomyces cerevisiae* pheromone response pathway. *Sem. Dev. Biol.* **1**, 151–158.

Kurjan, J., Hirsch, J. P., and Dietzel, C. (1991). Mutations in the guanine nucleotide binding domains of a yeast Gα protein confer a constitutive or uninducible state to the pheromone response pathway. *Genes Dev.* **5**, 475–483.

Lee, E., Taussig, R., and Gilman, A. G. (1992). The G226A mutant of $G_{s\alpha}$ highlights the requirement for dissociation of G protein subunits. *J. Biol. Chem.* **267**, 1212–1218.

Lefkowitz, R. J., Stadel, J. M., and Caron, M. G. (1983). Adenylate cyclase-coupled β-adrenergic receptors, structure and mechanisms of activation and desensitization. *Ann. Rev. Biochem.* **52**, 159–186.

McCaffrey, G., Clay, F. J., Kelsay, K., and Sprague, G. F., Jr. (1987). Identification and regulation of a gene required for cell fusing during mating of the yeast *Saccharomyces cerevisiae*. *Mol. Cell. Biol.* **7**, 2680–2690.

McGrath, J. P., and Varshavsky, A. (1989). The yeast *STE6* gene encodes a homologue of the mammalian multidrug resistance P-glycoprotein. *Nature (London)* **340**, 400–404.

MacKay, V. L., and Manny, T. R. (1974). Mutations affecting sexual conjugation and related processes in *Saccharomyces cerevisiae* II. Genetic analysis of non-mating mutants. *Genetics* **76**, 273–288.

Magee, A. I., and Hanley, M. R. (1988). Stickey fingers and CAAX boxes. *Nature (London)* **335**, 114–115.

Marsh, L., and Herskowitz, I. (1988). STE2 protein of *Saccharomyces kluyveri* is a member of the rhodopsin/β-adrenergic receptor family and is responsible for recognition of the peptide ligand α-factor. *Proc. Natl. Acad. Sci. U.S.A.* **85**, 3855–3859.

Marsh, L., Neiman, A. M., and Herskowitz, I. (1991). Signal transduction during pheromone response in yeast. *Ann. Rev. Cell Biol.* **7**, 699–728.

Meluh, P. B., and Rose, M. D. (1990). *KAR3*, a kinesin-related gene required for yeast nuclear fusion. *Cell* **60**, 1029–1041.

Miller, R. T., Masters, S. B., Sullivan, K. A., Beiderman, B., and Bourne, H. R. (1988). A mutation that prevents GTP-dependent activation of the α chain of G_s. *Nature (London)* **334**, 712–715.

Miyajima, I., Nakafuku, M., Nakayama, N., Brenner, C., Miyajima, A., Kaibuchi, K., Arai, K., Kaziro, Y., and Matsumoto, K. (1987). *GPA1*, a haploid essential gene, encodes a yeast homolog of mammalian G protein which may be involved in mating factor signal transduction. *Cell* **50**, 1011–1010.

Miyajima, I., Arai, K., and Matsumoto, K., (1989). *GPA1*$^{\text{Val-50}}$ mutation in the mating-factor signaling pathway in *Saccharomyces cerevisiae. Mol. Cell. Biol.* **9**, 2289–2297.

Morgan, N. G. (1989). "Cell Signalling." Open University Press, Milton Keynes, England.

Mueller, C. G. F., and Nordheim, A. (1991). A protein domain conserved between yeast MCM1 and human SRF directs ternary complex formation. *EMBO J.* **10**, 4219–4229.

Nadin-Davis, S. A., and Nasim, A. (1988). A gene which encodes a predicted protein kinase can restore some functions of the *ras* gene in fission yeast. *EMBO J.* **7**, 985–993.

Nakafuku, M., Itoh, H., Nakamura, S., and Kaziro, Y. (1987). Occurrence in *Saccharomyces cerevisiae* of a gene homologous to the cDNA coding for the subunit of mammalian G proteins. *Proc. Natl. Acad. Sci. U.S.A.* **84**, 2140–2144.

Nakayama, N., Miyajima, A., and Arai, K. (1985). Nucleotide sequences of *STE2* and *STE3*, cell type specific sterile genes from *Saccharomyces cerevisiae. EMBO J.* **4**, 2643–2648.

Nakayama, N., Miyajima, A., and Arai, K. (1987). Common signal transduction system shared by *STE2* and *STE3* in haploid cells of *Saccharomyces cerevisiae*: Autocrine cell-cycle arrest results from forced expression of *STE2*. *EMBO J.* **6**, 249–254.

Nakayama, N., Arai, K-I., and Matsumoto, K. (1988a). Role of *SGP2*, a suppressor of a *gpa1* mutation, in the mating-factor signaling pathway of *Saccharomyces cerevisiae. Mol. Cell. Biol.* **8**, 5410–5416.

Nakayama, N., Kaziro, Y., Arai, K., and Matsumoto, K. (1988b). Role of *STE* genes in the mating factor signaling pathway mediated by *GPA1* in *Saccharomyces cerevisiae. Mol. Cell. Biol.* **8**, 3777–3783.

Nash, R., Tokiwa, G., Anand, S., Erickson, K., and Futcher, A. B. (1988). The *WHI1*$^+$ gene of *Saccharomyces cerevisiae* tethers cell division to cell size and is a cyclin homolog. *EMBO J.* **13**, 4335–4346.

Nasmyth, K., and Dirick, L. (1991). The role of SWI4 and SWI6 in the activity of G1 cyclins in yeast. *Cell* **66**, 995–1013.

Nasmyth, K., and Shore, D. (1987). Transcriptional regulation in the yeast life cycle. *Science* **237**, 1162–1170.

Neer, E. J., and Clapham, D. E. (1988). Roles of G protein subunits in transmembrane signalling. *Nature (London)* **333**, 129–134.

Neiman, A. M., Chang, F., Komachi, K., and Herskowitz, I. (1990). *CDC36* and *CDC39* are negative elements in the signal transduction pathway of yeast. *Cell Reg.* **1**, 391–401.

Nishimoto, T., Eilen, E., and Basilico, C. (1978). Premature chromosome condensation in a ts DNA-mutant of BHK cells. *Cell* **15**, 475–483.

Nomoto, S., Nakayama, N., Arai, K.-I. and Matsumoto, K. (1990). Regulation of the yeast pheromone response pathway by G protein subunits. *EMBO J.* **9**, 691–696.

Ogas, J., Andrews, B. J., and Herskowitz, I. (1991). Transcriptional activation of *CLN1*, *CLN2*, and a putative new G1 cyclin (HCS26) by SWI4, a positive regulator of G1-specific transcription. *Cell* **66**, 1015–1026.

Ohtsubo, M., Kai, R., Furuno, N., Sekiguchi, T., Sekiguchi, M., Hayashida, H., Kuma, K., Miyata, T., Fukushige, S., Murotsu, T., Matsubara, K., and Nishimoto, T. (1987). Isolation and characterization of the active cDNA of the human cell cycle gene (*RCC1*) involved in the regulation of onset of chromosome condensation. *Genes Dev.* **1**, 585–593.

Ohtsubo, M., Yoshida, T., Seino, H., Nishitani, H., Clark, K. L., Sprague, G. F., Jr., Frasch, M., and Nishimoto, T. (1991). Mutation of the hamster cell cycle gene *RCC1* is complemented by the homologous genes of *Drosophila* and *S. cerevisiae. EMBO J.* **10**, 1265–1273.

Pardee, A. B. (1989). G_1 events and regulation of cell proliferation. *Science* **246**, 603–608.

Passmore, S., Elble, R., and Tye, B.-K. (1989). A protein involved in minichromosome maintenance in yeast binds a transcriptional enhancer conserved in eukaryotes. *Genes Dev.* **3**, 921–935.

Payette, P., Gossard, F., Whiteway, M., and Dennis, M. (1990). Expression and pharmacological characterization of the human M1 muscarinic receptor in *Saccharomyces cerevisiae. FEBS Lett.* **266**, 21–26.

Powers, S., Kataoka, T., Fasano, O., Goldfarb, M., Strathern, J., Broach, J., and Wigler, M. (1984). Genes in *S. cerevisiae* encoding proteins with domains homologous to the mammalian ras proteins. *Cell* **36**, 607–612.

Powers, S., Michaelis, S., Broek, D., Santa Anna-A. S., Field, J., Herskowitz, I., and Wigler, M. (1986). *RAM*, a gene of yeast required for a functional modification of RAS proteins and for production of mating pheromone **a**-factor. *Cell* **47**, 413–422.

Primig, M., Winkler, H., and Ammerer, G. (1991). The DNA binding and oligomerization domain of MCM1 is sufficient for its interaction with other regulatory proteins. *EMBO J.* **10**, 4209–4218.

Reed, S. I. (1980). The selection of *S. cerevisiae* mutants defective in the start event of cell division. *Genetics* **95**, 561–566.

Reneke, J. E., Blumer, K. J., Courchesne, W. E., and Thorner, J. (1988). The carboxy-terminal segment of the yeast α-factor receptor is a regulatory domain. *Cell* **55**, 221–234.

Rhodes, N. (1991). The STE11 kinase and its role in pheromone induced signal transduction. Ph.D. thesis. University of North Carolina, Chapel Hill.

Rhodes, N., Connell, L., and Errede, B. (1990). STE11 is a protein kinase required for cell-type-specific transcription and signal transduction in yeast. *Genes Dev.* **4**, 1862–1874.

Richardson, H. E., Wittenberg, C., Cross, F., and Reed, S. I. (1989). An essential G1 function for cyclin-like proteins in yeast. *Cell* **59**, 1127–1133.

Ross, E. M. (1989). Signal sorting and amplification through G protein-coupled receptors. *Neuron* **3**, 141–152.

Shepherd, G. M. (1991). Sensory transduction: Entering the mainstream of membrane signaling. *Cell* **67**, 845–851.

Simon, M. I., Strathmann, M. P., and Gautam, N. (1991). Diversity of G proteins in signal transduction. *Science* **252**, 802-808.

Song, O.-K., Dolan, J. W., Yuan, Y. O., and Fields, S. (1991). Pheromone-dependent phosphorylation of the yeast STE12 protein correlates with transcriptional activation. *Genes Dev.* **5**, 741–750.

Sprague, G. F., Jr. (1990). Combinatorial associations of regulatory proteins and the control of cell type in yeast. *In* "Advances in Genetics. Genetic Regulatory Hierarchies in Development" (J. G. Scandalios and T. R. F. Wright, eds.), Vol. 27, pp. 33–62. Academic Press, Orlando, Florida.

Sprague, G. F., Jr. (1991). Signal transduction in yeast mating. *Trends Genet.* **7**, 393–398.

Stevenson, B. J., Rhodes, N., Errede, B., and Sprague, G. F. Jr. (1992). Constitutive mutants of the protein kinase STE11 activate the yeast pheromone response pathway in the absence of the G protein. *Genes Dev.* **6**, 1293–1304.

Stone, D. E., and Reed, S. I. (1990). G protein mutants that alter the pheromone response in *Saccharomyces cerevisiae. Mol. Cell. Biol.* **10**, 4439–4446.

Stryer, L., and Bourne, H. R. (1986). G proteins: A family of signal transducers. *Ann. Rev. Cell Biol.* **2**, 391–419.

Sullivan, K. A., Miller, R. T., Masters, S. B., Beiderman, B., and Bourne, H. R. (1987). Identification of receptor contact site involved in receptor-G protein coupling. *Nature (London)* **330**, 758–760.

Takahashi, S., Negishi, M., and Ichikawa, A. (1991). Cytosol promotes the guanine nucleotide-induced release of the α subunit of G_{i2} from the membrane of mouse mastocytoma P-815 cells. *J. Biol. Chem.* **266**, 5367–5370.

Taylor, S. S., Buechler, J. A., and Yonemoto, W. (1990). cAMP-dependent protein kinase: Framework for a diverse family of regulatory enzymes. *Ann. Rev. Biochem.* **59**, 971–1005.

Teague, M. A., Chaleff, D. T., and Errede, B. (1986). Nucleotide sequence of the yeast regulatory gene *STE7* predicts a protein homologous to protein kinases. *Proc. Natl. Acad. Sci. U.S.A.* **83**, 7371–7375.

Toda, T., Shimanuki, T., and Yanagida, M. (1991). Fission yeast genes that confer resistance to staurosporine encode an AP-1-like transcription factor and a protein kinase related to the mammalian ERK1/MAP2 and budding yeast FUS3 and KSS1 kinases. *Genes Dev.* **5**, 60–73.

Trueheart, J., Boeke, J. D., and Fink, G. R. (1987). Two genes required for cell fusion during yeast conjugation: Evidence for a pheromone-induced surface protein. *Mol. Cell. Biol.* **7**, 2316–2328.

Ullrich, A., and Schlessinger, J. (1990). Signal transduction by receptors with tyrosine kinase activity. *Cell* **61**, 203–212.

van Arsdell, S. W., and Thorner, J. (1987). Hormonal regulation of gene expression in yeast. *In* "Transcriptional Control Mechanisms" (D. Granner, M. G. Rosenfeld, and S. Chang, eds.), pp. 325–332. Liss, New York.

Wang, Y., Xu, H., Riggs, M., Rodgers, L., and Wigler, M. (1991). *byr2*, a *Schizosaccharomyces pombe* gene encoding a protein kinase capable of partial suppression of the *ras1* mutant phenotype. *Mol. Cell. Biol.* **11**, 3554–3563.

Whiteway, M. S., Hougan, L., and Thomas, D. Y. (1988). Expression of *MFα1* in *MAT***a** cells supersensitive to α-factor leads to self-arrest. *Mol. Gen. Genet.* **214**, 85–88.

Whiteway, M. S., Hougan, L., Dignard, D., Thomas, D. Y., Bell, L., Saari, G. C., Grant, F. J., O'Hara, P., and MacKay, V. L. (1989). The *STE4* and *STE18* genes of yeast encode potential β and γ subunits of the mating factor receptor-coupled G protein. *Cell* **56**, 467–477.

Whiteway, M., Hougan, L., and Thomas, D. Y. (1990). Overexpression of the *STE4* gene leads to mating response in haploid *Saccharomyces cerevisiae*. *Mol. Cell. Biol.* **10**, 217–222.

Wittenberg, C., Sugimoto, K., and Reed, S. I. (1990). G1-specific cyclins of *S. cerevisiae*: Cell cycle periodicity, regulation by mating pheromone, and association with the $p34^{CDC28}$ protein kinase. *Cell* **62**, 225–237.

Yu, L., Blumer, K. J., Davidson, N., Lester, H. A., and Thorner, J. (1989). Functional expression of the yeast α-factor receptor in *Xenopus* oocytes. *J. Biol. Chem.* **264**, 20847–20850.

Yuan, Y. O., and Fields, S. (1991). Properties of the DNA-binding domain of the *Saccharomyces cerevisiae* STE12 protein. *Mol. Cell. Biol.* **11**, 5910–5918.

Zhou, Z., Gartner, A., Cade, R., Ammerer, G., and Errede, B. (1993). Pheromone-induced signal transduction in *Saccharomyces cerevisiae* requires the sequential function of three protein kinases. *Mol. Cell. Biol.* **13**, 2069–2080.

Chapter 9

Lipopeptide Pheromones of Yeast and Basidiomycetes Involved in Mating

WOLFGANG DUNTZE, RICHARD BETZ, and BERT PÖPPING

Abteilung für Biochemie der Differenzierung
Ruhr-Universität Bochum
D-4630 Bochum, Germany

I. Overview

Mating pheromones play important roles in the life cycles of several species of heterothallic fungi, in which they control the transition from the haploid to the diploid growth phase. The cells of the haploid phase usually reproduce vegetatively by mitotic division and exhibit either of two alternative mating types. Haploid cells of opposite mating type can conjugate by cell and nuclear fusion to initiate the diploid growth phase. The conjugation reaction is a complex process composed of a sequence of distinct steps. The establishment of specific cell–cell contact is followed by cellular fusion, leading to a heterokaryotic stage. In most fungi, the haploid nuclei fuse immediately to form a diploid nucleus; however, in some fungi, the heterokaryon persists for some time before karyogamy occurs.

In many fungi, initiation of the conjugation process and coordination of its stages involves extracellular mating pheromones that are secreted specifically by each mating type and act exclusively on the opposite mating type to induce the cellular responses necessary for conjugation. The role of mating pheromones in the life cycle of the yeast *Saccharomyces cerevisiae*, which is described in detail in Chapter 8, may be viewed as a general paradigm for pheromone action in many fungal life cycles.

The lipopeptide mating pheromones constitute a novel class of substances that have been found to control and coordinate the conjugation reaction between haploid cells in fungi. These pheromones combine the versatility of peptide structures with the hydrophobicity of an isoprenoid lipid. The extremely hydrophobic nature of the lipopeptide pheromones hampered their isolation and characterization. The initial discovery of this novel form of pheromone was made in the heterobasidiomycetous yeasts. Rhodotorucine A, one of the mating pheromones produced by *Rhodosporidium toruloides*, was found to be a peptide with a highly lipophilic amino acid at its C terminus (Kamiya *et al.*, 1978). Further characterization identified this residue as *S-trans,trans*-farnesyl cysteine (Kamiya *et al.*, 1979a,b). The pheromones produced by haploid strains of the closely related heterobasidiomycetes *Tremella mesenterica* and *Tremella brasiliensis* were identified as lipopeptides containing C-terminal farnesyl cysteine residues (Sakagami *et al.*, 1978,1979,1981a,b). Two of these pheromones, tremerogen A-10 and tremerogen A-9291-I, were shown to carry a methyl ester on the terminal carboxyl group (Sakagami *et al.*, 1979; Ishibashi *et al.*, 1984). This same modification also was encountered in a less eccentric organism when **a**-factor, the pheromone produced by *MAT***a** strains of *S. cerevisiae*, was shown to be a farnesyl cysteine-containing lipopeptide with a *C*-terminal methyl ester (Anderegg *et al.*, 1988).

Recently, a lipopeptide pheromone which is probably *S*-farnesylated has also been identified in the fission yeast *Schizosaccharomyces pombe* (Davey, 1992). In addition, genetic evidence has been provided which suggests the existence of similar pheromones in the smut fungus *Ustilago maydis* (Bölker *et al.*, 1992). Farnesylated lipopeptides thus appear to be rather common as signaling substances in fungal mating. The widespread occurrence of farnesylated peptides in eukaryotic microorganisms suggested the possibility that isoprenylation and C-terminal methylation might represent a more general type of protein modification in higher organisms. This view was confirmed subsequently by the discovery in several mammalian cells of a family of biologically important proteins, including Ras, all of which contain a C-terminal farnesyl cysteine (reviewed by Glomset *et al.*, 1990). The farnesyl residue in these proteins is believed to provide a hydrophobic anchor that acts to attach the protein to membranes.

The functional significance of farnesylation of the mating pheromones is uncertain. The extremely hydrophobic nature of these lipopeptides reduces their solubility and, therefore, their effective range of action in aqueous media. This seeming disadvantage may be compensated for if the hydrophobic residue increases the affinity of the lipopeptides for their receptors. The hydrophobic pheromones also have been proposed to remain bound to the plasma membrane of the producing cell and, therefore, to be transmitted more directly and efficiently to the mating partner at the region of intercellular contact (Jackson and Hartwell, 1990). Finally, the hydrophobic residue may promote intracellular membrane attachment and secretion by a novel mechanism (McGrath and Varshavsky, 1989; Kuchler *et al.*, 1990). Because the fungal systems, particularly *S. cerevisiae*, are readily amenable to biochemical and genetic analysis, they have been used widely as experimental systems in which to elucidate the significance and the biochemical mechanisms of hydrophobic modification of proteins. These studies have provided exciting new results with interesting prospects for cell biology and medicine.

II. Biochemistry

A. Structure and Function

1. Rhodotorucine A In the mating reaction of the heterobasidiomycetous fungus *R. toruloides*, cells of the yeast-like haploid growth phase of opposite mating types (*A* and *a*) conjugate to form a filamentous dikaryotic mycelium (Banno, 1967). This process is initiated by a mating pheromone secreted by type *A* cells that arrest cell division and induces the formation of conjugation tubes in type *a* cells (Abe *et al.*, 1975). The *a* cells then secrete a pheromone that acts on *A* cells in a similar manner.

The formation of mating tubes was used as a biological assay for the purification of the pheromone, designated rhodotorucine A, from culture filtrates of type *A* cells (Kamiya *et al.*, 1975). Purified rhodotorucine A binds specifically to mating type *a* but not to mating type *A* cells of *R. toruloides* (Tsuchiya and Fukui, 1978). The binding capacity is approximately 5×10^7 molecules per cell.

Rhodotorucine A was characterized as a very hydrophobic undecapeptide with a lipophilic C terminus. Proton magnetic resonance and mass spectrometry identified the C-terminal residue as *S*-farnesyl cysteine, a previously unknown derivative of cysteine (Kamiya *et al.*, 1979a). Stereochemical analysis revealed that the farnesyl residue is in the *trans,trans* configuration, one of four possible stereoisomeric forms (Fig. 9.1A; Kamiya *et al.*, 1979b). The pheromone induced by rhodotorucine A in *a* cells has not been characterized.

Rhodotorucine A is degraded rapidly by proteolytic cleavage of the peptide moiety after exposure to type *a* cells (Kamiya *et al.*, 1980). A cell-associated thiol protease, called trigger peptidase, has been characterized that is specific for rhodotorucine A as a substrate and has a strict requirement for

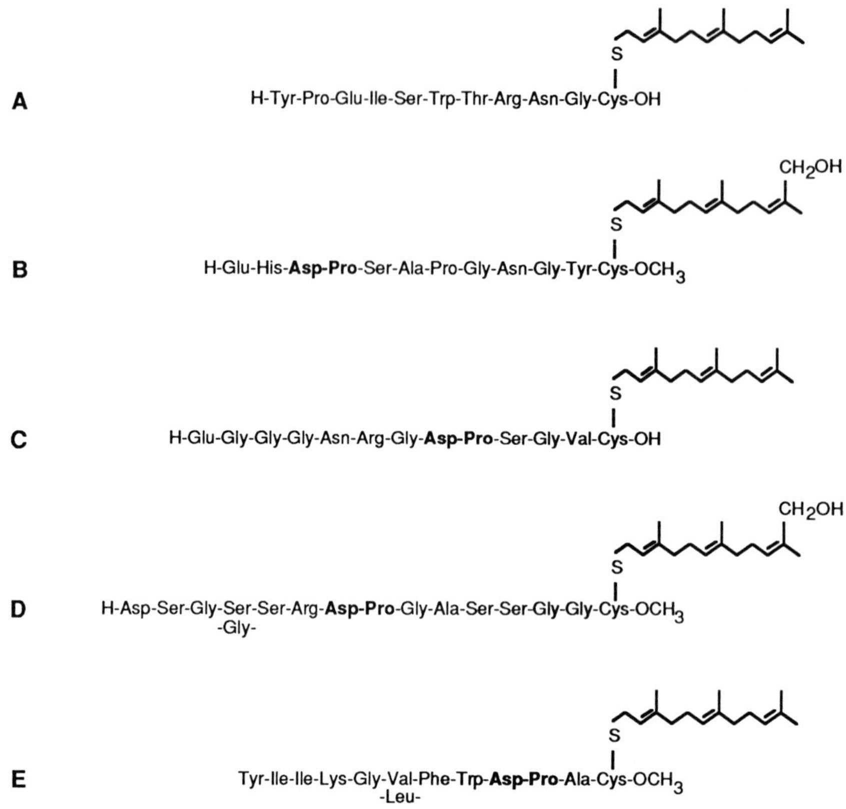

Figure 9.1 Primary structure of lipopeptide mating pheromones. (A) Rhodotorucine A from *Rhodosporidium toruloides* mating type *A* strain M 919. (B) Tremerogen A-10 isolated from *Tremella mesenterica* mating type *A* strain IFO 9310. (C) Tremerogen *a*-13 isolated from *Tremella mesenterica* mating type *a* strain IFO 9313. (D) Tremerogen A-9291-I, the major pheromone isolated from *Tremella brasiliensis* type *A* strain IFO 9291. Two forms of the pheromone that differ in an exchange at position 4 of a serine for a glycine residue have been found. (E) **a**-Factor from *Saccharomyces cerevisiae* (*MAT***a** strain X2180-A). Two forms of the pheromone differ in an exchange at position 6 of a valine for a leucine. The sequence Asp–Pro found in four of the five pheromones is shown in boldface.

the farnesyl moiety (Miyakawa *et al.*, 1987a). Trigger peptidase activity is not found in cells of *A* mating type. Trigger peptidase requires Ca^{2+} and specific phospholipids for *in vitro* activity, and cleaves the Arg–Asn sequence in the peptide moiety of the pheromone.

Several results suggest that cleavage of rhodotorucine A by trigger peptidase may be involved in transduction of the pheromone signal. A sterile mutant of *a* mating type that is defective in response to rhodotorucine *A* is also defective in trigger peptidase activity (Miyakawa *et al.*, 1982). Additional information about this mutant is required to determine whether the response defect results specifically from the trigger peptidase defect. Pheromone response is sensitive to thiol reagents that do not penetrate the membrane, suggesting that cell-surface protein sulfhydryls are involved in signal transduction (Miyakawa *et al.*, 1985a). Sulfhydryl blocking reagents inhibit both the biological effect of pheromone and the hydrolysis of pheromone; dithiothreitol (DTT) enhances both reactions (Miyakawa *et al.*, 1985a). Based on the correlation of biological response to rhodotorucine *A* and trigger peptidase activity, endoproteolytic cleavage has been suggested to play an essential role in triggering the mating process (Miyakawa *et al.*, 1982,1985a). Additional support for this conclusion requires information about whether the activity of cell-surface components other than trigger peptidase are affected by the sulfhydryl blocking and enhancing reagents. A role for a pheromone hydrolyzing peptidase in signal transmission would be quite novel, particularly in comparison with the *S. cerevisiae* system in which α-factor degradation by the barrier protease is required for recovery from pheromone arrest (Chan and Otte, 1982; MacKay *et al.*, 1988). A defect in the barrier protease leads to increased sensitivity to pheromone rather than to an absence of response to pheromone.

Another interesting observation is that Ca^{2+} may play a role in pheromone signaling. Rhodotorucine *A* causes a rapid transient increase in intracellular Ca^{2+} levels in its target cells (Miyakawa *et al.*, 1985c). Physiological concentrations of rhodotorucine *A* also inhibit the activity of a membrane-bound Ca^{2+}-ATPase (Miyakawa *et al.*, 1987b). The finding that both this Ca^{2+}-ATPase and the trigger peptidase activity are in the same particulate fraction has been suggested to indicate a close interaction of these proteins in the transduction of the pheromone signal. Additional structural and functional information about the trigger peptidase, the Ca^{2+}-ATPase, and the pheromone receptor is required to elucidate the molecular mechanism of transduction of the rhodotorucine *A* signal.

2. *Tremerogens* Lipopeptide pheromones also have been found to play an important role in conjugation in the closely related heterobasidiomycetous yeasts *T. mesenterica* (Sakagami *et al.*, 1981a) and *T. brasiliensis* (Ishibashi *et al.*, 1983) which, like *R. toruloides*, have a complex life cycle with dimorphic developmental phases. The cells of the yeast-like haplophase exist in two mating types, *A* and *a*, which can conjugate to produce dikaryotic mycelia (Flegel, 1981). Cells of *A* and *a* mating type each secrete a specific mating pheromone (originally designated *A*-factor and *a*-factor, respectively) that induce mating tube formation in the complementary cell type (Sakagami *et al.*, 1981a). Several pheromones were purified to homogeneity from culture filtrates of haploid

strains of both species using mating tube formation as an assay. Unfortunately, the multitude of compounds with pheromonal activity that was detected in the isolation from different strains gave rise to a somewhat confusing nomenclature.

A-factor and *a*-factor first were isolated from the complementary haploid strains IFO 9310 (type *A*) and IFO 9313 (type *a*) of *T. mesenterica* and renamed tremerogen *A*-10 and tremerogen *a*-13, respectively (Sakagami *et al.*, 1978,1981a; Yoshida *et al.*, 1981). Tremerogen *A*-10 and tremerogen *a*-13 were identified as peptides of 12 and 13 amino acids, respectively, containing C-terminal *S*-farnesyl cysteine residues (Fig. 9.1B and C; Sakagami *et al.*, 1981a). The farnesyl moiety of tremerogen *A*-10 is hydroxylated at one of the terminal methyl groups; the carboxyl group of farnesyl cysteine is blocked by a methyl ester (Sakagami *et al.*, 1979). The farnesyl residue of tremerogen *a*-13 is unmodified and the C terminus is free, as in rhodotorucine A (Sakagami *et al.*, 1981b). Alkylation of the cysteine residue of tremerogen *a*-13 has been shown to be required for biological activity (Kitada *et al.*, 1981). Interestingly, pheromone activity was increased by amidation of the terminal carboxyl group, which is unblocked in the native pheromone.

Eight compounds with pheromone activity were isolated from strain IFO 9291 (type *A*) of *T. brasiliensis*; these were designated tremerogens *A*-9291-I–*A*-9291-VIII (Ishibashi *et al.*, 1984). Structural analysis demonstrated that these pheromones are farnesylated lipopeptides in which the farnesyl residue is hydroxylated and the terminal carboxyl group is methylated (Fig. 9.1D). These pheromones exhibit microheterogeneity at the fourth residue of the peptide moiety, which may be either glycine or serine. The heterogeneity of the peptide moiety and hydrolysis of the terminal methyl ester appear to account for some of the multiplicity of the tremerogens isolated from this strain.

3. Saccharomyces cerevisiae ***a****-Factor* Because farnesylated lipopeptide pheromones previously had been found only in *Rhosdosporidium* and *Tremella*, this unusual type of compound seemed to be characteristic of the heterobasidiomycetous yeasts. Therefore, when the pheromone **a**-factor was isolated from haploid cultures of **a** cells of the ascomycetous yeast *S. cerevisiae* (Wilkinson and Pringle, 1974; Betz *et al.*, 1977) and characterized as an extremely hydrophobic peptide (Betz and Duntze, 1979), the molecule was thought to be acylated by a palmityl residue (Powers *et al.*, 1986; Becker *et al.*, 1987). The finding that **a**-factor, which is a mixture of closely related peptides that differ by the interchange of valine and leucine residues at position 6, contains an S-alkylated C-terminal cysteine residue with an esterified carboxyl group was rather surprising (Betz *et al.*, 1987). The detailed structure of **a**-factor remained elusive until analysis by nuclear magnetic resonance (NMR) and mass spectrometry revealed that it contains a C-terminal *S*-farnesyl cysteine residue that is blocked by a methyl ester, as seen in the tremerogens *A*-10 and *A*-9291-

I (Fig. 9.1E; Anderegg *et al.*, 1988). The farnesyl moiety was shown to be in the *trans,trans* stereochemical configuration. This structure subsequently was confirmed by total synthesis of the modified peptide (Xue *et al.*, 1989). The mechanism of signal transduction by **a**-factor and α-factor, the unacylated peptide pheromone produced by *S. cerevisiae* α cells, is discussed in Chapter 8.

4. *Schizosaccharomyces pombe M-Factor* Although the existence of diffusible mating pheromones in the fission yeast *S. pombe* has been indicated by several observations (Friedman and Egel, 1978; Leupold *et al.*, 1989), it was only recently that the pheromone M-factor could be isolated from M-type cells (Davey, 1991). Structural analysis (Davey, 1992) revealed that the M-factor is a nonapeptide in which the carboxyl-terminal cysteine residue is carboxymethylated and probably S-farnesylated having the structure: Tyr–Thr–Pro–Lys–Val–Pro–Tyr–Met–Cys-(*S*-farnesyl)-OCH_3. As in *S. cerevisiae* **a**-factor and tremerogen A-9291-I, the carboxyl-terminal methyl ester appears to be essential for biological activity of the M factor.

5. *Mating Pheromones of Ustilago maydis* The basidiomycete *Ustilago maydis*, the causative agent of corn smut disease, has a complex sexual life cycle which is initiated by the fusion of two compatible yeast-like haploid cells leading to the formation of a filamentous dikaryotic mycelium. The mating reaction is controlled by the *a* mating type locus, which exists in two allelic forms *a*1 and *a*2 defined by nonhomologous DNA sequences. Both alleles have been sequenced recently and were shown to contain short open reading frames coding for polypeptide products of 40 and 38 amino acids, respectively, which share at their carboxyl-terminus a C–A–A–X sequence. As this sequence motif is a characteristic feature of prenylated proteins, it has been suggested that the alleles of the mating type locus encode precursors of lipopeptide mating pheromones which control the mating reaction of the *U. maydis* (Bölker *et al.*, 1992).

B. Structure–Function Relationships

All the characterized fungal lipopeptide pheromones are composed of short peptide components 11–15 amino acids in length and terminating with a long hydrophobic farnesyl moiety bound in thioether linkage to the sulfur atom of the C-terminal cysteine residue. A comparison of the amino acid sequences of the peptide moieties reveals no similarities in primary sequence except for the sequence Asp–Pro, which is present in all the pheromones except rhodotorucine A and M factor (Fig. 9.1). No functional significance of this conserved dipeptide is known, but the peptide bond between these two amino acids is more acid labile than the standard peptide bond.

In all these pheromones, the *S-trans,trans*-farnesyl cysteine residue that constitutes the C terminus forms a very hydrophobic pole. In the tremerogens *A*-9291 and *A*-10 and in *S. cerevisiae* **a**-factor, this pole becomes even more hydrophobic by esterification of the terminal carboxyl group. Oxidation of the farnesyl moiety on C-13 in the tremerogens *A*-10 and *A*-9291 would not be predicted to have a significant influence on the lipophilic nature of these pheromones. This modification is not essential for pheromonal activity of tremerogen *A*-10 (Tsuchiya and Fukui, 1978; Fujino *et al.*, 1980). In the tremerogens, the overall hydrophobicity is reduced by more hydrophilic amino acids in the peptide portion of the molecule, but for **a**-factor an accumulation of rather apolar amino acids in the peptide increases hydrophobicity.

As predicted from the dissimilarity in amino acid sequences (Fig. 9.1), the models for the three-dimensional structures of the lipopeptide pheromones show no obvious resemblance to one another (Fig. 9.2). However, some similarity in the basic molecular design of all of the pheromones exists; the peptide portions form variable, probably flexible, structures from which the C-terminal farnesyl residues project as rigid rods.

The lipid and the peptide components of the pheromone molecules seem likely to contribute to different aspects of pheromonal activity. Specificity is suggested to be associated with the peptide portions of the molecules, as indicated by the observation that no cross-reactivity is seen between the different pheromones, including the pheromones of the two very closely related *Tremella* species, *T. mesenterica* and *T. brasiliensis* (Tsuchiya *et al.*, 1978; Ishibashi *et al.*, 1983). Further, proteolytic cleavage abolishes the activity of all the pheromones, leaving no residual activity in the C-terminal fragments containing the farnesyl cysteine residues (Betz *et al.*, 1977; Tsuchiya *et al.*, 1978; Sakagami *et al.*, 1981a). For tremerogen *A*-10, the removal of only the N-terminal amino acid is sufficient for complete loss of activity (Fujino *et al.*, 1980).

Studies of structure–function relationships show variable requirements for the farnesyl residue for pheromonal activity. Unsubstituted peptide analogs of rhodotorucine A and tremerogen A-10 are biologically inactive (Tsuchiya *et al.*, 1978). These analogs do not act as competitive inhibitors of the native pheromones, indicating an absolute requirement for farnesylation for receptor binding as well as activity. For tremerogens *A*-10 and *a*-13, substitution of the cysteine sulfur with alkyl groups other than farnesyl restores only partial activity (Fujino *et al.*, 1980; Kitada *et al.*, 1981). For alkyl residues shorter than farnesyl, the specific biological activity of the analogs tends to increase with increasing length of the alkyl residue. Tremerogen analogs with longer and more lipophilic side chains, for example, tetraprenyl and pentaprenyl groups, do not present a consistent picture; such tremerogen *a*-13 analogs are less potent (Kitada *et al.*, 1981) whereas such tremerogen *A*-10 analogs are more potent (Fujino *et al.*, 1980) than the native pheromones. For *S. cerevisiae* **a**-factor, the

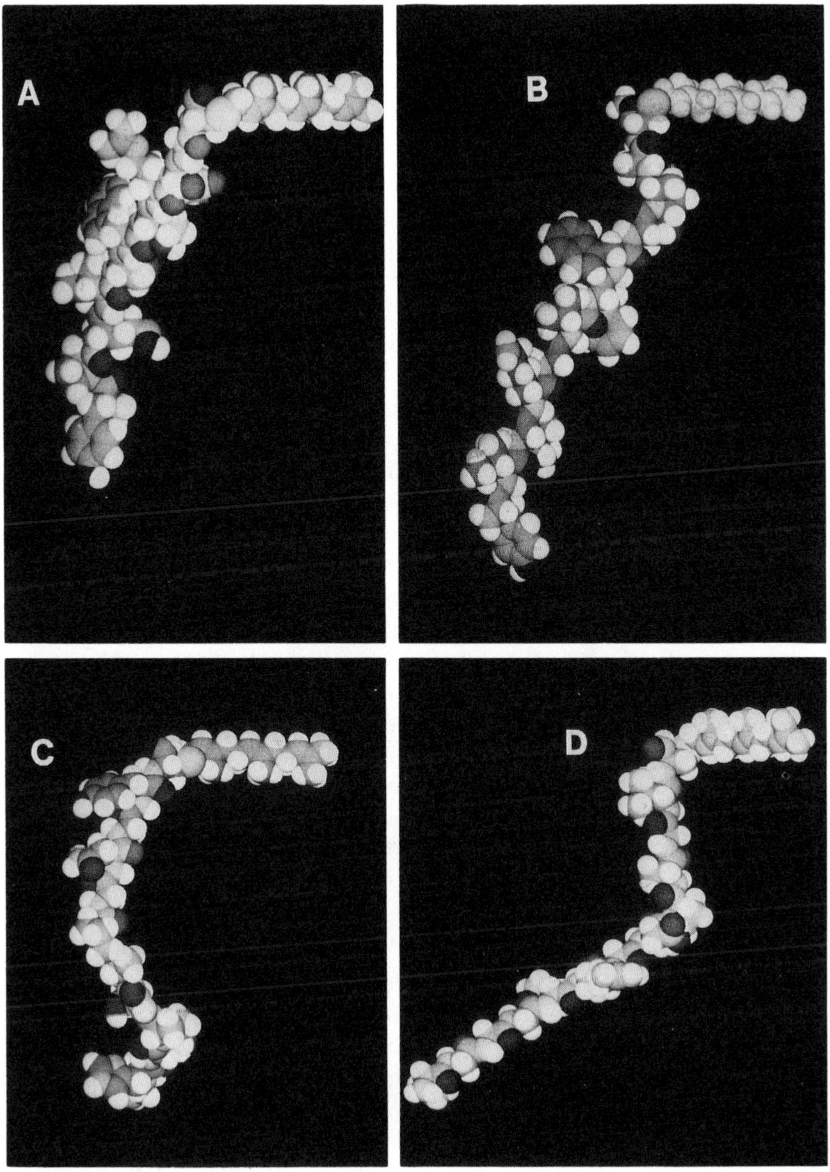

Figure 9.2 Computer-generated images of possible three-dimensional structures of lipopeptide pheromones. (A) Rhodotorucine A. (B) *Saccharomyces cerevisiae* **a**-factor. (C) Tremerogen A-10. (D) Tremerogen *a*-13. The models are arranged to show the farnesyl moiety on the upper right.

alkyl group is important but not absolutely essential for activity; the **a**-factor analog lacking an alkyl group shows reduced activity (Xue *et al.*, 1990). Intracellular modification of the lipopeptide pheromone by farnesylation also may play a role in secretion (see subsequent discussion).

In *S. cerevisiae* **a**-factor, *S. pombe* M-factor, and the tremerogens A-10 and A-9291, the blockage of terminal carboxyl group seems to be essential for pheromonal activity. Hydrolysis of the methyl ester abolishes the biological activity of the tremerogens A-10 and A-9291 (Fujino *et al.*, 1980; Ishibashi *et al.*, 1984) and M-factor (Davey, 1992). Hydrolysis of the methyl ester abolishes biological activity of **a**-factor (Anderegg *et al.*, 1988), but a synthetic farnesylated **a**-factor analog lacking C-terminal methylation has been reported to show low activity, about 800-fold lower than completely modified **a**-factor (Xue *et al.*, 1990). For tremerogen *a*-13, which has a free terminal carboxyl group in the native form, blocking of the C terminus which an amide group results in approximately an eightfold increase in specific activity (Kitada *et al.*, 1981), again indicating the importance of this modification for activity. Similarly, an octapeptide analog of M-factor acts as an inhibitor of pheromone activity when the C terminus is blocked, but not when it is free (Davey, 1992).

In summary, these structure–function studies suggest several conclusions: (1) pheromonal specificity depends primarily on the peptide portion of the pheromone molecules; (2) lipophilic substitution of the C-terminal cysteine, most favorably by farnesylation, is a requirement for high specific activity; and (3) for those pheromones that carry a terminal methyl ester, a block of the terminal carboxyl group is essential for high specific activity.

III. Biosynthesis and Secretion

During the past few years, alkylation and methylation of C-terminal cysteine residues have become recognized as common posttranslational modifications of many membrane-associated proteins, including the γ subunit of heterotrimeric G proteins (Backlund *et al.*, 1990; Finegold *et al.*, 1990; Lai *et al.*, 1990), Ras proteins (Hancock *et al.*, 1989; Guiterrez *et al.*, 1989), and nuclear lamins (Wolda and Glomset, 1988; Farnsworth *et al.*, 1989). The observation that farnesylation is required for the transforming activity of oncogenic Ras proteins has suggested possible approaches for the therapy of Ras-dependent tumors (Schafer *et al.*, 1989).

Because *S. cerevisiae* is readily amenable to biochemical and genetic analysis, it provides an excellent system in which to study the maturation of farnesylated proteins. Two genes encoding **a**-factor, *MFA1* and *MFA2*, were isolated using synthetic oligonucleotide probes corresponding to the amino acid sequence of the pheromone (Brake *et al.*, 1985). In addition, several **a**-specific sterile mutants were found to be defective in the production of **a**-factor (Fuller *et al.*, 1986; Wilson and Herskowitz, 1987; Michaelis and Powers, 1988).

MFA1 and *MFA2* are each responsible for production of about half the **a**-factor secreted (Michaelis and Herskowitz, 1988). These genes encode precursors of 36 and 38 amino acids, respectively, each of which contains a single copy of the **a**-factor peptide sequence (Brake *et al.*, 1985). These precursors terminate with the sequence Cys–Aaa–Aaa–Xaa, where Cys is the C-terminal cysteine residue of the pheromone, Aaa is an aliphatic amino acid, and Xaa is any amino acid. Because this sequence, often referred to as the CAAX box, is found at the C termini of all known farnesylated proteins, it appears to represent a general signal for protein farnesylation.

Investigation of the order of the processing and modification steps involved in **a**-factor production has been facilitated greatly by the availability of mutants defective in **a**-factor production. The order of steps determined by these studies is: (1) transfer of the farnesyl moeity from farnesyl pyrophosphate to the cysteine residue of the CAAX box, (2) removal of the terminal AAX tripeptide, and (3) transfer of a methyl group from *S*-adenosyl methionine to the C-terminal farnesyl cysteine. Two genes, *DPR1/RAM1* and *RAM2*, are required for farnesylation and probably encode farnesyltransferase subunits (Schafer *et al.*, 1989,1990; Goodman *et al.*, 1990). The *STE14* gene is required for carboxymethylation and is likely to encode the methyltransferase (Hrycyna and Clarke, 1990; Marr *et al.*, 1990). The order of these steps with respect to proteolytic removal of the N-terminal sequences is not known. Genes involved in proteolytic processing of both the C- and N-terminal sequences have not been identified.

Secretion of **a**-factor utilizes a route that is distinct from the typical eukaryotic secretory pathway. In the course of studying a ubiquitin-activating gene in yeast, McGrath and Varshavsky (1989) fortuitously characterized the adjacent *STE6* gene, which is required for **a**-factor secretion. The nucleotide sequence of the *STE6* gene predicts a protein with multiple putative membrane-spanning segments that is similar to a set of bacterial permeases. The deduced amino acid sequence suggests that the protein contains two ATP-binding sites. Therefore **a**-factor synthesis and processing has been proposed to occur in the cytoplasm; the *STE6* gene product functions as a permease that actively translocates **a**-factor through the plasma membrane by an ATP-driven process. Farnesylation of **a**-factor may facilitate interactions with the membrane and with the *STE6* product. Intriguingly, the *STE6* product also exhibits striking similarity to the mammalian *MDR1* gene product, which encodes a plasma membrane protein (P-glycoprotein) that can confer multidrug resistance to tumor cells. Therefore the normal role of the *MDR1* protein has been suggested to be to secrete peptides by a mechanism independent of the normal secretory pathway (Kuchler *et al.*, 1989).

Processing in *Rhodosporidium* and *Tremella* species has been characterized less extensively. Three genes (*RHA1*, *RHA2*, and *RHA3*) encoding the peptide moiety of rhodotorucine *A* have been cloned (Akada *et al.*, 1987,1989a,b); these genes contain four, five, and three tandem repeats of the rhodotorucine *A* peptide sequence, respectively, all separated by spacers. The C termini of the putative precursors end with a CAAX sequence. The spacers between rhodotorucine *A* repeats consist of the sequence (Cys)–Thr–Val–Ser/Ala–Lys. The individual pheromone peptides are likely to be liberated by proteolytic cleavage of the Ser/Ala bond, creating a terminal CAAX box that can serve as a signal for farnesylation. In a study of tremerogen *A* synthesis, putative precursors of 28 kDa, 12 kDa, and 2.8 kDa were observed (Miyakawa *et al.*, 1985b). Compactin, an inhibitor of mevalonic acid synthesis, blocked tremerogen *A* maturation, consistent with the farnesylation of the pheromone. Tunicamycin, which inhibits attachment of core *N*-linked carbohydrate in the endoplasmic reticulum, and monensin, which inhibits secretion at the level of the Golgi complex, also inhibited maturation. These results seem inconsistent with the results in *S. cerevisiae*, indicating that processing and secretion of the lipopeptide pheromone **a**-factor does not occur in the classical secretory pathway.

These studies on lipopeptide pheromone biosynthesis have important implications not only for cell biology but for medicine as well. Because the biochemical mechanisms of farnesylation and carboxy-methylation are likely to be similar in mammalian cells and in yeast, elucidation of the molecular mechanisms of these reactions may lead to the development of specific inhibitors that have potential therapeutic value for Ras-related tumors. In addition, analysis of the mechanism of **a**-factor translocation by the *STE6* gene product may help elucidate the physiological role and mechanism of *MDR1*, and therefore facilitate the development of permease inhibitors that are beneficial in cancer therapy.

Acknowledgments

The original research reported in this paper was supported by a grant from the Deutsche Forschungsgemeinschaft. We are also grateful to D. Wunsch for his help in constructing the computer images.

References

Abe, K., Kusaka, I., and Fukui, S. (1975). Morphological change in the early stages of the mating process of *Rhodosporidium toruloides*. *J. Bacteriol.* **122**, 710–718.

Akada, R., Minomi, K., Yamashita, I., Miyakawa, T., and Fukui, S. (1987). Cloning of a gene coding for rhodotorucine A, a farnesyl peptide mating pheromone of *Rhodosporidium toruloides*. *Agric. Biol. Biochem.* **51**, 1211–1215.

Akada, R., Minomi, K., Kai, J., Yamashita, I., Miyakawa, T., and Fukui, S. (1989a). Multiple genes coding for precursors of rhodotorucine A, a farnesyl peptide mating pheromone of the basidiomycetous yeast *Rhodosporidium toruloides*. *Mol. Cell. Biol.* **9**, 3491–3498.

Akada, R., Kai, J., Yamashita, I., Miyakawa, T., and Fukui, S. (1989b). Genomic organization of multiple genes coding for rhodotorucine A, a lipopeptide mating pheromone of the basidiomycetous yeast *Rhodosporidium toruloides*. *Arch. Microbiol.* **152**, 484–487.

Anderegg, R. J., Betz, R., Carr, S. A., Crabb, J. W., and Duntze, W. (1988). Structure of *Saccharomyces cerevisiae* mating hormone **a**-factor. *J. Biol. Chem.* **263**, 18236–18240.

Backlund, P. S., Jr., Simonds, W. F., and Spiegel, A. M. (1990). Carboxyl methylation and COOH-terminal processing of the brain G-protein gamma-subunit. *J. Biol. Chem.* **265**, 15572–15576.

Banno, I. (1967). Studies on the sexuality of *Rhodotorula*. *J. Gen. Appl. Microbiol.* **13**, 167–196.

Becker, J. M., Marcus, S., Kundu, B., Shenbagamurthi, P., and Naider, F. (1987). Towards determination of the structure of the *Saccharomyces cerevisiae* **a**-factor: An acylated pentadecapeptide blocks **a**-factor activity. *Mol. Cell. Biol.* **7**, 4122–4124.

Betz, R., and Duntze, W. (1979). Purification and partial characterization of **a**-factor, a mating hormone produced by mating-type-**a** cells from *Saccharomyces cerevisiae*. *Eur. J. Biochem.* **95**, 469–475.

Betz, R., MacKay, V. L., and Duntze, W. (1977). **a**-Factor from *Saccharomyces cerevisiae*: Partial characterization of a mating hormone produced by cells of mating type **a** *J. Bacteriol.* **132**, 462–472.

Betz, R., Crabb, J. W., Meyer, H. E., Wittig, R., and Duntze, W. (1987). Amino acid sequences of **a**-factor mating peptides from *Saccharomyces cerevisiae*. *J. Biol. Chem.* **262**, 546–548.

Bölker, M., Urban, M., and Kahmann, R. (1992). The *a* mating type locus of *U. maydis* specifies cell signaling components. *Cell* **68**, 441–450.

Brake, A., Brenner, C., Najarian, R., Laybourn, P., and Merryweather, J. (1985). Structures of genes encoding precursors of the yeast mating pheromone **a**-factor. *In* "Protein Transport and Secretion" (M.-J. Gething, ed.), pp. 103–108. Cold Spring Harbor Laboratory Press, Cold Spring Harbor, New York.

Chan, R. K., and Otte, C. A. (1982). Isolation and genetic analysis of *Saccharomyces cerevisiae* mutants supersensitive to G_1 arrest by **a** factor and α-factor pheromones. *Mol. Cell. Biol.* **2**, 11–20.

Davey, J. (1991). Isolation and quantitation of M-factor, a diffusible mating factor from the fission yeast *Schizosaccharomyces pombe*. *Yeast* **7**, 357–366.

Davey, J. (1992). Mating pheromones of the fission yeast *Schizosaccharomyces pombe*: purification and structural characterization of M-factor and isolation and analysis of two genes encoding the pheromone. *EMBO J.* **11**, 951–960.

Farnsworth, C. C., Wolda, S. L., Gelb, M. H., and Glomset, J. A. (1980). Human lamin B contains a farnesylated cysteine residue. *J. Biol. Chem.* **264**, 20422–20429.

Finegold, A. A., Schafer, W. R., Rine, J., Whiteway, M., and Tamanoi, F. (1990). Common modification of trimeric G-proteins and ras protein: Involvement of polyisoprenylation. *Science* **249**, 165–169.

Flegel, T. W. (1981). The pheromonal control of mating in yeasts and its phylogenetic implication: A review. *Can. J. Microbiol.* **27**, 373–389.

Friedmann, K., and Egel, R. (1978). Protein patterns during sporulation in fission yeast. *Z. Naturforsch.* **33c**, 84–91.

Fujino, M., Kitada, C., Sakagami, Y., Isogai, A., Tamura, S., and Suzuki, A. (1980). Biological activity of synthetic analogs of tremerogen A-10. *Naturwiss.* **67**, 406–408.

Fuller, R., Brake, A., Sterne, R., Kunisawa, R., Barnes, D., Flessel, M., and Thorner, J. (1986). Post-translational processing events in the maturation of yeast pheromone precursors. *In* "Yeast Cell Biology" (J. Hicks, ed.), pp. 461–476. Liss, New York.

Glomset, J. A., Gelb, M. H., and Farnsworth, C. C. (1990). Prenyl proteins in eukaryotic cells: A new type of membrane anchor. *Trends Biochem. Sci.* **15**, 139–142.

Goodman, L. E., Judd, S. R., Farnsworth, C. C., Powers, S., Gelb, M. H., Glomset, J. A., and Tamanoi, F. (1990). Mutants of *Saccharomyces cerevisiae* defective in the farnesylation of Ras proteins. *Proc. Natl. Acad. Sci. U.S.A.* **87**, 9665–9669.

Gutierrez, L., Magee, A. I., Marshall, C. J., and Hancock, J. F. (1989). Post-translational processing of $p21^{ras}$ is two-step and involves carboxyl-methylation and carboxyl-terminal proteolysis. *EMBO J.* **8**, 1093–1098.

Hancock, J. F., Magee, A. I., Childs, J. E., and Marshall, J. E. (1989). All ras proteins are polyisoprenylated but only some are palmitoylated. *Cell* **57**, 1167–1177.

Hrycyna, C. A., and Clarke, S. (1990). Farnesyl cysteine C-terminal methyltransferase activity is dependent upon the *STE14* gene product in *Saccharomyces cerevisiae*. *Mol. Cell. Biol.* **10**, 5071–5076.

Ishibashi, Y., Sakagami, Y., Isogai, A., Suzuki, A., and Bandoni, R. J. (1983). Isolation of tremerogens A-9291-I and A-9291-II, novel sex hormones of *Tremella brasiliensis*. *Can. J. Biochem. Cell. Biol.* **61**, 796–801.

Ishibashi, Y., Sakagami, Y., Isogai, A., and Suzuki, A. (1984). Structures of tremerogens A9291-I and A9291-VIII: Peptidyl sex hormones of *Tremella brasiliensis*. *Biochemistry* **23**, 1399–1404.

Jackson, C. L., and Hartwell, L. H. (1990). Courtship in *Saccharomyces cerevisiae*: An early cell–cell interaction during mating. *Mol. Cell. Biol.* **10**, 2202–2213.

Kamiya, Y., Yamamoto, K., Sakurai, A., Tamura, S., Abe, K., and Fukui, S. (1975). Isolation of a peptidyl factor controlling mating tube formation in *Rhodosporidium toruloides*. *Proc. Japan. Acad.* **51**, 571–576.

Kamiya, Y., Sakurai, A., Tamura, S., Takahashi, N., Abe, K., Tsuchiya, E., and Fukui, S. (1978). Isolation of rhodotorucine A, a peptidyl factor inducing the mating tube formation in *Rhodosporidium toruloides*. *Agric. Biol. Chem.* **42**, 1239–1243.

Kamiya, Y., Sakurai, A., Tamura, S., Takahashi, N., Tsuchiya, E., Abe, K., and Fukui, S. (1979a). Structure of rhodotorucine A, a peptidyl factor, inducing mating tube formation in *Rhodosporidium toruloides*. *Agric. Biol. Chem.* **43**, 363–369.

Kamiya, Y., Sakurai, A., Tamura, S., and Takahashi, N. (1979b). Structure confirmation of *S-trans-trans*-farnesylcysteine in rhodotorucine A by application of sulfonium salt cleavage reaction. *Agric. Biol. Chem.* **43**, 1049–1053.

Kamiya, Y., Sakurai, A., and Takahashi, N. (1980). Metabolites of mating pheromone rhodotorucine A, by a cells of *Rhodosporidium toruloides*. *Biochem. Biophys. Res. Commun.* **94**, 855–860.

Kitada, C., Fujino, M., Sakagami, Y., Yoshida, M., Isogai, A., and Suzuki, A. (1981). Synthesis and structure–activity relation of tremerogen *a*-13, a peptidal sex hormone of *Tremella mesenterica*. *Agric. Biol. Chem.* **45**, 1049–1051.

Kuchler, K., Sterne, R. E., and Thorner, J. (1989). *Saccharomyces cerevisiae* STE6 gene product: A novel pathway for protein export in eukaryotic cells. *EMBO J.* **8**, 3973–3984.

Lai, R. K., Perez-Sala, D., Canada, F. J., and Rando, R. R. (1990). The gamma subunit of transducin is farnesylated. *Proc. Natl. Acad. Sci. U.S.A.* **87**, 7673–7677.

Leupold, U., Sipicki, M., and Egel, R. (1991). Pheromone production and response in sterile mutants of fission yeast. *Curr. Genet.* **20**, 79–85.

McGrath, J. P., and Varshavsky, A. (1989). The yeast *STE6* gene encodes a homologue of the mammalian multidrug resistance P-glycoprotein. *Nature (London)* **340**, 400–404.

MacKay, V. L., Welch, S. K., Insley, M. Y., Manney, T. R., Holly, J., Saari, G. C., and Parker, M. L. (1988). The *Saccharomyces cerevisiae BAR1* gene encodes an exported protein with homology to pepsin. *Proc. Natl. Acad. Sci. U.S.A.* **85**, 55–59.

Maltese, W. A., and Robishaw, J. D. (1990). Isoprenylation of C-terminal cysteine in a G-protein gamma subunit. *J. Biol. Chem.* **265**, 18071–18074.

Marcus, S., Caldwell, G. A., Xue, C.-B., Naider, F., and Becker, J. M. (1990). Total in vitro maturation of the *Saccharomyces cerevisiae* **a**-factor lipopeptide mating pheromone. *Biochem. Biophys. Res. Commun.* **172**, 1310–1316.

Marr, R. S., Blair, L. C., and Thorner, J. (1990). *Saccharomyces cerevisiae STE14* gene is required for COOH-terminal methylation of **a**-factor mating pheromone. *J. Biol. Chem.* **265**, 20057–20060.

Michaelis, S., and Herskowitz, I. (1988). The **a**-factor pheromone of *Saccharomyces cerevisiae* is essential for mating. *Mol. Cell. Biol.* **8**, 1309–1318.

Michaelis, S., and Powers, S. (1988). Biogenesis of yeast mating pheromone **a**-factor and RAS-proteins. *In* "Molecular Biology of Intracellular Protein Sorting and Organelle Assembly" (R. A. Bradshaw *et al.*, eds), pp. 193–202. Liss, New York.

Miyakawa, T., Nishihara, M., Tsuchiya, E., and Fukui, S. (1982). Role of metabolism of the mating pheromone in sexual differentiation of the heterobasidiomycete *Rhodosporidium toruloides*. *J. Bacteriol.* **151**, 1184–1194.

Miyakawa, T., Kaji, M., and Yasutake, T., Jeong, Y.-K., Tsuchiya, E., and Fukui, S. (1985a). Involvement of protein sulfhydryls in the trigger reaction of rhodotorucine A, a farnesyl peptide mating pheromone of *Rhodosporidium toruloides*. *J. Bacteriol.* **162**, 294–299.

Miyakawa, T., Tabata, M., Tsuchiya, E., and Fukui, S. (1985b). Biosynthesis and secretion of tremerogen A-10, a polyisoprenyl peptide mating pheromone of *Tremella mesenterica*. *Eur. J. Biochem.* **147**, 489–493.

Miyakawa, T., Tachikawa, T., Jeong, Y.-K., Tsuchiya, E., and Fukui, S. (1985c). Transient increase of Ca^{2+} uptake as a signal for mating pheromone-induced differentiation in the heterobasidiomycetous yeast *Rhodosporidium toruloides*. *J. Bacteriol.* **162**, 1304–1306.

Miyakawa, T., Kaji, M., Jeong, Y.-K., Tsuchiya, E., and Fukui, S. (1987a). Purification and characterization of a Ca^{2+}-dependent membrane peptidase involved in the signaling of mating pheromone in *Rhodosporidium toruloides*. *J. Bacteriol.* **169**, 1626–1631.

Miyakawa, T., Tachikawa, T., Jeong, Y.-K., Tsuchiya, E., and Fukui, S. (1987b). Inhibition of membrane Ca^{2+}-ATPase *in vitro* by mating pheromone in *Rhodosporidium toruloides*, a heterobasidiomycetous yeast. *Biochem. Biophys. Res. Commun.* **143**, 893–900.

Powers, S., Michaelis, S., Broek, D., Santa Anna, A. S., Field, J., Herskowitz, I., and Wigler, M. (1986). *RAM*, a gene of yeast required for a functional modification of RAS proteins and for production of mating pheromone **a**-factor. *Cell* **47**, 413–422.

Sakagami, Y., Isogai, A., Suzuki, A., Tamura, S., Tsuchiya, E., and Fukui, S. (1978). Isolation of a novel sex hormone, tremerogen A-10, controlling conjugation tube formation in *Tremella mesenterica* Fries. *Agric. Biol. Chem.* **42**, 1093–1094.

Sakagami, Y., Isogai, A., Suzuki, A., Tamura, S., Kitada, C., and Fujino, M. (1979). Structure of tremerogen A-10, a peptidal hormone inducing conjugation tube formation in *Tremella mesenterica*. *Agric. Biol. Chem.* **43**, 2643–2645.

Sakagami, Y., Yoshida, M., Isogai, A., and Suzuki, A. (1981a). Peptidal sex hormones inducing conjugation tube formation in compatible mating-type cells of *Tremella mesenterica*. *Science* **212**, 1525–1527.

Sakagami, Y., Yoshida, M., Isogai, A., and Suzuki, A. (1981b). Structure of tremerogen a-13, a peptidal sex hormone of *Tremella mesenterica*. *Agric. Biol. Chem.* **45**, 1045–1047.

Schafer, W. R., Kim, R., Sterne, R., Thorner, J., Kim, S.-H., and Rine, J. (1989). Genetic and pharmacological suppression of oncogenic mutations in *ras* genes of yeast and humans. *Science* **245**, 379–385.

Schafer, W. R., Trueblood, C. E., Yang, C.-C., Mayer, M. P., Rosenberg, S., Poulter, C. B., Kim, S.-H., and Rine, J. (1990). Enzymatic coupling of cholesterol intermediates to a mating pheromone precursor and to the Ras protein. *Science* **249**, 1133–1139.

Tsuchiya, E., and Fukui, S. (1978). Binding of rhodotorucine A, a lipopeptidyl mating hormone to *a* cells of *Rhodosporidium toruloides* for induction of sexual differentiation. *Biochem. Biophys. Res. Commun.* **85**, 473–479.

Tsuchiya, E., Fukui, S., Kamiya, Y., Sakagami, Y., and Fujino, M. (1978). Requirements of chemical structure for hormonal activity of lipopeptidyl factors inducing sexual differentiation in vegetative cells of heterobasidiomycetous yeasts. *Biochem. Biophys. Res. Commun.* **85**, 459–463.

Wilkinson, L. E., and Pringle, J. R. (1974). Transient G_1 arrest of *S. cerevisiae* cells of mating type a by a factor produced by cells of mating type **a**. *Expt. Cell Res.* **89**, 175–187.

Wilson, K. L., and Herskowitz, I. (1987). *STE16*, a new gene required for pheromone production by **a** cells of *Saccharomyces cerevisiae*. *Genetics* **155**, 441–449.

Wolda, S. L., and Glomset, J. A. (1988). Evidence for modification of lamin B by a product of mevalonic acid. *J. Biol. Chem.* **263**, 5997–6000.

Xue, C.-B., Caldwell, G. A., Becker, J. M., and Naider, F. (1989). Total synthesis of the lipopeptide **a**-mating factor of *Saccharomyces cerevisiae*. *Biochem. Biophys. Res. Commun.* **162**, 253–257.

Xue, C.-B., Ewenson, A., Becker, J. M., and Naider, F. (1990). Solution phase synthesis of *Saccharomyces cerevisiae* **a**-mating factor and its analogs. *Int. J. Peptide Protein Res.* **36**, 362–373.

Yoshida, M., Sakagami, Y., Isogai, A., and Suzuki, A. (1981). Isolation of tremerogen a-13, a peptidal sex hormone of *Tremella mesenterica*. *Agric. Biol. Chem.* **45**, 1043–1044.

Chapter 10

Signal Transduction during Fertilization in *Chlamydomonas*

WILLIAM J. SNELL

Department of Cell Biology and Neuroscience
University of Texas Southwestern Medical Center
Dallas, Texas 75235

I. Introduction

Over the past several years, the mating reaction of *Chlamydomonas* has become an important model system for studying signal transduction induced by specific cell–cell interactions that occur during fertilization. This unicellular biflagellated green alga already has proved invaluable for research on flagellar motility. Much of our knowledge about the structure and function of the eukaryotic flagellum was discovered through studies on *Chlamydomonas* (reviewed by Luck, 1984; Lefebvre and Rosenbaum, 1986). In fact, this organism is featured in *The* Chlamydomonas *Handbook* (Harris, 1989), to which the reader is referred for more information about the many other biological processes that have been studied in *Chlamydomonas*, including photosynthesis, organelle regeneration, and phototaxis (Hegemann *et al.*, 1990).

During the interactions between gametes of this organism, cells of opposite mating type adhere to each other via complementary cell adhesion molecules located on the surfaces of their flagella. This adhesive interaction induces a sexual signal that, within minutes, leads to several subsequent events, including release of the cell wall, formation of an actin-filled protuberance at the apical end of the cell, increased adhesiveness of the flagella, and, finally, gamete fusion (Fig. 10.1). Fusion itself generates signals that initiate a new developmental pathway, leading to formation of a quiescent zygote that is well protected by a highly impervious cell wall. Under the appropriate environmental conditions, the zygote undergoes germination and reinitiates the vegetative life cycle.

II. *Chlamydomonas* Fertilization—Interaction between *mt*⁺ and *mt*⁻ Gametes Induces a Sexual Signal That Leads to a Set of Complex Responses

Mating type plus (mt^+) and mating type minus (mt^-) gametes have complementary adhesion molecules, called agglutinins, on their flagellar surfaces that

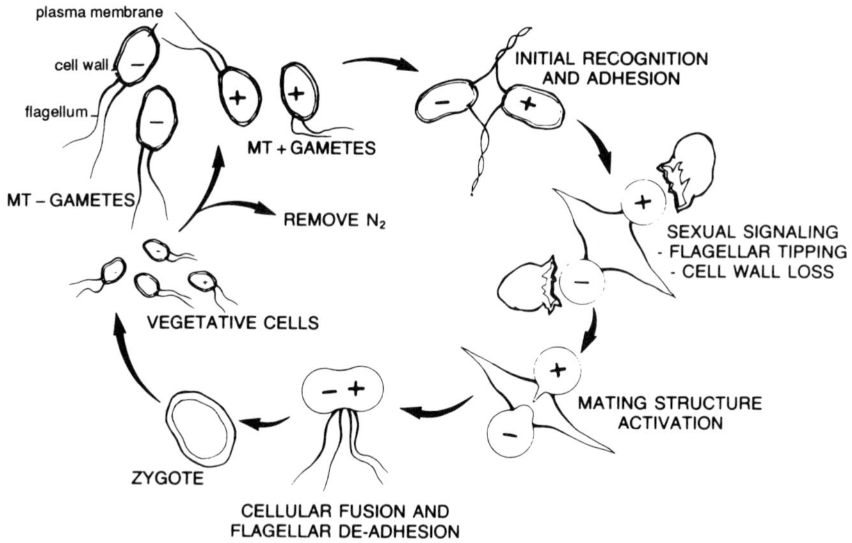

Figure 10.1 Life cycle of *Chlamydomonas reinhardtii*. (From Hunnicutt, 1989.)

are responsible for the initial adhesive interaction (Adair *et al.*, 1982,1983; Adair, 1985). The interaction between agglutinin molecules induces an almost immediate (2–3 min) 10-fold increase in levels of intracellular cAMP, followed by a return to baseline levels after cell fusion (approximately 20 min after mixing) (Pijst *et al.*, 1984; Pasquale and Goodenough, 1987,1988). The increased concentration of intracellular cAMP generates several cellular responses, most or all of which can be induced by addition of dibutyryl cAMP to gametes of a single mating type (Pasquale and Goodenough, 1987). These responses include collapse of a functional barrier (see subsequent text) between the cell body and flagella, permitting migration and activation of a pre-existing pool of agglutinins from the cell body membrane to the flagellar membrane (Goodenough, 1989; Hunnicutt *et al.*, 1990); movement of sites of adhesive interactions from random locations along the flagella to the flagellar tips (Goodenough *et al.*, 1980); alterations in the morphology of the flagellar tips, accompanied by elongation of axonemal microtubules (Mesland *et al.*, 1980); polymerization of actin filaments within a newly formed 4-μm-long, acrosome-like extension called the fertilization tubule that initiates cell fusion (Friedmann *et al.*, 1968; Goodenough and Jurivich, 1978; Goodenough and Weiss, 1978); and regulated secretion of a serine protease (see subsequent text) into the periplasm that converts an inactive prometalloprotease to its lower M_r, active form that degrades the cell wall (Buchanan *et al.*, 1988,1989; Snell *et al.*, 1989).

Cell fusion itself generates intracellular signals that initiate a new developmental pathway. Within minutes of gamete fusion to form zygotes, the flagellar agglutinins become inactive *in situ* (although they still can be detected by an *in vitro* assay) and ultimately are shed into the medium (Hunnicutt and Snell, 1991; see subsequent text). Within a few hours, flagella are resorbed and a new, highly impermeable cell wall is formed inside which the subsequent stages of zygote maturation occur. Under the proper regimen of light, germination ensues within a few days and yields two mt^+ and two mt^- daughter cells, which are the haploid cell products of meiosis. These processes have been reviewed elsewhere (Goodenough *et al.*, 1985a; Snell, 1985; Goodenough and Ferris, 1987; Musgrave and van den Ende, 1987).

This chapter focuses on the signals that are generated in the first few seconds after gametes adhere to each other and on the cellular mechanisms of the responses to the signals. Although occasional reference will be made to work in *Chlamydomonas eugametos*, most of the chapter centers on studies of *Chlamydomonas reinhardtii*. The reader is referred to other reviews for more information on *C. eugametos* (Musgrave and van den Ende, 1987; van den Ende, 1992).

III. Signal Transduction Induced by Interactions between Cell Surface Adhesion Molecules

The two flagella of *C. reinhardtii* play a central role in the life cycle of this highly motile organism. In both the vegetative and gametic phases of its life cycle, the flagella are used to propel the organism through its environment at ~100 μm/sec (Goodenough, 1983). During the gametic phase of the cycle, the flagella take on a new importance for the cells because of their display of the agglutinins (Adair *et al.*, 1982,1983; Adair, 1985; Goodenough and Ferris, 1987), which are glycoproteins that permit mt^+ gametes to adhere to mt^- gametes. No evidence exists for chemotaxis during the mating reaction. Instead, random collisions between the flagella of gametes of opposite mating type result in immediate adhesion, leading to the formation of large clumps of agglutinating gametes.

A. *Chlamydomonas* Agglutinins Are High Molecular Weight Rod-like Glycoproteins

1. Characterization of the Agglutinins The agglutinins are very high M_r glycoproteins that appear to extend orthogonally from the surface of the flagella when examined using rapid-freeze, deep-etch electron microscopy (Adair, 1985). These molecules are highly asymmetric (~225 nm × 4 nm) rodlike structures containing four distinct domains: a globular head, a rigid rodlike section, a flexible region, and a terminal hook. Both mt^+ and mt^- agglutinins are rich in hydroxyproline and serine and bear striking similarity to glycoproteins of the cell wall (Adair *et al.*, 1983; Cooper *et al.*, 1983; Goodenough *et al.*, 1985b). With the exception of sensitivity to chymotrypsin and reducing agents (Collin-Osdoby *et al.*, 1984; Goodenough, 1986; Saito and Matsuda, 1986), mt^+ and mt^- agglutinins essentially are indistinguishable by biochemical criteria, although they do have slight differences in morphology when isolated molecules are examined by rapid-freeze, deep-etch electron microscopy (Goodenough *et al.*, 1985b). Although evidence suggests that the agglutinins are arranged in a small number of rows (Goodenough, 1989), the nature of their association with the cell membrane remains elusive.

The high apparent molecular weight of the agglutinins made their identification by SDS–PAGE difficult because they were too large to enter a standard running gel but remained in the stacking gel (Adair *et al.*, 1982). More recently, the mt^+ agglutinin has been resolved in SDS–PAGE gradient gels with no stacking gel (Homan *et al.*, 1987) or using AcrylAide (FMC Bioproducts, Rockland, Maine) rather than bis(acrylamide) as cross-linker. Using 3–8% acrylamide gels cross-linked with AcrylAide, the agglutinin has an R_f of about 0.1 relative to the tracking dye, whereas the 220,000-M_r marker myosin has an R_f of .7–.9 (V. Kurvari, personal communication). Collectively, these results indicate that the agglutinin has an apparent molecular weight as high as 2×10^6. Although less work has been done on the *mt*$^-$ agglutinin, it has similar properties.

2. *Bioassay for Agglutinin Activity* *In vivo* assays of cell-free *Chlamydomonas* agglutinins in solution have been unsuccessful; even high concentrations of agglutinins do not block adhesion between gametes of opposite mating type, nor do they cause gametes of the opposite mating type to isoagglutinate. Instead, agglutinins are assayed using an *in vitro* bioassay (Adair *et al.*, 1982; Collin-Osdoby *et al.*, 1984). Soluble fractions containing agglutinin are diluted serially and portions are dried onto a glass slide. The slide is overlayed with a solution of gametes of the opposite mating type, which can adhere to the adsorbed agglutinin. The most dilute solution to which gametes adhere is considered the end point for the assay, which is used to calculate the units of agglutinin in the sample.

The ability to detect agglutinin activity in the glass slide assay but not in solution may provide information about agglutinin function. Possibly, the agglutinins have a very low affinity or are in an inactive conformation in solution. Adsorption onto glass might permit the molecules to assume an active conformation. Another possibility is that multivalent interactions are required for adhesion. Several molecules adsorbed to a small area on a glass surface or exposed on the surface of a flagellum could interact with several complementary agglutinins on the flagellum of a live gamete of the opposite mating type. These multiple, highly localized interactions, even if each were of low affinity, could be sufficient to permit cell adhesion. The tipping of agglutinins described next may reflect the need for such a multivalent interaction (Goodenough and Jurivich, 1978).

3. *Flagellar Attachment* Agglutinins originally were isolated from detergent extracts of flagella (Adair *et al.*, 1982). Later, incubation of gametes in ethylenediamine tetraacetic acid (EDTA) was discovered to lead to release of agglutinins into the medium, suggesting that the agglutinins were peripheral membrane proteins whose association with the membrane was dependent on divalent cations (Adair *et al.*, 1982,1983; Saito and Matsuda, 1984a). The efficacy of this extraction procedure depends on the method of cell culture (W. J. Snell, unpublished results). After several weeks of asynchronous vegetative growth on agar, the cells use up their nitrogen source and form gametes. Such gametes express flagellar agglutinins and are fully and immediately agglutinable when mixed with gametes of the opposite mating type. Extraction of these gametes with EDTA leads to release of all the detectable cellular agglutinins (Adair *et al.*, 1983). When vegetative cells are grown in liquid culture, however,

with the imposition of light–dark cycles, synchrony is induced. Gametes prepared from vegetative cells grown under these conditions do no release their agglutinins when incubated with EDTA (W. J. Snell *et al.*, unpublished results). One interpretation of these results is that the agglutinin is a transmembrane protein and EDTA induces an active release mechanism in cells grown on agar that does not occur in gametes grown in liquid culture. Consistent with this hypothesis, we have been unable to release agglutinins from cells grown in liquid culture using methods traditionally used to solubilize peripheral membrane proteins. In addition, phosphatidyl inositol-dependent phospholipase C does not release agglutinin from cells, suggesting that agglutinin is not associated with the membrane via a glycan phosphatidyl inositol (GPI) linkage (Y. Zhang and W. J. Snell, unpublished observations). On the other hand, after cell disruption by sonication, French Press, or extraction with octyl glucoside, agglutinin behaves as a soluble protein (Saito and Matsuda, 1984b; Saito *et al.*, 1985; Hunnicutt *et al.*, 1990) with an S value of 12 (Adair *et al.*, 1982). These observations indicate that active agglutinin can be isolated in a soluble form and suggest that it is a peripheral protein. The difference observed between gametes formed under the two different growth conditions remains paradoxical. Cloning and sequencing of the agglutinin genes should provide information about their membrane association.

More information is available about the nature of the interaction of the *C. eugametos* agglutinins with the flagellar membrane. The agglutinin in this species is associated with a 125,000-M_r wheat germ agglutinin (WGA) binding protein (Kooijman *et al.*, 1989). Although the agglutinin itself does not interact with WGA, treatment of *C. eugametos* gametes with WGA induces dramatic increases in intracellular cAMP levels and leads to several events associated with fertilization in this species. Moreover, the WGA binding proteins and agglutinin are mobilized to the flagellar tips under these conditions. Incubation of gametes with an anti-agglutinin antibody also induces migration of both the WGA binding protein and the agglutinin. In addition, both molecules bind to an anti-agglutinin affinity column in low concentrations of detergent, whereas only the agglutinin binds to the column in high detergent concentrations. These results suggest that the agglutinin and the WGA binding protein form a complex that is disrupted only in high concentrations of detergent (Kooijman *et al.*, 1989).

B. Interactions between Agglutinin Molecules Induce Increases in Intracellular Levels of cAMP

The prevailing view of the early events in the mating reaction is that direct interactions between mt^- and mt^+ flagellar agglutinins are responsible for flagellar adhesion and that this molecular interaction activates a signal transduction pathway. Possibly, the binding of agglutinins to each other alters their conformation, which in turn affects the conformation of nearby or associated proteins. These alterations generate an increase in the concentration of intracellular cAMP, which could occur through activation of a flagellar adenylyl cyclase or through inhibition of a cAMP phosphodiesterase. The second messenger directly or indirectly can activate protein kinases or protein phosphatases, which bring about the signaling events described earlier. The critical role of cAMP in signal transduction in *Chlamydomonas* is confirmed by the induction of events that normally are induced preceding cell fusion by incubation of gametes of a single mating type in dibutyryl cAMP (Pasquale and Goodenough, 1987; Goodenough, 1989; Hunnicutt *et al.*, 1990).

Adenylyl cyclase and phosphodiesterase activities have been found in flagella from both *C. reinhardtii* and *C. eugametos* (Pijst *et al.*, 1984; Pasquale and Goodenough, 1987,1988; Goodenough, 1989). Early experiments indicated that the levels of cAMP increased about 10-fold during the mating reaction in *C. eugametos*; preliminary studies showed that flagella contain an adenylyl cyclase activity (Pijst *et al.*, 1984). Similar results were found for *C. reinhardtii*. Interestingly, the *C. reinhardtii* adenylyl cyclase does not appear to be regulated by G proteins. Its *in vitro* activity was not activated by A1F, forskolin, GTP, GTPγS, or NaF (Pasquale and Goodenough, 1987). In addition, neither cholera toxin nor pertussis toxin inhibited or induced sexual signaling (Pasquale and Goodenough, 1987; W. J. Snell, unpublished results). These results imply that G proteins may not play a role in this signaling system and that other mechanisms regulate the adenylyl cyclase or phosphodiesterase. Using isolated mt^+ and mt^- flagella adhesion leads to a 3-fold increase in flagellar adenylyl cyclase, which again does not appear to be regulated by G proteins (Y. Zhang and W. J. Snell, unpublished observations). Interestingly, the flagellar adenylyl cyclase in gametes but not in vegetative cells of *C. reinhardtii* seems to be regulated by phosphorylation and dephosphorylation (Zhang *et al.*, 1991; Zhang and Snell, 1993).

In this system, adenylyl cyclase may be regulated by calmodulin (Pasquale and Goodenough, 1987,1988), which has been shown to be present in the cell body and the flagella of *Chlamydomonas* (Gitelman and Witman, 1980). Calmodulin stimulates flagellar adenylyl cyclase activity *in vitro*; trifluoperazine (TFP) blocks sexual signaling and the rise in cAMP that normally accompanies mating (Detmers and Condeelis, 1986). The TFP block to signaling is overcome by addition of exogenous cAMP (Pasquale and Goodenough, 1987).

Attempts to implicate other signal transduction systems in this process have been unsuccessful. Incubation of gametes in Ca^{2+} and the ionophore A23187 does not induce or inhibit signaling (Bloodgood and Levin, 1983; Pasquale and Goodenough, 1987). A possible role for Ca^{2+}, however, is suggested by the addition of 75 μM EDTA (W. J. Snell, unpublished observations) or the local anesthetic, lidocaine, which interferes with the interaction of Ca^{2+} with membranes (Snell *et al.*, 1982). Both these compounds inhibit signaling without blocking adhesion. Phorbol esters do not induce mating responses, suggesting that protein kinase C-dependent phosphorylations may not play a role in this signaling system (Pasquale and Goodenough, 1987).

IV. Cellular Responses to the Sexual Signal

The cellular response to the increased levels of intracellular cAMP described earlier is rapid and complex. This response includes alterations in the morphology of the tips of the flagella, erection of an actin-filled fertilization tubule, release of the cell wall, and recruitment of agglutinin molecules from the cell body to the flagella. The mechanisms that regulate flagellar tip morphology and formation of the mating structures have been reviewed (Goodenough *et al.*, 1985a; Snell, 1985). The subsequent discussion concentrates on information about cell wall loss and movement of agglutinin molecules on the surface of the cell body.

A. Release of the Cell Wall Is Initiated by a Protease Cascade

Because the cell wall acts a a barrier to cell fusion during fertilization, it must be removed early during the mating reaction. As might be expected, wall loss is the culmination of a series of cellular and biochemical processes initiated by the adhesion-induced increases in cAMP. The structure, composition, and degradation of the *Chlamydomonas* cell wall are described briefly before the regulated appearance of the enzymes that are responsible for its release are discussed. (See Adair and Snell, 1990, for a more comprehensive review.)

The cell wall is a multilayered extracellular matrix containing 20–30 proteins and glycoproteins, many of which are enriched in hydroxyproline. The majority (80–90%) of the wall protein is located in a crystalline portion of the wall that is soluble in sarkosyl–urea solutions as well as in chaotropic salts (Roberts *et al.*, 1972,1985; Hills *et al.*, 1973,1975; Goodenough and Heuser, 1985,1988a,b; Imam *et al.*, 1985; Adair, 1988; Imam and Snell, 1988; Adair and Appel, 1989). The wall has no cellulose, but contains a noncrystalline, highly insoluble framework portion that is responsible for the integrity and overall shape of the wall (Roberts *et al.*, 1972; Hills *et al.*, 1975). The framework is composed of cross-linked polypeptides and retains its shape even after boiling in SDS–PAGE sample buffer (Imam and Snell, 1988). This retention of shape even after separation from cells suggests that the cell wall polypeptides are linked covalently to form a single extremely large macromolecule. In addition, the framework contains the flagellar collars, which are striated channels in the wall through which the flagella protrude (Ringo, 1967; Snell, 1983; Imam and Snell, 1988). Both the wall-shaped macromolecule and the collar are substrates for the wall-degrading metalloprotease g-lysin, whereas the soluble crystalline portion of the wall is resistant to g-lysin (Goodenough and Heuser, 1985; Buchanan and Snell, 1988; Adair and Snell, 1990).

Although the cell wall is of interest in its own right, its primary relevance to signal transduction is that its loss is a marker for adhesion-induced release of the enzymes that degrade the wall. Early work indicated that mating gametes released g-lysin into their medium (Claes, 1971). This enzyme has been studied in some detail (reviewed by Snell, 1985; Matsuda, 1988), and its gene has been cloned and sequenced (Kinoshita *et al.*, 1992). g-Lysin activity appears concomitant with a dramatic increase in intracellular levels of cAMP and can be induced by addition of exogenous cAMP and phosphodiesterase inhibitors (Pasquale and Goodenough, 1987).

By analogy to other secretory systems, g-lysin was anticipated to be stored in intracellular vesicles, the exocytosis of which was induced by cAMP. However, experiments described here have shown that g-lysin is stored as an inactive higher M_r precursor in the periplasm, the space between the cell wall and the plasma membrane, rather than in the cytoplasm. The molecule is converted to the lower M_r active form as a consequence of sexual signaling.

1. Characterization of g-Lysin g-Lysin is a polypeptide with an M_r of ~60,000 whose cell wall degrading activity is inhibited by EDTA and phenanthroline, a zinc chelator typically used as an inhibitor of metalloproteases (Matsuda *et al.*, 1985,1987; Jaenicke *et al.*, 1987; Buchanan and Snell, 1988; Buchanan *et al.*, 1988). Two groups reported that phenylmethylsulfonyl fluoride (PMSF), *N*-tosyl-L-phenylalanine chloromethyl ketone (TPCK), 1-chloro-3-tosylamido-7-amino-2-heptone (TLCK), and diisopropylfluorophosphate (DFP) have little if any inhibitory activity on the purified enzyme (Matsuda *et al.*, 1985; Matsuda, 1988; Snell *et al.*, 1989), whereas a third group reported significant inhibition by PMSF and DFP (Jaenicke *et al.*, 1987). Cloning and sequencing of the g-lysin gene has shown that it has several functional domains that are similar to those of metalloproteases that degrade the extracellular matrix in animal cells (Kinoshita *et al.*, 1992).

2. g-Lysin Is Stored in the Periplasm as an Inactive Higher M_r Precursor Several approaches led to information on the localization of g-lysin. Carbohydrate-containing material within the periplasm was identified by staining with the lectin concanavalin A (Con A) and was found to be released during mating (Milliken and Weiss, 1984). Cell wall-less mutants are defective both in this periplasmic staining and in release of g-lysin activity during mating. Because g-lysin reacts with Con A, it has been suggested to correspond to the periplasmic material. Consistent with this idea, removal of the cell wall from mating gametes before mixing eliminated release of g-lysin and suggested that the enzyme is stored in the periplasm (Matsuda *et al.*, 1987). These results were surprising because storage of g-lysin in the periplasm should allow intimate association with its substrate. This paradox could be explained if the activity of g-lysin were regulated.

Immunoblots of material from nonmating gametes using anti-g-lysin antisera identify a higher 62,000-M_r polypeptide, pro-g-lysin (Buchanan *et al.*, 1989). As soon as the cells begin to mate and undergo sexual signaling, however, all the antigen is released from the cells as the lower M_r (60,000) mature polypeptide. Pro-g-lysin is localized to the periplasm rather than to an intracellular site (Buchanan *et al.*, 1989), a result that correlates well with the previous results (Millikin and Weiss, 1984; Matsuda *et al.*, 1987). Removal of cell walls from mt^+ gametes by treatment with g-lysin results in complete loss of cell-associated g-lysin antigen (Buchanan *et al.*, 1989). A substantial amount of pro-g-lysin is released from gametes by two cycles of freezing and thawing. This material has no g-lysin activity in a standard wall-loss assay. These results indicate that g-lysin is stored in the periplasm as an inactive precursor in gametes, and that sexual signaling induces conversion of this molecule to the lower M_r active enzyme.

3. Regulated Secretion of a Serine Protease that Converts Pro-g-lysin to the Active Wall-Degrading Enzyme g-Lysin Storage of g-lysin as an inactive periplasmic precursor suggests that the mating reaction might induce release of an extracellular converting enzyme (Snell *et al.*, 1989). To test this idea, mt^+ and mt^- gametes whose walls (and therefore endogenous pro-g-lysin) had been removed were mixed together to permit agglutination and sexual signaling. The cells were then removed by centrifugation and the resulting supernatant evaluated for converting activity by incubation with pro-g-lysin. Immunoblotting indicated that all the pro-g-lysin had been converted to the 60,000-M_r form, whereas the control sample contained the 62,000-M_r precursor. The 60,000-M_r g-lysin produced by the mating supernatant was active in a wall-loss assay. These results indicate that mating gametes release an activity that is distinct from g-lysin that converts inactive pro-g-lysin to active g-lysin. The results also demonstrate that this converting activity, which is designated p-lysinase, remains with the cells after wall removal, indicating that p-lysinase is stored in an intracellular compartment (Snell *et al.*, 1989).

4. Converting Activity Is Released on Incubation of Gametes in Dibutyryl cAMP and Papaverine and Has the Properties of a Serine Protease of ~300,000 M_r Further characterization of p-lysinase shows that it has the properties of a serine protease; that is, it is inhibited by PNSF and DFP but not by inhibitors of metalloproteases, aspartic proteases, or sulfhydryl proteases. Molecular sieve chromatography shows that the native molecule has an apparent M_r of approximately 300,000. Incubation of de-walled gametes in dibutyryl cAMP and papaverine (a phosphodiesterase inhibitor) releases p-lysinase activity into the medium (Snell *et al.*, 1989). This release represents one of the few examples of regulated secretion in a eukaryotic microorganism (Maihle and Satir, 1986; Vilmart Seuwen *et al.*, 1986). Obtaining mutants that are defective in p-lysinase secretion should be possible by screening for cells that do not lose their walls on sexual signaling. Such mutants could allow the characterization of cellular and molecular events that are important in regulated secretion, much like the yeast *sec* mutants have been instrumental in delineating new steps in constitutive secretion (Novick *et al.*, 1980).

Intriguing similarities exist between the system for regulating degradation of the extracellular matrix in *Chlamydomonas* and the system used in higher organisms. The widely distributed animal cell metalloprotease collagenase also is stored as an inactive higher M_r precursor in the pericellular space (Unemori and Werb, 1988). The enzyme(s) that converts it to the active form *in vivo* has not yet been identified, although serine proteases have been shown to convert it *in vitro*. Equally interesting in this regard are the observations that signal transduction through the fibronectin receptor induces secretion of procollagenase by fibroblasts (Unemori and Werb, 1986). Future studies should reveal whether these functional similarities extend to the molecular level.

B. Signaled Recruitment of Agglutinins from the Cell Body to the Flagella

1. More than 90% of Total Cellular Agglutinins Are Associated with the Cell Body Over 30 years ago, Forster and Wiese (1954) reported that gametes had the ability to inactivate adhesion molecules. Since then, one immediate consequence of interactions between mt^+ and mt^- flagellar agglutinins has been established to be their rapid inactivation (Snell and Roseman, 1979; Snell and Moore, 1980; Snell, 1981; Cooper *et al.*, 1983; Collin-Osdoby *et al.*, 1984; Saito *et al.*, 1985; Goodenough, 1986; Hunnicutt *et al.*, 1990). Addition of mt^- gametes to isolated mt^+ gametic flagella results in a rapid aggregation as cells adhere to the flagella (Snell and Roseman, 1979). This event is followed rapidly by disaggregation of the cells. Addition of fresh flagella to the disaggregated cells allows reaggregation. When the isolated flagella from the initial aggregation mixture are retested with fresh gametes, however, no aggregation occurs. Identical results are obtained if mt^- flagella are mixed with mt^+ gametes. These results indicate that the process of adhesion results in a loss of activity of the flagellar agglutinins.

Although only the added flagella lose their adhesiveness in these experiments, the observation that similar results occur with flagella isolated from either mating type suggests that the live gametes in the mixture also might be losing surface agglutinin function. An impotent mt^+ mutant *imp-1* (Goodenough *et al.*, 1976; Goodenough and Ferris, 1987), which can agglutinate but not fuse, was used to demonstrate that live gametes also lose their agglutinins when they adhere to gametes of the opposite mating type (Snell and Moore, 1980). Wild-type cells normally fuse within 5–15 min and then disaggregate, but the mt^+ *imp-1* gametes agglutinate with mt^- gametes for hours or even days without fusing (Goodenough *et al.*, 1976). Addition of inhibitors of protein synthesis (Snell and Moore, 1980) or protein glycosylation (Snell, 1981) leads to the loss of adhesiveness of aggregating wild-type mt^- and impotent mt^+ gametes after about 50 min. These results indicate that, during the process of fertilization, cells lose flagellar agglutinin activity, but this activity is regained rapidly.

These results led to the investigation of the reservoir of adhesion molecules that could be mobilized during adhesion. Because isolated flagella lose their adhesiveness within 5 min, whereas live gametes require 50–60 min to lose their adhesiveness, the reservoir would be predicted to represent about 90% of the total cellular agglutinins. Over 90% of total cellular agglutinins are found to be associated with cell bodies and are proposed to be the pool identified in the kinetic studies (Saito *et al.*, 1985). Comparison of flagellar and cell body agglutinin by SDS–PAGE and HPLC hydrophobic interaction chromatography shows that they are indistinguishable; the S value of cell body agglutinin is the same as the value previously reported for flagellar agglutinin (Adair *et al.*, 1982; Hunnicutt *et al.*, 1990). The cell body agglutinins also are similar to flagellar agglutinins because they can be released from the cell body by mechanical disruption and do not require detergent for solubility.

The size of the pool at various times after mixing *imp-1* mt^+ and wild-type mt^- gametes was measured (Snell and Moore, 1980; Saito *et al.*, 1985). In the presence of protein synthesis inhibitors, the pool is depleted completely in gametes that aggregate. The pool also is depleted significantly in the absence of protein synthesis inhibitors; in this case, however, as the cells continued to aggregate, the pool is replenished by protein synthesis before it becomes completely depleted. Thus, the cells are able to regulate the size of the pool both in resting gametes and when the pool is mobilized during prolonged adhesion.

2. *Cell Body Agglutinins Are on the Surface of the Cell* The lack of adherence of cell bodies to flagella or to cell bodies of gametes of the opposite mating type, even after removal of the cell walls, suggested that the pool of agglutinins associated with the cell body is intracellular (Saito *et al.*, 1985). In addition, rapid-freeze, deep-etch microscopy showed no evidence of agglutinins on the surface of the cell body plasma membrane (Goodenough, 1989). Subsequent work using exogenously added proteases and surface labeling shows that

these earlier interpretations of the evidence about the intracellular location of the agglutinins require reevaluation (Hunnicutt *et al.*, 1990). A brief protease treatment of deflagellated gametes destroys all detectable cell body agglutinin activity. In addition, the cell body agglutinin becomes radioactive after surface labeling with ^{125}I using the lactoperoxidase method. The accessibility of the cell body agglutinins to protease and surface labeling indicates that they are located on the cell surface. Comparisons of the estimated surface area of cell bodies and flagella to the amounts of agglutinin associated with each domain suggest that the agglutinins are distributed proportionately.

The presence of cell surface agglutinins on the cell body is surprising since adhesion is limited to the flagella. Posttranslational modification could account for the difference in *in vivo* activity of agglutinins from the two domains. Alternatively, interactions between agglutinins could be necessary for their activity and could be restricted to the flagella. Electron microscopic evidence indicates that the flagellar agglutinins are in a linear array (Goodenough, 1989), whereas cell body agglutinins have not been detected by rapid-freeze, deep-etch electron microscopy (Adair, 1985; Goodenough, 1989), suggesting that they do not have the rodlike structure of flagellar agglutinins. The arrays formed by the flagellar agglutinins may reflect an interaction of agglutinins that is necessary for their activity.

3. *Agglutinins Move from the Cell Body to the Flagella during Flagella Regeneration* On removal of *Chlamydomonas* flagella, assembly of two new full-length flagella begins immediately (reviewed by Lefebvre and Rosenbaum, 1986). Two half-length flagella can be regenerated in the presence of protein synthesis inhibitors, indicating that a significant pool of presynthesized flagellar molecules is available. Although the matrix and axonemal proteins in regenerated flagella must be derived from cytoplasmic precursors, no information is available on whether the pool of flagellar membrane proteins is in the form of intracellular membrane vesicles or is on the cell surface. In *Euglena*, some high molecular weight flagellar proteins also are found on the cell body plasma membrane (Geetha-Habib and Bouck, 1982; Rogalski and Bouck, 1982).

To determine whether the cell body agglutinins in *C. reinhardtii* could move to the flagella, gametes were de-walled with exogenous g-lysin, surface labeled with ^{125}I, and permitted to regrow flagella (Hunnicutt *et al.*, 1990). The newly grown flagella were isolated and analyzed by SDS–PAGE and autoradiography. The agglutinins on the regenerated flagella were iodinated, indicating that the agglutinins had moved from the cell bodies to the flagella during regeneration. Although the cell body agglutinins may have been internalized and reinserted into the regenerating flagella, the simplest explanation for the results is that the agglutinin molecules move directly from the cell body surface to the flagellar surface.

4. Agglutinins Move from the Cell Body to the Flagella during Sexual Signaling How the movement of agglutinins from the cell body to the flagella during flagellar regeneration relates to mobilization of flagellar agglutinins during adhesion is not clear. Two questions are of interest. (1) Are the agglutinins mobilized from the cell body to the flagella during aggregation? (2) Is agglutinin replacement stimulated by the loss of agglutinins itself or is agglutinin replacement one of the many responses to the sexual signal?

In *C. eugametos*, the flagellar adhesiveness of gametes is increased several-fold during the first few minutes of aggregation without a significant increase in agglutinin levels (Tomson *et al.*, 1990a). These experiments are interpreted to indicate that the increased adhesiveness cannot be accounted for by increased levels of agglutinins. Instead, microaggregation of agglutinins in the plane of the flagellar membrane is the most likely explanation for increased flagellar adhesiveness. Different conclusions have been drawn based on experiments with *C. reinhardtii* (Goodenough, 1989). The amount of cell body and flagellar agglutinin in mt^+ gametes was measured before and after mimicking sexual signaling by addition of dibutyryl cAMP. The flagellar agglutinin levels increased over 8-fold and the cell body agglutinins were depleted. Thus, in contrast to *C. eugametos*, the increased flagellar adhesiveness occurring during adhesion in *C. reinhardtii* can be accounted for by an increased number of flagellar agglutinins that are most likely relocalized from the cell body. In addition, this result demonstrates that the apparent requirement of adhesion for mobilization of the cell body agglutinins can be bypassed by addition of a second messenger, indicating that mobilization is a cellular response to the sexual signal generated during adhesion.

5. A Functional Barrier Normally Prevents Movement of Agglutinins between the Cell Body and the Flagella The experiments on movement of agglutinins between the cell body and the flagella raise new questions about the possibility of a barrier between the two domains that normally prevents movement and about whether the agglutinins are freely diffusible within each domain. Several possibilities exist for movement of agglutinins within and between the two domains. (1) Free diffusion of the agglutinin with its putative anchor protein may occur between the two domains in both resting (nonsignaling) gametes and gametes that are undergoing sexual signaling. Only when flagellar agglutinins are lost (for example, during adhesion) are they replaced by diffusion of agglutinins from the cell body. (2) Agglutinins may diffuse freely within each domain, but a barrier to movement exists between the domains. This barrier collapses during sexual signaling. (3) Diffusion of agglutinin may not occur because the protein is always associated via an anchor protein with elements of the cytoskeleton. Movement is an active process that can be regulated by sexual signaling. (4) Agglutinins may diffuse freely in one domain, but may be associated with cytoskeletal elements in the other domain. For example, cell body agglutinins may be associated with the submembranous array of microtubules that originates from the region around the basal bodies (Ringo, 1967). Because all the microtubules have the same polarity, agglutinins could be delivered (on generation of the appropriate signals) to the base of the flagella by motor molecules that associate with the putative anchor protein and with microtubules. Alternatively, agglutinins might diffuse freely in the plasma membrane of the cell body and those agglutinins that are near the base of the flagellum could associate with flagellar microtubules and migrate up the flagellum.

The flagellum has a surface motility system that is identified experimentally by its ability to move latex microspheres up or down the flagellum at over 2 μm/sec (Bloodgood *et al.*, 1979; Hoffman and Goodenough, 1980; Lewin, 1982; Snell *et al.*, 1982; Bloodgood and Salomonsky, 1990). Although the molecular mechanisms underlying flagellar surface motility have not been elucidated, this system may play a role during adhesion (Hoffman and Goodenough, 1980). Thus, once agglutinins are localized to the flagella, this system could allow any movement that is necessary for function.

In *C. eugametos*, a functional barrier to movement of agglutinins from the plasma membrane of the cell body to the flagella has been reported (Musgrave *et al.*, 1986). Under several circumstances, flagella agglutinins are not detectable although the cell body agglutinins are present in normal amounts. *Chlamydomonas eugametos* gametes have a cytoplasmic pool of agglutinins as well as the agglutinins on the cell membrane; either pool could be mobilized during aggregation (Musgrave *et al.*, 1986; Tomson *et al.*, 1990a,b).

An immunological approach based on a monoclonal antibody directed against the mt^+ agglutinin has been used to address the question of a barrier between the cell body and flagella in *C. reinhardtii* (Hunnicutt *et al.*, 1990). This antibody blocks adhesion of mt^+ gametes (Snell *et al.*, 1986). When the antibody is immobilized on Sepharose beads, mt^+ gametes bind to the beads via their flagella, whereas neither mt^- gametes nor vegetative cells of either mating type bind. Agglutinins on the cell body surface do not show any interaction with the antibody; that is, cells bind to the beads only via their flagella. Interestingly, binding of the mt^+ gametes to the beads is transient; cells adhere and then de-adhere, much like the *in vivo* interaction between mt^+ and mt^- gametes. During this interaction of the cells with the beads, the cells transfer their agglutinins to the beads. Thus, the interaction of the agglutinin with the antibody leads to release of the agglutinin from the flagella (Snell *et al.*, 1986).

These properties of the monoclonal antibody have been used to determine whether depletion of flagellar agglutinins in the absence of a sexual signal leads to movement of agglutinins from the cell body to the flagella (Hunnicutt *et al.*, 1990). When gametes are incubated with the antibody for several hours in the presence of the protein synthesis inhibitor cycloheximide, the cells remain fully flagellated and motile but their flagellar agglutinins are lost, as documented by an aggregation assay and by measurement of agglutinin activity using the *in vitro* assay. Surprisingly, the cell bodies of these cells still contain normal levels of cell body agglutinins. Thus, depleting flagellar agglutinins without affecting the level of cell body agglutinins is possible. These results indicate that, in nonmating nonsignaling gametes, cell body agglutinins are not freely exchangeable with flagellar agglutinins, indicating that a functional barrier exists between these two domains. The trivial explanation that the cell wall restricts movement is ruled out by showing that the functional barrier is maintained in gametes that are de-walled by addition of exogenous g-lysin.

To determine whether the cell body agglutinins in cells with reduced flagellar agglutinins still can be recruited to the flagella, gametes whose flagellar agglutinins have been depleted are induced to undergo signaling by incubating them in dibutyryl cAMP and papaverine. Within 30 min of addition, the cells reacquire the ability to agglutinate with mt^- gametes. Moreover, their levels of cell body agglutinins fall dramatically, concomitant with an increase in the levels of flagellar agglutinins. Incubation of isolated agglutinin-depleted flagella with dibutyryl cAMP does not lead to reacquisition of agglutinability, consistent with the idea that the agglutinins are moving from the cell body to the flagella. These results indicate that, in nonmating gametes, little if any movement of preexisting agglutinins from the cell body to the flagella occurs, presumably because of a functional barrier that prevents movement between these two domains (Hunnicutt *et al.*, 1990). This adhesion-induced movement of surface adhesion molecules is not restricted to *Chlamydomonas* gametes. Adhesion molecules on sperm of several mammalian species have been shown to undergo dramatic changes in surface localization during sperm–egg adhesion (Myles and Primakoff, 1984; Lopez and Shur, 1987; reviewed by Snell, 1990).

Thin-section and freeze–cleave electron microscopic methods have identified a structure called the flagellar or ciliary "necklace" in the membrane at the base of the flagellum in cells with cilia or flagella (Weiss *et al.*, 1977). This necklace has been proposed to prevent movement of membrane molecules from the cell body to the flagella (Weiss *et al.*, 1977). Whether this structure represents the barrier described here is unknown. In addition, the mechanism by which sexual signaling overcomes the functional barrier remains to be elucidated.

V. Avenues for Future Studies of Cell Fusion-Induced Signals

Studies on cellular events induced by cell fusion have involved a variety of approaches. Cell fusion induces two waves of novel gene transcription, one beginning within 5–10 min of fusion and the other appearing about 90 min after fusion (Ferris and Goodenough, 1987). Appearance of the early transcripts is not affected by incubation of the fusing gametes in cycloheximide, whereas induction of the late transcripts is blocked by this inhibitor. No gene induction is seen in mixtures of *imp-1* mt^+ and wild-type mt^- gametes, indicating that fusion is required for novel gene transcription. Characterization of the induced genes may allow identification of new components of the signaling pathway.

One specific aspect of the response is the fusion-induced decrease in cAMP levels mentioned earlier. The level of cAMP returns to the resting level after cell fusion, whereas it remains high in mixtures of *imp-1* mt^+ and wild-type mt^- gametes (Pasquale and Goodenough, 1987). The reduction in cAMP levels after fusion could result from a reduction in the activity of the adenylyl cyclase or from an activation of phosphodiesterase. Thus, adhesion would induce one signal that would lead to an increase in cAMP and cell fusion would induce a specific signal that reverses this effect. Alternatively, the reduction in cAMP could come about indirectly because the cells lose their flagellar adhesiveness within minutes of fusion. If interactions between mt^+ and mt^- agglutinins are required for activation of the adenylyl cyclase, then cessation of adhesion would lead to a reduction of adenylyl cyclase activity.

A rapid and complete loss of flagellar adhesiveness occurs concomitant with cell fusion (Hunnicutt and Snell, 1991). Surprisingly, the flagella isolated from these nonadhesive zygotes contain nearly the same amount of agglutinin activity as those of the nonfused control cells. The cell body agglutinins also remain high in these de-adhered cells. Within 1–2 hrs, both the cell body and the flagellar agglutinins begin to decline and finally become undetectable. An interpretation of these results is that cell fusion induces a signal that prevents activation of the agglutinins, but that the inactive agglutinins continue to move onto the flagella where eventually they are sloughed into the medium. The molecular mechanisms underlying this loss of activity *in situ* remains unknown. A similar loss of active flagellar agglutinins, in this case regulated by light, has been reported in *C. eugametos* (Kooijman *et al.*, 1988).

VI. Perspectives

The *Chlamydomonas* mating system has begun to yield new insights into the mechanisms of signal transduction induced by cell contact. The agglutinin molecule is a central and intriguing player in this process. This protein is both an adhesion molecule and a signaling molecule, yet the nature of its association with the flagellar and cell body membranes remains unclear. Agglutinin exists in an inactive form on the cell body and can migrate, presumably within the plane of the membrane, onto the flagellar membrane. Until the gametes have fused, migration of the agglutinins is accompanied by their conversion into a form that is active *in vivo*. Fusion, however, generates a signal that appears to prevent this activation. Once the agglutinin gene has been cloned and sequenced, the nature of the association of this protein with the membrane should be clearer. In addition, sequence information should make it possible to generate new antibodies that could be used to probe the location and movements of the agglutinin on the cell surface. New insights into the mechanism of signal transduction should be forthcoming, once more is learned about the interaction of agglutinin with adenylyl cyclase. Studies of the adenylyl cyclase itself should yield new information about the molecules (protein kinases and phosphatases?) that regulate this central signaling molecule.

Although *Chlamydomonas* is amenable to genetic manipulations, this aspect of the organism has not been exploited fully. Certainly the few mating mutants that have been well characterized (Goodenough and Ferris, 1987) have been invaluable for important discoveries. Investigators of signaling in *Chlamydomonas* probably would be well advised to pay attention to the central role that mutants have played in dissecting the molecular basis of flagellar motility in *Chlamydomonas* and to devote even more energy and resources to obtaining mutants blocked at all stages of the mating reaction. Such mutants, in combination with newly developed methods for transfecting *Chlamydomonas* with foreign DNA (Diener *et al.*, 1990; Kindle, 1990; Brown *et al.*, 1991; Mitchell and Kang, 1991), should make it possible to make important new advances in understanding signal transduction in this organism.

References

Adair, W. S. (1985). Characterization of *Chlamydomonas* sexual agglutinins. *J. Cell Sci. Suppl.* **2**, 233–260.

Adair, W. S. (1988). Organization and *in vitro* assembly of the *Chlamydomonas reinhardtii* cell wall. *In* "Self Assembling Architecture" (J. E. Varner, ed.), pp. 25–41. A. R. Liss, New York.

Adair, W. S., and Appel, H. (1989). Identification of a highly conserved hydroxyproline-rich glycoprotein in the cell walls of *Chlamydomonas reinhardtii* and two other Volvocales. *Planta* **179**, 381–386.

Adair, W. S., and Snell, W. J. (1990). The *Chlamydomonas reinhardtii* cell wall: Structure, biochemistry, and molecular biology. *In* "Organization and Assembly of Plant and Animal Extracellular Matrix" R. P. Mecham and W. S. Adair, eds.), pp. 15–84. Academic Press, Orlando, Florida.

Adair, W. S., Monk, B. C., Cohen, R., Hwang, C., and Goodenough, U. W. (1982). Sexual agglutinins from the *Chlamydomonas* flagellar membrane. Partial purification and characterization. *J. Biol. Chem.* **257**, 4593–4602.

Adair, W. S., Hwang, C., and Goodenough, U. W. (1983). Identification and visualization of the sexual agglutinin from the mating-type plus flagellar membrane of *Chlamydomonas*. *Cell* **33**, 183–193.

Bloodgood, R. A., and Levin, E. N. (1983). Transient increase in calcium efflux accompanies fertilization in *Chlamydomonas reinhardtii*. *J. Cell Biol.* **97**, 397–404.

Bloodgood, R. A., and Salomonsky, N. L. (1990). Calcium influx regulates antibody-induced glycoprotein movements within the *Chlamydomonas* flagellar membrane. *J. Cell. Sci.* **96**, 27–34.

Bloodgood, R. A., Leffler, E. M., and Bojczuk, A. T. (1979). Reversible inhibition of *Chlamydomonas reinhardtii* flagellar surface motility. *J. Cell Biol.* **82**, 664–674.

Brown, L. E., Sprecher, S. L., and Keller, L. R. (1991). Introduction of exogenous DNA into *Chlamydomonas reinhardtii* by electroporation. *Mol. Cell. Biol.* **11**, 2328–2332.

Buchanan, M. J., and Snell, W. J. (1988). Biochemical studies on lysin, a cell wall degrading enzyme released during fertilization in *Chlamydomonas*. *Exp. Cell Res.* **179**, 181–193.

Buchanan, M. J., Imam, S. H., and Snell, W. J. (1988). The *Chlamydomonas* wall degrading enzyme, lysin, is stored as a higher M_r precursor. *J. Cell Biol.* **105**, 306a.

Buchanan, M. J., Imam, S. H., Eskue, W. A., and Snell, W. J. (1989). Activation of the cell wall degrading protease, lysin, during sexual signalling in *Chlamydomonas*: The enzyme is stored as an inactive, higher relative molecular mass precursor in the periplasm. *J. Cell Biol.* **108**, 199–207.

Claes, H. (1971). Autolyse der Zellwand bei den Gameten von *Chlamydomonas reinhardii*. *Arch. Mikrobiol.* **78**, 180–188.

Collin-Osdoby, P., Adair, W. S., and Goodenough, U. W. (1984). *Chlamydomonas* agglutinin conjugated to agarose beads as an *in vitro* probe of adhesion. *Exp. Cell Res.* **150**, 282–291.

Cooper, J. B., Adair, W. S., Mecham, R. P., Heuser, J. E., and Goodenough, U. W. (1983). *Chlamydomonas* agglutinin is a hydroxy-proline-rich glycoprotein. *Proc. Natl. Acad. Sci. U.S.A.* **80**, 5898–5901.

Detmers, P. A., and Condeelis, J. (1986). Trifluoperazine and W-7 inhibit mating in *Chlamydomonas* at an early stage of gametic interaction. *Exp. Cell Res.* **163**, 317–326.

Diener, D. R., Curry, A. M., Johnson, K. A., Williams, B. D., Lefebvre, P. A., Kindle, K. L., and Rosenbaum, J. L. (1990). Rescue of a paralyzed-flagella mutant of *Chlamydomonas* by transformation. *Proc. Natl. Acad. Sci. U.S.A.* **87**, 5739–5743.

Ferris, P. J., and Goodenough, U. W. (1987). Transcription of novel genes, including a gene linked to the mating-type locus, induced by *Chlamydomonas* fertilization. *Mol. Cell. Biol.* **7**, 2360–2366.

Forster, H., and Wiese, L. (1954). Gamonwirkungen bei *Chlamydomonas eugametos*. *Z. Naturforsch.* **9b**, 548–550.

Friedmann, I., Colwin, A. L., and Colwin, L. H. (1968). Fine-structural aspects of fertilization in *Chlamydomonas reinhardi*. *J. Cell Sci.* **3**, 115–128.

Geetha-Habib, M., and Bouchk, G. B. (1982). Synthesis and mobilization of flagellar glycoproteins during regeneration in *Euglena*. *J. Cell Biol.* **93**, 432–441.

Gitelman, S. E., and Witman, G. B. (1980). Purification of calmodulin from *Chlamydomonas*: Calmodulin occurs in cell bodies and flagella. *J. Cell Biol.* **87**, 764–770.

Goodenough, U. W. (1983). Motile detergent-extracted cells of *Tetrahymena* and *Chlamydomonas*. *J. Cell Biol.* **96**, 1610–1621.

Goodenough, U. W. (1986). Experimental analysis of the adhesion reaction between isolated *Chlamydomonas* flagella. *Exp. Cell Res.* **166**, 237–246.

Goodenough, U. W. (1989). Cyclic AMP enhances the sexual agglutinability of *Chlamydomonas* flagella. *J. Cell Biol.* **109**, 247–252.

Goodenough, U. W., and Ferris, P. (1987). Genetic regulation of development in *Chlamydomonas*. *In* "Genetic Regulation of Development" (W. Loomis, ed.), pp. 171–189. Liss, New York.

Goodenough, U. W., and Heusser, J. E. (1985). The *Chlamydomonas* cell wall and its constituent glycoproteins analyzed by the quick-freeze, deep-etch technique. *J. Cell Biol.* **101**, 1550–1568.

Goodenough, U. W., and Heuser, J. E. (1988a). Molecular organization of cell-wall crystals from *Chlamydomonas reinhardtii* and *Volvox carteri*. *J. Cell Sci.* **90**, 717–733.
Goodenough, U. W., and Heuser, J. E. (1988b). Molecular organization of the cell wall and cell-wall crystals from *Chlamydomonas eugamatos*. *J. Cell Sci.* **90**, 735–750.
Goodenough, U. W., and Jurivich, D. (1978). Tipping and mating structure activation induced in *Chlamydomonas* gametes by flagellar membrane antisera. *J. Cell Biol.* **79**, 680–693.
Goodenough, U. W., and Weiss, R. L. (1978). Interrelationships between microtubules, a striated fiber, and the gametic mating structures of *Chlamydomonas reinhardtii*. *J. Cell Biol.* **76**, 430–438.
Goodenough, U. W., Hwang, C., and Martin, H. (1976). Isolation and genetic analysis of mutant strains of *Chlamydomonas reinhardtii* defective in gametic differentiation. *Genetics* **82**, 169–186.
Goodenough, U. W., Adair, W. S., Caligor, B., Forest, C. L., Hoffman, J. L., Mesland, D. A., and Speth, S. (1980). Membrane-membrane and membrane-ligand interactions in *Chlamydomonas* mating. *In* "Membrane–Membrane Interactions" (N. B. Gilula, ed.), pp. 131–152. Raven Press, New York.
Goodenough, U. W., Adair, W. S., Collin-Osdoby, P., and Heuser, J. E. (1985a). *Chlamydomonas* cells in contact. *In* "The Cell in Contact: Adhesions and Junctions as Morphogenetic Determinants" (G. M. Edelman and J.-P. Thiery, eds.) pp. 111–135. Neurosciences Research Foundation.
Goodenough, U. W., Adair, W. S., Collin-Osdoby, P., and Heuser, J. E. (1985b). Structure of the *Chlamydomonas* agglutinin and related flagellar surface proteins *in vitro* and *in situ*. *J. Cell Biol.* **101**, 924–941.
Harris, E. H. (1989). "The Chlamydomonas Sourcebook: A Comprehensive Guide to Biology and Laboratory Use." Academic Press, San Diego.
Hegemann, P., Neumeier, K., Hegemann, U., and Kuehnle, E. (1990). The role of calcium in *Chlamydomonas* photomovement responses as analyzed by calcium channel inhibitors. *Photochem. Photobiol.* **52**, 575–583.
Hills, G. J., Gurney Smith, M., and Roberts, K. (1973). Structure, composition, and morphogenesis of the cell wall of *Chlamydomonas reinhardtii*. II. Electron microscopy and optical diffraction analysis. *J. Ultrastruct. Res.* **43**, 179–192.
Hills, G. J., Phillips, J. M., Gay, M. R., and Roberts, K. (1975). Self-assembly of a plant cell wall *in vitro*. *J. Mol. Biol.* **96**, 431–441.
Hoffman, J. L., and Goodenough, U. W. (1980). Experimental dissection of flagellar surface motility in *Chlamydomonas*. *J. Cell Biol.* **86**, 656–665.
Homan, W. L., van Kalshoven, H., Kolk, A. H. J. , Musgrave, A., Schuring, F., and van den Ende, H. (1987). Monoclonal antibodies to surface glycoconjugates in *Chlamydomonas eugametos* recognize strain-specific O-methylated sugars. *Planta* **170**, 328–335.
Hunnicutt, G. R. (1989). "Biogenesis and Fate of Cell–Cell Adhesion Molecule, Agglutinin, during Gametogenesis and Fusion of *Chlamydomonas reinhardtii*. Dissertation, University of Texas Southwestern Medical Center.
Hunnicutt, G. R., and Snell, W. J. (1991). Rapid and slow mechanisms for loss of cell adhesiveness during fertilization in *Chlamydomonas*. *Dev. Biol.*
Hunnicutt, G. R., Kosfiszer, M. G., and Snell, W. J. (1990). Cell body and flagellar agglutinins in *Chlamydomonas reinhardtii*: The cell body plasma membrane is a reservoir for agglutinins whose migration to the flagella is regulated by a functional barrier. *J. Cell Biol.* **111**, 1605–1616.
Imam, S. H., and Snell, W. J. (1988). The *Chlamydomonas* cell wall degrading enzyme, lysin, acts on two domains within the framework of the wall. *J. Cell Biol.* **106**, 2211–2221.
Imam, S. H., Buchanan, M. J., Shin, H.-C., and Snell, W. J. (1985). The *Chlamydomonas* cell wall: Characterization of the wall framework. *J. Cell Biol.* **101**, 1599–1607.

Jaenicke, L., Kuhne, W., Spessert, R., Wahle, U., and Waffenschmidt, S. (1987). Cell-wall lytic enzymes (autolysins) of *Chlamydomonas reinhardtii* are (hydroxy)proline-specific proteases. *Eur. J. Biochem.* **170**, 485–491.

Kindle, K. L. (1990). High-frequency nuclear transformation of *Chlamydomonas reinhardtii. Proc. Natl. Acad. Sci. U.S.A.* **87**, 1228–1232.

Kinoshita, T., Fukuzawa, H., Shimoda, T., and Matsuda, Y. (1992). Primary structure and expression of a gamete lytic enzyme in *Chlamydomonas reinhardtii*: Similarity of functional domains to matrix metalloproteases. *Proc. Natl. Acad. Sci. U.S.A.* **89**, 4693–4697.

Kooijman, R., de Wildt, P., Homan, W. L., Musgrave, A., and van den Ende, H. (1988). Light affects flagellar agglutinability in *Chlamydomonas eugametos* by modification of the agglutinin molecules. *Plant Physiol.* **86**, 216–223.

Kooijman, R., de Wildt, P., Beumer, S., van der Vliet, G., Homan, W., Kalshoven, H., Musgrave, A., and van den Ende, H. (1989). Wheat germ agglutinin induces mating reactions in *Chlamydomonas eugametos* by cross-linking agglutinin-associated glycoproteins in the flagellar membrane. *J. Cell Biol.* **109**, 1677–1687.

Lefebvre, P. A., and Rosenbaum, J. L. (1986). Regulation of the synthesis and assembly of ciliary and flagellar proteins during regeneration. *Ann. Rev. Cell Biol.* **2**, 517–546.

Lewin, R. A. (1982). A new kind of motility mutant (non-gliding) in *Chlamydomonas. Experientia* **38**, 348–349.

Lopez, L. C., and Shur, B. D. (1987). Redistribution of mouse sperm surface galactosyltransferase after the acrosome reaction. *J. Cell Biol.* **105**, 1663–1670.

Luck, D. J. (1984). Genetic and biochemical dissection of the eucaryotic flagellum. *J. Cell Biol.* **98**, 789–794.

Maihle, N. J., and Satir, B. H. (1986). Protein secretion in *Tetrahymena thermophila*. Characterization of the major proteinaceous secretory proteins. *J. Biol. Chem.* **261**, 7566–7570.

Matsuda, Y. (1988). The *Chlamydomonas* cell walls and their degrading enzymes. *Jpn. J. Phycol.* **36**, 246–264.

Matsuda, Y., Saito, T., Yamaguchi, T., and Kawase, H. (1985). Cell wall lytic enzyme released by mating gametes of *Chlamydomonas reinhardtii* is a metalloprotease and digests the sodium percholorate-insoluble component of cell wall. *J. Biol. Chem.* **260**, 6373–6377.

Matsuda, Y., Saito, T., Yamaguchi, T., Koseki, M., and Hayashi, K. (1987). Topography of cell wall lytic enzyme in *Chlamydomonas reinhardtii*: Form and location of the stored enzyme in vegetative cell and gamete. *J. Cell Biol.* **104**, 321–330.

Mesland, D. A. M., Hoffman, J. L., Caligor, E., and Goodenough, U. W. (1980). Flagellar tip activation stimulated by membrane adhesions in *Chlamydomonas* gametes. *J. Cell Biol.* **84**, 599–617.

Millikin, B. E., and Weiss, R. L. (1984). Distribution of concanavalin a binding carbohydrates during mating in *Chlamydomonas. J. Cell Sci.* **66**, 223–239.

Mitchell, O. R., and Kang, Y. (1991). Identification of *odab* as a *Chlamydomonas* dynein mutant by rescue with the wild-type gene. *J. Cell Biol.* **113**, 835–842.

Musgrave, A., and van den Ende, H. (1987). How *Chlamydomonas* court their partners. *Trends Biochem. Sci.* **12**, 470–473.

Musgrave, A., de Wildt, P., van Etten, I., Pijst, H., Scholma, C., Kooyman, R., Homan, W., and van den Ende, H. (1986). Evidence for a functional membrane barrier in the transition zone between the flagellum and cell body of *Chlamydomonas eugametos* gametes. *Planta* **167**, 544–553.

Myles, D. G., and Primokoff, P. (1984). Localized surface antigens of guinea pig sperm migrate to new regions prior to fertilization. *J. Cell Biol.* **99**, 1634–1641.

Novick, P., Field, C., and Schekman, R. (1980). Identification of 23 complementation groups required for post-translational events in the yeast secretory pathway. *Cell* **21**, 205–215.

Pasquale, S. M., and Goodenough, U. W. (1987). Cyclic AMP functions as a primary sexual signal in gametes of *Chlamydomonas reinhardtii*. *J. Cell Biol.* **105**, 17a.

Pasquale, S. M., and Goodenough, U. W. (1988). Calmodulin sensitivity of the flagellar membrane adenylate cyclase and signaling of motile responses by cAMP in gametes of *Chlamydomonas reinhardtii*. *Bot. Acta* **101**, 118–122.

Pijst, H. L. A., van Driel, R., Janssens, P. M. W., Musgrave, A., and van den Ende, H. (1984). Cyclic AMP is involved in sexual reproduction of *Chlamydomonas eugametos*. *FEBS Lett.* **174**, 132–136.

Ringo, D. L. (1967). Flagellar motion and fine structure of the flagellar apparatus in *Chlamydomonas*. *J. Cell Biol.* **33**, 543–571.

Roberts, K., Gurney Smith, M., and Hills, G. J. (1972). Structure, composition and morphogenesis of the cell wall of *Chlamydomonas reinhardi*. I. Ultrastructure and preliminary chemical analysis. *J. Ultrastruct. Res.* **40**, 599–613.

Roberts, K., Grief, C., Hills, G. J., and Shaw, P. J. (1985). Cell wall glycoproteins: Structure and function. *J. Cell Sci. Suppl.* **2**, 105–127.

Rogalski, A. A., and Bouck, G. B. (1982). Flagellar surface antigens in *Euglena*: Evidence for an external glycoprotein pool and its transfer to the regenerating flagellum. *J. Cell Biol.* **93**, 758–766.

Saito, T., and Matsuda, Y. (1984a). Sexual agglutinin of mating type minus gametes in *Chlamydomonas reinhardii*: 1. Loss and recovery of agglutinability of gametes treated with EDTA. *Exp. Cell Res.* **152**, 322–330.

Saito, T., and Matsuda, Y. (1984b). Sexual agglutinin of mating-type minus gametes in *Chlamydomonas reinhardii*: 2. Purification and characterization of minus agglutinin and comparison with plus agglutinin. *Arch. Microbiol.* **139**, 95–99.

Saito, T., and Matsuda, Y. (1986). The *Chlamydomonas reinhardtii* plus and minus agglutinin difference in sensitivity to dithiothreitol. *Arch. Microbiol.* **146**, 25–29.

Saito, Y., Tsubo, Y., and Matsuda, Y. (1985). Synthesis and turnover of cell body agglutinin as a pool of flagellar surface agglutinin in *Chlamydomonas reinhardtii* gamete. *Arch. Microbiol.* **142**, 207–210.

Snell, W. J. (1981). Flagellar adhesion and deadhesion in *Chlamydomonas* gametes: Effects of tunicamycin and observations on flagellar tip morphology. *J. Supramol. Struct. Cell. Biochem.* **16**, 371–376.

Snell, W. J. (1983). Characterization of the *Chlamydomonas* flagellar collar. *Cell Motil.* **3**, 273–280.

Snell, W. J. (1985). Cell–cell interactions in *Chlamydomonas*. *Ann. Rev. Plant Physiol.* **36**, 287–315.

Snell, W. J. (1990). Adhesion and signalling during fertilization in multicellular and unicellular organisms. *Curr. Opin. Cell Biol.* **2**, 821–832.

Snell, W. J., and Moore, W. S. (1980). Aggregation-dependent turnover of flagellar adhesion molecules in *Chlamydomonas* gametes. *J. Cell Biol.* **84**, 203–210.

Snell, W. J., and Roseman, S. (1979). Kinetics of adhesion and de-adhesion of *Chlamydomonas* gametes. *J. Biol. Chem.* **254**, 10820–10829.

Snell, W. J., Kosfiszer, M. G., Clausell, A., Perillo, N., Imam, S., and Hunnicutt, G. (1986). A monoclonal antibody that blocks adhesion of *Chlamydomonas* mt^+ gametes. *J. Cell Biol.* **103**, 2449–2456.

Snell, W. J., Buchanan, M., and Clausell, A. (1982). Lidocaine reversibly inhibits fertilization in *Chlamydomonas*: A possible role for calcium in sexual signalling. *J. Cell Biol.* **94**, 607–612.

Snell, W. J., Eskue, A. E., and Buchanan, M. J. (1989). Regulated secretion of a serine protease that activates an extracellular matrix-degrading metalloprotease during fertilization in *Chlamydomonas*. *J. Cell Biol.* **109**, 1689–1694.

Tomson, A. M., Demets, R., Musgrave, A., Kooijman, R., Stegwee, D., and van den Ende, H. (1990a). Contact activation in *Chlamydomonas* gametes by increased binding capacity of sexual agglutinins. *J. Cell Sci.* **95**, 293–301.

Tomson, A. M., Demets, R., van Spronsen, E. A., Brakenhoff, G. J., Stegwee, D., and van den Ende, H. (1990b). Turnover and transport of agglutinins in conjugating *Chlamydomonas* gametes. *Protoplasma* **155**, 200–209.

Unemori, E. N., and Werb, Z. (1986). Reorganization of polymorized actin: A possible trigger for induction of procollagenase in fibroblasts cultured in and on collagen gels. *J. Cell Biol.* **103**, 1021–1031.

Unemori, E. N., and Werb, Z. (1988). Collagenase expression and endogenous activation in rabbit synovial fibroblasts stimulated by the calcium ionophore A23187. *J. Biol. Chem.* **263**, 16252–16259.

van den Ende, H. (1992). Sexual signalling in *Chlamydomonas. In* "Perspectives in Plant Recognition" (J. A. Callow and J. R. Green, eds.), pp. 1–19. Cambridge University Press, Cambridge.

Vilmart Seuwen, J., Kersken, H., Sturzl, R., and Plattner, H. (1986). ATP keeps exocytosis sites in a primed state but is not required for membrane fusion: An analysis with *Paramecium* cells *in vivo* and *in vitro. J. Cell Biol.* **103**, 1279–1288.

Weiss, R. L., Goodenough, D. A., and Goodenough, U. W. (1977). Membrane particle arrays associated with the basal body and with contractile vacuole secretion in *Chlamydomonas. J. Cell Biol.* **72**, 133–143.

Zhang, Y., and Snell, W. J. (1993). Differential regulation of adenylyl cyclase in vegetative and gametic flagella of *Chlamydomonas. J. Biol. Chem.*

Zhang, Y., Ross, E. M., and Snell, W. J. (1991). ATP-dependent regulation of flagellar adenylyl cyclase in gametes of *Chlamydomonas reinhardtii. J. Biol. Chem.* **266**, 22954–22959.

Chapter 11

Photoreception in *Chlamydomonas*

PETER HEGEMANN and HARTMANN HARZ

Max-Planck-Institut für Biochemie
D-8033 Martinsried, Germany

I. Introduction

The unicellular photosynthetic alga *Chlamydomonas*, a representative of the Chlorophyceae, serves as a model organism for the understanding of general principles of eukaryotic cell behavior. One of the best studied characteristics of *Chlamydomonas* is its ability to orient in light and to find the optimal light conditions for photosynthesis (Pfeffer, 1904; Buder, 1917; Nultsch and Häder, 1988). Most experiments have been carried out with the species *C. reinhardtii*.

Although much biochemistry has been done in this system, and molecular genetic methods, including chloroplast and nuclear transformation (Kindle, 1990, and literature therein) are now well established, the mechanisms of signal transduction processes in *Chlamydomonas* are far from understood. The major reasons for this lack of understanding are (1) that the photobehavior depends on many environmental parameters, which makes reproducible measurements of behavioral responses rather difficult; (2) that investigations with selected mutants have begun only very recently; and (3) that biophysical methods that allow the detection of fast flagellar responses or electrical phenomena were developed very slowly in *Chlamydomas*, in contrast to systems such as *Paramecium* (for review see Martinac *et al.*, 1988).

Chlamydomonas normally swims with the flagella in the forward direction (Fig. 11.1); the flagella beat with a frequency of between 20 and 50 Hz. A lateral component of the three-dimensional flagellar beating causes the cell to rotate with a frequency of about 2 Hz (Bessen *et al.*, 1980; Rüffer and Nultsch, 1985,1987). In contrast to bacteria, *Chlamydomonas* can detect not only the light intensity but also the light direction. This ability requires two key features that are not present in bacteria. First, the amount of absorbed light should vary with the orientation of the cell relative to the light source. For this purpose, the photoreceptor molecules must be localized to selected areas of the cell or, at least, must be distributed unevenly over the cell. On the basis of theoretical arguments about the function of the eyespot as an optical device for photoreception, and based on the high density of intramembrane particles in eyespot overlaying membranes, the photoreceptor has been proposed to be localized to one of the eyespot overlaying membranes (Foster and Smyth, 1980; Melkonian and Robenek, 1984). This localization of the photoreceptor allows the lay-

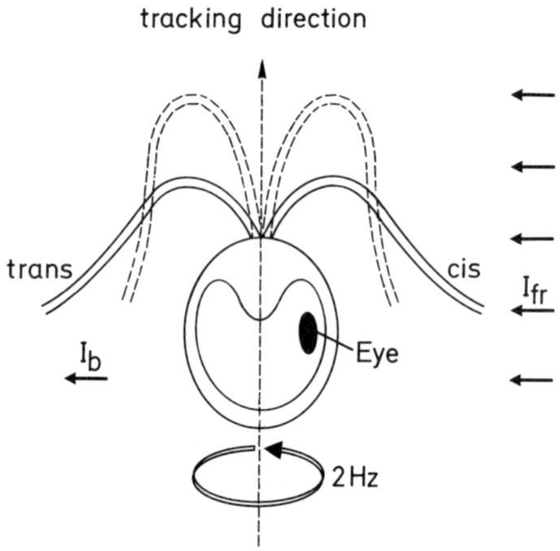

Figure 11.1 *Chlamydomonas* cell with the flagella in the forward beating mode. For the description of light antenna effectiveness, the "front to back contrast" has been defined by the modulation function as $m = (I_{fr} - I_b)/(I_{fr} + I_b)$ (Foster and Smyth, 1980). I_{fr} is the photon irradiance seen by the photoreceptor when the eye is facing the light. I_b is the irradiance after extinction through the cell and the eyespot, which is seen when the eyespot is facing away from the light.

ered structure of the eyespot granula to serve as a directional antenna. This antenna, in conjunction with the photoreceptor molecules, forms an intact eye (Fig. 11.1; Foster and Smyth, 1980). Second, the cell must be able to make directional changes with a distinct preference to one site relative to the light incidence. In *Chlamydomonas* cells, this asymmetry is expressed by an uneven responsiveness of the two flagella when light is perceived by the eye. Although these principles seem very simple, understanding their mechanism presents a problem and a challenge for future research.

II. Behavior in Light

A. Responses to Flashes and Step-up Stimulation

When *Chlamydomonas* is faced with changes in direction, wavelength, or photon irradiance (photons m^{-2} sec^{-1}) of the ambient light, the organism exhibits distinct behavioral responses that depend on the size and duration of the change. Weak flashes or low increases in stimulation cause alterations of the swimming direction (Fig. 11.2A), whereas more intense flashes cause the cells to switch transiently to a slow backward swimming mode (stop response or phobic response) (Fig. 11.2B and C; Schmidt and Eckert, 1976; Boscov and Feinleib, 1979; Hegemann and Marwan, 1988). On stimulation with repetitive flashes above 5 Hz or increased stimulation with continuous light of high photon irradiance, the cells first show a stop response, which is followed by directional changes that become less and less pronounced; the cells finally show almost smooth helical swimming toward or away from the light (positive or negative phototaxis) (Fig. 11.2D). Phototaxis of unicellular algae was reported as early as 1817 by Trevianus and has been under intense investigation since that time (Famitzins, 1867; Buder, 1917; Nultsch and Häder, 1988).

Phototaxis has been characterized by three main approaches: the accumulation of cells in a selected area of a measuring cuvette (Feinleib and Curry, 1967), the average orientation of a cell population with respect to the light incidence (directedness; Häder and Lebert, 1985), or the net movement of a single cell or a cell population to or away from a light source (phototactic rate; Foster and Smyth, 1980; Hegemann *et al.*, 1988). Another important parameter is the absolute light sensitivity of the cells, which is defined as 1/threshold photon irradiance (for phototaxis) or 1/threshold photon exposure (for flash-induced responses).

In principle, a directional change may be caused by (1) activation or inactivation of the flagellum nearer to the eyespot (the *cis* flagellum), (2) inactivation or activation of the *trans* flagellum, (3) a combination of *cis*- and *trans*-flagellar activation or inactivation, or (4) alteration of the three-dimensional flagellar

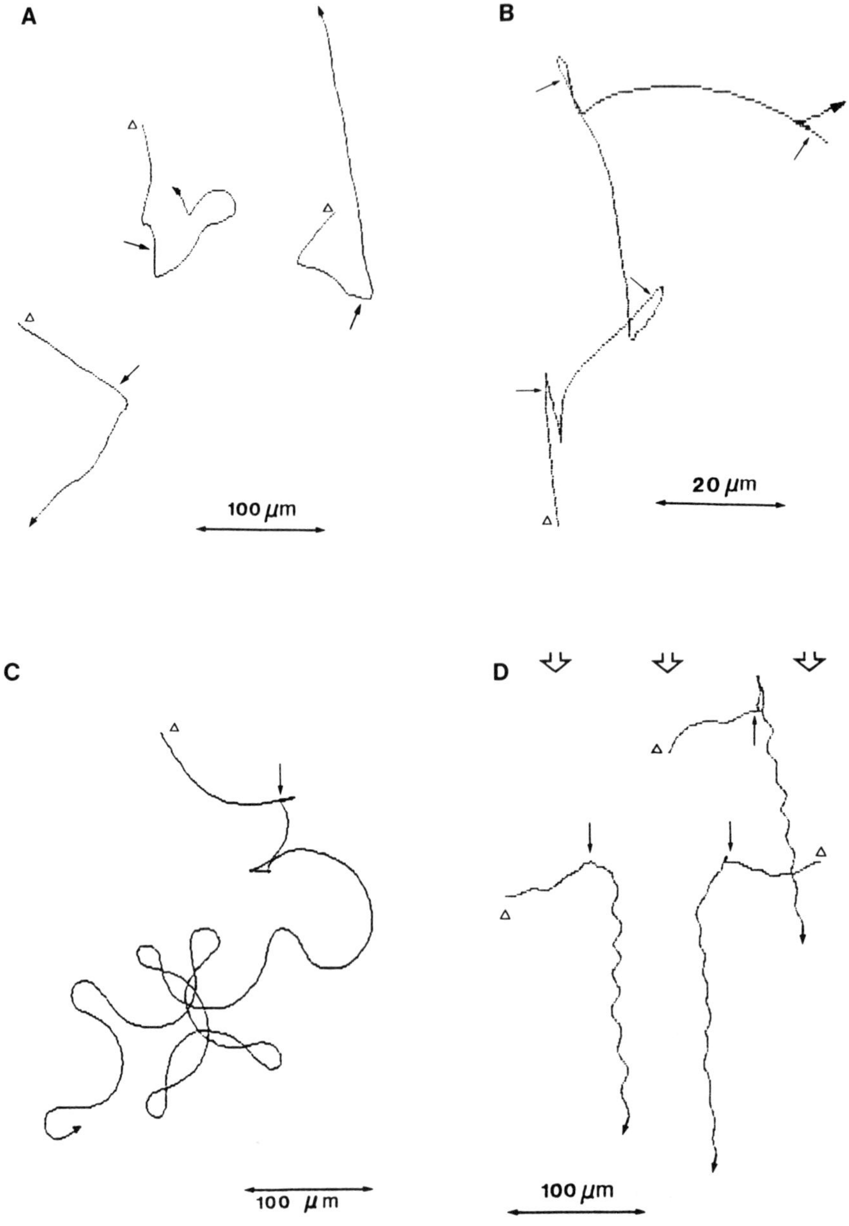

Figure 11.2 Typical light responses of strain 806 cells. (A) 3-sec tracks of cells that were stimulated at low ($2 \times 10^{-7}M$) calcium with green flashes of 50-msec duration (500 nm). The cells perform only direction changes. (B) 4-sec track of a cell stimulated at high calcium ($1 \times 10^{-4}M$) with flashes as in A. The cells respond to every flash with a stop response. (C) Track of a cell that

beating modes. Flagellar beating in living *Chlamydomonas* cells has been subjected to very detailed analysis using high speed cinematography. Because of the high time resolution, these experiments all were carried out in the presence of intense red background irradiance (Rüffer and Nultsch, 1985,1987,1990); therefore, the responses to stimulation with light pulses cannot be compared with those measured in other behavioral experiments. During a stop response in *Chamydomonas*, the flagella transiently switch from the forward "breast stroke style" to undulation, resulting in slow backward swimming (Schmidt and Eckert, 1976). Microscopic studies in the *Chlamydomonas* relative *Haematococcus* indicate that, during response to green flashes, the frequency of beating of both flagella transiently increases, but to a slightly different extent (Sineshchekov, 1991). More indirect conclusions have been drawn from experiments with detergent-extracted *Chlamydomonas* cells and isolated flagella (Kamiya and Witman, 1984). In these permeabilized cells, a slight increase in calcium concentration leads to *cis*-flagellum inactivation and *trans*-flagellum activation, whereas a larger increase in calcium concentration causes the cells to switch from forward to slow backward swimming. The authors suggest that these calcium responses resemble the light-induced directional changes and stop responses of living cells.

We are convinced that phototaxis is the result of individual directional changes that are very pronounced at low light, under which photon absorption occurs relatively rarely. At high photon irradiance, when light is perceived almost continuously, the directional changes are very small and almost indistinguishable from smooth swimming behavior. The higher probability for photon absorption when the eye is facing the light rather than pointed away from the light causes an uneven distribution of directional changes relative to the light, resulting in a biased walk toward or away from the light source (Hegemann and Marwan, 1988). The strength of the bias depends on the difference of the photon absorption (front to back contrast), which can be expressed by the contrast function $m = (I_{fr} - I_b)/(I_{fr} + I_b)$ (Fig. 11.1; Foster and Smyth, 1980). At low light levels, the contrast should be proportional to the phototactic rate. At high photon irradiance, the cells should receive a continuous light pattern that is modulated depending on the orientation of the cells with respect to the light incidence. The cells correct their swimming until they receive an unmodulated

was pulsed repetitively with a frequency of 1 Hz (500 msec light on and off) at $1 \times 10^{-6} M$ calcium shows a few stops (two in this case) followed by step-down loops. (D) Cells stimulated with 5 Hz flashing (shown here) or continuous light are negatively phototactic. Triangles (△) indicate the beginning of tracking. Small arrows indicate the position of the cells when flashes were applied (A,B) or the beginning of repetitive flashing (C,D). Large open arrows indicate the direction of the light incidence in D. In all cases, light of 500 nm and 2.5×10^{19} photons $m^{-2} sec^{-1}$ was used. Reprinted with permission from Hegemann and Bruck (1989).

signal, which occurs only if they swim directly toward or away from the light source (Foster and Smyth, 1980). Under these conditions, the phototactic rate approaches the swimming speed.

According to the elegant theoretical explanation of Foster and Smyth (1980), significant contrast is caused by absorption, reflection, and constructive interference at the layers of carotenoid-containing vesicles in the eyespot and only to a minor extent by shading caused by photosynthetic pigments. Therefore, the contrast should depend strongly on the number of these layers, which varies from one (quarterwave plate) in *Haematococcus* to seven (quarterwave stack) in *Chlamydomonas* (Foster and Smyth, 1980; Staehlin, 1986). The correlation between reflectivity and constructive interference and the number of layers has been demonstrated for three unicellular algae—*Tetraselmis chui, Chlamydomonas eugametos, Hafniomonas reticulata*—using confocal laser scanning microscopy (Kreimer and Melkonian, 1990). The absolute contrast at the site of the photoreceptor can be calculated from the shift of the stimulus–response curves for flash-induced photocurrents under two different orientations of the eyespot, either facing the light directly or facing away from it. The stimulus–response curves are shifted by a factor of 3 for *Haematococcus* (single layered eyespot; Sineshchekov, 1991) and by a factor of 8 for *Chlamydomonas* (multilayered eyespot; Harz and Hegemann, 1991).

The function of the contrast for phototaxis has been clarified further by the finding that, in eyespot-less mutants that are missing the optical device but not the photoreceptor molecules, the phototactic rate is very low at all stimulus photon irradiances, whereas the absolute light sensitivity remains close to that of wild-type cells (Morel-Laurence and Feinleib, 1983; Foster *et al.*, 1984). The absolute light sensitivity is determined by the number of photoreceptor molecules in the cells (total absorption cross section for photon capture) and by the efficiency of the signal transduction process, and is therefore fairly independent of the cell contrast. Because directional changes in dark-adapted cells are caused by individual photons—either by one (Hegemann and Marwan, 1988) or by two (R. Uhl, unpublished results)—the amplification is not limiting and the number of photoreceptor molecules should be the dominant parameter that determines the absolute light sensitivity of the cell.

Pairs formed during the mating of gametes of different mating type (mt^+ or mt^-) are also phototactic. The pair is pushed forward by the flagella belonging to the mt^+ parent, whereas the flagella belonging to the mt^- are paralyzed. This behavioral difference was actually the original definition that distinguished mt^+ and mt^- cells. Experiments with *C. eugametos* have shown that no communication occurs between the photoreceptor of the mt^- cell and the flagella of the mt^+ partner; phototaxis is controlled completely by the mt^+ partner. In response to increased stimulation, slow backward swimming followed by a characteristic spiralling period occurs using only one active flagellum (Fig. 11.3; Hegemann and Musgrave, 1991; Musgrave and Häder, 1991).

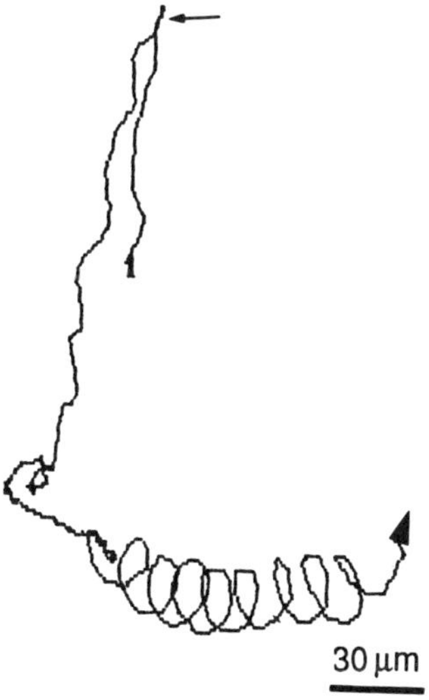

Figure 11.3 Trace of a dark-adapted *Chlamydomonas eugametos* mating pair during 20 sec of irradiation with green light. The start of irradiation is indicated by a small arrow. Reprinted with permission from Hegemann and Musgrave (1991).

B. Adaptation Phenomena

Chlamydomonas cells are able to adapt to longer light pulses or repetitive stimulation. This adaptation results in a change in the behavioral responses to particular stimuli, that is, cells can respond positively or negatively to the identical stimulus depending on the irradiance to which the cells have been exposed previously (Mayer, 1968). The adaptation level (for definition see Delbrück and Reichert, 1956) depends on the duration and the photon irradiance of the adapting stimulus and on the calcium concentration of the extracellular medium. At high calcium concentrations (~1 m*M*), the final adaptation level is high; on decrease of the photon irradiance (step-down stimulation), the cells perform stop responses, step-down loops (Fig. 11.2C), or direction changes. At low calcium levels, when the adaptation level is lower, the step-down responses to the same stimulation program are only very small (Hegemann and Bruck, 1989).

Chlamydomonas can orient phototactically to photon irradiances covering at least five orders of magnitude. The cells change from positive to negative phototaxis at certain light levels; the absolute light level of the switch strongly depends on the environmental conditions and the strain (Nultsch, 1977; Morel-Laurence, 1987). Several experiments have shown that white light causes a strong desensitization of the cells to stimulation by green light (Marwan and Hegemann, 1988; Uhl and Hegemann, 1990a). Full sensitivity slowly recovers within a period of minutes (up to an hour) after the cells are returned to darkness. Green light that is phototactically active causes desensitization when given as background light. However, in contrast to white light, green light desensitization is rapidly reversible within a few minutes. These experiments, in conjunction with earlier phototaxis experiments with photosynthesis inhibitors (Feinleib and Curry, 1979), indicate that adaptation to white or red light is caused primarily by photosynthesis and is probably a result of hyperpolarization of the plasma membrane (Sineshchekov *et al.*, 1990). Adaptation to green light instead occurs by photoreceptor inactivation, which results from bleaching or modification of the receptor or from a decrease in signal transduction efficiency (Uhl and Hegemann, 1990b). These phenomena indicate a complex adaptation system that has not been elucidated on a molecular level. Adaptation has been overlooked widely because most phototaxis experiments have been carried out in continuous white light, under which photosynthesis influences photobehavior to an unknown degree.

III. A Rhodopsin Is the Photoreceptor for Photomovement Responses

A. Identification of Rhodopsin from Behavioral Studies

Until a few years ago, the nature of the photoreceptor that controls photomovement responses in *Chlamydomonas* had been discussed exclusively on the basis of action spectroscopy. When action spectra are constructed from phototaxis measurements at high photon irradiance, they exhibit maxima at 503 and 440 nm (Nultsch *et al.*, 1971), leading to the suggestion that flavins or carotenoids may function as the photoreceptor (discussed by Smyth *et al.*, 1988). However, low intensity phototaxis action spectra reveal the absorption of the functional photoreceptor much better than do high intensity spectra. Using experimental data from Nultsch and colleagues (1971), Foster and Smyth (1980) constructed threshold action spectra for phototaxis. These spectra were rhodopsin shaped and peaked at 503 nm, as do absorption spectra of rhodopsin from higher animals. These results led to the first suggestion that a rhodopsin might serve as the primary photoreceptor for phototaxis in *Chlamydomonas*. (The term "rhodopsin" is used here to mean a retinal-containing protein photoreceptor.) The absence of a maximum at 440 nm in the threshold action spectrum can be explained either by a second photoreceptor present in much smaller amounts, thus contributing only at high irradiances, or by photochromism of only one photoreceptor (P. Hegemann and R. Uhl, unpublished observation).

The action spectra of green cells with intact eyes (Fig. 11.4) are significantly narrower than the Dartnall nomogram, which is the standard spectrum

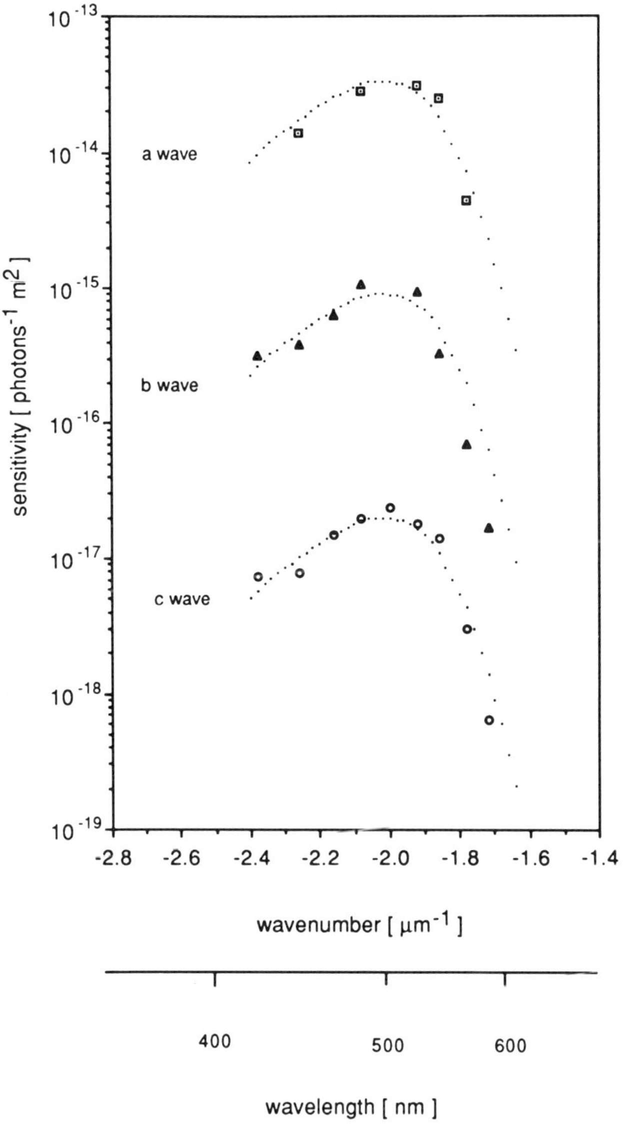

Figure 11.4 Action spectra for three different flash-induced light scattering transients. The b and c waves reflect direction changes and stop responses, respectively. The a wave reflects photon absorption, which is so low that no behavioral responses can be detected. Reprinted with permission from Uhl and Hegemann (1990).

of animal rhodopsins. The small half-width of the spectrum can be explained by the small half bandwidth of the rhodopsin absorption itself (discussed subsequently) and by optical properties of the eyespot. As determined by detection of flash-induced photoreceptor currents (H. Harz and P. Hegemann, unpublished observations), the front-to-back contrast produced by the multilayered eyespot in *Chlamydomonas* is wavelength dependent with a maximum also near 500 nm (see also Foster and Smyth, 1980; Foster and Saranak, 1989). The contrast contributes to different extents to phototaxis recorded in continuous light, to flash-induced responses of free-swimming cells, and to electrical responses of fixed cells. These variables explain some differences of the shape and the precise maxima of the various action spectra (Nultsch, 1971; Foster *et al.*, 1984; Uhl and Hegemann, 1991; Harz and Hegemann, 1991).

The existence of rhodopsins in other algae also has been suggested on the basis of action spectroscopy of their movement responses (Foster *et al.*, 1984). Based on the rhodopsin absorption drawn from action spectroscopy, the rhodopsins in these algae show maxima between 457 and 493 nm (summarized by Foster *et al.*, 1984). These algal photosensory receptors can be regarded as a newly identified class of lower plant rhodopsins.

Additional evidence for the rhodopsin-like nature of the *Chlamydomonas* photoreceptor is derived from experiments with FN68 mutant cells, which are almost completely blind if grown in darkness (Table I). Full phototactic sensitivity of these blind cells can be reconstituted by supplementation of the cells with exogenous retinal or retinal analog compounds (Foster *et al.*, 1984). More than 30 analogs have been tested for their ability to restore phototaxis in this mutant. The compounds 11-*cis* and all-*trans* retinal were found to result in action spectra with maxima very similar to those of wild-type cells (505 and 501 nm), whereas the action spectrum in the presence of 9-*cis* retinal was shifted slightly toward blue (488 nm). Retinal analogs with shorter conjugated polyene chains (dihydro compounds) showed a more pronounced blue shift. The retinal dependence of phototactic sensitivity, in conjunction with the shift of the action spectra maxima by retinal analogs, led to the final conclusion that a rhodopsin is the functional photoreceptor for phototaxis (Foster *et al.*, 1984).

In later experiments, phototaxis was tested using retinal analogs that cannot isomerize around certain double bonds or have very short polyene chains (Foster *et al.*, 1988b,1989). Surprisingly, most of these compounds readily restored phototaxis. These reconstitution experiments led to several conclusions about the properties of *Chlamydomonas* rhodopsin and its light activation: (1) The aldehyde function of a retinal analog is essential for the formation of an active rhodopsin. (2) If isomerization is required for rhodopsin activation, it must be nonregiospecific. (3) Neither detachment of the chromophore from the binding site (bleaching) nor protonation of a C=N bond is required for restoration of phototaxis. (4) Phototaxis can be restored by aldehydes, which

TABLE I
Strains that Have Been Used for the Analysis of Photoreception in *Chlamydomonas*[a]

Strain	Cell wall	Retinal	Phototaxis	Use
CC137 *mt*⁺	++	++	++ pos/neg	Phototaxis
622E	++	++	++ pos/neg	Phototaxis; cinematography
CW806	+	++	++ neg	Phototoxis; rhodopsin
CW2	−	++	++ pos/neg	Electrophysiology
FN68(F1)	+	−	+ neg/pos[b]	Phototaxis
CC2359	++	− −	++ neg/pos[b]	Phototaxis

[a]++, Wild-type level; +, below wild-type; −, poor; − −, not detectable.
[b]After addition to retinal.

have only one or even no C=C bond, for example, hexenal or hexanal. Specific C=C double bond isomerization accompanies ion pumping in bacteriorhodopsin and halorhodopsin, photosignaling by bacterial sensory rhodopsins, and signal transduction in the visual processes of higher animals (Oesterhelt *et al.*, 1986; Yan *et al.*, 1990). The possibility that mechanisms other than C=C isomerization may activate *Chlamydomonas* rhodopsin made the *Chlamydomonas* system especially appealing. *Chlamydomonas* rhodopsin could be activated by C=N bond isomerization or directly by charge redistribution in the excited state of the chromophore (Nakanishi *et al.*, 1989). This concept, however, is not generally accepted because theoretical arguments have led to the prediction that drastic changes of the polyene system also should influence the charge redistribution in the excited state (Birge, 1990) and lead to a distortion of the activation mechanism and an altered absolute light sensitivity of the cell.

Originally, 11-*cis* retinal was favored as the natural retinal isomer of *Chlamydomonas* rhodopsin because the final phototactic sensitivity of FN68 cells (Table I) was about three times higher with 11-*cis* than with all-*trans* retinal (Foster *et al.*, 1984). However, when retinal analogs were used all-*trans* isomers showed the same and sometimes even higher sensitivity than 11-*cis* isomers (Nakanishi *et al.*, 1989); therefore, the question of the natural isomer cannot be answered from phototaxis measurements alone.

Phototaxis measurements also show that the rhodopsin is easily accessible from the extracellular medium. Hydroxylamine reversibly suppresses phototaxis in wild-type cells in a dose-dependent manner. After washing of the cells, added retinal can restore the full phototactic sensitivity within a few minutes (Hegemann *et al.*, 1988). Therefore, hydroxylamine as well as action spectroscopy can be used to assay rhodopsin in other algal species for which blind mutants are not available.

In addition to phototaxis, flash-induced behavioral responses, that is, directional changes or stop responses, have been found to fulfill the criteria for rhodopsin-regulated processes. They are retinal dependent, show rhodopsin action spectra with a maximum at 495 nm, and are sensitive to hydroxylamine. These results suggest that flash-induced responses and phototaxis are controlled by the same or very similar rhodopsin-type photoreceptors (Hegemann and Marwan, 1988; Lawson *et al.*, 1990; Uhl and Hegemann, 1990b).

A retinal-deficient mutant CC2359 (Table I), which, like the FN68 mutant, is almost completely blind when grown in the dark, has been used to monitor incorporation of opsin. By detection of flash-induced responses as light-scattering transients, *in vivo* all-*trans* retinal has been demonstrated to incorporate into the opsin within 1 min, raising the sensitivity of the cells by a factor of more than 1000. The compound 11-*cis* retinal reconstitutes at least 50 times slower when low concentrations of the retinal isomers (between 1 and 10 n*M*) are used. The relative efficiency of the isomers is

all-*trans* >> 13-*cis* > 11-*cis* > 9-*cis* > 7-*cis*

The same graduated efficiency also has been found by detection of stop responses in isomer-reconstituted CC2359 cells by motion analysis (Zacks *et al.*, 1991). These reconstitution kinetics do not prove the natural isomer configuration but we think they are a strong argument for all-*trans*. Flash experiments with truncated or locked retinal analogs indicate that some structural features of the retinal are essential, that once had been regarded as nonessential based on previous phototaxis experiments. Competition experiments have shown that 13-demethyl-retinal also incorporates into *Chlamydomonas* rhodopsin without mediating function, suggesting that the 13-methyl group is essential for retinal–protein coupling (Hegemann *et al.*, 1991). A 13-locked retinal has been reported to be fully active in the phototaxis population assay with FN68 cells (Foster *et al.*, 1989). However, with the phenotypically similar CC2359 cells, the 13-locked retinal entered the rhodopsin binding site but could not induce stop responses (Zacks *et al.*, 1991), suggesting that a "free" 13-double bond is critical for rhodopsin function. Experiments with short-chain retinal analogs with 3, 2, 1 or no C=C bonds such as octa-trienals, octa-dienals, hexenal, and hexanal revealed that at least three double bonds in conjunction with the aldehyde function are necessary for the formation of a rhodopsin that can trigger stop responses. These experiments suggest that the mechanism of *Chlamydomonas* rhodopsin activation without double-bond isomerization cannot be generalized or extended to stop responses. From the analysis of flash-induced responses, a more conservative view of the retinal binding site in *Chlamydomonas* rhodopsin has been favored (Hegemann *et al.*, 1991).

The physiological experiments described establish rhodopsin as the functional photoreceptor for photomovement responses in *Chlamydomonas*. The properties of this rhodopsin are summarized in Table II. Several questions about the mechanism of rhodopsin activation remain unanswered.

1. Why was the phototactic sensitivity of FN68 cells after reconstitution with naphthyl retinals higher than in the presence of normal retinals? Why was the sensitivity with the short-chained analogs as high as with retinal itself?
2. Does retinal isomerize in *Chlamydomonas* rhodopsin under physiological conditions despite the fact that this isomerization seems not to be essential for phototaxis in the early experiments with FN68 cells?
3. Are stop response and phototaxis controlled by the same rhodopsin or by two different rhodopsin species?
4. Why do some retinal analogs regenerate phototaxis in FN68 cells but not light responses in CC2359 cells?

B. Retinal Analysis

Because wild-type *Chlamydomonas* cells contain large amounts of carotenoids in the eyespot and in the photosynthetic membranes (Krinsky and Levine, 1964), the identification of retinoids in wild-type cells has been difficult. The blind mutant FN68, however, is virtually free of carotenoids. The retinoid deficiency is thought to be caused by a defect in the carotenoid pathway at an early stage. Phototactic sensitivity can be recovered by irradiation with green light. The action spectrum for sensitization is similar to those for phototaxis or flash-induced responses, suggesting that the very tiny amount of functional rhodopsin that is present in dark-grown cell induces retinal synthesis (Foster *et al.*, 1988a). Because these sensitized cells are still devoid of most of the carotenoids, they are ideal for retinal extraction. All-*trans* retinal has been identified in extracts of these resensitized cells, but no 11-*cis* isomer has been isolated (Beckmann and Hegemann, 1991). Retinal also has been identified in green strain 806 cells that have been grown in low light to reach the maximal phototactic sensitivity (Fig. 11.5; Beckmann and Hegemann, 1991). The HPLC diagram of wild-type cells was distorted by large amounts of carotenoids, but

TABLE II
Properties of *Chlamydomonas* Rhodopsin

Maximal absorbance	495 nm
Extinction coefficient	50,000 M^{-1} cm^{-1}
Functional retinal isomer	All *trans*
Retinal C6–C7 bond	Planar and *trans*
Negative counterion provided by the protein	Close to C=N bond
Molecular mass	32 kDa

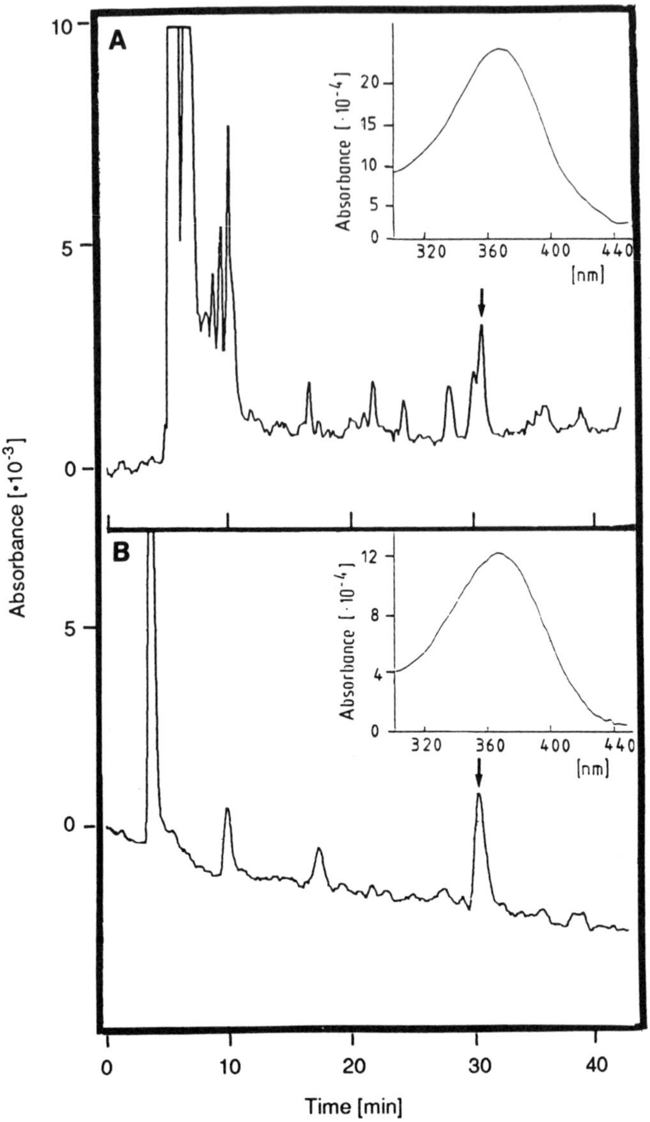

Figure 11.5 Pigment HPLC analysis from strain 806 whole cells (A) and rhodopsin-containing membranes (B). Cells or membranes were extracted with organic solvent and pigments were analyzed by polar phase HPLC. Inserts show the absorption spectra of the compounds eluting at the position of the arrows. These spectra are identical to those of all-*trans* retinal. Standard all-*trans* retinal also elutes at this position (Beckmann, 1991; Beckmann and Hegemann, 1991).

all-*trans* and small amounts of 13-*cis* retinal were detectable by on-line absorption spectroscopy. The amount of retinal in sensitized white cells as well as in green cells reached 30,000 per cell, which is in the estimated range of the number of rhodopsin molecules (Melkonian and Robenek, 1984; Smyth *et al.*, 1988; Beckmann and Hegemann, 1991). In both strains, retinol was also detectable; the amount increased with the photon irradiance to which the cells had been exposed during growth and gametogenesis. Retinol might serve as a storage form for retinoids in the cell.

C. *In Vitro* Identification of Rhodopsin

For the spectroscopic identification of rhodopsin, membranes free of carotenoids and photosynthetic material were prepared from a 20-liter cell culture (Beckmann and Hegemann, 1991). The absorption spectrum of these membranes was rhodopsin-shaped and showed a maximum at 495 nm (Table II). The absorption spectrum (Fig. 11.6A) was fine structured, which is extremely unusual for a rhodopsin. Fine-structured absorption spectra have been observed for the β-lactoglobulin–retinol complex (Fugate and Song, 1980) and for phoborhodopsin (SR II), a photoreceptor for the repellent photomovement responses to blue-green light in halobacteria) (Takahashi *et al.*, 1990). Fine-structured absorption spectra can explain the high resolution action spectra for photomovement responses of various algae (summarized by Foster *et al.*, 1984) and for the depolarization on light exposure in the marine alga *Acetabularia* (Gradmann, 1984), whereas bell-shaped absorption spectra are not consistent with these phenomena.

The rhodopsin in *Chlamydomonas* membranes can be bleached in continuous green light (500 nm). Bleaching correlates with the appearance of free retinal. The corresponding bleaching difference spectra (Fig. 11.6B) still exhibit fine structure. The rhodopsin absorption spectrum in detergent-washed membranes is also unstable in the dark and decays over a few days, even at 4°C. However, absorption can be reconstituted by addition of all-*trans* retinal. From these reconstitution experiments, the maximum extinction coefficient is determined to be near 50,000 $M^{-1}cm^{-1}$. Reconstitution with [^{3}H]retinal followed by reduction with cyanoborohydride identifies only one labeled retinal protein with a molecular mass of 32 kDa in the final membrane preparation. This protein identifies the rhodopsin that functions as the photoreceptor for photomovement responses in *Chlamydomonas* (Beckmann and Hegemann, 1991).

D. Properties of the Rhodopsin Chromophore

Fine-structured absorption spectra allow certain conclusions to be drawn about the retinylidene conformation and its protein environment that cannot

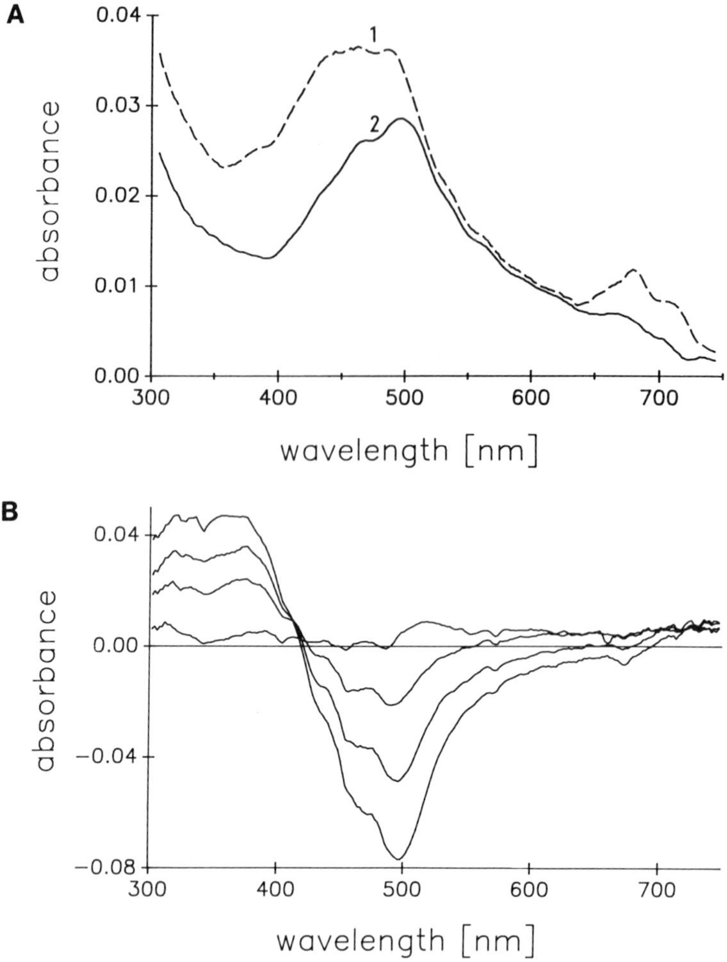

Figure 11.6 (A) Absorption spectra of crude eyespot extracts (1) and of *Chlamydomonas* rhodopsin in detergent-treated membranes (2) from strain 806 cells. (B) Absorption difference spectra of bleached minus unbleached rhodopsin-containing membranes after different times of bleaching. The absorption of free retinal appears around 360 nm (Beckmann, 1991; Beckmann and Hegemann, 1991).

be drawn from bell-shaped absorption spectra. A detailed analysis has been presented for phoborhodopsin (SR II) (Takahashi *et al.*, 1990). The primary arguments made by these investigators should be applicable also to *Chlamydomonas* rhodopsin. The major causes of spectroscopic broadening (no fine structure) in retinal and related compounds in solution are thought to result from a distribution of vibronic activity among a large number of excited-state vibrational modes or from a conformational heterogeneity of the ground state. Both effects arise primarily from a distorted *cis* conformation around the C6–C7 single bond (Warshel and Karplus, 1974). Therefore the development of fine structure in the *Chlamydomonas* rhodopsin spectrum indicates a forced coplanar conformation around the C6–C7 bond. A planar conformation is expected to be 6-s-*trans*, because a steric hindrance argues against a planar 6-s-*cis* geometry (Fig. 11.7). A planar 6-s-*trans* configuration for *Chlamydomonas* rhodopsin has been predicted already from *in vivo* incorporation and phototaxis that occurs on addition of 6–7 locked retinal analogs in white mutants (Foster *et al.*, 1987) and from stop response measurements (Zacks *et al.*, 1991). Additional support for the ring/chain coplanarity may be derived from phototaxis experiments with naphthyl retinal (3,7-dimethyl-12,4,6-heptatrienal), in which the bond corresponding to the C6–C7 bond in retinal is fixed into a planar conformation. The planar binding site may facilitate naphthyl retinal incorporation and explains the high light sensitivity of FN68 cells in the presence of this compound. However, the fast reconstitution of rhodopsin with regular all-*trans* retinal demonstrates that planarization of the retinal ring is not a rate limiting step. In visual pigments, additional distortion of the chromophore planarity beyond the 6-s-*cis* conformation arises from the steric interaction between the C13 methyl group and the C10 proton resulting from the C11=C12 *cis* retinal configuration (Warshel and Karplus, 1974; Smith *et al.*, 1987). Therefore, fine-structured absorption spectra support the concept of a retinal chromophore with all-*trans* conformation in *Chlamydomonas*. This argument lends additional support to the notion that the natural chromophore of *Chlamydomonas* is all-*trans*, as already concluded from retinal analysis and from the *in vivo* reconstitution experiments.

In a comparison of the λ_{max} value of an all-*trans* protonated Schiff base (PSB^+) in solution (440 nm) with the absorption maximum of 6–7-locked retinal, which is fixed in an all-*trans* configuration (465 nm; van der Steen *et al.*,

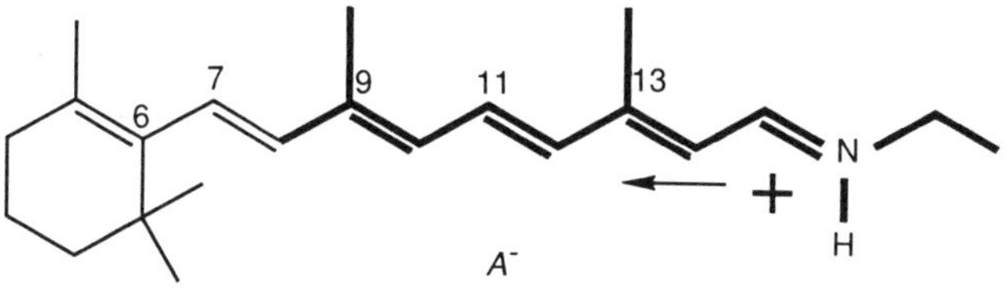

Figure 11.7 Structure of the *Chlamydomonas* rhodopsin chromophore as derived from behavioral and biochemical studies. The bold lines indicate the shortest rhodopsin chromophore that triggers phototaxis *and* stop responses in CC2359 cells (Hegemann *et al.*, 1991). A^- indicates the negative counterion provided by a negatively charged amino acid.

1986), planarization can account for a 1200–1300 cm^{-1} red shift. In *Chlamydomonas* rhodopsin, only an additional 1300 cm^{-1} remains for the shift induced by the protein environment, indicating that the negative counterion is close to the protonated Schiff base and that the interaction with other protein charges is relatively weak.

Although an all-*trans* and 6-s-*trans* conformation seems to be a prerequisite for a rigid chromophore, such a structure does not necessarily imply the existence of a fine-structured absorption spectrum. Bacteriorhodopsin (BR) meets the two initial requirements (van der Steen *et al.*, 1986) but does not have a fine-structured spectrum. For three halobacterial proteins—BR (max = 568 nm), the light-driven chloride pump halorhodopsin (HR; max = 578 nm), and sensory rhodopsin (SR; max = 588 nm)—the strong red shift (opsin shift) indicates heavy chromophore–opsin interactions. These interactions are responsible for the inhomogeneous broadening of the absorption spectra in BR, HR, and SR. Similar conclusions have been drawn from absorption measurements of organic molecules in solvents of different polarity. Solutions of the least polar solvents, with the lowest solvating power, have the least effect on spectra and are least effective in inhibiting vibrational fine structure (Jaffe and Orchin, 1962). Because the opsin shift of *Chlamydomonas* rhodopsin is much lower than that of BR, the protein–chromophore interaction is much weaker and similar to that of phoborhodopsin (λ_{max} = 487 nm; Takahashi *et al.*, 1990). The location of negative charges in *Chlamydomonas* rhodopsin has been probed by action spectroscopy using a series of dihydroretinals for the reconstitution of blind cells (Foster *et al.*, 1988b). The sharp change in the opsin shift observed between trienals and dienals or 7,8- and 9,10-dihydro-retinal (two or three C=C conjugated bonds) led to the conclusion that only one negative charge is near the C10 or C11 position of the retinal and that no charge is located near the C5 position of the retinal ring (Foster and Saranak, 1989). However, the failure of short-chained analogs to restore function in CC2359 cells precludes a clear interpretation. The location of the counterion will have to be analyzed in more detail by FTIR or resonance Raman spectroscopy.

IV. Multiple Roles of Calcium

A. Calcium Is Essential for Signal Transduction

Light-induced behavioral responses are dependent on calcium, indicating that calcium plays an important role in photoreception in *Chlamydomonas* (Stavis and Hirschberg, 1973; Schmidt and Eckert, 1976; Nultsch, 1979; Dolle *et al.*,

1986; Hegemann and Bruck, 1989). A role for the intracellular calcium concentration in the control of the beating pattern of the flagella has been shown by analyzing the beating of the isolated flagellar apparatus (Hyams and Borisy, 1978) or of permeabilized cells (Kamyia and Witman, 1984). Mechanical microinjection of Mg^{2+} and Ca^{2+} into the cytoplasm of intact cells gave rise to changes in flagellar beating (Nichols and Rickmanspoel, 1978). The role of calcium in signal transduction also has been indicated by the suppression of phototaxis and stop responses by calcium channel inhibitors under conditions of fairly constant cell motility (Nultsch *et al.*, 1986; Hegemann *et al.*, 1990). In addition to its direct role in controlling signal transduction and flagellar beating, calcium also regulates adaptation of the cells to continuous light or to repetitive flash stimulation (Hegemann and Bruck, 1989).

Direct measurements of light-induced ^{45}Ca uptake have been precluded in wild-type cells by the cell wall, but have been possible in the cell-wall defective 806 mutant (Hegemann *et al.*, 1990). Verapamil, a selective inhibitor of L-type calcium channels, completely suppresses calcium uptake to a level even lower than the dark level. ^{45}Ca measurements, however, suffer from poor time resolution; the calcium uptake during a single stop response is too small to be detected. From studies with tritiated verapamil, nimodipine, and diltiazeme, several binding sites in the plasma membrane and the flagellar membrane have been postulated (Dolle, 1988; Dolle and Nultsch, 1988).

B. Electrical Recording of Ion Channel Activation

Light-regulated ion currents have been recorded directly from Chlorophyceae by application of the suction pipette technique. These experiments were done first in the large *Chlamydomonas* relative *Haematococcus* (Litvin *et al.*, 1978) and later in *Chlamydomonas* itself (Harz and Hegemann, 1991). The photocurrents in both algal species are very similar. The minor differences can be explained by the different experimental conditions used.

Stimulation with light flashes or onset of actinic light initiates a sequence of clearly distinguishable inward currents, all of which are likely to be carried, at least to some extent, by calcium (Fig. 11.8A). The first current component appears in the eyespot region within less than 500 μsec after the flash. With increasing photon exposure of the actinic light flash, the peak amplitude increases and the current decay gradually accelerates in parallel (Fig. 11.8B). In both algal species, blue-green light near 500 nm is most effective. In *Haematococcus*, the action spectrum shows six maxima, suggesting that several photoreceptor pigments are involved. For *Chlamydomonas*, the action spectrum is clearly rhodopsin shaped with the maximum at 494 nm, a result that is very similar to the rhodopsin absorption detected in purified membrane

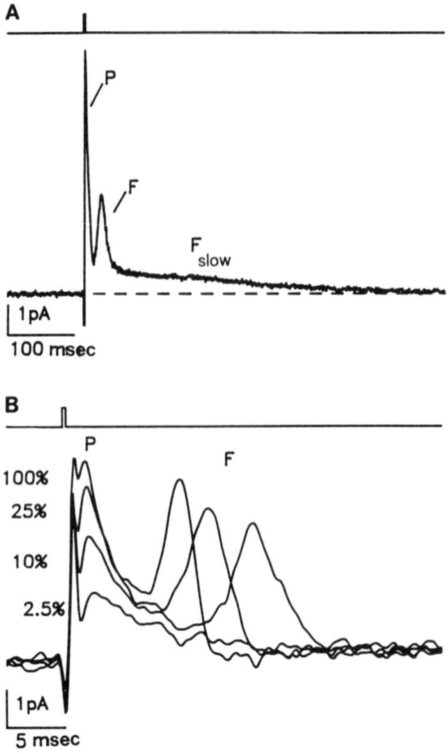

Figure 11.8 (A) Flash-induced photocurrents from a strain CW2 cell recorded under a configuration with the eyespot and flagella outside the pipette (0.3-msec flashes of 495 nm). The photoreceptor current P originates from a conductivity increase in the eyespot region, whereas the flagellar currents F and F_{slow} are localized in the flagellar region, the former presumably near the flagellar base and the latter within the flagella. (B) Photoreceptor and flagellar currents of a cell that was stimulated with flashes of various photon exposure. The top trace indicates the light stimulation. The sharp negative and positive signals appearing immediately after the flash are flash artifacts (conditions as described by Harz and Hegemann, 1991).

preparations discussed earlier. This first signal originally was termed primary potential difference (PPD) because of the rapid appearance and the measuring configuration that was used for *Haematococcus*. To express the function and location of this current, we call it the photoreceptor current (P).

When the photoreceptor current oversteps a certain threshold value, a transient inward current in the flagellar region is triggered that has been called the regenerative response (RR) or, more descriptively, the flagellar current (F) (Fig. 11.8A). The time interval between flash stimulation and the flagellar current increases when the photon exposure is lowered (Fig. 11.8B). Interestingly, the integral of the photoreceptor current between the flash and the beginning of the flagellar current changes only very little over a wide range of photon exposure. The all-or-none appearance of the flagellar current is similar to that of action potentials observed in animal neurons or the marine alga *Acetabularia* (Gradmann, 1984). The flagellar current in *Chlamydomonas* lasts only a few msec and its increase and decrease have similar kinetics.

A third flash-induced calcium current component has been detected in *Chlamydomonas* (Fig. 11.8A). This current also is localized in the flagellar region (F_{slow}), but very little is known about its nature and function. The amplitude is very small (<1pA) and decays slowly within 300–500 msec of the flash. This third current can be detected only when the two faster current components have ceased. This long-lasting current, similar to the first flagellar current component, shows an all-or-none appearance and its duration is independent of the flash length. The hypothesis is that the fast flagellar current is localized in the flagellar roots, whereas the slowly decaying component is a current into the flagella themselves. In *Haematococcus*, another long-lasting component (late potential difference; LPD) has been detected in continuous light. This slow current is localized in the eyespot region and, in contrast to the slow flagellar current, its duration is correlated strictly with the duration of the light pulse. The ion specificity of this slow photoreceptor current is also unknown (Sineshchekov, 1991).

The primary potential difference in *Haematococcus* has been recorded with high time resolution after stimulation of cells with a laser flash (Sineshchekov *et al.*, 1990). Under these conditions, the rise and the decay are at least biphasic. The smaller first component with a delay of less than 30 μsec is interpreted as a light-induced charge movement within the photoreceptor molecules and might be comparable to the early receptor potential of the visual process of higher animals (Cone 1967).

A causal connection between electrical phenomena and behavioral light responses has not been shown directly. In addition, simultaneous recording of electrical processes and flagellar beating from a single cell has not been done. Even so, the photocurrents are likely to be part of the rhodopsin-regulated signal transduction chain, based on several observations.

1. The action spectra for the photoreceptor current and for the physiological responses are very similar (shape and maxima).
2. The all-or-none appearance of the flagellar current and its delay after the flash closely resemble the all-or-none appearance and the delay of the stop response (Uhl and Hegemann, 1990b).
3. The duration of the slow flagellar current (F_{slow}) is in the same time range as the duration of the stop response.

4. All substances that suppress the photocurrents, that is, low calcium, pimozide, or verapamil, also suppress the behavioral responses.

The activation mechanism of the photoreceptor channel, the relationship between the currents, and the regulation of flagellar beating by the entering Ca^{2+} has not been determined yet.

V. Ionic Composition of the Medium Controls Cellular Sensitivity in a Complex Manner

In addition to the direct current measurements, more indirect conclusions about the ionic requirements for movement responses can be drawn from light scattering measurements that detect structural changes in the cell and alterations of the flagellar beating pattern with microsecond resolution (Uhl and Hegemann, 1990b). Using this assay, membrane-impermeable cations such as NMG (*N*-methylglucamine) can be demonstrated to shift the stimulus–response curve for flash-induced behavioral responses (direction changes and stops) to higher stimulus photon irradiances. This effect is more drastic with multivalent ions such as gentamycin. Cations that normally are membrane permeable, for example, potassium and, to a lesser extent sodium, slightly shift the stimulus–response curve to lower photon irradiance, that is sensitize the cells; at higher concentrations, they reduce the amplitude of the light response.

These experiments suggest that the influence of extracellular cations on the excitability of a *Chlamydomonas* cell is very similar to that seen in *Paramecium* (Eckert and Brehm, 1979). Any cation reduces the surface potential of the cell by screening the surface charges of the cell wall and the plasma membrane lipids. The portion of the membrane potential that drops across the lipid bilayer (transmembrane potential) is increased without changing the resting potential. This effect leads to cell desensitization because of apparent hyperpolarization. The membrane potential is carried mainly by potassium, so addition of potassium depolarizes and thereby sensitizes the cell. If the cell is depolarized below a critical level, it is no longer excitable (P. Hegemann and R. Uhl, unpublished results). These conclusions explain the old finding that potassium plays a role in phototaxis (Stavis and Hirschberg, 1973; Nultsch, 1979) and that at least 10 μM potassium must be present during cell growth to keep the cells excitable.

VI. The Signal Transduction Chain

As described earlier, much has been learned about the physical principles of the function of the *Chlamydomonas* eye. The photoreceptor in this system has been identified as a rhodopsin, both *in vivo* and *in vitro*. The detection of rhodopsin-regulated currents in *Haematococcus* and *Chlamydomonas* has increased our knowledge about light-regulated signaling in unicellular algae. However, we have not yet characterized any protein involved in this transduction pathway in molecular detail.

At low photon irradiance, when only single photons are absorbed, the change of the flagellar beating is caused by only one or two activated rhodopsin molecules, which activate the photoreceptor channel in less than 1 msec. Photoreceptor cells of vertebrates show a much slower amplification cascade that involves several biochemical steps including G protein activation, phosphodiesterase activation, cGMP degradation, and channel inactivation (Stryer, 1988). Photoreception in *Chlamydomonas* is probably much simpler and more closely related to the signaling processes of *Paramecium* or *Dictyostelium*.

In *Chlamydomonas*, G proteins have been regarded as likely signal transducing components, as in other eukaryotic systems. A putative G protein with a molecular mass of 24 kDa has been detected in eyespot preparations using antibodies against G protein α sites (Korolkov *et al.*, 1990). In addition, two GTP-binding proteins of 50 and 70 kDa can be labeled in eyespot preparations with [^{32}P]GTP (Fig. 11.9A). These two proteins also can be ADP-ribosylated in presence of pertussis toxin (Fig. 11.9B). However, no light sensitivity of GTP binding or ATP ribosylation has been found (P. Hegemann, unpublished observation). A gene encoding a G protein β subunit-like polypeptide has been identified and sequenced (Schloss, 1990), but the absence of light sensitivity of GTP labeling, GTPase activity, or ATP ribosylation precludes an implication for a role of these putative G protein subunits in photoreception. Another putative second messenger, inositol triphosphate (IP_3), has been identified in the cell body and the flagella of *C. reinhardtii* and *C. eugametos* (Irvine *et al.*, 1989). However, its putative role in photoreception has not yet been tested.

Because of the rapid ion channel activation and the small number of ions sufficient for membrane depolarization, we do not see the need for the involvement of a G protein or second messenger other than calcium in rhodopsin-regulated processes. In *Paramecium*, enormous efforts have failed to identify a G protein involved in receptor-initiated signaling (Schultz and Schade, 1989). Photoreception in *Chlamydomonas* also may not involve G proteins. G proteins or IP_x compounds might, however, participate in the much slower adaptation processes.

A 50-kDa protein that binds dihydropyridine-type calcium channel inhibitors was identified by photolabeling with the dihydropyridine derivative azidopine (Hegemann *et al.*, 1990). This protein is a candidate for a rhodopsin-regulated calcium channel in *Chlamydomonas*. However, in higher plant systems,

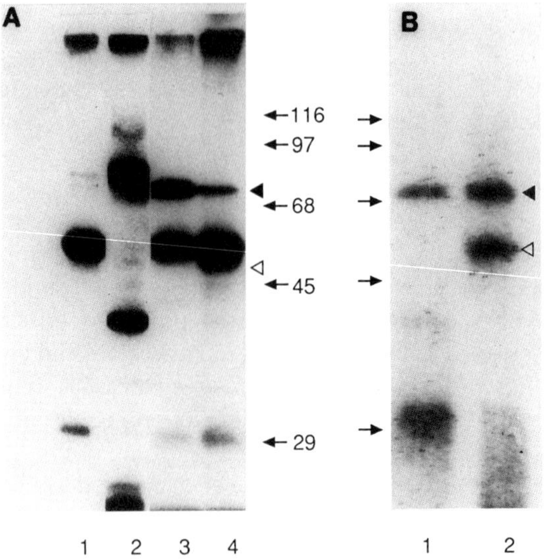

Figure 11.9 Identification of G protein candidates. (A) Identification of [^{32}P]GTP (lanes 1,3, and 4) and [^{32}P]ATP (lane 2) binding proteins in total membranes fractions (lanes 1 and 2) and eyespot membrane preparations (lanes 3 and 4) in the presence (lanes 1, 2, and 3) or absence (lane 4) of Mg^{2+} by direct photoaffinity labeling, according to Maruta and Korn (1981). No ATP binding protein was detectable in eyespot fractions. The labeling of the 70-kDa protein (◀) is enhanced by Mg^{2+} whereas that of the 50-kDa (◁) is reduced. The intensely labeled protein at 52 kDa is tubulin and is not specific for eyespots. (B) NAD-Ribosylation of total membrane fractions (lane 1) or eyespot preparations (lane 2). Membranes with a protein concentration of 50 μg were incubated for 2 hr at 30°C in the dark with 2 μg pertussis toxin and 2 μCi [^{32}P]NAD. Precipitated protein was analyzed by 11% SDS–PAGE.

dihydropyridines have been found to influence photosystem II, which makes the interpretation of inhibitor experiments problematic as long as electrophysiological data are not available.

The pathway proposed to be involved in photoreception in *Chlamydomonas*, based on experiments described in this chapter, is summarized in Fig. 11.10. On light absorption, the excited rhodopsin activates a photoreceptor calcium current via an unknown signal transduction pathway, causing one flagellum to be activated or inactivated transiently, probably directly via calcium. At high photon exposure, the photoreceptor current is followed by two flagellar currents. Under these conditions, the elevated calcium concentration leads to

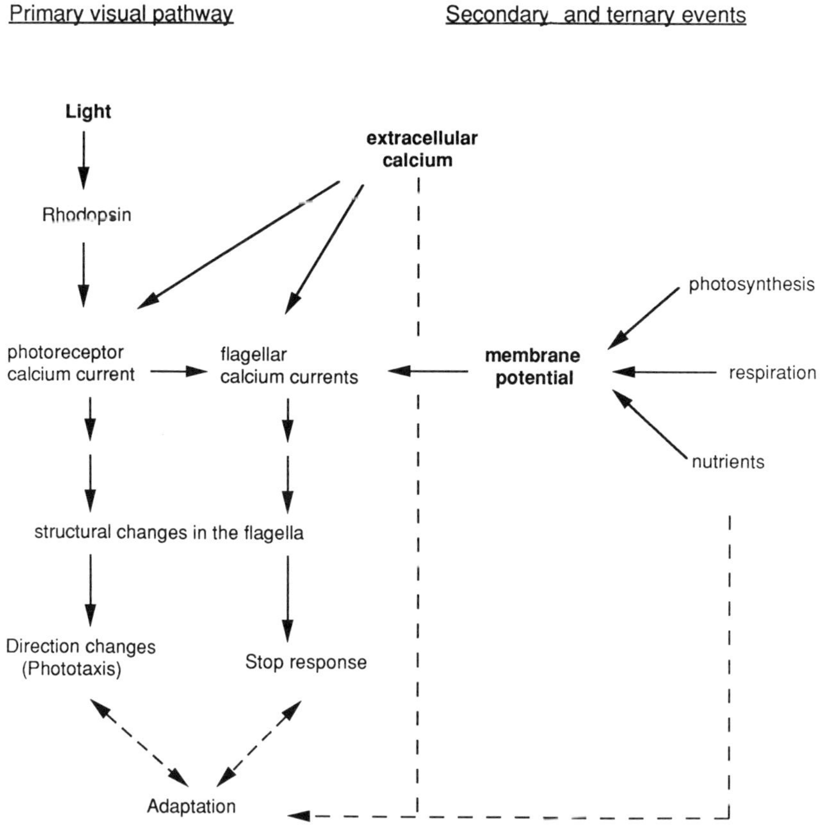

Figure 11.10 Photoreception in *Chlamydomonas*. On light absorption, the excited rhodopsin activates a photoreceptor calcium current via an unknown signal transduction pathway, causing one flagellum to be activated or inactivated transiently, probably directly by calcium. At high photon exposure, the photoreceptor current is followed by two flagellar currents. Under these conditions, the elevated calcium concentration leads to a transient backward swimming (stop response). The currents are initiated by light and depend on extracellular calcium and on the membrane potential. Photosynthesis, respiration, and nutrients indirectly influence photoreception by affecting the membrane potential or chemically affecting the adaptation process.

a transient backward swimming (stop response). The currents are initiated by light and depend on extracellular calcium and on the membrane potential. The identification of additional proteins involved in this process and the isolation of their genes should allow further elucidation of this system.

Acknowledgments

We thank J. L. Spudich for critical reading of the manuscript.

References

Beckmann, M. (1991). Isolierung und spektroskopische Charakterisierung eines Rhodopsins aus der einzelligen Alge *Chlamydomonas reinhardtii*. Dissertation. Universität München, Germany.

Beckmann, M., and Hegemann, P. (1991). *In vitro* identification of rhodopsin in the green alga *Chlamydomonas*. *Biochemistry* **30**, 3692–3697.

Bessen, M. R., Fay, B., and Witman, G. B. (1980). Calcium control of waveform in isolated flagellar axonemes of *Chlamydomonas*. *J. Cell Biol.* **86**, 446–455.

Birge, R. R. (1990). Nature of the primary photochemical events in rhodopsin and bacteriorhodopsin. *Biochem. Biophys. Acta* **1016**, 293–327.

Boscov, J. S., and Feinleib, M. E. (1979). Phototactic response of *Chlamydomonas* to flashes of light. II. Response of individual cells. *Photochem. Photobiol.* **30**, 499–505.

Buder, J. (1917). Zur Kenntnis der phototaktischen Richtungsbewegung. *Jahrb. Wiss. Bot.* **58**, 105–220.

Cone, R. A. (1967). Early receptor potential: Photoreversible charge displacement in rhodopsin. *Science* **155**, 1128–1131.

Delbrück, M., and Reichert, W. (1956). System analysis for the light growth reactions of *Phycomyces*. *In* "Cellular Mechanisms in Differentiation and Growth" pp. 3–44. Princeton University Press, Princeton, New Jersey.

Dolle, R. (1988). Isolation of the plasma membrane and binding of the Ca^{2+} agonist nimodipine in *Chlamydomonas reinhardtii*. *Physiol. Plant.* **73**, 7–14.

Dolle, R., and Nultsch, W. (1988). Specific binding of the calcium channel blocker [^{3}H]verapamil to membrane fractions of *Chlamydomonas reinhardtii*. *Arch. Microbiol.* **144**, 393–397.

Dolle, R., Pfau, J., and Nultsch, W. (1986). Role of calcium ions on motility and phototaxis on *Chlamydomonas reinhardtii*. *J. Plant Physiol.* **126**, 467–473.

Eckert, R., and Brehm, P. (1979). Ionic mechanisms of excitation in *Paramecium*. *Ann. Rev. Biophys. Bioeng.* **8**, 353–383.

Famitzins, A. (1867). Die Wirkung des Lichtes auf Algen und einige andere ihnen nahe verwandten Organismen. *Jahrb. Wiss. Bot.* **6**, 1–44.

Feinleib, M. E., and Curry, G. M. (1967). Methods for measuring phototaxis of cell populations and individual cells. *Physiol. Plant. (Kopenh.)* **20**, 1083–1095.

Feinleib, M. E., and Curry, G. M. (1979). The nature of the photoreceptor in phototaxis. *In*: "Handbook of Physiology. I. Receptor mechanisms" pp. 366–395.

Foster, K. W., and Saranak, J. (1989). The *Chlamydomonas* (Chlorophyceae) eye as a model of cellular structure, intracellular signaling and rhodopsin activation. *In* "Algae as an Experimental System" (A. W. Coleman, L. J. Goff, and J. R. Stein-Taylor, eds.), pp. 215–230. Liss, New York.

Foster, K. W., and Smyth, R. D. (1980). Light antennas in phototactic algae. *Microbiol. Rev.* **44**, 572–630.

Foster, K. W., Saranak, J., Patel, N., Zarrilli, G., Okabe, M., Kline, T., and Nakanishi, K. (1984). A rhodopsin is the functional chromophore for phototaxis in the unicellular alga *Chlamydomonas*. *Nature (London)* **311**, 756–759.

Foster, K. W., Saranak, J., van der Steen, R., and Lugtenburg, J. (1987). Retinal in *Chlamydomonas* rhodopsin is in a planar 6-*S-trans* configuration as shown by in vivo incorporation of 6–7 locked retinal analogs. *Invest. Ophthal. Vis. Sci. Suppl.* S253.

Foster, K. W., Saranak, J., and Zarrilli, G. R. (1988a). Autoregulation of rhodopsin synthesis in *Chlamydomonas reinhardtii. Proc. Natl. Acad. Sci. U.S.A.* **85**, 6379–6883.

Foster, K. W., Saranak, J., Derguini, F., Rao, J., Zarrilli, G. R., Okabe, M., Fang, J.-M., Shimizu, N., and Nakanishi, K. (1988b). Rhodopsin activation: A novel view suggested by *in vivo Chlamydomonas* experiments. *J. Am. Chem. Soc.* **110**, 6588–6589.

Foster, K. W., Saranak, J., Derguini, F., Zarrilli, G., Johnson, R., Okabe, M., and Nakanishi, K. (1989). Activation of *Chlamydomonas* rhodopsin in vivo does not require isomerisation of retinal. *Biochemistry* **28**, 819–824.

Fugate, R. D., and Song, P.-S. (1980). Spectroscopic characterization of β-lactoglobulin-retinal complex. *Biochim. Biophys. Acta* **625**, 28–42.

Gradmann, D. (1984). Electrogenic Cl^- pump in the marine alga Acetabularia. *In* "Chloride Transport Coupling in Biological Membranes and Epithelia" (G. A. Gerencser, ed.), pp. 14–61. Elsevier, Amsterdam.

Häder, D.-P., and Lebert, M. (1985). Real time computer controlled tracking of motile microorganisms. *Photochem. Photobiol.* **42**, 509–514.

Harz, H., and Hegemann, P. (1991). Rhodopsin-regulated ion channel activation in *Chlamydomonas. Nature (London)* **351**, 489–491.

Hegemann, P., and Bruck, B. (1989). The light-induced stop response in *Chlamydomonas*. Occurrence and adaptation phenomena. *Cell Motil.* **14**, 501–515.

Hegemann, P., and Marwan, P. (1988). Single photons are sufficient to trigger movement responses in *Chlamydomonas reinhardtii. Photochem. Photobiol.* **48**, 99–106.

Hegemann, P., and Musgrave, A. (1991). Phototaxis of *Chlamydomonas eugametos* pairs is controlled by the mt^+ partner. *J. Plant Physiol.* **138**, 285–291.

Hegemann, P., Hegemann, U., Foster, K. W. (1988). Reversible bleaching of *Chlamydomonas reinhardtii* rhodopsin *in vivo. Photochem. Photobiol.* **48**, 123–128.

Hegemann, P., Neumeier, K., Hegemann, U., and Kuehnle, E. (1990). The role of calcium in *Chlamydomonas* movement responses as analyzed by calcium channel inhibitors. *Photochem. Photobiol.* **52**, 575–583.

Hegemann, P., Gärtner, W., and Uhl, R. (1991). All-*trans* retinal constitutes the functional chromophore in *Chlamydomonas* rhodopsin. *Biophys. J.* **60**, 1477–1489.

Hyams, J. S., and Borisy, G. G. (1978). Isolated flagella apparatus of *Chlamydomonas*: Characterization of forward swimming and alteration of waveform and reversal of motion by calcium ions in vitro. *J. Cell Sci.* **33**, 235–253.

Irvine, R. F., Letcher, A. J., Lander, D. J., Drøbak, B. K., Dawson, A. P., and Musgrave, A. (1989). Phosphatidylinositol(4,5)bisphosphate and phosphatidylinositol(4)phosphate in plant tissues. *Plant. Physiol.* **89**, 888–892.

Jaffe, H. H., and Orchin, M. (1962). "Theory and Application of Ultraviolet Spectroscopy." Wiley, New York.

Kamiya, R., and Witman, G. (1984). Submicromolar levels of calcium control the balance of beating between the two flagella in demembranated models of *Chlamydomonas. J. Cell Biol.* **98**, 97–107.

Kindle, K. (1990). High-frequency nuclear transformation of *Chlamydomonas reinhardtii. Proc. Natl. Acad. Sci. U.S.A.* **87**, 1228–1232.

Korolkov, S. N., Garnovskaya, M. N., Basov, A. S., Chunaev, A. S., and Dumler, I. L. (1990). The detection and characterization of G-proteins in the eyespot of *Chlamydomonas reinhardtii. FEBS Lett.* **270**, 132–134.

Kreimer, G., and Melkonian, M. (1990). Reflection confocal laser scanning microscopy of eyespots in flagellated green algae. *Eur. J. Cell Biol.* **53**, 101–111.

Krinski, N. I., and Litvin, R. P. (1964). Carotenoids of wild type and mutant strains of the green alga *Chlamydomonas reinhardtii. Plant Physiol.* **48**, 680–687.

Lawson, M., Derguini, F., Nakanishi, K., and Spudich, J. L. (1990). Computerised cell tracking and motion analysis of photophobic responses of *Chlamydomonas reinhardtii*. Paper presented at the International Conference of Cell Molecular Biology of *Chlamydomonas*. Delavan, Wisconsin.

Litvin, F. F., Sineshchekov, O. A., and Sineshchekov, V. A. (1978). Photoreceptor electric potential in the phototaxis in the alga *Haematococcus pluvialis. Nature (London)* **271**, 476–477.

Martinac, B., Saimi, Y., Gustin, M. C., Culbertson, M. R., Adler, J., and Kung, C. (1988). Ion channels in microbes. *Period. Biol.* **90**, 375–384.

Maruta, H., and Korn, A. A. (1981). Direct photoaffinity labeling by nucleotides of the apparent catalytic site on the heavy chain of the smooth muscle and *Acanthamoeba* myosins. *J. Biol. Chem.* **256**, 499–502.

Mast, S. O. (1927). Structure and function of the eye-spot in unicellular and colonial organisms. *Arch. Protistenkd.* **60**, 197–220.

Mayer, A. M. (1968). *Chlamydomonas*: Adaptation phenomena in phototaxis. *Nature (London)* **217**, 875–876.

Melkonian, M., and Robenek, H. (1984). The eyespot apparatus of flagellated algae: A critical review. *Progr. Phycol. Res.* **3**, 193–268.

Morel-Laurence, N. (1987). Calcium control of phototactic orientation in *Chlamydomonas reinhardtii*: Sign and strength of response. *Photochem. Photobiol.* **45**, 119–128.

Morel-Laurence, N., and Feinleib, M. E. (1983). Photomovement in an "eyeless" mutant of *Chlamydomonas. Photochem. Photobiol.* **37**, 189–194.

Musgrave, E., and Häder, D.-P. (1991). Phototactic responses of *Chlamydomonas eugametos* gametes before and after fusion. *Gen. Physiol. Biophys.* **10**, 189–199.

Nakanishi, K., Derguini, F., Rao, J., Zarrilli, G., Okabe, M., Lien, T., Johnson, R., Foster, K. W., and Saranak, J. (1989). Theory of rhodopsin activation: Probable charge redistribution of excited state chromophore. *Pure Appl. Chem.* **61**, 361–364.

Nichols, K. M., and Rikmenspoel, R. (1978). Control of flagellar motion in *Chlamydomonas* and *Euglena* by mechanical microinjection of Mg^{2+} and Ca^{2+} and by electrical current injection. *Cell Sci.* **29**, 233–247.

Nultsch, W. (1977). Effect of external factors on phototaxis of *Chlamydomonas reinhardtii*. Carbon dioxide, oxygen and pH. *Arch. Microbiol.* **112**, 179–185.

Nultsch, W. (1979). Effect of external factors on phototaxis of *Chlamydomonas reinhardtii*. III. Cations. *Arch. Microbiol.* **123**, 93–99.

Nultsch, W., and Häder, D.-P. (1988). Photomovement in motile microorganisms. II. *Photochem. Photobiol.* **47**, 837–869.

Nultsch, W., Throm, G., and von Rimscha, I. (1971). Phototaktische Untersuchungen an *Chlamydomonas reinhardtii* Dangeard in homokontinuierlicher Kultur. *Arch. Microbiol.* **80**, 351–369.

Nultsch, W., Pfau, J., and Dolle, R. (1986). Effect of calcium channel blockers on phototaxis and motility of *Chlamydomonas reinhardtii. Arch. Microbiol.* **144**, 393–397.

Oesterhelt, D., Hegemann, P., Tavan, P., and Schulten, K. (1986). *Trans–cis* isomerisation of retinal and a mechanism for ion translocation in halorhodopsin. *Eur. J. Biophys.* **14**, 123–129.

Pfeffer, W. (1904). "Pflanzenphysiologie." Leipzig.

Rüffer, U., and Nultsch, W. (1985). High speed cinematographic analysis of the movement of *Chlamydomonas. Cell Motil.* **5**, 251–263.

Rüffer, U., and Nultsch, W. (1987). Comparison of the beating of *cis* and *trans* flagella of *Chlamydomonas* cells held on micropipettes. *Cell. Motil.* **7**, 87–93.

Rüffer, U., and Nultsch, W. (1990). Flagellar photoresponses of *Chlamydomonas* cells held on micropipettes: I. Change in flagellar beat frequency. *Cell Motil.* **15**, 162–167.

Schloss, J. A. (1990). A *Chlamydomonas* gene encodes a G protein β subunit-like polypeptide. *Mol. Gen. Genet.* **221**, 443–452.

Schmidt, J. A., and Eckert, R. (1976). Calcium couples flagellar reversal to photostimulation in *Chlamydomonas reinhardtii. Nature (London)* **262**, 713–715.

Schultz, J., and Schade, M. (1989). Calcium channel activation and inactivation in *Paramecium* biochemically measured by cyclic GMP production. *J. Membrane Biol.* **109**, 251–267.

Sineshchekov, O. A. (1991). "Photoreception in unicellular flagellates: Bioelectric phenomena in phototaxis. *In* "Light in Biology and Medicine" (R. H. Douglas, ed.), Vol. II, pp. 523–532. Plenum Press, New York.

Sineshchekov, O. A., Litvin, F. F., and Keszthelyi, L. (1990). Two components of photoreceptor potential in phototaxis of the flagellated green alga *Haematococcus pluvialis. Biophys. J.* **57**, 33–39.

Smith, S. O., Paling, I., Copie, V., Raleigh, D. P., Courtin, J., Pardoen, J. A., Lugtenburg, J., Mathies, R. A., and Griffin, R. G. (1987). Low temperature solid state ^{13}C-NMR studies of retinal chromophore in rhodopsin. *Biochemistry* **26**, 1606–1611.

Smyth, R. D., and Berg, H. C. (1982). Change in flagellar beat frequency of *Chlamydomonas* in response to light. *Cell Motil. Suppl.* **1**, 211–215.

Smyth, R. D., Saranak, J., and Foster, K. W. (1988). Algal visual systems and their photoreceptor pigments. *Progr. Physiol. Res.* **6**, 254–286.

Stavis, R. L., and Hirschberg, R. (1973). Phototaxis in *Chlamydomonas reinhardtii. J. Cell Biol.* **59**, 367–377.

Staehlin, L. A. (1986). *In* "Enzyclopedia of Plant Physiology, 19, Photosynthesis III." Springer, Berlin.

Stryer, L. (1988). Molecular basis of visual excitation. *Cold Spring Harbor Symp. Quant. Biol.* **53**, 283–294.

Takahashi, T., Yan, B., Mazur, P., Derguini, F., Nakanishi, K., and Spudich, J. L. (1990). Color regulation in the archebacterial phototaxis receptor phoborhodopsin (sensory rhodopsin II). *Biochemistry* **29**, 8467–8474.

Trevianus, L. C. (1817). Fernere Beobachtungen über die Bewegung der grünen Materie im Pflanzenreiche. *In* "Vermischte Schriften Anatomischen und Physiologischen Inhalts" (G. R. Trevianus und L. C. Trevianus, eds.), Vol. II, pp. 72–92. Heyse, Bremen.

Uhl, R., and Hegemann, P. (1990a). Adaptation of *Chlamydomonas* phototaxis: I. A light scattering apparatus for measuring the phototactic rate of microorganisms with high time resolution. *Cell Motil.* **15**, 230–244.

Uhl, R., and Hegemann, P. (1990b). Probing visual transduction in a plant cell. Optical recording of rhodopsin-induced structural changes from *Chlamydomonas. Biophys. J.* **58**, 1295–1302.

Ulehla, V. (1911). Ultramikroskopische Studien über Geisselbewegung. *Biol. Zbt.* **25**, 689–705.

van der Steen, R., Biesheuvel, P. L., Mathies, R. A., and Lugtenburg, J. (1986). Retinal analogs with locked 6–7 conformation show that bacteriorhodopsin requires the 6-*s-trans* conformation. *J. Am. Chem. Soc.* **108**, 6410–6411.

Warshel, A., and Karplus, M. (1974). Calculation of ° excited state conformations and vibronic structure of retinal and related molecules. *J. Am. Chem. Soc.* **96**, 5677–5689.

Yan, B., Takahashi, T., Johnson, R., Derguini, F., Nakanishi, K., and Spudich, J. (1990). All-*trans*/13-*cis* isomerisation of retinal is required for phototaxis signaling by sensory rhodopsins in *Halobacterium halobium. Biophys. J.* **57**, 807–814.

Zacks, D., Lawson, M., Derguini, F., Nakanishi, K., and Spudich, J. L. (1991). *Chlamydomonas* retinal dependent photoreception: Evidence the native chromophore polyene configuration for the stop response is all-*trans. Biophys. J.* **59**, 329a.

Chapter 12

Paramecium Chemosensory Transduction

JUDITH VAN HOUTEN
Department of Zoology
University of Vermont
Burlington, Vermont 05405

I. Introduction

It was observed long ago that paramecia can be attracted to or repelled from chemical stimuli (Jennings, 1906; see Van Houten and Preston, 1987, for review). *Paramecium* swimming is observed easily, so the fact that stimuli affect swimming behavior of individual cells to bring about the population response was quickly obvious. However, the underlying signal transduction pathways that were set in motion by chemical stimuli were poorly understood until the right combination of electrophysiology, genetics, biochemistry, and behavioral analysis became available. Although many puzzles about these pathways remain unsolved, chemical sensing in *Paramecium* clearly shares common elements with olfaction, taste, and other sensory transduction and may have some novel aspects as well (Van Houten, 1990,1991,1992). This chapter describes our current knowledge about chemosensory signal transduction in *Paramecium*.

II. Stimuli

In general, attractant stimuli seem to indicate that bacteria (*Paramecium* food) are available and repellents signify nonoptimal ionic and pH conditions (Van Houten, 1978; Van Houten *et al.*, 1981; Van Houten and Preston, 1988). Distinct species differences exist in the stimuli recognized: For example, *Paramecium tetraurelia* is attracted to acetate, folate, lactate, glutamate, and other bacterial products but is not responsive to alcohols, which constitute a set of bacterial fermentation products that repel the species *Paramecium caudatum* (Dryl, 1959; Van Houten, 1978; Preston and Usherwood, 1988b). Both species are repelled by high salts and extremes of pH, but the pH optima differ (Van Houten and Preston, 1988).

Paramecia appear to release neither mating pheromones from mating competent cells (Van Houten *et al.*, 1981) nor chemical signals that inform the predator *Didinium* about its prey. *Didinium* finds paramecia to devour by seeking out the same bacteria that produce chemical stimuli and serve as food for the paramecia (Antipa *et al.*, 1983).

Most attractant stimuli are organic compounds that are effective at neutral pH, whereas repellents include extremes of pH, high ionic strength, and "membrane active" compounds that elicit action potentials (Van Houten and Preston, 1988). Therefore, the sensory transduction systems for attractants and repellents can be expected to differ. Chemosensory receptors are likely to have evolved as part of chemosensory transduction for organic attractants, whereas repellent stimuli generally do not appear to require cell surface receptors (Van Houten and Preston, 1988). Most of the following discussion centers on the receptor-mediated sensory transduction pathway of folate, cAMP, and similar attractant compounds.

III. Chemosensory Behavior

The end point of chemosensory transduction in *Paramecium* is the accumulation or dispersal of a population of cells. Since paramecia are holotrichous ciliates, that is, covered with cilia, the alteration in ciliary motility in response to stimuli underlies this change in population behavior. To understand this connection, we must delve into *Paramecium* physiology.

Cilia beat toward the posterior of the cell, driving the cell forward through its natural habitat of pond water or through laboratory media and buffers. Small changes in resting membrane potential cause the cilia to change their beat: faster and more directly toward the posterior with hyperpolarization or more slowly and more toward the side with depolarization (for reviews, see

Machemer, 1989; Preston, 1990). These changes result in speeding up and slowing down the cell, respectively. Sufficiently large depolarizations trigger a calcium action potential, which transiently increases intraciliary calcium above 10^{-6} *M* and briefly changes the timing of the power stroke of the cilia (Eckert, 1972). The result is that, for each action potential, the cell makes a jerky little turn in swimming path. The new direction of swimming is almost random.

Chemical stimuli of *P. tetraurelia* affect swimming paths of the cells. In general, attractants induce smooth, slightly faster swimming than that in controls and repellents induce the converse (Van Houten, 1978; Van Houten and Preston, 1988). The cells are attracted or repelled by an indirect means, a kinesis or change in turning frequency that creates a biased random walk. Similar to bacteria, the cells accumulate or disperse by biasing turning frequency but, unlike bacteria, paramecia modulate both turning and speed to achieve a smooth long path in attractants and a jerky short path in repellents. Adaptation is a necessary ingredient in this kinetic mechanism (Schnitzer *et al.*, 1990). Paramecia do, indeed, adapt to unchanging chemical environments (Van Houten and Van Houten, 1982; Van Houten *et al.*, 1982). The biochemical basis of this adaptation is not yet understood.

The smooth long mean free paths of cells in attractants such as acetate are characteristic of hyperpolarization of the cells; the jerky slow paths of repellents are characteristic of depolarization. Indeed, when electrodes are used to record the membrane potential of cells, attractants generally hyperpolarize the cells and repellents depolarize them (Van Houten, 1979). However, on a more extensive inventory of stimuli, repellents are found that appear to make the cells swim very quickly and extremely smoothly; attractants (such as high $BaCl_2$, which probably does not occur in the natural habitat) are found that cause such slow and jerky motions that the cells are trapped in place and thus are attracted to the stimulus (Van Houten, 1978; Van Houten and Preston, 1988). These behaviors are in keeping with extreme hyperpolarization and depolarization, respectively, which has been confirmed by direct electrical recordings (Van Houten, 1979; Van Houten and Preston, 1988). Therefore, based on behavioral criteria, two classes of attractants and two of repellents exist for *Paramecium*. Attractants I hyperpolarize slightly and cause slightly increased mean free paths; attractants II depolarize strongly and cause such short mean free paths that the cells are trapped. Repellents I slightly depolarize, causing slightly shorter mean free paths than in controls, repellents II cause extreme hyperpolarization and such fast smooth swimming that cells streak out of a region and thus are repelled. All this seemingly complex behavior is, nonetheless, in keeping with *Paramecium* physiology.

With hindsight, the physiology underlying these behaviors seems obvious, but mutants were required to clarify that, indeed, two mechanisms of attraction and repulsion existed, one that relied most significantly on turning frequency and another that relied on speed modulation (Van Houten, 1978). The mutant Pawn, named for the chess piece, can go only forward because of the absence of a functional calcium channel and the calcium action potential that initiates turns (Kung *et al.*, 1975; Kung and Saimi, 1982; Saimi and Kung, 1987). Pawns are not attracted or repelled by stimuli such as acetate or folate that require turning in the response, but pawns can respond normally to other stimuli. Another mutant, fast-2, is pawnlike in Na^+ solutions but normal in K^+(Kung *et al.*, 1975; Kung and Saimi, 1982; Saimi and Kung, 1987). In Na^+ solutions, these cells are not attracted to stimuli such as acetate that require turning frequency control, but they show normal attraction to K^+ salts of these same stimuli.

An important mutant, d4-530, is repelled by acetate under conditions that attract wild-type (Van Houten, 1977). Whereas normal cells respond to acetate with smooth swimming, these mutants are repelled by smooth fast swimming to an extreme. These mutants provided the first hint that a second mechanism of chemoresponse exists, and that paramecia differ from the bacterial model that relies solely on turning modulation (Schnitzer *et al.*, 1990).

An effective assay of attraction and repulsion is a simple T-maze assay in which cells distribute between two arms of a three-way glass stopcock (Van Houten *et al.*, 1982; Fig. 12.1). Cells are loaded into a two-way plug through the third arm of the stopcock. The plug is then turned to connect the two arms with test and control solutions that are balanced for inorganic salts and pH to measure the responses of the cells to acetate (OAc^-) and Cl^-, for example. Once the plug is turned to connect the two arms of the T, the cells enter an arm, then leave and reenter many times. When the cells come to a pseudo-steady state, the number of cells in each arm is counted to generate an index of chemokinesis. [I_{che} = (number of cells in test arm/total number of cells in test and control arms); $I_{che} > 0.5$ indicates attraction; $I_{che} < 0.5$ indicates repulsion.] The number of cells in an arm is proportional to the time spent there;

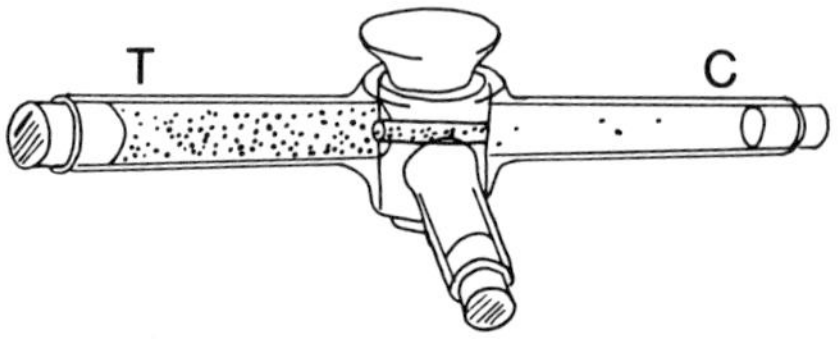

Figure 12.1 The T maze used to assay chemoresponse. Cells in control buffer are loaded into the two-way plug in the three-way stopcock. The plug is turned to connect the two arms of the T containing buffer with attractant or control buffer containing a comparable amount of salt. (For example, K acetate in buffer is tested against KCl in buffer with monovalent ions and pH kept constant.) After 30 min, the plug is turned, arms emptied, and cells counted to calculate the index of chemokinesis as a measure of attraction or repulsion. The openings in the arms are 4 mm, sufficiently large to allow turning and free access and egress of the cells. T, test arm; C, control arm.

accumulation is not caused by the oriented motility of a subset of cells or by trapping cells in an arm of the T that they first entered. For example, if cells are put into the stopcock plug between two buffers, one with attractant NH_4Cl and the other with the control salt NaCl, more than 85% of the cells will be found in the arm with NH_4Cl after 5 min. If these cells in attractant are removed, the arm refilled with fresh attractant solution, and the cells left in the NaCl arm allowed to redistribute, 85% of these cells will be found in the arm with NH_4Cl (Van Houten *et al.*, 1982). This process can be repeated over and over as the cells enter and reenter arms.

The distribution in the T maze is a result of the changes in motility that the cells experience, most likely right at the boundary between the arm and the plug. In the rest of the arm, where little gradient of attractant or repellent exists, the cells adapt to the relatively unchanging chemical environment. Without this adaptative return of swimming behavior to a basal level, the cells would not accumulate or disperse by these kinetic mechanisms (Fraenkel and Gunn, 1961; Van Houten and Van Houten, 1982; Van Houten *et al.*, 1982; Schnitzer *et al.*, 1990).

IV. Signal Transduction Underlying Chemokinesis

A. Receptors

Abundant evidence suggests the presence of membrane receptors for attractants in *Paramecium* chemoresponse. First, single-locus mutants have lost one specific chemoresponse while retaining all others tested and also have lost specific binding of ^3H-labeled ligand (DiNallo *et al.*, 1982; Schulz *et al.*, 1984; Smith *et al.*, 1987). Second, competition experiments in which cells detect attractants in the saturating and uniform distribution of other attractants (e.g., a gradient of acetate in the presence of lactate) imply a specificity of response to the stimulus that could be at the cell-surface level (Van Houten, 1978). Saturating amounts of lactate and acetate are additive in the hyperpolarizations that they generate, also implying separate sensory transduction pathways, at least at the receptor level (J. Van Houten, unpublished observations). Third, the covalent linkage of ligand to its receptor should render the receptor useless in detection of gradients of attractants. Exposure of cells to *N*-hydroxylsuccimide "activated" folate covalently and specifically links folate to the cell-surface sites (presumably including receptors). As a result, the cells cannot respond to folate, but other responses remain intact (Sasner and Van Houten, 1989). Similarly, cells labeled with the photoaffinity probe N_3-cAMP are specifically unable to detect and respond to cAMP (Van Houten *et al.*, 1991). Therefore, the simplest interpretation is that specific cell-surface receptors mediate chemoreception. Additionally, only a small percentage of receptors need to be inactivated to compromise the ability of the cell to detect a stimulus. This effect may underlie the success in the use of antibodies to disrupt chemoresponse to cAMP (see subsequent text) and the slight rightward shift in dose–response curves for the electrophysiology and behavioral responses relative to the binding curves for folate and cAMP (Schulz *et al.*, 1984; Preston and Van Houten, 1987a; Smith *et al.*, 1987). Fourth, ^{3}H-labeled ligand binding studies provide measures of the number and affinity of potential receptor sites on the *Paramecium* cell surface. The affinites are low (K_d = 29 μM for K_2 folate, 250 μM for cAMP; Schulz *et al.*, 1984; Smith *et al.*, 1987), but are in keeping with the half-maximal concentrations for behavioral response and hyperpolarization. Binding sites per cell are numerous, but no more dense than on *Dictyostelium*, for example. Binding of ^{3}H-labeled ligands is specific and reflects the specificity shown in the behavioral assays.

Interestingly, for folate and cAMP almost all (~99%) the binding sites are on the cell body rather than on the cilia, which are covered with approximately 50% of the cell-surface membrane (Dunlap, 1977). The cell body binding sites demonstrated by ^{3}H-labeled ligand binding is supported by electrical recording from deciliated cells. Cells can be depleted of most cilia relatively gently with only some oral cilia left behind (Fig. 12.2). Attractant-induced hyperpolarizations clearly are intact in deciliated cells (Preston and Van Houten, 1987a) and do not require the regeneration of cilia to be recovered, whereas voltage gated Ca^{2+} channel activity does require regeneration of cilia (Ogura and Takahashi, 1976; Dunlap, 1977). Therefore, the sensory transduction pathway for most attractants from receptor to membrane potential are found within the cell body and do not require the ciliary compartment.

The chemosensory pathway for glutamate may prove to be an interesting exception to this trend. [^{3}H]Glutamate binds specifically to cilia (Preston and Usherwood, 1988a). Additionally, the sensory transduction pathway may include a K^+ channel conductance (Preston and Usherwood, 1988b), which is distinct from the sensory transduction pathway for other attractants such as acetate, folate, and cAMP, as discussed in a subsequent section.

Electrophysiological studies demonstrate a gradient of responsiveness to folate from anterior to posterior along the ventral side of the cell (Preston and Van Houten, 1987b). The leading edge of the cell is the first to encounter a change in chemical milieu and, therefore, the cell would benefit from concentrating receptors in the anterior to facilitate a rapid response to changing chemical stimuli. *Paramecium* does not require an even distribution of receptors along its surface because it does not seem to evaluate chemical stimuli by comparing occupancy of receptors over the length of the cell. Like bacteria, paramecia are capable of responding to temporal shifts in which the cells are

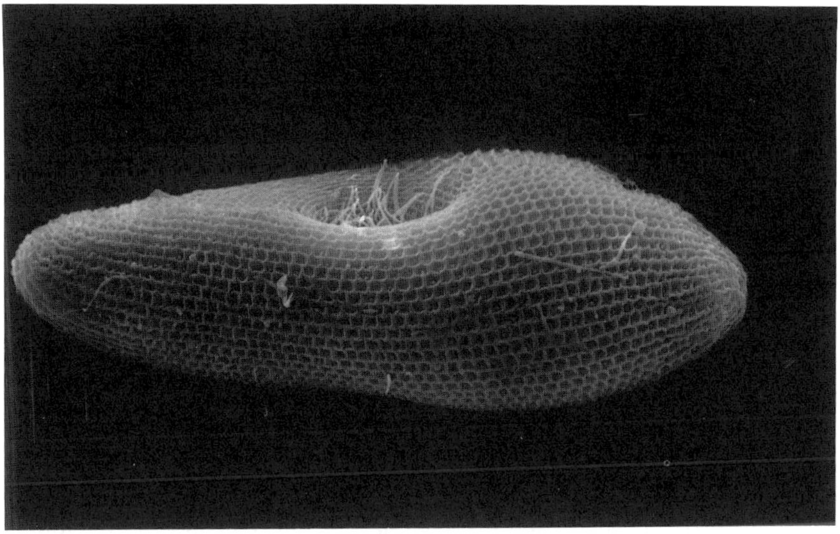

Figure 12.2 *Paramecium tetraurelia* deciliated with ethanol treatment. Such cells show normal hyperpolarization when exposed to attractant stimuli such as acetate, folate, or cAMP. The voltage-gated calcium gate function is removed with the cilia.

moved from one uniform milieu lacking a stimulus to another with stimulus. A physical gradient of stimuli resulting in a gradient of occupied receptors along the length of the cell is not necessary for a response (Van Houten, 1978). Therefore, from these observations, one would predict a gradient of receptors or other sensory transduction components from anterior to posterior, reminiscent of the gradient of depolarizing mechanoreceptors (Ogura and Machemer, 1980).

Encouraged by indirect evidence suggesting the existence of cell-surface receptors for attractants such as acetate, folate, and cAMP, we solubilized proteins of the complex cell-surface membranes (pellicles or cortex; Keryer *et al.*, 1990) in Triton X-100 and applied them to folate and cAMP affinity columns. Although the folate binding proteins have been narrowed down to five candidates, none of these has been identified definitively as the folate receptor (Schulz, 1987). However, only one major doublet of cAMP binding protein of 48 kDa can be purified partially with cAMP–agarose (Van Houten *et al.*, 1991). The doublet can be eluted from the column with cAMP and AMP (a partial agonist) and slightly with KOAc (which slightly interferes with attraction) but not with cGMP. The proteins of the doublet are inseparable by reverse phase HPLC, but can be resolved by two-dimensional electrophoresis (pI 6.6 and 9.1; A. Freshler, B. Cote, and J. Van Houten, unpublished results). Both are glycoproteins. CNBr cleavage shows that they share all peptides except one

(Fig. 12.3). The N-terminal sequences of these proteins differ; presumably, the peptide that is not shared is from the N terminus (B. Cote and J. Van Houten, unpublished results). Perhaps the most persuasive evidence that this doublet corresponds to the cAMP receptor is provided by the use of blocking antibodies. Polyclonal antibodies made against the doublet render cells specifically unable to respond to cAMP in T mazes while responses to other attractants are intact (Van Houten *et al.*, 1991). Currently, a battery of monoclonal antibodies is being evaluated for an ability to distinguish between the two bands of the doublet and, subsequently, to block chemoresponse.

Mutants with specific defects in folate and cAMP binding are of no practical assistance in identifying the cAMP or folate receptors because each is conditional. Null mutants would be most useful in detecting differences in protein profiles between wild-type and mutants. Why all these behavioral mutants are conditional is not clear, but this characteristic apparently is not uncommon among *Paramecium* behavioral mutants in general (Kung and Saimi, 1982; Saimi and Kung, 1987; Preston, 1990).

The cAMP receptor of 48 kDa is not the regulatory subunit of the protein kinase A (Hochstrasser and Nelson, 1989; Van Houten *et al.*, 1991), a binding protein from an internal membrane (Van Houten, 1990), or a fragment of the immobilization antigen (Van Houten, 1990; Van Houten *et al.*, 1991). The *Dictyostelium* cAMP receptor doublet (40–43 kDa) does not share an epitope recognized by monoclonal antibodies (Klein *et al.*, 1988; Devreotes, 1989; Van Houten *et al.*, 1991) and primers from the *Dictyostelium* sequence do not suc-

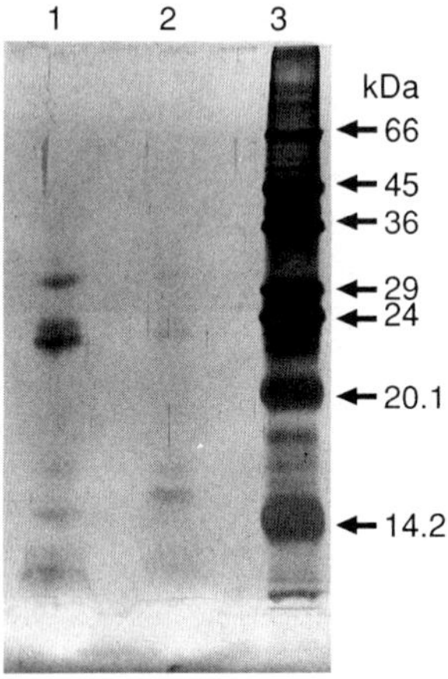

Figure 12.3 Peptides from the upper and lower bands of the cAMP receptor doublet. The peptides were generated by CNBr cleavage (Matsudaira, 1989).

cessfully amplify genomic sequences by polymerase chain reaction (PCR) or detect *Paramecium* sequences on Southern blots (I. M. Schlichtherle and J. Van Houten, unpublished results). Determining the structure of the *Paramecium* receptor and the relationship to *Dictyostelium* awaits the cloning of the *Paramecium* gene.

B. Role of Membrane Potential

The characteristic hyperpolarizations associated with attractant stimuli such as acetate and folate are sufficient to trigger the remainder of the sensory transduction pathway. This conclusion can be demonstrated by manipulating the membrane potential directly through external cations in the absence of the usual organic stimuli. The membrane potential of the cell changes as a function of external K^+ or Na^+ in the bathing buffer. Therefore, using pairs of salt buffers in which cells have known membrane potentials for the test and control arms of the T maze, the relative hyperpolarization or depolarization that the cells experience when they swim between arms of the T maze with attractant or repellent were simulated. The receptors thus could be bypassed and the sensory transduction pathway intersected at the level of membrane potential. The responses in the T maze were entirely predictable based on manipulated changes in membrane potential alone (Schulz *et al.*, 1983). Therefore, the attractant-induced hyperpolarization is sufficient to communicate with the axoneme to produce the recorded changes in ciliary beating and the altered behavior of the cell.

How does membrane potential control the ciliary beating pattern? The microtubule doublets of the axoneme slide relative to each other to generate the ciliary power stroke; the timing of the sliding determines the direction of the power stroke and the beat frequency (Satir, 1985). Membrane potential affects this timing (Machemer, 1989), at least partially through the control of internal levels of cyclic nucleotides (Hennessey *et al.*, 1985; Bonini *et al.*, 1986; Bonini and Nelson, 1988). For example, a hyperpolarization caused by a decrease in external K^+ results in faster beating and in increased levels of cAMP (Schultz *et al.*, 1984; Nakaoka and Ooi, 1985; Bonini *et al.*, 1986). However, the effects of this change in membrane potential cannot be attributed completely to the cyclic nucleotide levels (Bonini *et al.*, 1986). The nucleotide-dependent protein kinases of the cilia (Hochstrasser and Nelson, 1989; Mason and Nelson, 1989) and their substrates are being studied (Bonini and Nelson, 1990; Hamasaki *et al.*, 1991). The study of ciliary beat control is now entering a new phase of biochemistry. However, the actual control mechanisms still remain to be elucidated.

Chemical stimuli affect V_m, and attractants hyperpolarize cells. Although a hyperpolarization from a decrease in K_o clearly generates cAMP, the hyperpolarization from attractant stimuli does not alter cAMP levels (M. V. Wright and J. Van Houten, unpublished results). The hyperpolarization in attractants is generally smaller than that used to demonstrate robust changes in cAMP levels (8–12 mV vs 20) but, nonetheless, cAMP does not appear to be a second messenger in chemoresponse. Some evidence suggests that adenylyl cyclase is a K^+ channel that controls resting V_m and that hyperpolarization is sensed by this channel and is translated into activation of adenylyl cyclase activity (Schultz *et al.*, 1992). How this adenylyl cyclase distinguishes between membrane potentials of different origins, for example, changes in K_o vs attractant stimuli, must be determined.

What produces the change in membrane potential? For the attractants acetate and folate, standard electrophysiological measurements have uncovered no dependence on resting membrane voltage, no reversal potential, and no dependence on K_o or Na_o. (Removing all Ca_o without compromising the cells is more difficult.) Therefore, the change in membrane potential in acetate, for example, is caused by a 0.2 nA current, but the origin of this conductance apparently is not a channel (Preston and Van Houten, 1987a). Mutants with alterations in various conductances helped rule out the predictable mechanisms of hyperpolarization by specific channel activity or surface potential (Preston and Van Houten, 1987a). Most likely, the hyperpolarizing current is from a pump such as a cell-surface pump that produces the efflux of a divalent cation such as Ca^{2+}.

What is the evidence for a calcium pump? Perhaps the best correlative evidence for a calcium pump is provided with the use of lithium to perturb chemoresponse. The original intent was to explore the involvement of phosphoinositol lipids in chemoresponse using lithium to block their recycling through the inositol phosphate cycle (Berridge, 1987). Inositol phosphate and phosphoinositol lipids do not seem to change much with chemical stimulation, whereas lithium profoundly affects chemoresponse (Wright, 1990; Wright *et al.*, 1992). In addition to affecting chemoresponse, lithium inhibits normal calcium homeostasis efflux by 50% and displaces about 50% of K_i (Wright, 1990; Wright *et al.*, 1992). Additionally, the K-shy mutant appears to have defects in calcium homeostasis (Evans *et al.*, 1987) and does not respond to many attractants (Van Houten, 1990). Collectively, these observations suggest calcium pumping in chemoresponse.

To better understand the role of calcium pumps in chemoresponse, we isolated a Ca^{2+}-ATPase activity from the cell body membranes of *Paramecium*. This activity has all the hallmarks of a cell-surface calcium pump: K_m for calcium of 95 nM; fastidious requirement for ATP and no other nucleotide triphosphate; inhibition by calmodulin inhibitors, and vanadate; and lack of effect of inhibitors of endoplasmic reticulum and mitochondrial calcium pumps (Wright and Van Houten, 1990). This Ca^{2+}-ATPase activity does not come from contaminating ATPases such as dyneins of the cilia, the endoplasmic reticulum, or the alveolar sacs that sequester calcium (Stelly *et al.*, 1991), because pH optima, Mg^{2+} requirements, and pharmacological agents distinguish between these possibilities and the putative plasma membrane Ca^{2+}-ATPase (Wright and Van Houten, 1990). The alveolar sacs that underlie the plasma membrane sequester calcium; this uptake can be measured by ^{45}Ca, incorporation (Wright, 1990; Stelly *et al.*, 1991; Wright *et al.*, 1992). Calmodulin inhibitors are not nearly as effective in inhibiting this uptake as they are in reducing the Ca^{2+}-ATPase activity (Wright *et al.*, 1992). Thapsigargin is a specific inhibitor of the calcium pump of the endoplasmic reticulum (Papp *et al.*, 1991). Although thapsigargin very effectively inhibits alveolar uptake (Table I), it has little effect on the Ca^{2+}-ATPase activity (Table II). Interestingly, this agent affects alveolar calcium uptake since there is speculation that these sacs function much like an endoplasmic reticulum in *Paramecium* (Satir and Wissig, 1982; Stelly *et al.*, 1991). Therefore, by process of elimination, the plasma membrane calcium pump and the Ca^{2+}-ATPase activity seem to be the same molecule.

A characteristic of calcium pumps is the transient covalent phosphorylation of the protein by [^{32}P]ATP as part of the catalytic mechanism (Carafoli, 1991). One major protein of the *Paramecium* pellicle preparation becomes covalently labeled with [^{32}P]ATP. This labeling is hydroxylamine labile, which distinguishes it from kinase substrates (Wright and Van Houten, 1990). This protein of 133-kDa apparent molecular mass is appropriate in size for a pump with a calmodulin binding domain, as expected for a plasma membrane calcium pump, and is clearly distinguishable from the endoplasmic reticulum pump. (Dynein, although an ATPase, will not form this phosphoenzyme intermediate.) This phosphoenzyme intermediate is diminished greatly in ciliary membranes, as is Ca^{2+}-ATPase activity (Wright, 1990). By process of elimination, the phosphoenzyme intermediate is most likely the protein correlated with Ca^{2+}-ATPase activity. Therefore, the protein pump and the Ca^{2+}-ATPase activity appear to be the same.

How does the calcium pump relate to chemosensory transduction? Again, indirect evidence implicates calcium pumping in chemoresponse. Direct demonstration of stimulation of Ca^{2+}-ATPase activity by chemical stimuli is not feasible because only a small percentage increase in the pumping activity would be required to generate the small current associated with the attractant-induced hyperpolarization (Wright, 1990). The interference of lithium in the chemoresponse can be explained by its displacement of K_i and the concomitant loss of Ca^{2+}-ATPase activity and measurable decreased rate of ^{45}Ca efflux (Wright, 1990; Wright and Van Houten, 1990; Wright *et al.*, 1992). Although Li^+ replaces K_i, it does not function as K^+ does to activate the Ca^{2+}-ATPase activity (Wright and Van Houten, 1990). If the calcium cell-surface pump were calmodulin regulated, extant calmodulin mutants could be of some assistance in sorting out the role of the calcium pump in chemoresponse. Indeed, the

TABLE I
^{45}Ca Uptake into Alveoli[a]

Treatment	Uptake of ^{45}Ca (%)	
	60 sec	120 sec
None	100	100
Vanadate (20 μ*M*)	52.7	—
Thapsigargin (50 n*M*)	15.4	22.8
Thapsigargin (100 n*M*)	13.5	36.0

[a]Uptake comes to equilibrium by 120 sec. The data are averages of two experiments. All measurements done in duplicate for each experiment. (Methods in Wright and Van Houten, 1990; Wright *et al.*, 1992.)

TABLE II
Ca–ATPase Activity in *Paramecium* Pellicles[a]

Treatment	Activity (%)
None	100
Vanadate (2 μ*M*)	54.9 ± 2.9
Thapsigargin (100 n*M*)	100.3 ± 13.9
Thapsigargin (50 n*M*)	85.8 ± 20.0
Thapsigargin (25 n*M*)	105

[a]The data are averages of three experiments with standard deviation shown with the exception of the 25 n*M* thapsigargin condition which was done once. All measurements done in duplicate for each experiment. (Methods in Wright and Van Houten, 1990; Wright *et al.*, 1992.)

calmodulin mutants *fnaP* and *fna*, which up- and down-regulate calmodulin-regulated conductances, respectively (Preston, 1990; Preston *et al.*, 1991), show defects in chemoresponse in Na^+ solutions, that is, under conditions in which the mutant phenotype is most prevalent, and almost normal chemoresponse in K^+ solutions, in which the phenotype is close to wild-type (Table III; Wright *et al.*, 1992). Under similar conditions, efflux of calcium from the cells, a crude measure of calcium pumping, is divergent from normal in Na^+ solutions but close to normal in K^+ solutions (Table IV). Collectively, these observations are consistent with a calmodulin-regulated calcium pump participating in chemosensory transduction.

TABLE III
Chemoresponse of Calmodulin Mutants

	Treatment[a]			
	K acetate	Na acetate	K lactate	Na lactate
Normal	0.84 ± 0.07	0.82 ± 0.11	0.83 ± 0.06	0.73 ± 0.04
fna	0.87 ± 0.09	0.51 ± 0.08	0.79 ± 0.11	0.48 ± 0.03[b]
fnaP	0.77 ± 0.02	0.55 ± 0.02	nd[c]	nd

[a]Data are averages of three or more T-maze assays ± one standard deviation. Solutions are all buffers with 5 m*M* of attractant salt in test solution and 5 m*M* of control salt in control solution.
[b]90 min instead of the standard 30 min for the assay.
[c]Not done.

C. Coupling the Receptor to the Pump

To date we have no evidence for a mechanism that couples receptors such as the cAMP receptor to a calcium pump. No heterotrimeric G proteins have been identified in *Paramecium*, but the small Ras-like proteins that have been identified (Peterson, 1991; D. Fraga and R. Hinrichsen, personal communication) may serve this function. Other second messengers such as cAMP and inositol triphosphate (IP_3) have been investigated and not found to change on stimulation. The mechanism of coupling awaits clarification and probably will become more evident with the cloning of the Ca^{2+}-ATPase gene and the cAMP receptor genes (B. Cote, N. Elwess, and J. Van Houten, unpublished results).

TABLE IV
^{45}Ca Efflux as a Measure of Pumping Activity[a]

	Efflux	
Strain	in KCl	in NaCl
fnaP	1.37	2.5
Normal	1.0	1.0
fna	1.95	0.65

[a]Cells are loaded with ^{45}Ca to equilibrium. The flux of calcium out of the cells is measured in a filtration assay (Wright, 1990; Wright *et al.*, 1992). Data are averages from three experiments, normalized to the normal cell values.

D. Other Transduction Pathways

In addition to coupling to the calcium pump, two other transduction pathways exist in *Paramecium*. The hyperpolarization in response to L-glutamate can be attributed to a K^+ conductance (Preston and Usherwood, 1988b). This hyperpolarization is not sustained and therefore differs from the hyperpolarizations in response to acetate or folate in this respect as well. Yet other attractant stimuli create a sustained hyperpolarization, but do not appear to involve a calcium pump in the generation of the hyperpolarizing current. For example, LiCl has profound effects on chemoresponse to folate or acetate but none on behavior in NH_4Cl. Likewise, the K-shy mutant is not able to accumulate in folate or acetate but does respond normally to NH_4Cl. K-shy is thought to have defects in calcium homeostasis. These observations therefore imply that at least two chemosensory pathways are distinguished by LiCl: one involving the calcium pump and the other involving a conductance that is pH sensitive. We speculate that the cells in NH_4Cl experience a sudden acidification on leaving the NH_4Cl attractant buffer and entering a NaCl solution. This acidification may alter conductances or pump activity to depolarize the cell in the NaCl buffer and thereby cause repulsion from this solution. These hypotheses are being tested with calcium- and pH-sensitive fluorescent dyes.

V. Context of Other Chemosensory Transduction Systems

The field of sensory transduction, that is, the transformation of stimulus energy into electrical energy of the receptor cell, has come of age (Shepherd, 1991). Sensory transduction is a subset of the larger study of signal transduction that includes systems from neurotransmission to mitogenesis. The basic transduction components seem to translate well across systems and phyla, and include receptors, G proteins, soluble second messengers (cyclic nucleotides, Ca^{2+}, IP_3, and so on), ion channels including those that are nucleotide gated, and kinases (Shepherd, 1991). For example, olfactory stimuli appear to bind to specific G-coupled receptors and induce the production of cAMP or IP_3 in a very rapid and GTP-dependent manner. These second messengers then modulate channel activity to affect the impulses from these receptor cells to the central nervous system, where the information from across the entire olfactory epithelium is decoded (for reviews, see Vogt *et al.*, 1990; Anholt, 1991; Shepherd, 1991). In taste, receptors for sweet stimuli seem to use the same basic arrangement to stimulate cAMP production, which subsequently results in closure of K^+ channels, perhaps involving protein kinase A. Closure of these channels leads to depolarization and opening of voltage-dependent calcium channels, bringing in the calcium needed for neurotransmitter release. Alternatively, Na^+ and H^+ are detected in taste by direct channel interaction rather than by receptors. Bitter stimuli appear to activate phospholipase C with the expected effects on IP_3 and Ca^{2+}; some of these activations may be receptor mediated (for reviews, see Kinnamon, 1988; Roper, 1989; Shepherd, 1991).

Paramecium, Dictyostelium, yeast, and sea urchin spermatozoa are single cell chemosensory systems that have been useful in elucidating sensory transduction pathways (for reviews, see Kincaid, 1991; Van Houten, 1991,1992). In *Dictyostelium* in particular, many details of two parallel pathways, one for chemotaxis and one for cAMP secretion, have been described from receptor to second messengers and their effects on cytoskeleton (Devreotes, 1989; Firtel, 1991). The components have remarkable similarity to those in mammalian systems with seven transmembrane receptors and G proteins, but also have some distinct differences, for example, in IP metabolism (Van Haastert *et al.*, 1991). Yeast similarly have "classical" seven transmembrane receptors for a and α pheromone, but the coupling of the receptors to the G protein becomes more interesting because the G protein β and γ subunits rather than the α subunit seem to carry on the main sensory transduction pathway (Hoekstra *et al.*, 1991; Sprague, 1991; Chapter 8). Sea urchin receptors for egg peptides likewise demonstrate an interesting deviation from the G protein-mediated systems. Here the peptide receptor is the guanylyl cyclase of the surface membrane. In mammals, the receptors for peptides such as atrial natriuretic factor and endotoxin similarly are membrane guanylyl cyclases (Schulz *et al.*, 1991).

All the eukaryotic chemoreceptor proteins isolated and all the genes cloned to date have been isolated from unicellular organisms. The putative vertebrate olfactory receptor genes that have been cloned appear promising (Buck and Axel, 1991; see also Dahmen *et al.*, 1991; Wang *et al.*, 1991) but function remains to be demonstrated. An advantage of *Paramecium* or the other unicellular systems is that function can be confirmed through simple behavioral assays, mating assays, biochemistry, or electrophysiology of normal cells or by mutants. (Kincaid, 1991).

Paramecium has been useful in the isolation of a receptor protein. The simple behavioral assay provides an effective means to verify the function of this protein in chemoresponse. However, until the gene that encodes the protein is cloned we will not know whether *Paramecium* receptors will follow the superfamily models for G protein-coupled or other receptors such as those with kinase, channel, or guanylyl cyclase activity. *Paramecium* differs from many of the systems described in this section through its apparent lack of heterotrimeric G proteins and its protein coupling to a calcium pump or internal pH regulation. However, this organism shares with sensory neurons the modulation of electrical properties as an integral step in the chemosensory transduction pathway. Olfactory receptor neurons can depolarize or hyperpolarize in the presence of chemical stimuli (McClintock and Ache, 1989). Therefore, when a mixture of excitatory and inhibitory stimuli is applied, the time of onset and other integrated aspects of the resulting electrical impulses must contribute to the coding information used by the brain to assess quantity and quality of the odor. *Paramecium* likewise depolarizes or hyperpolarizes in chemical stimuli. Action potentials are generated with depolarization but, unlike olfactory neurons, the cell population is relatively uniform in its response. No specific cell population responds to only one of many stimuli.

VI. Summary

Paramecia are handy chemosensory cells that have parallels with and also distinct differences from metazoan chemosensory transduction systems. At the moment, these differences are suggested to be novel mechanisms, but they may actually direct us to common transduction mechanisms. Significant progress is being made in the molecular biology of olfactory receptors, and expression of receptor clones now is possible (Raming *et al.*, 1993).

References

Anholt, R. R. H. (1991). Odor recognition and olfactory transduction: the new frontiers. *Chem. Senses* **16**, 421–427.

Antipa, G., Martin, K., and Rintz, M. (1983). A note on the possible ecological significance of chemotaxis in certain ciliated protozoa. *J. Protozool.* **30**, 55–58.

Berridge, M. J. (1987). Inositol trisphosphate and diacyl glycerol: Two interacting second messengers. *Ann. Rev. Biochem.* **56**, 159–193.

Bonini, N. M., and Nelson, D. (1988). Differential regulation of *Paramecium* ciliary motility by cAMP and cGMP. *J. Cell Biol.* **106**, 1615–1623.

Bonini, N. M., and Nelson, D. (1990). Phosphoproteins associated with cyclic nucleotide stimulation of ciliary motility in *Paramecium. J. Cell Sci.* **95**, 219–230.

Bonini, N. M., Gustin, M. C., and Nelson, D. (1986). Regulation of ciliary motility by membrane potential in *Paramecium*: A role for cAMP. *Cell Motil. Cytoskel.* **6**, 256–272.

Buck, L., and Axel, R. (1991). A novel multigene family may encode odorant receptors: A molecular basis for odor recognition. *Cell* **65**, 175–187.

Carafoli, E. (1991). The calcium pumping ATPase of the plasma membrane. *Ann. Rev. Physiol.* **53**, 531–547.

Dahmen, N., Wang, H., Getchell, T. V., and Margolis, F. L. (1991). Expression of rat olfactory receptors in *Xenopus* oocytes. *Soc. Neurosci. Abstr.* **17**, 634.

Devreotes, P. (1989). *Dictyostelium discoideum*: A model system for cell–cell interactions in development. *Science* **245**, 1054–1058.

DiNallo, M., Wohlford, M., and Van Houten, J. (1982). Mutants of *Paramecium* defective in chemoreception of folate. *Genetics* **102**, 149–158.

Dryl, S. (1959). Chemotactic and toxic effects of lower alcohols in *Paramecium caudatum. Acta Biol. Exp.* **19**, 95–104.

Dunlap, K. (1977). Localization of calcium channels in *Paramecium caudatum. J. Physiol.* **271**, 119–133.

Eckert, R. (1972). Bioelectric control of ciliary activity. *Science* **176**, 473–481.

Evans, T. C., Hennessey, T., and Nelson, D. L. (1987). Electrophysiological evidence suggests a defective Ca^{2+} control mechanism in a new *Paramecium* mutant. *J. Membrane Biol.* **98**, 275–283.
Firtel, R. A. (1991). Signal transduction pathways controlling multicellular development in *Dictyostelium. Trends Genet.* **7**, 381–388.
Fraenkel, G. S., and Gunn, D. L. (1961). "Orientation of Animals." Dover, New York.
Hamasaki, T., Barkalow, K., Richmond, J., and Satir, P. (1991). cAMP-stimulated phosphorylation of an axonemal polypeptide that copurifies with the 22S dynein arm regulates microtubule translocation velocity and swimming speed in *Paramecium. Proc. Natl. Acad. Sci. U.S.A.* **88**, 7918–7922.
Hennessey, T., Machemer, H., and Nelson, D. L. (1985). Injected cyclic AMP increases ciliary beat frequency in conjunction with membrane hyperpolarization. *Eur. J. Cell Biol.* **36**, 153–156.
Hochstrasser, M., and Nelson, D. L. (1989). Cyclic AMP-dependent protein kinase in *Paramecium tetraurelia. J. Biol. Chem.* **254**, 14510–14518.
Hoekstra, M. F., Demaggio, A. J., and Dhillon, N. (1991). Genetically identified protein kinases in yeast I: Transcription, translation, transport and mating. *Trends Genet.* **7**, 256–261.
Jennings, H. S. (1906). "Behavior of the Lower Organisms." Columbia University Press, New York.
Keryer, G., Adoutte, A., Ng, S. F., Cohen, J., de Loubresse, N. G., Rossignol, M., Stelly, N., and Beisson, J. (1990). Purification of the surface membrane–cytoskeleton complex (cortex) of *Paramecium* and identification of several of its protein constituents. *Eur. J. Protistol.* **25**, 209–225.
Kincaid, R. L. (1991). Signaling mechanisms in microorganisms: Common themes in the evolution of signal transduction pathways. *In* "Advances in Second Messenger and Phosphoprotein Research" (P. Greengard and G. A. Robison, eds), pp. 165–184. Raven Press, New York.
Kinnamon, S. (1988). Taste transduction: A diversity of mechanisms. *Trends Neurosci.* **11**, 491–496.
Klein, P. S., Sun, T. J., Saxe, C. L., Kimmel, A. R., Johnson, R. L., and Devreotes, P. N. (1988). A chemoattractant receptor controls development in *Dictyostelium discoideum. Science* **241**, 1467–1472.
Kung, C., and Saimi, Y. (1982). The physiological basis of taxes in *Paramecium. Ann. Rev. Physiol.* **44**, 519–534.
Kung, C., Chang, S-Y., Satow, Y., Van Houten, J., and Hansma, H. (1975). Genetic dissection of behavior in *Paramecium. Science* **188**, 898–904.
McClintock, T. S., and Ache, B. W. (1989). Hyperpolarizing receptor potentials in lobster olfactory receptor cells: Implications for transduction and mixture suppression. *Chem. Senses.* **14**, 637–647.
Machemer, H. (1989). Cellular behavior modulated by ions: Electrophysiological implications. *J. Protozool.* **36**, 463–487.
Mason, P. A., and Nelson, D. L. (1989). Cyclic AMP-dependent protein kinases of *Paramecium.* I. Chromatographic and physical properties of the enzymes from cilia. *Biochim. Biophys. Acta* **1010**, 108–115.
Matsudaira, P. T. (1989). "A Practical Guide to Protein and Peptide Purification for Microsequencing." Academic Press, San Diego.
Nakaoka, Y., and Ooi, H. (1985). Regulation of ciliary reversal in Triton-extracted *Paramecium* by calcium and cyclic adenosine monophosphate. *J. Cell Sci.* **77**, 185–195.
Ogura, A., and Machemer, H. (1980). Distribution of mechanoreceptor channels in the *Paramecium* surface membrane. *J. Comp. Physiol.* **135**, 233–242.
Ogura, A., and Takahashi, K. (1976). Artificial deciliation causes loss of Ca-dependent responses in *Paramecium. Nature (London)* **264**, 170–172.
Papp, B., Enyedi, A., Kovacs, T., Sarkadi, B., Wuytack, F., Thastrup, O., Gardos, G., Bredoux, R., Levy-Toledano, S., and Enouf, J. (1991). Demonstration of two forms of calcium pumps with thapsigargin inhibition and radioimmunoblotting in platelet membrane vesicles. *J. Biol. Chem.* **266**, 14593–14596.

Peterson, J. B. (1991). Small GTP-binding proteins associated with secretory vesicles of *Paramecium. J. Protozool.* **38**, 495–501.

Preston, R. R. (1990). Genetic dissection of Ca^{2+}-dependent ion channel function in *Paramecium. BioEssays* **12**, 273–281.

Preston, R. R., and Usherwood, P. N. R. (1988a). Characterization of a specific L-[^{3}H]glutamic acid binding site on cilia isolated from *Paramecium tetruarelia. J. Comp. Physiol. B* **158**, 345–351.

Preston, R. R., and Usherwood, P. N. R. (1988b). L-Glutamate-induced membrane hyperpolarization and behavioural responses in *Paramecium tetraurelia. J. Comp. Physiol. A* **164**, 75–84.

Preston, R. R., and Van Houten, J. (1987a). Chemoreception in *Paramecium*: Acetate- and folate-induced membrane hyperpolarization. *J. Comp. Physiol. A* **160**, 525–536.

Preston, R. R., and Van Houten, J. (1987b). Localization of chemoreceptive properties of the surface of *Paramecium. J. Comp. Physiol. A* **160**, 537–541.

Preston, R. R., Kink, J. A., Hinrichsen, R. D., Saimi, Y., and Kung, C. (1991). Calmodulin mutants and Ca^{2+}-dependent channels in *Paramecium. Ann. Rev. Physiol.* **53**, 309–319.

Raming *et al.* (1993). *Nature (London)* **361**, 353–356.

Roper, S. D. (1989). The cell biology of vertebrate taste receptors. *Ann. Rev. Neurosci.* **12**, 329–353.

Saimi, Y., and Kung, C. (1987). Behavioral genetics of *Paramecium. Ann. Rev. Genet.* **21**, 47–65.

Sasner, M. J., and Van Houten, J. (1989). Evidence for a *Paramecium* folate receptor. *Chem. Senses* **14**, 241–251.

Satir, P. (1985). Switching mechanisms in the control of ciliary motility. *Modern Cell Biol.* **4**, 1–46.

Satir, P., and Wissig, S. L. (1982). Alveolar sacs of *Tetrahymena*: Ultrastructural characteristics and similarities to subsurface cisterns of muscle and nerve. *J. Cell Sci.* **55**, 13–33.

Schnitzer, M. J., Block, S. M., Berg, H. C., and Purcell, E. M. (1990). Strategies for chemotaxis. *In* "Biology of the Chemotactic Response" (J. P. Armitage and J. M. Lackie, eds.), pp. 15–34. Cambridge University Press, Cambridge.

Schultz, J. E., Grunemund, R., von Hirschhausen, R., and Schonefeld, U. (1984).Ionic regulation of cyclic AMP levels in *Paramecium tetraurelia. FEBS Lett.* **167**, 113–116.

Schultz, J. E., Pohl, T., and Klumpp, S. (1986). Voltage-gated Ca^{2+} entry into *Paramecium* linked to intraciliary increase in cyclic GMP. *Nature (London)* **322**, 271–273.

Schultz, J. E., Klumpp, S., Benz, R., Schürhoff-Goeters, W. J. C., and Schmid, A. (1992). Regulation of adenylyl cyclase from *Paramecium* by an intrinsic potassium conductance. *Science* **255**, 600–603.

Schulz, S. (1987). Folate binding proteins of *Paramecium tetraurelia*: Relationship to chemoreception. Ph.D. Thesis. University of Vermont, Burlington, Vermont.

Schulz, S., Denaro, M., and Van Houten, J. (1983). Evidence for membrane potential control of chemoreception in *Paramecium tetraurelia. Soc. Neurosci. Abstr.* **9**, 1023.

Schulz, S., Denaro, M., and Van Houten, J. (1984). Relationship of folate binding and uptake to chemoreception in *Paramecium. J. Comp. Physiol.* **155**, 113–119.

Schulz, S., Yuen, P. S., and Garbers, D. L. (1991). The expanding family of guanylyl cyclases. *Trends Pharmacol. Sci.* **12**, 116–120.

Shepherd, G. M. (1991). Sensory transduction: Entering the mainstream of membrane signalling. *Cell* **67**, 845–851.

Smith, R., Preston, R. R., Schulz, S., and Van Houten, J. (1987). Correlation of cyclic adenosine monophosphate binding and chemoresponse in *Paramecium. Biochim. Biophys. Acta* **928**, 171–178.

Sprague, G. F. (1991). Signal transduction in yeast mating: Receptors, transcription factors, and the kinase connection. *Trends Genet.* **7**, 393–398.

Stelly, N., Mauger, J-P., Claret, M., and Adoutte, A. (1991). Cortical alveoli of *Paramecium*: A vast submembranous calcium storage compartment. *J. Cell Biol.* **113**, 103–112.

Van Haastert, P. J. M., Janssens, P. M. W., and Erneux, C. (1991). Sensory transduction in eukaryotes. A comparison between *Dictyostelium* and vertebrate cells. *Eur. J. Biochem.* **195**, 289–303.
Van Houten, J. (1977). A mutant of *Paramecium* defective in chemotaxis. *Science* **198**, 746–749.
Van Houten, J. (1978). Two mechanisms of chemotaxis in *Paramecium. J. Comp. Physiol.* A **127**, 167–174.
Van Houten, J. (1979). Membrane potential changes during chemokinesis in *Paramecium. Science* **204**, 1100–1103.
Van Houten, J. (1980). Chemosensory transduction in *Paramecium*: Role of membrane potential. *In* "Olfaction and Taste" (J. Atema, ed.), Vol. VIII, pp. 53–56. IRL Press, London.
Van Houten, J. (1990). Chemosensory transduction in *Paramecium. In* "Biology of the Chemotactic Response" (J. P. Armitage and J. M. Lackie, eds.), pp. 297–321. Cambridge University Press, Cambridge.
Van Houten, J. (1991). Signal transduction in chemoreception. *In* "Sensory Receptors and Signal Transduction." (J. L. Spudich and B. H. Satir, eds.), pp. 65–135. Wiley–Liss, New York.
Van Houten, J. (1992). Chemosensory transduction in eukaryotic microorganisms. *Ann. Rev. Physiol.* **54**, 639–664.
Van Houten, J., and Preston, R. R. (1987). Chemoreception in single-celled organisms. *In* "Neurobiology of Taste and Smell" (T. E. Finger, ed.), pp. 11–38. John Wiley and Sons, New York.
Van Houten, J., and Preston, R. R. (1988). Chemokinesis. *In* "Paramecium" (H.-D. Görtz, ed.), pp. 282–300. Springer-Verlag, Berlin.
Van Houten, J., and Van Houten, J. C. (1982). Computer simulation of *Paramecium* chemokinesis behavior. *J. Theor. Biol.* **98**, 453–468.
Van Houten, J., Hauser, D. C. R., and Levandowsky, M. L. (1981). Chemosensory behavior in protozoa. *In* "Biochemistry and Physiology of Protozoa" (S. Hutner and M. L. Levandowsky, eds.), 2d Ed., Vol. 4, pp. 67–124. Academic Press, New York, NY.
Van Houten, J., Martel, E., and Kasch, T. (1982). Kinetic analysis of chemokinesis in *Paramecium. J. Protozool.* **29**, 226–230.
Van Houten, J. L., Cote, B., Zhang, J., Baez, J., and Gagnon, M. L. (1991). Studies of the cyclic adenosine monophosphate chemoreceptor of *Paramecium. J. Membrane Biol* **119**, 15–24.
Vogt, R. G., Tybczynski, R., and Lerner, M. R. (1990). The biochemistry of odorant reception and transduction. *In* "Chemosensory Information Processing" (D. Schild, ed.), pp. 33–76. Springer-Verlag, Berlin.
Wang, H., Dahmen, N., Kodukula, K., Yan, N., Teeter, J., and Margolis, F. L. (1991). Expression of catfish L-amino acid olfactory receptors in *Xenopus* oocytes injected with in vitro transcribed RNA. *Soc. Neurosci. Abstr.* **17**, 634.
Wright, M. V. (1990). Signal transduction in *Paramecium* chemoresponse: The role of calcium and lithium. Ph.D. Thesis. University of Vermont, Burlington, Vermont.
Wright, M. V., and Van Houten, J. (1990). Characterization of a putative Ca^{2+}-transporting Ca^{2+}-ATPase in the pellicles of *Paramecium tetraurelia. Biochim. Biophys. Acta* **1029**, 241–245.
Wright, M. V., Elwess, N., Frantz, M., and Van Houten, J. (1992). Lithium fluxes in *Paramecium* and their relationship to chemoresponse. *Biochim. Biophys. Acta* **1107**, 223–230.

Chapter 13

Biochemistry and Genetics of Sensory Transduction in *Dictyostelium*

PETER J. M. VAN HAASTERT

Department of Biochemistry
University of Groningen
9747 AG Groningen, The Netherlands

PETER N. DEVREOTES

Department of Biological Chemistry
The Johns Hopkins University School of Medicine
Baltimore, Maryland 21205-2185

I. Introduction

The organization of multicellular organisms depends on cell–cell communication. The signals are often soluble molecules in the extracellular fluid, but also include odors and light. A large array of surface receptors is involved in the detection of these signals. The receptors transduce the signals across the plasma membrane so enzymes on the inner face of the membrane are activated, producing second messengers, which activate target proteins or genes by a complex network of interactions.

Vertebrate cells have been used to study hormone and neurotransmitter action, vision, the regulation of cell growth, and differentiation. By comparing sensory transduction in lower and higher eukaryotes, general principles may be recognized that are found in all organisms and deviations may be found that are present in specialized systems. These comparative studies also may help us understand the differences between cell types within one organism and the importance of a particular sensory transduction pathway.

During *Dictyostelium discoideum* development, an initially nearly homogeneous population of cells aggregates and differentiates into two final cell states (Loomis, 1985). About 80% of the cells form a ball of spores supported by a dead, highly vacuolized, cellulose stalk composed of the remaining 20% of the cells. Triggered by starvation, this differentiation program begins with the expression of genes that constitute a cAMP sensory transduction system. This apparatus controls aggregation as well as early and late gene expression. Its components include surface cAMP receptors, G proteins, and adenylyl cyclase. During aggregation, cAMP is secreted from central points of the cell population at 5 min intervals. This molecule diffuses to surrounding cells, which move up the cAMP concentration gradient toward the aggregation center and synthesize and secrete additional cAMP. This process results in the outward propagation of cAMP concentration waves, which direct periodic inward chemotactic movement steps, leading to the formation of a multicellular structure (Fig. 13.1).

In this review, we describe sensory transduction in *D. discoideum* with emphasis on the biochemistry and genetics of the events that occur at the plasma membrane. The transduction pathways that lead to chemotaxis and cell type-specific gene expression are discussed briefly. A more detailed discussion of gene regulation by cellular signalling is found in Chapter 14.

II. Components of Sensory Transduction Pathways

A. Signal Molecules

All *Dictyostelium* species use pterin or folic acid as a chemoattractant during growth. Chemotaxis toward these compounds may function as a food seeking behavior since bacteria secrete pterins (see De Wit and Konijn, 1983b). Most slime mold species become chemotactically sensitive to specific compounds, called acrasins, that are secreted during cell aggregation. Species-specific

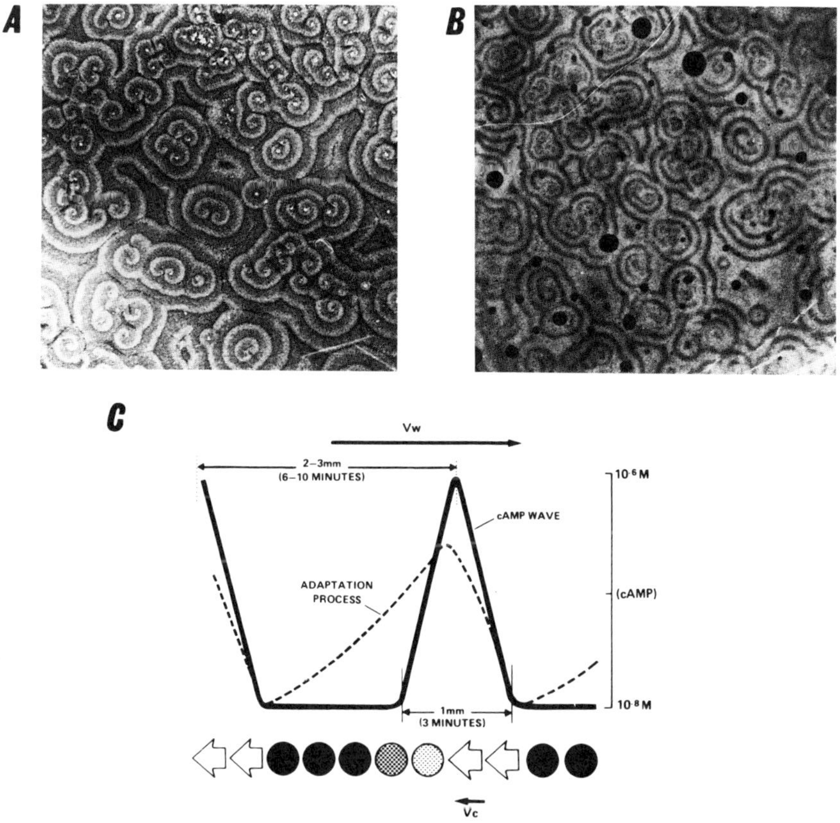

Figure 13.1 Chemotaxis and cAMP signaling mediate aggregation in *Dictyostelium* (A) Dark-field photograph of aggregating cells at 5 hr in the developmental program. Territories containing ~1 million cells are 1–2 cm in diameter. (B) Print of fluorographic image of cAMP waves within monolayer of aggregating cells. Waves were detected by a solid-phase isotope dilution technique. (C) Dynamics of signal relay and chemotaxis. The boldface line representing the cAMP concentration is drawn from analyses of scans of the fluorographic images of the cAMP waves. Symbols under the graph represent a single radial line of cells: open arrows, cells moving toward the center; shaded circles, randomly oriented cells; arrow vectors, the speed and direction of motion of the cAMP wave (V_w = 300 μm/min) and the moving cells (V_c = 20 μm/min).

acrasins provide a mechanism for the segregation of species during cell aggregation. The acrasin of *Polysphondilium violaceum* is a circular dipeptide called glorin (Shimomura *et al.*, 1982), the acrasin of *Dictyostelium lacteum* is a derivative of pterin (Van Haastert *et al.*, 1982a), and the acrasin of *Dictyostelium minutum* is a derivative of folic acid (De Wit and Konijn, 1983a). The first acrasin identified was cAMP, which mediates cell aggregation in at least four *Dictyostelium* species: *D. discoideum, D. mucoroides, D. rosarium*, and *D. purpureum* (Konijn *et al.*, 1967).

In addition to the chemoattractants, other compounds that regulate differentiation have been defined, including differentiation-inducing factor (DIF; Morris *et al.*, 1987), ammonia, and adenosine (Schaap and Wang, 1986), as well as several unidentified factors, such as conditioned medium factor and slug-turning factor. The role of these molecules during differentiation is discussed in more detail in Chapter 14.

B. Signal Degrading Enzymes

Degradation of chemoattractants by enzymes exposed on the cell surface and secreted into the medium insures a low background concentration, which improves signal-to-noise ratio of chemoattractant gradients. Indeed, mutants unable to degrade extracellular cAMP fail to aggregate (Barra *et al.*, 1980).

In most species, including *D. discoideum*, folic acid and pterins are degraded by a hydrolytic deaminase that removes the amino group at the 2 position. The products, 2-deamino-folic acid and lumazine, are inactive. In *D. minutum*, folic acid is degraded by a C9–N10 hydrolytic enzyme producing 6-hydroxymethylpterine and *p*-aminobenzoylglutamic acid; these products are inactive in *D. minutum*. Interestingly, some species respond to the breakdown products of other species. For example, 2-deamino-folic acid is active in *D. minutum*, whereas 6-hydroxymethylpterine is active in *D. discoideum* (see De Wit and Konijn, 1983b).

The degradation of extracellular cAMP is mediated by a cyclic nucleotide phosphodiesterase (PDE), yielding the inactive compound 5′-AMP. Although the metabolism of cAMP is simple, the regulation of PDE is complex (see Kessin, 1988). The enzyme is present both on the cell surface and in the extracellular medium. The extracellular enzyme is regulated by a heat-stable inhibitor (PDI). During starvation, PDE activity increases several fold because of enhanced *de novo* synthesis of PDE and repression of PDI. The genes encoding PDE (Lacombe *et al.*, 1986) and PDI (Wu and Franke, 1990) have been cloned, regulatory elements in the 5′-untranscribed regions have been identified. Expression of PDE is regulated by three promotors that are active during the growth, cell aggregation, or multicellular stage, respectively. These promotors are regulated by starvation and by extracellular cAMP (Faure *et al.*, 1990). The mutant defective in PDE activity can be rescued by expression of the PDE -cDNA (Faure *et al.*, 1989). Further, overexpression of PDE in wild-type cells disturbs cell aggregation and differentiation, indicating that PDE activity must be controlled optimally to provide the appropriate cAMP signals (Faure *et al.*, 1988).

C. Receptors

Growing cells possess relatively high levels of folic acid binding sites and relatively low levels of cAMP binding sites. On starvation, the number of folic acid binding sites declines from a few thousand to undetectable levels whereas the number of cAMP binding sites increases 10-fold, reaching a maximum of about 10^5/cell during the aggregation stage (see De Wit and Van Haastert, 1985; Janssens and Van Haastert, 1987). Multiple kinetic forms of folic acid and cAMP binding activity have been defined based on observed differences in affinities and rates of dissociation (Van Haastert and De Wit, 1984; De Wit and Van Haastert, 1985; Van Haastert *et al.*, 1986). Similar kinetic forms are found in purified membranes. In membranes, the proportions of the different forms can be regulated by guanine nucleotides, suggesting that folic acid and cAMP receptors interact with one or more GTP-binding proteins (Van Haastert, 1984; De Wit and Bulgakov, 1985).

The surface cAMP receptor protein has been purified and the gene has been cloned (Klein *et al.*, 1987,1988). The deduced amino acid sequence of the protein, designated cAR1, predicts a topological structure now characteristic of all receptors coupled to G proteins: an extracellular N-terminal domain, seven putative transmembrane spanning domains, and a cytoplasmic C-terminal domain (Fig. 13.2). Three other genes, designated *cAR2*, *cAR3*, and *cAR4*, have been cloned that encode proteins that are homologous with cAR1 in the transmembrane spanning domains and connecting loops (Saxe *et al.*, 1991a,b); however, the extracellular N-terminal domains and the cytoplasmic C-terminal domains differ markedly in length and sequence. The divergence of the cytoplasmic domains is intriguing because these regions are not required for binding of the ligand and presumably are involved in regulation of receptor function. cAR1 contains multiple serines that are heavily phosphorylated and presumably play a role in desensitization, as in β-adrenergic receptors and rhodopsin (see subsequent text).

cAR1, cAR2, and cAR3 have been expressed during the growth stage using a constitutive promotor (Klein *et al.*, 1988; Johnson *et al.*, 1991, 1992). These cells do not display abnormal phenotypes, and permit the study of the biochemistry of each individual cAR protein in the absence of the other cARs. In this context, each of the receptors displays multiple affinities. The average affinities for cAR1, cAR2, and cAR3 are 50 n*M*, 500 n*M*, and >2μ*M*, respectively. Reconstituting the cAMP signal transduction pathway during the growth phase by coexpression of other developmentally regulated components may be possible. The cyclic nucleotide specificity of cAR1, cAR2, and cAR3 was assessed (Johnson *et al.*, 1992) and compared with the specificity of binding sites displayed on wild-type cells during cell aggregation (Van Haastert and Kien, 1983; Van Ments-Cohen and Van Haastert, 1989). Although the specificity of

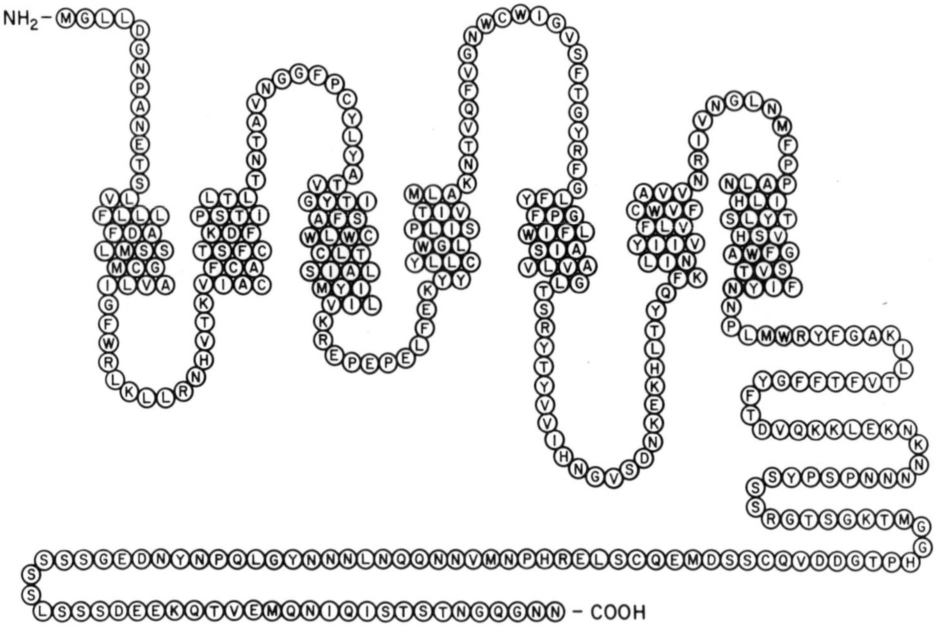

Figure 13.2 Deduced structure of the cell surface cAMP receptor cAR1.

each cAR is similar to the other cARs and to binding sites on wild-type cells, each cAR has a unique pharmacological profile. Assigning the individual receptor types to each of the kinetic forms observed on cells and in membranes has not been possible yet.

Cell lines with reduced expression of cAR1 were created by antisense RNA mutagenesis (Klein *et al.*, 1988; Sun *et al.*, 1990) or by gene disruption (Sun and Devreotes, 1991). cAMP stimulation of adenylyl cyclase, guanylyl cyclase, chemotaxis, cell aggregation, and differentiation is impaired in cells in which expression of cAR1 is low or absent. These results suggest that cAR1 mediates all those responses or that cAR1 is required for expression of other signal transduction components. The functions of cAR2, cAR3, and cAR4 are being investigated by inactivation of their genes. These receptors are expressed during the multicellular stages of development.

In vivo labeling of cells with [^{32}P]orthophosphate demonstrates that cAMP induces a fourfold increase of cAR1 phosphorylation over the basal ligand-independent level (Klein *et al.*, 1987; Vaughan and Devreotes, 1988). Phosphorylation causes a shift in the electrophoretic mobility of the receptor from 40 to 43 kD. The majority of phosphate is added to serine residues in the

C-terminal cytoplasmic domain that are grouped in four clusters. Preliminary evidence suggests that the cluster closest to the membrane is the major target of the receptor kinase.

Phosphorylation occurs in mutants that lack Gα2, the G protein that appears to interact with cAR1 (described subsequently), indicating that this G protein is not required for activation of the receptor kinase or phosphatase (Kesbeke *et al.*, 1988). Evidence suggests that a given receptor molecule need not be occupied to become phosphorylated. As long as an overall increase in occupancy occurs, the fraction of the phosphorylated receptors increases (Snaar-Jagalska *et al.*, 1988a). Thus, a global signal is transduced to the receptor kinase or phosphatase, leading to an increase in the steady-state level of phosphorylation.

Adaptation of adenylyl cyclase is correlated with phosphorylation of the C terminus of cAR1. Increments of stimulus concentration elicit proportionally larger transient activation of the enzyme. Pretreatment with a given concentration of stimulus abolishes the capacity to respond to the same or a lower concentration of stimulus. The extent and kinetics of phosphorylation are consistent with the extent and kinetics of adaptation of adenylyl cyclase. On removal of cAMP, the receptor is dephosphorylated with kinetics that match the return of sensitivity of the enzyme (Devreotes and Sherring, 1985; Knox *et al.*, 1986; Vaughan and Devreotes, 1988). The phosphorylated form of cAR1 appears to lead to persistent modification of a "downstream" component because adenylyl cyclase cannot be activated *in vitro* by 5′-γ thio-GTP (GTPγS) in the adapted system. This last observation suggests that the cAR system differs from two analogous systems, rhodopsin and β-adrenergic receptors, which display similar adaptation phenomena. Phosphorylation of these receptors allows the protein arrestin or α-arrestin to block interaction of the receptor with its respective G protein (see Deterre *et al.*, 1987).

D. G Proteins

Heterotrimeric G proteins transduce signals from cell-surface receptors to intracellular effector proteins. In vertebrates, G proteins are composed of a 38–52 kDa α subunit, a 35–36 kDa β subunit, and a 8–10 kDa γ subunit (see Neer and Clapham, 1988). In *D. discoideum*, genes encoding eight α subunit homologs and one β subunit homolog have been cloned. Gα1 and Gα2 were cloned by screening a cDNA library with oligonucleotides based on the sequence of the conserved GTP-binding domains (Pupillo *et al.*, 1989); Gα3–8 were cloned by polymerase chain reaction (PCR) using primers directed against these domains (Hadwiger and Firtel, 1991; Wu and Devreotes, 1991; M. Pupillo, S. Vogelmeier, G. Pitt, and P. N. Devreotes, unpublished results). Each of the α subunits is expressed transiently at a specific stage of development, so each stage is characterized by one or a few G proteins (Kumagai *et al.*, 1989). The gene encoding the β subunit is expressed during all developmental stages, with highest levels during growth, and probably serves all the α subunits (Lilly *et al.*, 1988).

Biochemical experiments support the idea that multiple G proteins are functional in *D. discoideum*; evidence has been presented for an adenylyl cyclase stimulatory G protein (Theibert and Devreotes, 1986; Van Haastert *et al.*, 1987b), an adenylyl cyclase inhibitory G protein that is sensitive to pertussis toxin (Van Haastert *et al.*, 1987b), and a G protein that stimulates phospholipase C (Europe-Finner and Newell, 1987; Van Haastert *et al.*, 1989; Bominaar *et al.*, 1991). The two putative G proteins that regulate adenylyl cyclase do not appear to be among the cloned Gα1–Gα5 subunits because disruption of Gα1, Gα2, and Gα3 does not affect this activity, and Gα4 and Gα5 are expressed only later in development. Gα6, Gα7, and Gα8 have not been tested yet. Finally, adenylyl cyclase may be regulated by the Bγ subunits of G proteins.

Gα2 is the most extensively characterized α subunit, and is most likely the G protein that regulates phospholipase C; this conclusion is based on the observation that *D. discoideum* mutant *fgdA* (Coukell *et al.*, 1983) shows defects in a Gα subunit (Kesbeke *et al.*, 1988; Snaar-Jagalska *et al.*, 1988), alterations of the Gα2 gene (Kumagai *et al.*, 1989) and a defect in GTPγS-mediated stimulation of phospholipase C (Bominaar *et al.*, 1991a; Bominaar and Van Haastert, unpublished observations). cAMP stimulates a transient phosphorylation of Gα2 on serine residues (Gundersen and Devreotes, 1990). The phosphorylation requires cAMP-binding to surface receptors and occurs at cAMP concentrations that elicit physiological responses. The role of Gα2 phosphorylation is unknown; the kinetics and transient nature suggest that it may be related to excitation responses that are also transient in these cells. The kinase responsible for Gα2 phosphorylation remains to be determined.

E. Adenylyl Cyclase Pathway

Adenylyl cyclase structural genes have been cloned from yeast and mammalian cells (see Krupinski *et al.*, 1989). Although the yeast enzyme is soluble, the deduced amino acid sequence of the mammalian enzyme shows features of a transporter protein: twelve putative membrane-spanning domains localized in two groups, each of which is followed by a large cytoplasmic domain. The two cytoplasmic domains show strong homology to the catalytic domains of yeast adenylyl cyclase and of vertebrate and sea urchin guanylyl cyclase. The *Dictyostelium* adenylyl cyclase that mediates cell–cell interaction during aggregation had a complex topography similar to that of the mammalian enzyme (Pitt *et al.*, 1992).

Receptor-mediated activation of adenylyl cyclase in *D. discoideum* is of special interest because the pathway may be different from that characterized in mammalian cells. Stimulatory and inhibitory G proteins are likely to be involved in this process, since GTP-mediated activation and inhibition of adenylyl cyclase in a cell-free system have been detected (Theibert and Devreotes, 1986; Van Haastert *et al.*, 1987b). Coupling of the surface receptor to the enzyme is diminished greatly in broken cell preparations, but stimulation with guanine nucleotides is robust. For maximal stimulation, GTPγS must be present during cell lysis and the lysate must be incubated for an additional 5 min at 0°C before GTPγS-mediated stimulation of adenylyl cyclase can be detected (Theibert and Devreotes, 1986). Alternatively, stimulation by GTPγS can be observed if the adenylyl cyclase assay with isolated membranes is performed at 0°C in the presence of a cytosolic protein (Van Haastert *et al.*, 1987b). This cytosolic protein, called GRP, appears to be essential for the loading of a G protein with GTP and is nonfunctional in the mutant *synag* 7 (Theibert and Devreotes, 1986; Van Haastert *et al.*, 1987b; Snaar-Jagalska and Van Haastert, 1988). *synag* 7 represents one of at least four complementation groups identified by mutants that are defective in guanine nucleotide regulation of the enzyme.

In *D. discoideum*, cAMP also acts intracellularly, as it does in vertebrates. cAMP binds to intracellular proteins including the cAMP-dependent protein kinase (PKA; Arents and Van Driel, 1982; De Gunzburg and Veron, 1982; Leichtling *et al.*, 1982) and the cAMP-binding protein cABP-1 (Tsang *et al.*, 1988). The gene that encodes the regulatory subunit of PKA has been cloned; the deduced amino acid sequence predicts a protein that is homologous to mammalian PKA and different from the surface cAMP receptor (Mutzel *et al.*, 1987). The strong homology to the vertebrate PKA is reflected also in the nearly identical cyclic nucleotide binding specificity of these proteins (De Wit *et al.*, 1982).

The function of the cAMP signal transduction pathway in *D. discoideum* has been investigated using several different approaches, including cAMP derivatives that specifically activate either surface cAMP receptors or intracellular PKA (Van Haastert, 1983; Van Haastert and Kien, 1983; Schaap and Van Driel, 1985; Harirabu and Dottin, 1986; Oyama and Blumberg, 1986; Theibert *et al.*, 1986; Kimmel, 1987; Mann and Firtel, 1987; Van Ments-Cohen and Van Haastert, 1989), mutants that are defective in the activation of adenylyl cyclase (*synag* 7) (Schaap *et al.*, 1986), transformants that overexpress the regulatory subunit of PKA (Simon *et al.*, 1989), or transformants that express antisense mRNA complementary to the mRNA encoding the surface cAMP receptor (Klein *et al.*, 1988; Sun *et al.*, 1990; Sun and Devreotes, 1991). These studies suggest that all functions of extracellular cAMP are mediated by surface receptors. Further, the activation of adenylyl cyclase is not required for cell growth, chemotaxis, or differentiation but is required for the production of the extracellular relay signal and for the fine regulation of the developmental program. On the other hand, overexpression of the regulatory subunit of PKA or of a regulatory subunit that cannot bind cAMP, both of which are expected to neutralize the activity of the catalytic subunit, block cell type-specific differentiation. Further, RdeC, a rapidly developing sporogenous mutant, has a mutation in the regulatory subunit that prevents interaction with the catalytic subunit. The mechanism of these intriguing phenotypes remains to be determined. PKA may transduce signals to developmental gene expression, or PKA may regulate signal transduction pathways indirectly. For example, PKA may be involved in cAMP-induced down-regulation of cAR1 (see subsequent text).

F. Inositol Phosphate/Ca^{2+} Pathway

In vertebrates, stimulation of a specific phospholipase C (PLC) leads to hydrolysis of phosphatidylinositol 4,5-bisphosphate [PI(4,5)P_2] to produce diacylglycerol and inositol 1,4,5-trisphosphate [I(1,4,5)P_3] (Fig. 13.3). I(1,4,5)P_3 releases Ca^{2+} from a nonmitochondrial pool, whereas diacylglycerol and Ca^{2+} stimulate the calcium/phospholipid-dependent protein kinase PKC (see Nishizuka, 1988; Beridge and Irvine, 1989). Calcium also may stimulate calmodulin-dependent enzymes and other calcium-dependent activities such as the polymerization of actin. Several genes encoding PLC have been cloned from mammalian cells and from *Drosophila* (see Rhee *et al.*, 1989). The family of PLC genes encodes a diverse set of proteins with different molecular weights, although all contain two small conserved sequences that probably form the catalytic domain. Sequence data from a *Dictyostelium* PLC shows that it contains the two conserved sequences and has the topography of PLC-δ (Drayer and Van Haastert, 1992).

In *Dictyostelium* cells and membranes cAMP and GTPγS, respectively, stimulate the formation of I(1,4,5)P_3 (Europe-Finner *et al.*, 1989; Van Haastert *et al.*, 1989). *In vivo*, the concentration of I(1,4,5)P_2 increases from about 3.3 μM to 5.5 μM (Van Haastert, 1989). Substantial secretion of I(1,4,5)P_3 occurs (Van Haastert, 1989). As in higher eukaryotic cells, I(1,4,5)P_3 stimulates the release of Ca^{2+} from nonmitochondrial pools (Europe-Finner and Newell, 1986).

In vertebrates, inositol and CDP-diacylglycerol are converted to PI, which is subsequently phosphorylated to PI(4)P and PI(4,5)P_2. These phospholipids are hydrolyzed by PLC activity (Rhee *et al.*, 1989); the products of these reactions are I(1)P, I(1,4)P_2, and I(1,4,5)P_3, respectively. I(1,4,5)P_3 is metabolized by two routes: direct dephosphorylation via I(1,4)P_2 and I(4)P to inositol or phosphorylation to $I(1,3,4,5)P_4$ followed by a complex series of dephosphorylations and phosphorylations (see Majerus *et al.*, 1990; Fig. 13.3).

In *Dictyostelium* cells, PI, PIP, and PIP_2 were identified based on the kinetics of label incorporation of [^{32}P]orthophosphate and [^{3}H]inositol (Van Haastert *et al.*, 1989). In membranes, the kinases that phosphorylate PI and PIP were characterized (Varela *et al.*, 1987). In both experiments, the polar head groups of the lipids were not identified and the fatty acid composition of inositol-containing phospholipids was not determined. *Dictyostelium discoideum* is reported to be devoid of arachidonic acid (MacDonald and Weeks 1985).

Dictyostelium cells possess the $I(1,4,5)P_3$ 5-phosphatase, the $I(1,4)P_2$/$I(1,3,4)P_3$ 1-phosphatase, and the IP monophosphatase that have been described in vertebrates. However, the major route of dephosphorylation in aggregating cells is by a 1-phosphatase to $I(4,5)P_2$ and subsequently to I(4)P (Van Lookeren Campagne *et al.*, 1988a). The presence of multiple dephosphorylation routes of $I(1,4,5)P_3$ suggests that these routes are regulated and that the IP_2 products may have specific functions. Indeed, the metabolism of $I(1,4,5)P_3$ changes during differentiation of *D. discoideum* cells (Bominaar *et al.*, 1991a).

Lithium ions have profound effects on the development of *D. discoideum* cells; lithium ions promote prestalk differentiation and inhibit prespore differentiation (Van Lookeren Campagne *et al.*, 1988b; Peters *et al.*, 1989). Lithium ions inhibit the activity of multiple inositol phosphate phosphatases: the inositol

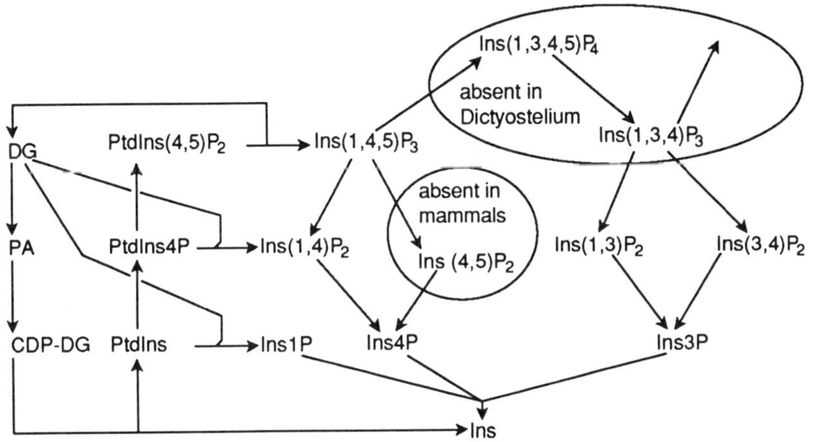

Figure 13.3 The inositol cycle of mammalian and *D. discoideum* cells. CDP-DG, CDP-diacylglycerol; PA, phosphatidic acid; DG, diacylglycerol; Ins, inositol; PtdIns, phosphatidyl inositol.

monophosphate phosphatase, the $I(1,4)P_2$ 1-phosphatase, and the $I(1,4,5)P_3$ 1-phosphatase (Van Lookeren Campagen *et al.*, 1988a). The overall effect of lithium treatment is a reduction of the concentration of $I(1,4,5)P_3$ (Peters *et al.*, 1989). The inhibition of prespore gene expression by lithium could be the result of the reduction of $I(1,4,5)P_3$ formation.

The stimulation of phospholipase C leads not only to the formation of inositol phosphates but also to the production of diacylglycerol which, in combination with Ca^{2+}, may stimulate PKC. In vertebrates, different PKC isoenzymes have been identified that are encoded by different genes. These isoenzymes show different sensitivity to the activators diacylglycerol, Ca^{2+}, and phosphatidylserine and the artificial activator phorbol ester. The isoenzymes are also likely to show different substrate specificities (Nishizuka, 1988). In *D. discoideum*, a PKC-like enzyme activity has been identified (Luderus *et al.*, 1989). The existence of this enzyme was predicted by the effects of phorbol ester, ATP, and Ca^{2+} on cAMP binding to surface receptors (Van Haastert *et al.*, 1985; Luderus *et al.*, 1988). Phorbol esters, in combination with $I(1,4,5)P_3$, have been reported to induce prespore gene expression in saponin-permeabilized *D. discoideum* cells (Ginsburg and Kimmel, 1989). A more complete picture of the function of PKC in *D. discoideum* awaits the cloning of the gene(s) and the reduction of transcription by antisense mRNA inactivation or gene disruption.

In yeast, Ras directly regulates the activity of adenylyl cyclase (Toda *et al.*, 1985; Chapter 7), but similar regulation does not appear to occur in other organisms, including *Dictyostelium* (Reymond *et al.*, 1986). However, Ras may play a role in the inositol phosphate pathway. *Dictyostelium discoideum* cells contain two genes that are highly homologous with the mammalian *ras* genes, one gene expressed during growth (Robbins *et al.*, 1989) and one during differentiation (Reymond *et al.*, 1984). Cell lines have been constructed that overexpress either the wild-type *RAS* gene or a *RAS* gene encoding a protein with a Gly-12 → Thr-12 mutation, which would confer oncogenic activity in mammalian *ras* genes (Reymond *et al.*, 1986). Overexpression of $RAS^{Thr\text{-}12}$ produces a biochemical phenotype that could be explained by constitutive activation of the inositol phosphate pathway (Van Haastert *et al.*, 1987a; Europe-Finner *et al.*, 1988; Luderus *et al.*, 1988; Klein *et al.*, 1989). $I(1,4,5)P_3$ levels are increased as are intracellular Ca^{2+} levels; as a result, guanylyl cyclase is inactivated more rapidly by Ca^{2+} after receptor-mediated activation. A PKC-like activity probably also is activated. A more detailed investigation of the inositol cycle suggests that increased $I(1,4,5)P_3$ levels in the mutated *RAS* transformant are not caused by increased phospholipase C activity, but by the enhanced conversion of PI to PIP, leading indirectly to increased levels of PIP_2 and $I(1,4,5)P_3$ (Van der Kaay *et al.*, 1990). The reason why PI kinase activity is increased in these $RAS^{Thr\text{-}12}$ transformants is presently unclear.

G. Guanylyl Cyclase Pathway

The cloning of genes encoding guanylyl cyclase in higher eukaryotes reveals that this transduction pathway shows unique characteristics (see Garbers, 1989). The membrane-bound guanylyl cyclase in sea urchin spermatozoa is stimulated by peptides secreted by sea urchin eggs. These peptides presumably induce chemoattraction of spermatozoa. In mammalian cells, a membrane-bound guanylyl cyclase is activated by atrial natriuretic peptide (ANP). In both organisms, membranous guanylyl cyclase activity and specific binding of the ligand appear to co-localize to one protein.

In *Dictyostelium*, guanylyl cyclase is activated within 2 sec of addition of cAMP to sensitive cells. The cGMP concentration peaks at 10 sec and the basal cGMP level is recovered after 30 sec (see Janssens and Van Haastert, 1987). The mechanism by which extracellular cAMP elicits these changes in cGMP is not known. Pharmacologically, the response appears to be mediated by the surface receptor (Van Haastert and Kien, 1983). However, the currently known cARs are related topologically to the G protein-coupled receptors and contain no guanylyl cyclase catalytic domain (Klein *et al.*, 1988). Therefore, cAMP is unlikely to bind directly to the enzyme. Based on observations of saponin-permeabilized cells, the enzyme may be activated by intracellular Ca^{2+} generated by $I(1,4,5)P_3$-induced release from internal stores (Europe-Finner and Newell, 1985; Small *et al.*, 1986). However, observations suggest that the enzyme is strongly inhibited by submicromolar Ca^{2+} concentrations (Janssens *et al.*, 1989; Valkema and Van Haastert, 1992). A mutant defective in the activation of phospholipase C shows normal activation of guanylyl cyclase (Bominaar *et al.*, 1991b). Guanylyl cyclase has been observed to be activated by GTPγS, suggesting that a G protein may be involved in the regulation of enzyme activity (Janssens *et al.*, 1989); however, stimulation by GTPγS is not inhibited by GDPβS, an accepted prerequisite for assigning G protein involvement.

The newly synthesized cGMP remains largely intracellular, where it binds to specific intracellular cGMP-binding proteins and is hydrolyzed by a cGMP-stimulated cGMP phosphodiesterase (see Janssens and Van Haastert, 1987). cGMP-binding proteins have been detected that bind to DNA (Parissenti and Coukell, 1989) and a cGMP-stimulated protein kinase was identified in *Dictyostelium* lysates (Wanner and Wurster, 1990).

H. Ion Channels

Investigation of the regulation of ion channels by extracellular signals in *Dictyostelium* has been impaired by a property of the plasma membrane that hinders the formation of tight seals required for patch-clamp analysis (Muller *et al.*, 1986). However, extracellular cAMP is known to increase the efflux of H^+ and

K^+ and the influx of Ca^{2+} (Malchow *et al.*, 1978; Bumann *et al.*, 1983; Aeckerle *et al.*, 1985). The transport of protons appears to involve an electrogenic H^+-ATPase and is an important determinant of the plasma membrane potential (Van Duijn and Vogelzang, 1989). The transport of Ca^{2+} is regulated by a Ca^{2+}-ATPase that pumps Ca^{2+} out of the cell and by a Ca^{2+} channel that is opened by receptor agonists, leading to the entrance of Ca^{2+} (Wick *et al.*, 1978; Boehme *et al.*, 1987; Milne and Coukell, 1988).

The efflux of K^+ may be stimulated by intracellular Ca^{2+} or cGMP (Aeckerle and Malchow, 1989; Van Duijn and Wang, 1990). The *stmF* mutant shows both enhanced cGMP formation and Ca^{2+} influx, suggesting that the receptor-stimulated Ca^{2+} influx may be stimulated by intracellular cGMP (Menz *et al.*, 1991). However, the receptor-stimulated influx of Ca^{2+} is normal in mutant *fgdA*, which does not respond to extracellular cAMP with an increase of cAMP, cGMP, or $I(1,4,5)P_3$ (Milne and Cookell, 1991).

The role of all these ion translocations for cellular functions is limited, since removal of extracellular Ca^{2+} with EGTA does not interfere with chemotaxis, cell aggregation, or differentiation (see Janssens and Van Haastert, 1987). Moreover, signal transduction and chemotaxis are essentially normal in electropermeabilized cells, in which free diffusion of ions and small metabolites through the plasma membrane is possible (Van Duijn *et al.*, 1990). Although these experiments establish that membrane potential and translocation of K^+ and Na^+ are not necessarily important for signal transduction and chemotaxis, regulation of intracellular Ca^{2+} still may be critical since EGTA abolishes the chemotactic response in electroporated cells (Van Duijn *et al.*, 1990; Van Duijn and Van Haastert, 1992).

I. Desensitization of Signal Transduction

Persistent stimulation of cells with high ligand concentrations generally leads to desensitization. This effect may be mediated by several mechanisms, including uncoupling of surface receptors from transducing G proteins (adaptation), removal of binding sites from the cell surface (loss of ligand binding), degradation of the receptors (down-regulation), regulation of receptor expression, and inhibition of downstream effects (see Lefkowitz and Caron, 1988). These processes often are associated with ligand-induced phosphorylation of the receptor, which occurs in virtually all G protein-coupled receptors including rhodopsin, the β-adrenergic receptor (Hausdorff *et al.*, 1989), the *D. discoideum* cAMP receptor (Klein *et al.*, 1987; Vaughan and Devreotes, 1988), and the yeast α-factor receptor (Reneke *et al.*, 1988).

Using a collection of signal transduction mutants, desensitization of sensory transduction in *Dictyostelium* was demonstrated to consist of at least four components (Van Haastert *et al.*, 1992).

1. Loss of ligand binding occurs so the receptor no longer binds cAMP, but still may be on the cell surface. This step occurs with a $t_{1/2}$ of about 2 min after addition of receptor agonists or antagonists and does not require the activation of any known second messenger or the phosphorylation of the receptor.
2. Phosphorylation of the receptor leads to uncoupling of the sensory transduction pathway. This step occurs with a $t_{1/2}$ of about 1 min and requires the activation of the receptor with agonists but does not require the activation of known second messenger enzymes.
3. Degradation of the receptor occurs on a time scale of about 30 min. This step presumably requires the activation of adenylyl cyclase and phospholipase C.
4. Disappearance of mRNA encoding cAR1 occurs on a time scale of a few minutes and may require the activation of phospholipase C.

In addition to these different mechanisms of desensitization of the receptor, desensitization of different effector systems is not mediated by one uniform mechanism. For instance, guanylyl cyclase stimulation desensitizes within 2 sec of stimulation whereas adenylyl cyclase desensitizes with a $t_{1/2}$ of 1 min. Further, desensitization and recovery from desensitization for the adenylyl and guanylyl cyclases display very different temperature dependencies (Van Haastert, 1987). Finally, desensitization of phospholipase C may not occur in *Dictyostelium*.

III. Genetics of Signal Transduction in *Dictyostelium discoideum*

In *D. discoideum*, the genetics of signal transduction have been investigated using several mutants that are defective in specific components of the network (Fig. 13.4). Signal transduction depends on the presence of surface cAMP receptors, as indicated by studies with cells in which the expression of the *cAR1* gene was inactivated (Klein *et al.*, 1988; Sun *et al.*, 1990; Sun and Devreotes, 1991). Preliminary observations suggest that cAR2 and cAR3 can substitute for some but not all of the functions of cAR1.

Activation of a specific G protein (Gα2) appears to be essential for multiple sensory responses, as indicated using mutant *fgdA*, which is defective in this G protein. Gα2 appears to be the G protein directly coupled to cAR1; in its absence, the affinity of cAR1 for cAMP is low and insensitive to regulation by guanine nucleotides. One of the effectors of Gα2 is most likely PLC because, in *fgdA, in vitro* GTPγS stimulation of PLC is absent, whereas GTPγS stimulation of adenylyl cyclase and guanylyl cyclase is unaltered. *In vivo*, however, cAMP

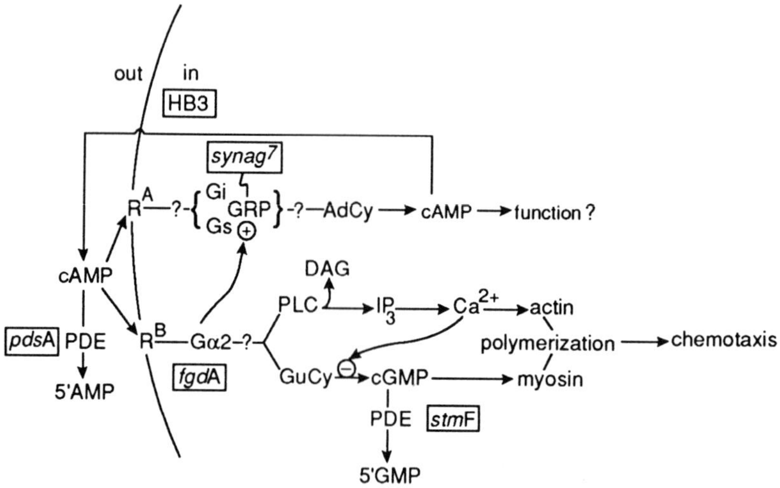

Figure 13.4 Model for sensory transduction in *Dictyostelium*. cAMP binds to two forms of surface receptors that activate the relay pathway (R^A) and the chemotaxis/differentiation pathway (R^B), respectively. R^A and R^B may be two forms of the same molecule. The relay pathway is composed of unidentified G proteins with stimulatory and inhibitory properties, GRP (an associated protein defective in mutant *synag* 7), and adenylyl cyclase. cAMP is secreted by an unknown mechanism that is defective in mutant *hb3*. The chemotaxis/differentiation pathway is composed of the G protein Gα2, guanylyl cyclase, and phospholipase C. The second messengers cGMP and DAG/Ins(1,4,5)P_3/Ca^{2+} are involved in actin and myosin polymerization, resulting in chemotaxis. DAG and Ca^{2+}, either alone or in combination, are likely to be involved in certain types of gene expression. Also several cross-interactions occur, notably the inhibition of guanylyl cyclase by Ca^{2+} and the requirement of Gα2 for the activation of adenylyl cyclase. Several mutants are indicated in boxes. PDE, cyclic nucleotide phosphodiesterase; DAG, diacylglycerol; AdCy, adenylyl cyclase; GuCy, guanylyl cyclase; PLC, phospholipase C; IP3, inositol(1,4,5)P_3; GRP, GTP reconstituting protein; +, activation; −, inhibition.

and cGMP as well as IP_3 are blocked. Why adenylyl and guanylyl cyclase are not stimulated by cAMP *in vivo* in $G\alpha 2^-$ cells is not known, but this result suggests a link between the signal transduction pathways.

A series of mutants designated *synag* are specifically defective in cell–cell signaling. The most well characterized is *synag* 7, described earlier. In these cells, cAMP fails to elicit cAMP production *in vivo* and GTPγS fails to activate adenylyl cyclase *in vitro*. Since these cells do not produce appropriate signaling responses, they fail to aggregate. However, when mixed with wild-type cells, they can respond to wild-type signals and form chimeric aggregates. This phenotype has been exploited to isolate multiple *synag* mutants. Many of these mutants appear to be specifically defective in GTPγS activation of adenylyl cyclase. Preliminary observations suggest that at least four complementation groups are involved in this response.

Two additional mutants have been used to investigate the role of cAMP. In mutant *pdsA*, extracellular phosphodiesterase activity is absent (Barra *et al.*, 1980); in mutant *hb3*, cAMP secretion is impaired (Barclay and Henderson, 1986; Kesbeke and Van Haastert, 1988). As in the *synag* mutants, the activation of guanylyl cyclase and PLC is essentially normal and cAMP induces normal chemotaxis and cell differentiation (Schaap *et al.*, 1986).

The function of intracellular cGMP has been investigated using mutant *stmF*, which appears to be defective in cGMP phosphodiesterase (Ross and Newell, 1981; Van Haastert *et al.*, 1982b). In this mutant, extracellular cAMP induces elevated and prolonged cGMP accumulation, which is associated with prolonged chemotactic movement. The enhanced accumulation of cGMP in mutant *stmF* has no effect on the stimulation of adenylyl cyclase or PLC. The importance of the cGMP pathway will be established more accurately when mutants lacking guanylyl cyclase are available.

Mutant *fgdC* (Coukell *et al.*, 1983) was characterized further. Extracellular cAMP induces the normal activation of adenylyl and guanylyl cyclase but no activation of phospholipase C occurs in this mutant. This defect is associated with the absence of cAMP-induced prespore and prestalk gene expression. These observations suggest that activation of phospholipase C is required for cAMP-mediated gene expression, but is not required for the activation of adenylyl or guanylyl cyclase, as was suggested previously (Bominaar *et al.*, 1991b). More mutants in the inositol phosphate–Ca^{2+} pathway are required to unravel the functions of the many components of the inositol cycle.

IV. Comparison of Signal Transduction in *Dictyostelium* and Vertebrates

The description of signal transduction provided in this chapter suggests that many sensory transduction pathways identified in vertebrates are also present in *Dictyostelium*, perhaps with the exception of sensory transduction via steroid receptors. Although the evolutionary distance between *D. discoideum* and vertebrates is nearly one billion years, surface receptors, G proteins, adenylyl cyclase, guanylyl cyclase, and the inositol cycle are remarkably well conserved.

The powerful genetic techniques that can be used in microorganisms are proving very useful in studying sensory transduction in these organisms. In the yeast *Saccharomyces cerevisiae*, many components of the sensory transduction pathways were identified by a combination of biochemistry and molecular genetics. *Dictyostelium discoideum* has the additional advantage of a tightly programmed development of identical free-moving amoeboid cells to a multicellular organism consisting of two or three cell types. In the coming years, we expect to learn in much more detail how simple extracellular signals regulate complex cellular functions such as chemotaxis, morphogenesis, cell differentiation, and pattern formation.

References

Aeckerle, S., and Malchow, D. (1989). Calcium regulates cAMP-induced potassium ion efflux in *Dictyostelium discoideum*. *Biochim. Biophys. Acta* **1012**, 196–200.

Aeckerle, S., Wurster, B., and Malchow, D. (1985). Oscillations and cAMP-induced changes of the K^+ concentration in *Dictyostelium discoideum*. *EMBO J.* **4**, 39–43.

Arents, J. C., and Van Driel, R. (1982). Soluble cytoplasmic cyclic AMP binding proteins of *D. discoideum*. *FEBS Lett.* **137**, 201–204.

Barclay, S. L., and Henderson, E. J. (1986). Altered cAMP receptor activity and morphogenesis in a chemosensory mutant of *Dictyostelium discoideum*. *Differentiation* **33**, 111–120.

Barra, J., Barrand, P., Blondelet, M. H., and Brachet, P. (1980). *pdsA*, a gene involved in the production of active phosphodiesterase during starvation of *Dictyostelium discoideum* amoebae. *Mol. Gen. Genet.* **177**, 607–613.

Beridge, M. J., and Irvine, R. F. (1989). Inositol phosphates and cell signalling. *Nature (London)* **341**, 197–205.

Boehme, R., Bumann, J., Aeckerle, S., and Malchow, D. (1987). A high affinity plasma membrane Ca^{2+}-ATPase in *Dictyostelium discoideum*: Its relation to cAMP-induced Ca^{2+} fluxes. *Biochim. Biophys. Acta* **904**, 125–130.

Bominaar, A. A., Van der Kaay, J., and Van Haastert, P. J. M. (1991a). Dynamics and function of the inositol cycle in *Dictyostelium discoideum*. *Dev. Gen.* **12**, 19–24.

Bominaar, A. A., Kesbeke, F., Snaar-Jagalska, B. E., Peters, D. J. M., Schaap, P., and Van Haastert, P. J. M. (1991b). Aberrant chemotaxis and differentiation in mutant *fgd* C with a defective regulation of receptor-stimulated phosphoinositidase C. *J. Cell Sci.* **100**, 825–831.

Bumann, J., Wurster, B., and Malchow, D. (1983). Attractant-induced changes and oscillations of the extracellular Ca^{2+} concentration in suspensions of differentiating *Dictyostelium* cell. *J. Cell Biol.* **98**, 173–178.

Coukell, M. B., Lappano, S., and Cameron, A. M. (1983). Isolation and characterization of cAMP unresponsive (frigid) aggregation-deficient mutants of *Dictyostelium discoideum*. *Dev. Gen.* **3**, 283–297.

De Gunzburg, J., and Veron, M. (1982). A cAMP-dependent protein kinase is present in differentiating *Dictyostelium discoideum* cells. *EMBO J.* **1**, 1063–1068.

Deterre, P., Pfister, C., Bigay, J., and Chabre, M. (1987). The retinal phototransduction process: Enzymatic cascade and regulation. *Biochimie* **69**, 365–370.

Devreotes, P. N., and Sherring, J. A. (1985). Kinetics and concentration dependence of reversible cAMP-induced modification of the surface cAMP receptor in *Dictyostelium*. *J. Biol. Chem.* **260**, 6378–6384.

De Wit, R. J. W., and Bulgakov, R. (1985). Guanine nucleotides modulate the ligand binding properties of cell surface folate receptors in *Dictyostelium discoideum*. *FEBS Lett.* **179**, 257–261.

De Wit, R. J. W., and Konijn, T. M. (1983a). Identification of the acrasin of *Dictyostelium minutum* as a derivative of folic acid. *Cell Diff.* **12**, 205–210.

De Wit, R. J. W., and Konijn, T. M. (1983b). Pterines and folates as extracellular signals for chemotaxis and differentiation in cellular slime molds. *In* "Biochemical and Clinical Aspects of Pteridines" (H. C. Curtius, W. Pfleiderer, and H. Wachter, eds.), Vol. 2, pp. 383–400. Walter De Gruyter, Berlin.

De Wit, R. J. W., and Van Haastert, P. J. M. (1985). Binding of folates to *Dictyostelium discoideum* cells. Demonstration of five classes of binding sites and their interconversion. *Biochim. Biophys. Acta* **814**, 199–213.

De Wit, R. J. W., Arents, J. C., and Van Driel, R. (1982). Ligand binding properties of the cytoplasmic cAMP-binding protein of *Dictyostelium discoideum. FEBS Lett.* **145**, 150–154.

Drayer, A. L., and Van Haastert, P. J. M. (1992). Molecular cloning and expression of a phosphoinositide-specific phospholipase C of *Dictyostelium discoideum. J. Biol. Chem.* **267**, 18387–18392.

Europe-Finner, G. N., and Newell, P. C. (1985). Inositol 1,4,5-triphosphate induces cyclic GMP formation in *Dictyostelium discoideum. Biochem. Biophys. Res. Commun.* **130**, 1115–1122.

Europe-Finner, G. N., and Newell, P. C. (1986). Inositol 1,4,5-trisphosphate induces calcium release from a non-mitochondrial pool in amoebae in *Dictyostelium. Biochim. Biophys. Acta* **887**, 335–340.

Europe-Finner, G. N., and Newell, P. C. (1987). GTP-analogues stimulate inositol trisphosphate formation transiently in *Dictyostelium. J. Cell Sci.* **87**, 513–518.

Europe-Finner, G. N., Luderus, M. E. E., Small, N. V., Van Driel, R., Reymond, C. D., Firtel, R. A., and Newel, P. C. (1988). Mutant RAS gene induces elevated levels of inositol tris- and hexakis phosphates in *Dictyostelium discoideum. J. Cell Sci.* **89**, 13–20.

Europe-Finner, G. N., Gammon, B., Wood, C. A., and Newell, P. C. (1989). Inositol tris- and polyphosphate formation during chemotaxis of *Dictyostelium. J. Cell Sci.* **93**, 585–592.

Faure, M., Podgorski, G. J., Franke, J., and Kessin, R. H. (1988). Disruption of *Dictyostelium discoideum* morphogenesis by overproduction of cAMP phosphodiesterase *Proc. Natl. Acad. Sci. U.S.A.* **85**, 8076–8080.

Faure, M., Podgorski, G. J., Franke, J., and Kessin, R. H. (1989). Rescue of a *Dictyostelium discoideum* mutant defective in phosphodiesterase. *Dev. Biol.* **131**, 366–372.

Faure, M., Franke, J., Hall, A. L., Podgorski, G. J., and Kessin, R. H. (1990). The cyclic nucleotide phosphodiesterase gene of *Dictyostelium discoideum* contains three promoters specific for growth, aggregation, and late development. *Mol. Cell. Biol.* **10**, 1921–1930.

Garbers, D. L. (1989). Guanylyl cyclase, a cell surface receptor. *J. Biol. Chem.* **264**, 9103–9106.

Ginsburg, G., and Kimmel, A. R. (1989). Inositol trisphosphate and diacylglycerol can differentially modulate gene expression in *Dictyostelium. Proc. Natl. Acad. Sci. U.S.A.* **86**, 9332–9336.

Gundersen, R. E., and Devreotes, P. N. (1990). In vivo receptor-mediated phosphorylation of a G protein in *Dictyostelium. Science* **248**, 591–593.

Hadwiger, J., and Firtel, R. (1991). Identification of *Dictyostelium* Gα genes expressed during molecular development. *Proc. Natl. Acad. Sci. U.S.A.* **88**, 8213–8217.

Harirabu, B., and Dottin, R. P. (1986). Pharmacological characterization of cyclic AMP receptors mediating gene regulation in *Dictyostelium discoideum. Mol. Cell Biol.* **6**, 2402–2408.

Hausdorff, W. P., Bouvier, M., O'Dowd, B. F., Irons, G. P., Caron, M. G., and Lefkowitz, R. J. (1989). Phosphorylation sites of two domains of the [beta]2-adrenergic receptor are involved in distinct pathways of receptor desensitization. *J. Biol. Chem.* **264**, 12657–12665.

Janssens, P. M. W., and Van Haastert, P. J. M. (1987). Molecular basis of transmembrane signal transduction in *Dictyostelium discoideum. Microbiol. Rev.* **51**, 396–418.

Janssens, P. M. W., De Jong, C. C. C., Vink, A. A., and Van Haastert, P. J. M. (1989). Regulatory properties of magnesium-dependent guanylyl cyclase in *Dictyostelium discoideum* membranes. *J. Biol. Chem.* **264**, 4329–4335.

Johnson, R., Vaughan, R., Caterina, M., Van Haastert, P. J. M., and Devreotes, P. N. (1991). Overexpression of the cAMP receptor (cAR1) in growing *Dictyostelium* cells. *Biochemistry* **30**, 6982–6986.

Johnson, R. L., Van Haastert, P. J. M., Kimmel, A. R., Saxe, C. L., III, Jastorff, B., and Devreotes, P. N. (1992). The cyclic nucleotide specificity of three cAMP receptors in *Dictyostelium. J. Biol. Chem.* **267**, 4600–4607.

Kesbeke, F., and Van Haastert, P. J. M. (1988). Reduced cAMP secretion in *Dictyostelium discoideum* mutant HB3. *Dev. Biol.* **130**, 464–470.

Kesbeke, F., Snaar-Jagalska, B. E., and Van Haastert, P. J. M. (1988). Signal transduction in *Dictyostelium* fgd A mutants with a defective interaction between surface cAMP receptors and a GTP-binding regulatory protein. *J. Cell Biol.* **107**, 521–528.

Kessin, R. H. (1988). Genetics of early *Dictyostelium discoideum* development. *Microbiol. Rev.* **52**, 29–49.

Kimmel, A. R. (1987). Different molecular mechanisms for cAMP regulation of gene expression during *Dictyostelium* development. *Dev. Biol.* **122**, 163–171.

Klein, P., Vaughan, R., Borleis, J., and Devreotes, P. (1987). The surface cyclic AMP receptor in *Dictyostelium*. Levels of ligand-induced phosphorylation, solubilization, identification of primary transcript, and developmental regulation of expression. *J. Biol. Chem.* **262**, 358–364.

Klein, P. S., Sun, T. J., Saxe, C. L., III, Kimmel, A. R., Johnson, R. L., and Devreotes, P. N. (1988). A chemoattractant receptor controls development in *Dictyostelium discoideum*. *Science* **241**, 1467–1472.

Klein, G., Martin, J.-B., Satre, M., and Reymond, C. D. (1989). Exocytosis and inositol hexakisphosphate levels in RAS transformants of *Dictyostelium discoideum* transformants. *Experientia* **45**, 365–367.

Knox, B. E., Devreotes, P. N., Goldbeter, A., and Segel, L. A. (1986). A molecular mechanism for sensory adaptation based on ligand-induced receptor modification. *Proc. Natl. Acad. Sci. U.S.A.* **83**, 2345–2349.

Konijn, T. M., Van de Meene, J. G. C., Bonner, J. T., and Barkley, D. S. (1967). The acrasin activity of adenosine-3′,5′-cyclic monophosphate. *Proc. Natl. Acad. Sci. U.S.A.* **58**, 1152–1154.

Krupinski, J., Coussen, F., Bakalyar, H. A., Tang, W. J., Feinstein, P. G., Orth, K., Slaughter, C., Reed, R. R., and Gilman, A. G. (1989). Adenylyl cyclase amino acid sequence: Possible channel- or transporter-like structure. *Science* **244**, 1558–1564.

Kumagai, A., Pupillo, M., Gundersen, R., Miake-Lye, R., Devreotes, P. N., and Firtel, R. A. (1989). Regulation and function of Gα protein subunits in *Dictyostelium*. *Cell* **57**, 265–275.

Lacombe, M.-L., Podgorski, G. J., Franke, J., and Kessin, R. H. (1986). Molecular cloning and developmental expression of the cyclic nucleotide phosphodiesterase gene of *Dictyostelium discoideum*. *J. Biol. Chem.* **261**, 16811–16817.

Lefkowitz, R. J., and Caron, M. G. (1988). Adrenergic receptors. Models for the study of receptors coupled to guanine nucleotide regulatory proteins. *J. Biol. Chem.* **263**, 4993–4998.

Leichtling, B. H., Majerfeld, I. H., Coffman, D. S., and Rickenberg, H. V. (1982). Identification of the regulatory subunit of the cAMP-dependent protein kinase in *Dictyostelium discoideum*. *Biochim. Biophys. Res. Commun.* **105**, 949–955.

Lilly, P., Klein, P., Theibert, A., Vaughan, R., Pupillo, M., Saxe, C., Kimmel, A., and Devreotes, P. (1988). Receptor G-protein interactions in the development of *Dictyostelium*. *Bot. Acta* **101**, 123–127.

Loomis, W. F. (1985). "*Dictyostelium discoideum*: A Developmental System." Academic Press, New York.

Ludérus, M. E. E., Reymond, C. D., Van Haastert, P. J. M., and Van Driel, R. (1988). Expression of a mutated ras gene in *Dictyostelium discoideum* alters the binding of cyclic AMP to its chemotactic receptor. *J. Cell Sci.* **90**, 701–706.

Ludérus, M. E. E., Van der Most, R. G., Otte, A. P., and Van Driel, R. (1989). A protein kinase C-related enzyme activity in *Dictyostelium discoideum*. *FEBS Lett.* **253**, 71075.

MacDonald, J. I. S., and Weeks, G. (1985). The biosynthesis and turnover of lipid during the differentiation of *Dictyostelium discoideum*. *Biochim. Biophys. Acta* **834**, 301–307.

Majerus, P. W., Ross, T. S., Cunningham, T. W., Caldwell, K. K., Bennett Jefferson, A., and Bansal, V. S. (1990). Recent insights in phosphatidylinositol signaling. *Cell* **63**, 459–465.

Malchow, D., Nanjundiah, V., and Gerisch, G. (1978). pH oscillations in cell suspensions on *Dictyostelium discoideum*, their relation to cAMP signal. *J. Cell Sci.* **30**, 319–330.

Mann, S. K. O., and Firtel, R. A. (1987). Cyclic AMP regulation of early gene expression in *Dictyostelium discoideum*: Mediation via the cell surface cyclic AMP receptor. *Mol. Cell. Biol.* **7**, 458–469.

Menz, S., Bumann, J., Jaworski, E., and Malchow, D. (1991). Mutant analysis suggests that cGMP mediates the cAMP-induced Ca^{2+} uptake in *Dictyostelium. J. Cell Sci.* **99**, 187–191.

Milne, J. L., and Coukell, M. B. (1988). Isolation and characterization of a plasma membrane calcium pump from *Dictyostelium discoideum. Biochem. J.* **249**, 223–230.

Milne, J. L., and Coukell, M. B. (1991). A Ca^{2+} transport system associated with the plasma membrane of *Dictyostelium discoideum* is activated by different chemotactic receptors. *J. Cell Biol.* **112**, 103–110.

Morris, H. R., Taylor, G. W., Masento, M. S., Jermyn, K. A., and Kay, R. R. (1987). Chemical structure of the morphogen differentiation inducing factor from *Dictyostelium discoideum. Nature (London)* **328**, 811–814.

Muller, U., Malchow, D., and Hartung, K. (1986). Single ion channels in the slime mold *Dictyostelium discoideum. Biochim. Biophys. Acta* **857**, 287–290.

Mutzel, R., Lacombe, M. L., Simon, M. N., De Gunzburg, J., and Veron, M. (1987). cAMP-dependent protein kinase from *Dictyostelium discoideum*: Cloning and cDNA sequence of the regulatory subunit. *Proc. Natl. Acad. Sci. U.S.A.* **84**, 6–10.

Neer, E. J., and Clapham, D. E. (1988). Roles of G protein subunits in transmembrane signalling. *Nature (London)* **333**, 129–134.

Nishizuka, Y. (1988). The molecular heterogeneity of protein kinase C and its implications for cellular regulation. *Nature (London)* **334**, 661–665.

Oyama, M., and Blumberg, D. D. (1986). Interaction of cAMP with the cell-surface receptor induces cell-type specific mRNA accumulation in *Dictyostelium discoideum. Proc. Natl. Acad. Sci. U.S.A.* **83**, 4819–4823.

Parissenti, A. M., and Coukell, M. B. (1989). Identification of a nucleic acid-regulated cyclic GMP-binding activity in *Dictyostelium discoideum. J. Cell Sci.* **92**, 291–301.

Peters, D. J. M., Van Lookeren-Campagne, M. M., Van Haastert, P. J. M., Spek, W., and Schaap, P. (1989). Lithium ions induce prestalk-associated gene expression and inhibit prespore gene expression in *Dictyostelium discoideum. J. Cell Sci.* **93**, 205–210.

Pitt, G. S., Milona, N., Borleis, J., Lin, K. C., Reed, R. R., and Devreotes, P. N. (1992). Structurally distinct and stage-specific adenylyl cyclase gene play different roles in *Dictyostelium* development. *Cell* **69**, 305–315.

Pupillo, M., Kumagai, A., Pitt, G. S., Firtel, R. A., and Devreotes, P. N. (1989). Multiple α subunits of guanine nucleotide-binding proteins in *Dictyostelium. Proc. Natl. Acad. Sci. U.S.A.* **86**, 4892–4896.

Reneke, J. E., Blumer, K. J., Courchesne, W. E., and Thorner, J. (1988). The carboxy-terminal segment of the yeast [alpha]-factor receptor is a regulatory domain. *Cell* **55**, 221–234.

Reymond, C. D., Gomer, R. H., Mehdy, M. C., and Firtel, R. A. (1984). Developmental regulation of a *Dictyostelium* gene encoding a protein homologous to mammalian ras protein. *Cell* **39**, 141–148.

Reymond, C. D., Gomer, R. H., Nellen, W., Theibert, A., Devreotes, P., and Firtel, R. (1986). Phenotypic changes induced by a mutated ras gene during the development of *Dictyostelium* transformants. *Nature (London)* **323**, 340–343.

Rhee, S. G., Suh, P.-G., Ruy, S.-H., and Lee, S. Y. (1989). Studies of inositol phospholipid-specific phospholipase C. *Science* **244**, 546–550.

Robbins, S. M., Williams, J. G., Jermyn, K. A., Spiegelman, G. B., and Weeks, G. (1989). Growing and developing *Dictyostelium* cells express different ras genes. *Proc. Natl. Acad. Sci. U.S.A.* **86**, 938–942.

Ross, F. M., and Newell, P. C. (1981). Streamers: Chemotactic mutants of *Dictyostelium discoideum* with altered cyclic GMP metabolism. *J. Gen. Microbiol.* **127**, 339–350.

Saxe, C. L., III, Johnson, R., Devreotes, P. N., and Kimmel, A. R. (1991a). Multiple genes for cell surface cAMP receptors in *Dictyostelium discoideum. Dev. Gen.* **12**, 6–13.

Saxe, C. L., III, Johnson, R. L., Devreotes, P. N., and Kimmel, A. R. (1991b). Expression of a cAMP receptor gene of *Dictyostelium* and evidence for a multigene family. *Genes Dev.* **5**, 1–8.

Schaap, P., and Van Driel, R. (1985). Induction of post-aggregative differentiation in *Dictyostelium discoideum* by cAMP. *Exp. Cell Res.* **159**, 388–398.

Schaap, P., and Wang, M. (1986). Interactions between adenosine and oscillatory cAMP signalling regulate seize and pattern in *Dictyostelium. Cell* **45**, 137–144.

Schaap, P., Van Lookeren Campagne, M. M., Van Driel, R., Spek, W., Van Haastert, P. J. M., and Pinas, J. (1986). Postaggregative differentiation induction by cyclic AMP in *Dictyostelium*: Intracellular transduction pathway and requirement for additional stimuli. *Dev. Biol.* **118**, 52–63.

Shimomura, O., Suthers, H. L. B., and Bonner, J. T. (1982). Chemical identity of the acrasin of the slime mold *Polysphondilium violaceum. Proc. Natl. Acad. Sci. U.S.A.* **79**, 7376–7379.

Simon, M. N., Driscol, D. M., Mutzert, R., Part, D., Williams, J. G., and Veron, M. (1989). Overproduction of the regulatory subunit of the cAMP-dependent protein kinase blocks the differentiation of *Dictyostelium discoideum. EMBO J.* **8**, 2039–2043.

Small, N. V., Europe-Finner, G. N., and Newell, P. C. (1986). Calcium inducescyclic GMP formation in *Dictyostelium. FEBS Lett.* **203**, 11–14.

Snaar-Jagalska, B. E., and Van Haastert, P. J. M. (1988). *Dictyostelium discoideum* mutant synac 7 with altered G-protein adenylyl cyclase interaction. *J. Cell Sci.* **91**, 287–294.

Snaar-Jagalska, B. E., Devreotes, P. N., and Van Haastert, P. J. M. (1988a). Ligand-induced modification of a surface cAMP receptor of *Dictyostelium discoideum* does not require its occupancy. *J. Biol. Chem.* **263**, 897–901.

Snaar-Jagalska, B. E., Kesbeke, F., Pupillo, M., and Van Haastert, P. J. M. (1988b). Immunological detection of G-protein α-subunits in *Dictyostelium discoideum. Biochem. Biophys. Res. Commun.* **156**, 757–761.

Sun, T. J., and Devreotes, P. N. (1991). Gene targeting of the aggregation stage cAMP receptor cAR1 in *Dictyostelium. Gene Dev.* **5**, 572–582.

Sun, T. J., Van Haastert, P. J. M., and Devreotes, P. N. (1990). Surface cAMP receptors mediate multiple responses during development in *Dictyostelium*: Evidenced by antisense mutagenesis. *J. Cell Biol.* **110**, 1549–1554.

Theibert, A., and Devreotes, P. (1986). Surface receptor-mediated activation of adenylyl cyclase in *Dictyostelium. J. Biol. Chem.* **261**, 15121–15125.

Theibert, A., Palmisano, M., Jastorff, B., and Devreotes, P. (1986). The specificity of the cAMP receptor mediating activation of adenylyl cyclase in *Dictyostelium discoideum. Dev. Biol.* **114**, 529–533.

Toda, T., Uno, I., Ishikawa, T., Powers, S., Kataoka, T., Broek, D., Cameron, S., Broach, J., Matsumoto, K., and Wigler, M. (1985). In yeast, RAS proteins are controlling elements of adenylyl cyclase. *Cell* **40**, 27–36.

Tsang, A., Grant, C., Kay, C., Bain, G., Greenwood, M., Noce, T., and Tasaka, M. (1988). Characterization of an unusual cAMP receptor and its related polypeptides in *Dictyostelium discoideum. Dev. Genet.* **8**, 237–245.

Valkema, R., and Van Haastert, P. J. M. (1992). Inhibition of receptor-stimulated guanylyl cyclase by intracellular calcium ions in *Dictyostelium* cells. *Biochem. Biophys. Res. Commun.* **186**, 263–268.

Van der Kaay, J., Draijer, R., and Van Haastert, P. J. M. (1990). Increased conversion of phosphatidylinositol to phosphatidylinositolphosphate in *Dictyostelium* cells expressing a mutated RAS gene. *Proc. Natl. Acad. Sci. U.S.A.* **87**, 9197–9201.

Van Duijn, B., and Van Haastert, P. J. M. (1992). Independent control of locomotion and chemotaxis during *Dictyostelium* chemotaxis. *J. Cell Sci.* **102**, 763–768.

Van Duijn, B., and Vogelzang, S. A. (1989). The membrane potential of the cellular slime mold *Dictyostelium discoideum* is mainly generated by an electrogenic proton pump. *Biochim. Biophys. Acta* **983**, 186–192.

Van Duijn, B., and Wang, M. (1990). Chemoattractant-induced membrane hyperpolarization in *Dictyostelium discoideum*. *FEBS Lett.* **275**, 201–204.

Van Duijn, B., Vogelzang, S. A., Ypey, D. L., Van der Molen, L. G., and Van Haastert, P. J. M. (1990). Normal chemotaxis in *Dictyostelium discoideum* with a depolarized plasma membrane potential. *J. Cell Sci.* **95**, 177–183.

Van Haastert, P. J. M. (1983). Binding of cAMP and adenosine derivatives to *Dictyostelium discoideum* cells. Relationships of binding, chemotactic, and antagonistic activities. *J. Biol. Chem.* **258**, 9643–9648.

Van Haastert, P. J. M. (1984). Guanine nucleotides modulate cell surface cAMP-binding sites in membranes from *Dictyostelium discoideum*. *Biochem. Biophys. Res. Commun.* **124**, 597–604.

Van Haastert, P. J. M. (1987). Differential effects of temperature on cAMP-induced excitation, adaptation and deadaptation of adenylyl and guanylyl cyclase in *Dictyostelium discoideum*. *J. Cell Biol.* **105**, 2301–2306.

Van Haastert, P. J. M. (1989). Determination of inositol 1,4,5-trisphosphate levels in *Dictyostelium* by isotope dilution assay. *Anal. Biochem.* **177**, 115–119.

Van Haastert, P. J. M., and De Wit, R. J. W. (1984). Demonstration of receptor heterogeneity and affinity modulation by nonequilibrium binding experiments. The cell surface cAMP receptor of *Dictyostelium discoideum*. *J. Biol. Chem.* **259**, 13321–13328.

Van Haastert, P. J. M., and Kien, E. (1983). Binding of cAMP derivatives to *Dictyostelium discoideum* cells. Activation mechanism of the cell surface cAMP receptor. *J. Biol. Chem.* **258**, 9636–9642.

Van Haastert, P. J. M., De Wit, R. J. W., Grijpma, Y., and Konijn, T. M. (1982a). Identification of a pterin as the acrasin of the cellular slime mold *Dictyostelium* lacteum. *Proc. Natl. Acad. Sci. U.S.A.* **79**, 6270–6274.

Van Haastert, P. J. M., Van Lookeren Campagne, M. M., and Ross, F. M. (1982b). Altered cCMP-phosphodiesterase activity in chemotactic mutants of *Dictyostelium discoideum*. *FEBS Lett.* **147**, 149–152.

Van Haastert, P. J. M., De Wit, R. J. W., and Van Lookeren Campagne, M. M. (1985). Ca^{2+}- or phorbolester-dependent effect of ATP on a subpopulation of cAMP cell-surface receptors in membranes from *D. discoideum*. A role for protein kinase C. *Biochem. Biophys. Res. Commun.* **128**, 185–192.

Van Haastert, P. J. M., De Wit, R. J. W., Janssens, P. M. W., Kesbeke, F., and DeGoede, J. (1986). G-Protein mediated interconversions of cell-surface cAMP receptors and their involvement in the activation and desensitization of guanylyl cyclase in *Dictyostelium discoideum*. *J. Biol. Chem.* **261**, 6904–6911.

Van Haastert, P. J. M., Kesbeke, F., Reymond, C. D., Firtel, R., Luderus, E., and Van Driel, R. (1987a). Aberrant transmembrane signal transduction in *Dictyostelium* cells expressing a mutated ras gene. *Proc. Natl. Acad. Sci. U.S.A.* **84**, 4905–4909.

Van Haastert, P. J. M., Snaar-Jagalska, B. E., and Janssens, P. M. W. (1987b). The regulation of adenylyl cyclase by guanine nucleotides in *Dictyostelium discoideum* membranes. *Eur. J. Biochem.* **162**, 251–258.

Van Haastert, P. J. M., De Vries, M. J., Penning, L. C., Roovers, E., Van der Kaay, J., Erneux, C., and Van Lookeren Campagne, M. M. (1989). Chemoattractant and guanosine 5′-[τ-thio]triphosphate induce the accumulation of inositol 1,4,5-triphosphate in *Dictyostelium* cells that are labelled with [^{3}H]inositol by electroporation. *Biochem. J.* **258**, 577–586.

Van Haastert, P. J. M., Wang, M., Bominaar, A. A., Devreotes, P. M., and Schaap, P. (1992). cAMP-induced desensitization of surface cAMP receptors in *Dictyostelium*: Different second messengers mediate receptor phosphorylation, loss of ligand binding, degradation of receptor, and reduction of receptor mRNA levels. *Mol. Biol. Cell* **3**, 603–612.

Van Lookeren Campagne, M. M., Erneux, C., Van Eijk, R., and Van Haastert, P. J. M. (1988a). Two dephosphorylation pathways of inositol 1,4,5-trisphosphate in homogenates of the cellular slime mould *Dictyostelium discoideum. Biochem. J.* **254**, 343–350.

Van Lookeren Campagne, M. M., Wang, M., Spek, W., Peters, D., and Schaap, P. (1988b). Lithium respecifies cyclic-AMP induced cell-type specific gene expression in *Dictyostelium. Dev. Gen.* **9**, 589–596.

Van Ments-Cohen, M., and Van Haastert, P. J. M. (1989). The cyclic nucleotide specificity of eight cAMP-binding proteins in *Dictyostelium discoideum* is correlated into three groups. *J. Biol. Chem.* **25**, 8717–8722.

Varela, I., Van Lookeren Campagne, M. M., Alvarez, J. F., and Mato, J. M. (1987). The developmental regulation of phosphatidylinositol kinase in *Dictyostelium discoideum. FEBS Lett.* **211**, 64–68.

Vaughan, R. A., and Devreotes, P. N. (1988). Ligand-induced phosphorylation of the cAMP receptor from *Dictyostelium discoideum. J. Biol. Chem.* **263**, 14538–14543.

Wanner, R., and Wurster, B. (1990). Cyclic GMP-activated protein kinase from *Dictyostelium discoideum. Biochim. Biophys. Acta* **1053**, 179.

Wick, U., Malchow, D., and Gerisch, G. (1978). Cyclic-AMP stimulated calciumin flux into aggregating cells of *Dictyostelium discoideum. Cell Biol. Int. Rep.* **2**, 71–79.

Wu, L., and Devreotes, P. N. (1991). *Dictyostelium* transiently expresses eight distinct G-protein α-subunits during its developmental program. *Biochem. Biophys. Res. Commun.* **179**, 1141–1147.

Wu, L., and Franke, J. (1990). A developmentally regulated and cAMP repressible gene of *Dictyostelium discoideum*: Cloning and expression of the gene encoding cyclic nucleotide phosphodiesterase inhibitor. *Gene* **91**, 51–56.

Chapter 14

Gene Regulation by Hormone-like Signals in *Dictyostelium*

PAULINE SCHAAP and DORIEN J. M. PETERS

Department of Biology
University of Leiden
2333 AL Leiden, The Netherlands

BODDULURI HARIBABU and ROBERT P. DOTTIN

Department of Biological Sciences
Hunter College, City University of New York
New York, New York 10021

I. Introduction

The social amoeba *Dictyostelium discoideum* offers unique possibilities for studying regulation of gene expression and cell differentiation by intercellular signaling. Several essential signaling molecules have been identified, and many components of their signal transduction pathways have been characterized. As are other microorganisms, *Dictyostelium* is easy to grow as single cells and easy to manipulate. In addition, cells can pass through a short well-defined multicellular developmental cycle that is accompanied by a highly regulated program of stage- and cell type-specific gene expression and therefore bears a resemblance to embryological development of higher organisms. However, since development involves nonproliferating cells, the stimuli that affect differentiation can be studied independent of signals that mediate proliferation.

Although the cells can be conditionally diploid, they are usually haploid, which eliminates the masking of mutations by dominant alleles and makes isolation of useful mutants by spontaneous mutation or by random mutagenesis relatively easy. Of particular importance is the fact that advances in DNA technology have made DNA transfection routine, and have facilitated overexpression of specific genes and elimination of gene products by expression of antisense RNA. Moreover, unlike in most other eukaryotes, specific transformation procedures in *Dictyostelium* can induce disruption or modification of individual endogenous genes at a relatively high frequency.

In the feeding stage, unicellular amoebas digest large amounts of bacteria. Depletion of food initiates the social phase; cells begin to communicate by means of chemotactic signals and aggregate to form multicellular structures. As described in Chapter 13, the chemoattractant is cAMP, the well-known ubiquitous intracellular second messenger, which acts in *Dictyostelium* as an extracellular hormone-like signal (Konijn *et al.*, 1967). A few cells in the population secrete cAMP in an oscillatory manner. Surrounding cells are stimulated and respond by moving toward the cAMP source. They in turn relay the initial cAMP pulse, which results in concentric or spiral patterns of cAMP waves and cell streaming (Roos *et al.*, 1975; Shaffer, 1975; Tomchik and Devreotes, 1981). The resulting aggregate transforms into a slug-shaped structure that can migrate over the substratum in response to signals such as light and temperature to find a favorable situation for fruiting body formation (Bonner *et al.*, 1950; see Fig. 14.1). The slug then points upward; cells at the tip move into a central cellulose tube, and differentiate into rigid stalk cells (Raper and Fennell, 1952). The newly formed stalk carries the remaining cell mass upward. These cells differentiate into spores. Differentiation is already evident shortly after aggregation; the presumptive spores and stalk cells are laid down in an antero–posterior pattern in the slug (Raper, 1940; Bonner *et al.*, 1955).

Studies performed over the last several decades have made it clear that intercellular signaling mediated by diffusible signal molecules is largely responsible for the control of cell differentiation in *Dictyostelium*. Several putative morphogenic signal molecules have been identified. A central role is played by cAMP which, in addition to its function as chemoattractant, also regulates the expression of several classes of genes in different stages of development. Cyclic AMP binds to cell-surface receptors that interact with G proteins (Van Haastert, 1984; Klein *et al.*, 1988), making transmembrane processing of this signal similar to transduction of well-known hydrophilic hormones in higher vertebrates, for example, epinephrine, dopamine, serotonin, histamine, luteinizing hormone, vasopressin, and many others (see Birnbaumer, 1990). A chlorinated hexanone called DIF (differentiation-inducing factor) specifically induces

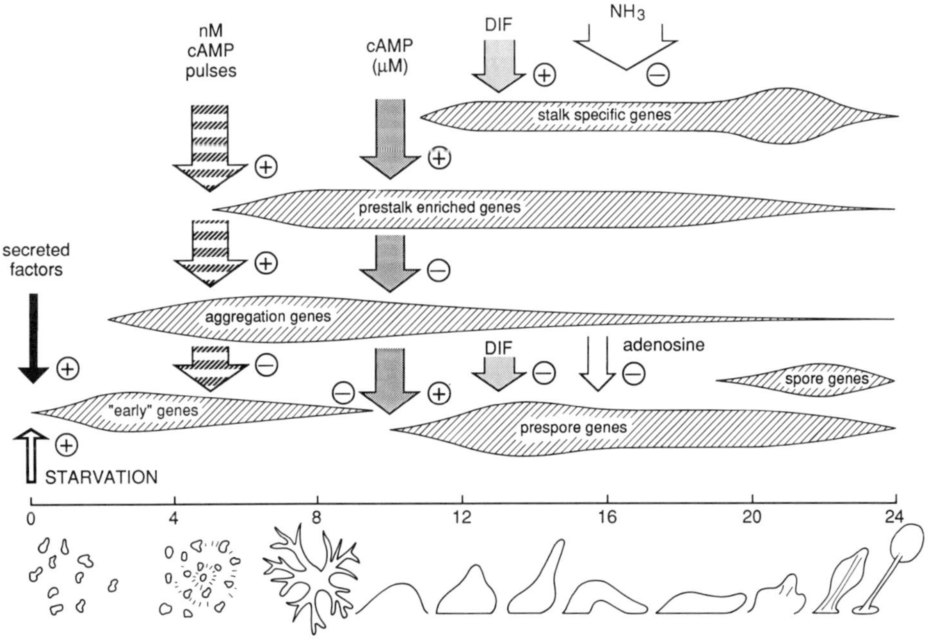

Figure 14.1 Classes of genes expressed during *Dictyostelium* development and extracellular signals regulating their expression.

stalk cell differentiation (Kay and Jermyn, 1983; Morris *et al.*, 1987). This lipophilic molecule can enter the cell, so its mode of action may be mechanistically similar to that of steroid hormones and retinoic acid, the vertebrate morphogen (Thaller and Eichele, 1987; Evans, 1988). Other identified signaling molecules are adenosine and ammonia, which act respectively as cAMP and DIF antagonists (Gross *et al.*, 1983; Weijer and Durston, 1985; Schaap and Wang, 1986), and CMF (conditioned-medium factor), an 80-kDa glycoprotein that regulates gene expression early in development (Mehdy and Firtel, 1985; Gomer *et al.*, 1991). Several other cellular secretion products that may optimize other aspects of gene expression during development also have been recognized, including the prespore inducing factor (PIF; Kumagai and Okamoto, 1986), the spore inducing factor (SPIF; Wilkinson *et al.*, 1985), and the prestarvation factor (PSF; Clarke *et al.*, 1987).

In this chapter, we focus our attention largely on the mechanisms involved in gene regulation by extracellular cAMP because transmembrane transduction of this signal has been investigated most thoroughly. However, we also give some attention to the processing of the other signals for which limited knowledge is available.

II. Gene Regulation during *Dictyostelium* Development

Gene expression during development has been a subject of investigation for the last several decades. At first, biochemical and immunological methods demonstrated the developmental regulation of specific enzyme activities or antigens (Bonner *et al.*, 1955; Takeuchi, 1963; Susmann and Osborn, 1964; Roth and Susmann, 1968; Loomis, 1969; Newell *et al.*, 1969; Telser and Sussman, 1971; Fishel *et al.*, 1982). Later, one- and two-dimensional SDS–PAGE analysis revealed patterns of changing protein synthesis during development (Alton and Lodish, 1977; Morrissey *et al.*, 1984; Cardelli *et al.*, 1985). Now many developmentally regulated genes have been cloned, and regulation of their expression has been analyzed (Barklis and Lodish, 1983; Mehdy *et al.*, 1983; Fishell *et al.*, 1985; Kopachik *et al.*, 1985a; Singleton *et al.*, 1987,1988; Williams *et al.*, 1987). Several signal molecules that regulate gene expression have been characterized and possible functions have been suggested for other molecules, some of which are partially purified. From these data, a general pattern of cascades of gene expression emerges in which cAMP and other hormones play critical roles (Fig. 14.1).

On starvation, expression of genes specific for the growth phase of development decreases (Kopachik *et al.*, 1985a; Singleton *et al.*, 1987,1988) and new classes of genes are transcribed, including the genes encoding discoidin I, α-mannosidase, the cAMP phosphodiesterase (PDE) inhibitor (Ma and Firtel, 1978; Wu and Franke, 1990; Schatzle *et al.*, 1991), and a class of genes called the I genes (Singleton *et al.*, 1988). The expression of these genes is considered to be induced by amino acid deprivation and by secreted factors such as PSF and CMF, which are used to sense cell density (Grabel and Loomis, 1978; Margolskee *et al.*, 1980; Clarke *et al.*, 1987; Mann and Firtel, 1989; Gomer *et al.*, 1991). Their expression is transient; repression is, in many cases, induced by cAMP (Bozzone and Berger, 1987; Singleton *et al.*, 1988; Mann and Firtel, 1989; Wu and Franke, 1990).

A second class of genes accumulates about 4 hr into starvation. Among them are genes encoding cAMP receptors, cyclic nucleotide PDE, adhesive contact sites A, and a G protein α subunit (Gα2) (Darmon *et al.*, 1975; Gerisch *et al.*, 1975; Lacombe *et al.*, 1986; Noegel *et al.*, 1986; Klein *et al.*, 1988; Kumagai *et al.*, 1989). Addition of nanomolar cAMP pulses to starved cells accelerates the expression of these aggregative genes, whereas infusion of constant levels of cAMP fails to stimulate expression (with the exception of the PDE gene).

During aggregation, a third class of genes is transcribed, all of which later are expressed preferentially at the anterior prestalk region. These prestalk-enriched genes are regulated positively by exogenous cAMP, added either in pulses at low (nanomolar) concentrations or as a continuous infusion at high (micromolar) concentrations. This class includes genes encoding cysteine proteases and one of the *ras* proto-oncogenes, as well as a number of genes encoding other proteins of unknown function (Barklis and Lodish, 1983; Mehdy *et al.*, 1983; Chisholm *et al.*, 1984; Reymond *et al.*, 1984; Mehdy and Firtel, 1985; Pears *et al.*, 1985 Peters *et al.*, 1989).

After aggregates have formed, expression of aggregative genes declines. At this stage, a major class of spore-specific genes as well as nonspecific genes is expressed (Alton and Lodish, 1977; Barklis and Lodish, 1983; Borth and Ratner, 1983; Mehdy *et al.*, 1983; Morrissey *et al.*, 1984; Cardelli *et al.*, 1985; Haribabu *et al.*, 1986). Accumulation of the mRNAs of most of these genes ceases when the aggregates are disrupted but can be restored by addition of high (micromolar) amounts of cAMP (Chung *et al.*, 1981; Blumberg *et al.*, 1982; Barklis and Lodish, 1983; Mehdy *et al.*, 1983; Haribabu *et al.*, 1986; Takemoto *et al.*, 1990). Two low molecular weight factors, PIF (a lipophilic molecule) and SPIF (a methionine analog), appear to be required additionally for synthesis of prespore-specific gene products (Kumagai and Okamoto, 1986; Gibson and Hames, 1988). At what level PIF affects spore gene expression is not yet known, but SPIF appears to be required for translation of spore coat-specific mRNAs (Gibson *et al.*, 1991). Later during development, a few spore-specific genes are expressed that are not responsive to extracellular cAMP (Richardson *et al.*, 1991) or to secretion products (Corney *et al.*, 1990; Richards *et al.*, 1990).

Two stalk-specific genes, *ecmA* and *ecmB*, encoding extracellular stalk matrix proteins also are transcribed during the formation of tight aggregates (Jermyn *et al.*, 1987; McRobbie *et al.*, 1988). Expression of these genes is induced by DIF (Williams *et al.*, 1987). At the onset of culmination, a large number of stalk-specific proteins is synthesized (Morrissey *et al.*, 1984; Cardelli *et al.*, 1985; Kopachik *et al.*, 1985b). DIF induces synthesis of most of these proteins (Kopachik *et al.*, 1985b).

DIF and other regulatory molecules have been found to be antagonistic to each other; DIF and adenosine, a cAMP hydrolysis product, inhibit cAMP induced expression of prespore genes (Weijer and Durston, 1985; Schaap and Wang, 1986; Wang *et al.*, 1986; Early and Williams, 1988). DIF-induced stalk cell differentiation and stalk-specific gene expression are, in turn, inhibited by ammonia and cAMP (Gross *et al.*, 1983; Berks and Kay, 1990; Wang and Schaap, 1989). Thus, complex regulatory loops dampen the effects of individual agonists. Spatial variations in extracellular levels of DIF, ammonia, cAMP, and adenosine have been hypothesized to control pattern formation in *Dictyostelium*.

III. Processing of Extracellular cAMP Signals

Efforts to elucidate cAMP signal processing in *Dictyostelium* soon followed its discovery as a chemoattractant. A role for cAMP as a putative morphogen motivated extensive research to identify transduction components and final targets of this signal molecule. Extracellular cAMP binds to highly specific cell-surface cAMP receptors (Green and Newell, 1975; Henderson, 1975).

At present four highly homologous surface cAMP receptors have been cloned, which show similar but not identical cyclic nucleotide specificity (Johnson *et al.*, 1992) and are expressed at different stages of development. cAR1 is expressed during aggregation, cAR3 between 5 and 15 hr, cAR2 from 12 to 24 hr, and cAR4 from 15 to 24 hr (Johnson *et al.*, 1993; Saxe *et al.*, 1993; P. N. Devreotes, personal communication). This pattern suggests that the four cARs may have different functions during development.

Binding of cAMP to surface receptors elicits a number of rapid responses; these receptors act through G proteins to activate several effectors. The components of these pathways are described in detail in Chapter 13 and diagrammed in Fig. 14.2.

Figure 14.2 cAMP-activated transmembrane signal processes in *Dictyostelium*. Four homologous cAMP receptors (CARs) have been identified (Klein *et al.*, 1988; Saxe *et al.*, 1991). Interaction of cAMP with these cell surface receptors acts through G proteins to activate a number of intracellular target enzymes such as adenylyl cyclase (Roos *et al.*, 1975; Shaffer, 1975), guanylyl cyclase (Mato *et al.*, 1977; Wurster *et al.*, 1977), and phospholipase C (Europe-Finner and Newell, 1987; Van Haastert *et al.*, 1989), resulting in increases in their associated second messengers (Europe-Finner and Newell, 1986; Abe *et al.*, 1988). Extracellular cAMP also induces an influx of Ca^{2+} (Wick *et al.*, 1978; Milne and Coukell, 1991) and an efflux of K^+ ions and H^+ ions (Malchow *et al.*, 1978; Aeckerle *et al.*, 1985). Activation of adenylyl cyclase and phospholipase C is mediated by different G proteins (Theibert and Devreotes, 1986; Van Haastert *et al.*, 1987,1989). In addition, adenylyl cyclase activation requires unidentified cytosolic factors, one of which is not active in the cAMP relay-defective mutant *synag7* (Theibert and Devreotes, 1986). Whether G proteins are involved in the activation of guanylyl cyclase remains unknown. Eight distinct G protein α subunits and one β subunit have been cloned (Lilly *et al.*, 1988; Kumagai *et al.*, 1989; Pupillo *et al.*, 1989; Hadwiger *et al.*, 1991; Wu and Devreotes, 1991). At present, the function of only one of these G proteins is clear. G2, a large heterotrimeric G protein, most likely couples the cAMP receptor to PLC; the α subunit of this G protein is absent or inactive in *fgdA* mutants (Coukell *et al.*, 1983; Kesbeke *et al.*, 1988; Kumagai *et al.*, 1989). Several intracellular target proteins for the potential intracellular messengers Ca^{2+}, cAMP, and cGMP have been identified. Intracellular cAMP can activate a cAMP-dependent protein kinase (A kinase) (De Gunzburg and Veron, 1982; Rutherford *et al.*, 1982; Leichtling *et al.*, 1984; Mutzel *et al.*, 1987), which consists of only one regulatory and one catalytic subunit, whereas the A kinase of higher eukaryotes consists of two regulatory and two catalytic subunits. Cells also contain another cAMP binding protein (cABP1) of unknown function (Tsang and Tasaka, 1986; Grant and Tsang, 1990). Ca^{2+} ions can interact with calmodulin (Clarke *et al.*, 1980; Marshak *et al.*, 1984) or activate a protein kinase C that is stimulated by

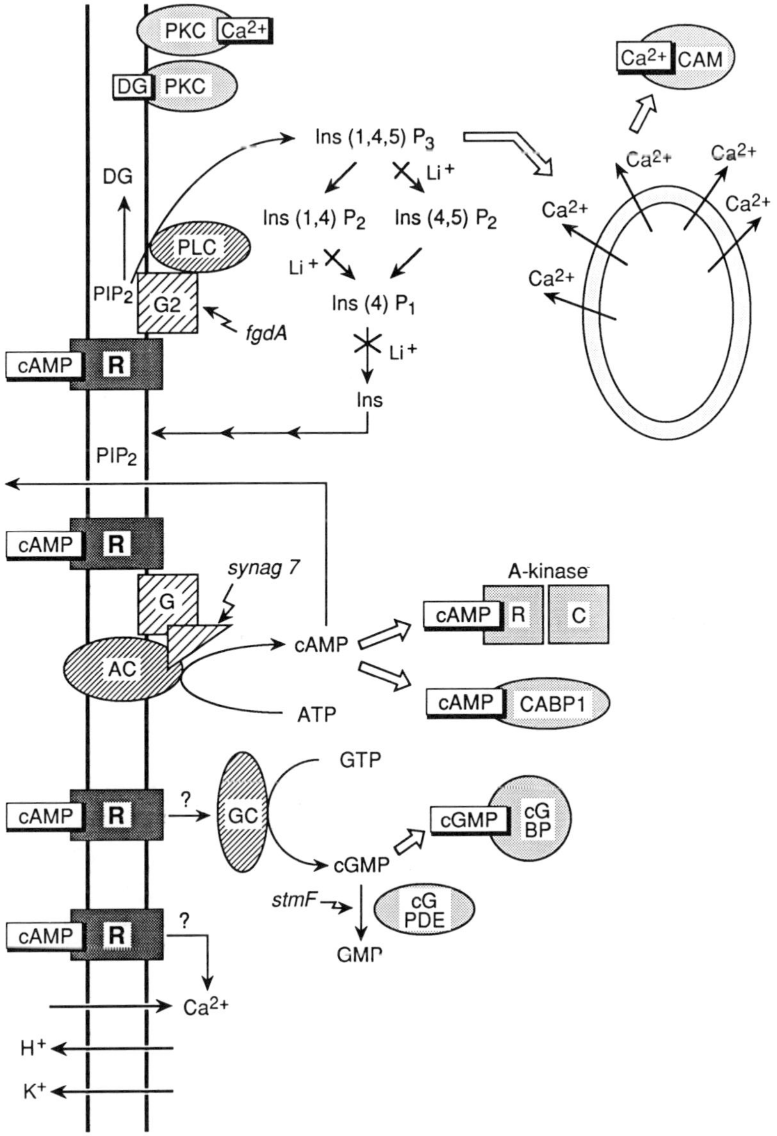

Ca^{2+}, phospholipids, or phorbol esters (Luderus *et al.*, 1989). A phospholipid-stimulated kinase activity that is not stimulated by Ca^{2+} also has been demonstrated (Jimenez *et al.*, 1989). R, surface cAMP receptor; G, guaninine nucleotide regulatory protein; AC, adenylyl cyclase; GC, guanylyl cyclase; PLC, phospholipase C; cABP1, cAMP binding protein 1; cGBP, cGMP binding protein; CAM, calmodulin; PKC, protein kinase C; cGPDE, cGMP specific phosphodiesterase; DG, diacylglycerol; PIP_2, phosphatidyl inositol bisphosphate.

The large repertory of intracellular responses mediated by extracellular cAMP (Fig. 14.2) comprises potential intermediates for its regulation of gene expression. We will explore the molecular genetic and pharmacological approaches that have been used to identify transduction pathways involved in gene regulation in the next section.

IV. Mechanisms of Gene Regulation by cAMP

Experiments dating from the early 1970s showed that cAMP affects not only cell movement but also cell differentiation (Bonner, 1970). However, because unphysiologically high (millimolar) concentrations of cAMP were required to induce differentiation (Town *et al.*, 1976; George, 1977; Kay, 1982), cAMP initially was thought to enter the cell and act as an intracellular messenger, bypassing undefined transmembrane transduction events, as is frequently the case in mammalian systems.

Careful dose-dependency studies showed that micromolar cAMP concentrations induced postaggregative gene expression (Mehdy and Firtel, 1985; Schaap and Van Driel, 1985; Haribabu and Dottin, 1986; Oyama and Blumberg, 1986). Pharmacological studies using systematically modified cAMP derivatives showed that the nucleotide specificities of the various cell-surface cAMP receptors are very similar but entirely distinct from those of the intracellular cAMP binding proteins PKA and cABP1 (De Wit *et al.*, 1982; Van Haastert and Kien, 1983; Van Ments-Cohen and Van Haastert, 1989). These data were followed by observations that the nucleotide specificity of cAMP induction of postaggregative prestalk and prespore gene expression is similar to the specificity of the surface receptor but distinct from that of the intracellular cAMP binding proteins (Schaap and Van Driel, 1985; Gomer *et al.*, 1986; Haribabu and Dottin, 1986; Oyama and Blumberg, 1986; Takemoto *et al.*, 1990). These results strongly suggest that cAMP is an extracellular first messenger for gene regulation and utilizes transmembrane signal processing systems.

Aggregation-related gene expression is induced by the chemotactic signal, that is, nanomolar cAMP pulses, which generally are assumed to be detected by chemotactic cAMP receptors. However, some data seem to contradict involvement of these receptors. Pulses of the two cAMP analogs, 8-bromo-cAMP and N^6-monobutyryl cAMP, which are relatively poor surface receptor agonists, were found to induce contact site A (*csA*) gene expression at only 10-fold higher concentrations than cAMP, suggesting that *csA* gene expression is induced by a different receptor (Ma and Siu, 1990). However, preferential resistance of these two agonists to PDE or the conversion of N^6-monobutyryl cAMP to cAMP also may account for these unexpected results.

Extracellular cAMP regulates the expression of 1000–2000 genes; some of these genes are aggregation related and others are prestalk or prespore specific. Therefore, the pathways from the receptors to these genes are likely to diverge, or multiple receptors may activate different pathways. cAMP, cGMP, diacylglycerol (DG), inositol triphosphate (IP_3) or its derivatives, and Ca^{2+} or other cAMP-induced ion fluxes (Fig. 14.2) all may function as intracellular intermediates. Some or all of these molecules may be involved in cAMP-mediated gene induction. One response to cAMP is the production of platelet-activating factor (PAF). Synthesis of PAF starts slowly and peaks after about 6 min of stimulation. Since the compound remains cell associated, it may have a function in intracellular signaling (Bussolino *et al.*, 1991).

In analogy to regulation of many mammalian genes, intracellular cAMP synthesized in response to activation of surface cAMP receptors seems a likely candidate to control gene regulation. The literature, however, is rather divided on the importance of this second messenger. Several studies showed that, when receptor-mediated activation of adenylyl cyclase is blocked by caffeine or by starvation in glucose–albumin medium, the pulse-induced aggregative genes as well as the cAMP-inducible prestalk and prespore genes continue to be induced by exogenous cAMP (Oyama and Blumberg, 1986; Schaap *et al.*, 1986; Kimmel, 1987; Haribabu *et al.*, 1989; Ma and Siu, 1990). These genes also are induced normally in a class of mutants (*synag*) that are defective in receptor-mediated activation of adenylyl cyclase (Schaap *et al.*, 1986; Theibert and Devreotes, 1986; Bozzaro *et al.*, 1987; Mann *et al.*, 1988; Ma and Siu, 1990). These mutants have normal cAMP receptors and activation of other cAMP-induced responses, as well as normal basal adenylyl cyclase activity. These results show that adenylyl cyclase activation is not required for the induction of these genes. The vegetative *M4-1* gene, whose expression normally is down-regulated in response to cAMP pulses, is not down-regulated by pulses in *synag* mutants, suggesting that intracellular cAMP may mediate its repression (Kimmel, 1987). Pharmacological efforts to either increase or decrease intracellular cAMP levels suggest that transcription of the *PDE* gene is repressed by intracellular cAMP (Riley and Barclay, 1990) whereas expression of one late spore-specific gene, as well as spore maturation, requires elevated intracellular cAMP levels (Kay, 1989; Riley and Barclay, 1990; Richardson *et al.*, 1991).

A completely different line of evidence suggests that intracellular cAMP has general importance in gene regulation. Overexpression of either the native *Dictyostelium* PKA regulatory subunit or of dominant negative subunits (lacking the cAMP binding sites) from *Dictyostelium* or mammals blocks development and inhibits expression of the cAMP-induced prestalk and prespore genes *CP2* and *D19* and the DIF-induced *ecmA* and *ecmB* genes (Simon *et al.*, 1989; Firtel and Chapman, 1990; Harwood *et al.*, 1992). Overexpression of the regulatory subunit does not affect pulse-induced genes, but the *PDE* gene appears to be overexpressed. Negative regulation of the discoidin I gene by cAMP does not occur (Firtel and Chapman, 1990). The last two observations are consistent with the pharmacological and *synag* mutant studies, and strongly suggest that intracellular cAMP mediates repression of the *M4-1* and discoidin genes and that this effect is induced by extracellular cAMP. Because overexpression of the regulatory subunit results in overexpression of the *PDE* gene, cAMP is unlikely to be an intermediate for the extracellular signal. Possibly, expression of this gene is under a secondary type of regulation that involves a subtle modulation of intracellular cAMP levels.

In the case of the late spore genes, intracellular cAMP is not an intermediate for extracellular cAMP because extracellular cAMP does not induce expression of these genes (Richards *et al.*, 1990; Richardson *et al.*, 1991). The largest discrepancy seems to be the regulation of the cAMP-inducible prestalk and prespore genes. Here, one set of data indicates that cAMP-induced synthesis of intracellular cAMP is not required, whereas the other set of data emphasizes that an active PKA is necessary. One explanation could be that PKA is activated by intracellular cAMP, the levels of which are increased by some signal other than exogenous cAMP. Alternatively, the PKA catalytic subunit in *Dictyostelium* may be active in the absence of cAMP because, in contrast to mammalian PKA, the regulatory and catalytic subunits of *Dictyostelium* PKA are not tightly associated (De Gunzburg *et al.*, 1984). Overexpression of the regulatory subunit may result in binding all unassociated catalytic subunits.

A role for cGMP as an intracellular second messenger remains unclear. *Streamer* F mutants which are defective in cGMP PDE show a strongly increased and prolonged cGMP response, but gene expression is unaffected (Ross and Newell, 1981; Schaap *et al.*, 1986). However, two lines of evidence yield correlative evidence for involvement of cGMP in aggregative gene expression. The partial cAMP antagonist 8-*p*-chlorophenylthio-cAMP (8CPT-cAMP) selectively induces the expression of aggregative genes but not of postaggregative genes. This molecule induces normal accumulation of cGMP and cAMP but decreases rather than increases IP_3 levels (Peters *et al.*, 1991a). LiCl promotes cAMP pulse-induced aggregative gene expression as well as prestalk gene expression, but inhibits cAMP-induced prespore gene expression (Peters *et al.*, 1991b; B. Haribabu and R. P. Dottin, unpublished results). At the second-messenger level, LiCl is found to inhibit IP_3 accumulation strongly, to reduce cAMP accumulation, and to promote guanylyl cyclase activation (Peters *et al.*, 1989; D. J. M. Peters *et al.*, unpublished observations). These data exclude IP_3 from mediating aggregative gene expression, and cAMP was excluded already by the experiments with *synag* and PKA mutants described earlier. These results therefore suggest an involvement of cGMP or an unknown effector as an intermediate of this response. Clearly, the causal relationship between intracellular cGMP levels and transcriptional activity of aggregative genes requires stronger evidence.

cGMP can interact with a cGMP-dependent protein kinase (PKG) (Wanner and Wurster, 1990) and a nuclear cGMP-binding protein whose kinetics change on interaction with oligonucleotides (Parissenti and Coukell, 1990; Butler and Coukell, 1992). The binding of this protein is altered most effectively by GC-rich DNA sequences and therefore may interact with regulatory DNA sequences, which are frequently GC rich in *Dictyostelium*.

Several experiments suggest the involvement of phospholipase C (PLC)-mediated pathways in gene induction. Addition of IP_3 or, more efficiently, IP_3 and DG to saponin-permeabilized cells results in negative regulation of a cAMP-inducible prestalk gene and a non-cell-type-specific gene, and in positive regulation of a cAMP-inducible prespore gene, apparently bypassing the requirement for receptor activation (Ginsburg and Kimmel, 1989). However, stimulation with IP_3 and DG is required prior to the stage at which cAMP-mediated induction occurs, which argues against direct involvement of these intermediates in cAMP-regulated transcription.

Frigid A mutants are defective in a G protein α subunit (Gα2) that activates PLC, and do not express cAMP-induced aggregative and postaggregative genes (Coukell *et al.*, 1983; Kesbeke *et al.*, 1988; Mann *et al.*, 1988; Kumagai *et al.*, 1989). Mutants with the same phenotype also have been obtained by disruption of the endogenous Gα2 gene (Kumagai *et al.*, 1991). This phenotype suggests that Gα2 is of major importance to gene induction, presumably by its action on PLC. The latter supposition is contradicted in part by the effects of LiCl and 8-CPT-cAMP on signal transduction and gene expression, because these compounds (respectively) inhibit or are unable to induce activation of PLC, but nevertheless promote or induce the expression of aggregative genes. Only in case of the prespore genes does both the pharmacological and the molecular genetic evidence indicate involvement of PLC activation. This conclusion is substantiated further by observations that prespore gene expression is particularly sensitive to inhibition by calcium antagonists (Schaap *et al.*, 1986; Blumberg *et al.*, 1989; Haribabu *et al.*, 1989).

In addition to the second messengers mentioned above cAMP-induced ion fluxes are putative intermediates for gene regulation. Much attention has been paid to a role for intracellular pH in gene regulation. cAMP induces both a transient and a sustained increase in intracellular pH (Aerts *et al.*, 1987; Van Lookeren Campagne *et al.*, 1989). Conditions leading to respective increases or decreases in intracellular pH (pH_i) were shown to promote the spore and stalk cell differentiation pathways, respectively (Gross *et al.*, 1983; Town, 1984; Dominov and Town, 1986). However, increasing or decreasing pH_i does not directly affect the expression of cAMP-inducible prestalk and prespore genes (Van Lookeren Campagne *et al.*, 1989; Peters *et al.*, 1991b). Decreasing pH_i

cannot induce expression of the DIF-regulated prestalk genes (Wang *et al.*, 1990). A direct role for pH_i changes as intermediates for cAMP- or DIF-regulated gene expression is therefore unlikely. Increasing pH_i does, however, inhibit DIF-induced gene expression (Wang *et al.*, 1990), which is probably responsible for the observed effects of pH_i changes on stalk and spore cell proportions.

Very little is known of possible roles of the cAMP-induced Ca^{2+} influx and K^+ efflux in gene regulation. Interestingly, Ca^{2+} influx may be involved in translational control of gene expression. Depletion of intracellular Ca^{2+} was reported to inhibit synthesis of cAMP–PDE at the post-transcriptional level (Coukell and Cameron, 1990). A later study indicated that Ca^{2+} depletion disrupted the secretory pathway, resulting in selective cAMP–PDE degradation (Coukell *et al.*, 1992).

Studies on mammalian systems suggest that phosphorylation cascades initiated by transmembrane signals play an important role in controlling gene regulation. A number of *Dictyostelium* protein kinase (Tan and Spudich, 1990; Haribabu and Dottin, 1991a; Devreotes, Veron, and Williams, personal communication) and protein phosphatase genes (Haribabu and Dottin, 1991b; Kincaid and Kessin, Mutzel, personal communication) have been isolated. It will be of interest to see if any of these have a role in mediating cAMP responses in *Dictyostelium*.

V. Mechanisms of Gene Regulation by Other Extracellular Signals

Although extensive knowledge about the action of cAMP as extracellular signal has been obtained, very little is known of transduction of other hormone-like signals in *Dictyostelium*. The stalk-inducing factor DIF is a small lipophilic molecule (Morris *et al.*, 1987) that has been shown to bind to an intracellular DIF binding protein. This putative DIF receptor appears shortly before the stalk genes become DIF inducible. The protein binds DIF with high affinity (2 n*M*) and is found in both cytosolic and nuclear fractions, suggesting a direct function as a transcription factor (Insall and Kay, 1990). Early data showed that induction of terminal stalk cell differentiation by DIF could be bypassed by agents causing cytoplasmic acidification, suggesting that the effects of DIF were mediated by a decrease in pH_i (Gross *et al.*, 1983; Town, 1984). However, DIF itself induces only a very minor decrease of pH_i (Inouye, 1985; Kay *et al.*, 1986) and cytoplasmic acidification cannot induce transcription of stalk genes directly (Wang *et al.*, 1990). However, a low pH_i enhances responsiveness of cells to DIF. The DIF antagonist ammonia most likely inhibits stalk cell differentiation by increasing pH_i, thereby interfering with DIF signal transduction.

The cAMP antagonist adenosine inhibits cAMP-induced responses such as adenylyl cyclase activation (Newell, 1982; Theibert and Devreotes, 1984) and prespore gene expression (Schaap and Wang, 1986). Adenosine also inhibits binding of cAMP to surface cAMP receptors (Newell and Ross, 1982) with specificity similar to that of its inhibition of adenylyl cyclase activation and prespore gene expression (Van Lookeren Campagne *et al.*, 1986). Therefore, adenosine is likely to act directly at the cell surface and inhibit cAMP-induced responses by inhibiting cAMP binding to a (sub)population of cAMP receptors.

A large field of research investigating transduction of these signals, as well as of other *Dictyostelium* hormones such as CMF, PSF, PIF, and SPIF is, as yet, completely unexplored. The identification of CMF may initiate studies into transduction of this novel class of glycoprotein hormones in *Dictyostelium*. Considering the observed DIF/ammonia and cAMP/adenosine antagonisms, investigating at what level the different signals can modulate each other's transduction is also of considerable interest.

VI. Signal Transduction Response Elements and *trans*-Acting Factors

In contrast with the detail in which the cell-surface events are being elucidated, our knowledge of the subsequent reactions that lead to transcription in response to cAMP and other hormones remains sparse. Several cDNAs corresponding to cAMP- and DIF-responsive genes have been cloned, but only a few genes have been studied with the goal of characterizing the promoters that mediate cAMP or DIF responsiveness. These genes include the early genes α-fucosidase (May *et al.*, 1989,1991), discoidin I (Vauti *et al.*, 1990), prestalk gene *DG17*, two cysteine protease genes *CP1* and *CP2* (Datta and Firtel, 1987,1988; Pears and Williams, 1987,1988), a developmentally regulated *ras* homolog (Esch and Firtel, 1991; Louvion *et al.*, 1991), prespore genes *SP60* and *E7* (Haberstroh and Firtel, 1990; Haberstroh *et al.*, 1991; Ramji *et al.*, 1991), a *UDPGP1* gene (Pavlovic *et al.*, 1988,1989), and the DIF-induced *ecmB* gene (Ceccarelli *et al.*, 1991). The fact that the *Dictyostelium* genome is AT rich (over 70% in the coding region and over 85% in noncoding DNA) has allowed identification of the relatively rare GC-rich regions in noncoding DNA as potential regulatory sequences. These GC-rich regions are hypersensitive to DNaseI, as are regulatory sites in several other eukaryotic genes (Reeves, 1984; Pavlovic *et al.*, 1986,1989). The regulatory elements have been characterized by transfecting cells with different deletion constructs coupled to reporter gene sequences and analyzing their transcripts. In general, these regulatory elements appear to contain a short (20-bp) GC-rich region adjacent to an AT-rich region, and are located 200–500 bp from the start of transcription. These elements are distinct from CAAT and TATA sequences, which often occur in *Dictyostelium* as well as in other eukaryotic genes. In the case of *UDPGP1*, the α-fucosidase gene, and the *CP1* and *CP2* genes, the GC-rich sequences appear to be required for induction of the genes in response to cAMP (Pears and Williams, 1987,1988; Datta and Firtel, 1988; Pavlovic *et al.*, 1989; May *et al.*, 1991), whereas AT-rich regions directly downstream of the GC-rich sequences have been shown to be essential for basal or optimal transcription of the *UDPGP1* and *CP2* genes, respectively (Datta and Firtel, 1988; Pavlovic *et al.*, 1989). Deletion of the GC-rich hormone-response element of the *UDPGP1* gene causes this gene to be down-regulated rather than induced during the multicellular stages of development (Pavlovic *et al.*, 1989), suggesting that cAMP also is involved in the temporal induction of the gene during multicellular development. Developmental induction of the discoidin I gene is mediated by a GC-rich element with a characteristic TGG motif at −347; an AT-rich region required for cAMP-induced down-regulation of discoidin I expression is located upstream of this region (Vauti *et al.*, 1990).

Use of promoters coupled to cell-autonomous markers such as specific antigen tags or β-galactosidase has allowed analysis of pattern formation as well as dissection of promoter elements that control gene regulation (Dingermann *et al.*, 1989; Jermyn *et al.*, 1989; Haberstroh and Firtel, 1990; Esch and Firtel, 1991; Haberstroh *et al.*, 1991; Jermyn and Williams, 1991). Such studies show that the promoter of the DIF-inducible prestalk gene *ecmB* contains three distinct elements, each of which activates or represses transcription in different parts of the culminating fruiting body (Ceccarelli *et al.*, 1991). Analysis of the promoter of the cAMP-inducible prespore gene *SP60* reveals that multiple CA-rich elements regulate the expression of this gene and that each of these elements functions in distinct regions of the slug (Haberstroh *et al.*, 1991). Two CA-rich and two C-rich regions were found in the promoter of the cAMP-inducible prespore specific gene *7E*; however, whether these elements are essential for cAMP inducibility is not yet known (Ramji *et al.*, 1991).

None of these elements shows significant homology to any other and no single consensus sequence for cAMP responsiveness has been delineated. These elements are not related to the mammalian cAMP response elements (Roesler *et al.*, 1988), a result that was expected because most of these genes are not regulated by intracellular cAMP. Different *Dictyostelium* genes that are regulated by exogenous cAMP may use distinct proteins that activate unique regulatory sequences. This possibility would require many regulatory proteins because thousands of genes are regulated by cAMP. A second explanation is that relatively few cAMP-responsive transcription factors exist and that each of them could recognize different cAMP-responsive sequence elements, as does the HAP protein in yeast (Pfeifer, 1987). Protein–protein interactions then may confer additional specificity. A heat-stable DNA binding activity termed GBF that interacts with a G box upstream from the *CP2* gene has been identified

(Hjorth *et al.*, 1989,1990). This activity also seems to interact with several other cAMP response elements *in vitro* and is itself regulated by exogenous cAMP. A third possibility is that several cAMP genes share the same regulatory sequences that have not been represented in the limited number of cloned cAMP-responsive genes. How multiple responses of cAMP are integrated at the level of transcription in *Dictyostelium* is not yet clear. Further analysis of specific promoters and *trans*-acting factors should help define the pathways between second messengers and gene transcription.

VII. Conclusions

The past 10 years have seen enormous advances in our understanding of regulation of developmental gene expression. Notably, the recognition of cAMP as an extracellular hormone that controls gene regulation and the discovery of its highly conserved G protein-mediated transduction pathway emphasize that even molecular mechanisms of vertebrate development may have originated in the primitive eukaryotes.

A progressive increase in our understanding of cAMP signal processing downstream from receptors and G proteins is to be expected within the next several years, considering the multitude of putative transduction components that are being cloned. However, the future usefulness of the *Dictyostelium* system will require further expansion of its molecular genetic potential to the opportunities represented by the yeast system. Use of reverse genetics with cloned genes to disrupt their homologs by recombination, suppressor genetics, and complementation of the signal transduction mutants then can be used to identify signal transduction elements. Also, the putative role of cascades of expression of gene regulatory proteins under control of morphogenic gradients, which are known to be of great importance in regulating insect and vertebrate development, remains to be explored in *Dictyostelium*. The observations that developmental gene expression often requires prior protein synthesis and that at least one *Dictyostelium* gene regulatory protein is cAMP inducible suggests that such mechanisms also are active in this organism. Hopefully, the next several years will provide novel molecular details on the mechanism of hormone action in *Dictyostelium*, which may elucidate related phenomena in mammals.

References

Abe, T., Maeda, Y., and Iijima, T. (1988). Transient increase of the intracellular Ca^{2+} concentration during chemotactic signal transduction in *Dictyostelium discoideum* cells. *Differentiation* **39**, 90–96.

Aeckerle, S., Wurster, B., and Malchow, D. (1985). Oscillations and cyclic AMP-induced changes of the K^+ concentration in *Dictyostelium discoideum*. *EMBO J.* **4**, 39–43.

Aerts, R. J., De Wit, R. J. W., and Van Lookeren Campagne, M. M. (1987). Cyclic AMP induces a transient alkalinization in *Dictyostelium*. *FEBS Lett.* **220**, 366–370.

Alton, T. H., and Lodish, H. F. (1977). Synthesis of developmentally regulated proteins in *Dictyostelium discoideum* which are dependent on continued cell-cell interaction. *Dev. Biol.* **60**, 207–216.

Barklis, E., and Lodish, H. F. (1983). Regulation of *Dictyostelium discoideum* mRNAs specific for prespore or prestalk cells. *Cell* **32**, 1139–1148.

Berks, M., and Kay, R. R. (1990). Combinatorial control of cell differentiation by cAMP and DIF1 during development of *Dictyostelium discoideum*. *Development* **110**, 977–984.

Birnbaumer, L., Abramwitz, J., and Brown, A. M. (1990). Receptor–effector coupling by G proteins. *Biochim. Biophys. Acta* **1031**, 163–224.

Blumberg, D. D., Margolskee, J. P., Barklis, E., Chung, S. N., Cohen, N. S., and Lodish, H. F. (1982). Specific cell–cell contacts are essential for induction of gene expression during differentiation of *Dictyostelium discoideum*. *Proc. Natl. Acad. Sci. U.S.A.* **79**, 127–131.

Blumberg, D. D., Comer, J. F., and Walton, E. M. (1989). Ca^{++} antagonists distinguish different requirements for cAMP-mediated gene expression in the cellular slime mold, *Dictyostelium discoideum*. *Differentiation* **41**, 14–21.

Bonner, J. T. (1970). Induction of stalk cell differentiation by cyclic AMP in the cellular slime mold *Dictyostelium discoideum*. *Proc. Natl. Acad. Sci. U.S.A.* **65**, 110–113.

Bonner, J. T., Clarke, W. W., Jr., Neely, C. L., Jr., and Slifkin, M. K. (1950). The orientation to light and the extremely sensitive orientation to temperature gradients in the slime mold *Dictyostelium discoideum*. *J. Cell. Comp. Physiol.* **36**, 149–158.

Bonner, J. T., Chiquoine, A. D., and Kolderie, M. Q. (1955). A histochemical study of differentiation in the cellular slime molds. *J. Exp. Zool.* **130**, 133–157.

Borth, W., and Ratner, D. (1983). Different synthetic profiles and developmental fates of prespore versus prestalk proteins of *Dictyostelium*. *Differentiation* **24**, 213–219.

Bozzaro, S., Hagmann, J., Noegel, A., Westphal, M., Calautti, E., and Bogliolo, E. (1987). Cell differentiation in the absence of intracellular and extracellular cyclic AMP pulses in *Dictyostelium discoideum*. *Dev. Biol.* **123**, 540–548.

Bozzone, D. M., and Berger, E. A. (1987). Distinct developmental regulation and properties of the responsiveness of different genes to cyclic AMP in *Dictyostelium discoideum*. *Differentiation* **33**, 197–206.

Bussolino, F., Sordano, C., Benfenati, E., and Bozzaro, S. (1991). *Dictyostelium* cells produce platelet-activating factor in response to cAMP. *Eur. J. Biochem.* **196**, 609–615.

Cardelli, J. A., Knecht, D. A., Wunderlich, R., and Dimond, R. L. (1985). Major changes in gene expression occur during at least four stages of development of *Dictyostelium discoideum*. *Dev. Biol.* **110**, 147–156.

Ceccarelli, A., Mahbubani, H., and Williams, J. G. (1991). Positively and negatively acting signals regulating stalk cell and anterior-like cell differentiation in *Dictyostelium*. *Cell* **65**, 983–989.

Chisholm, R. L., Barklis, E., and Lodish, H. F. (1984). Mechanism of sequential induction of cell type specific mRNAs in *Dictyostelium* differentiation. *Nature (London)* **310**, 67–69.

Chung, S., Landfear, S. M., Blumberg, D. D, Cohen, N. S., and Lodish, H. F. (1981). Synthesis and stability of developmentally regulated *Dictyostelium* mRNAs are affected by cell–cell contact and cAMP. *Cell* **24**, 785–797.

Clarke, M., Bazari, W. L., and Kayman, S. C. (1980). Isolation and properties of calmodulin from *Dictyostelium discoideum*. *J. Bacteriol.* **141**, 397–400.

Clarke, M., Kayman, S. C., and Riley, K. (1987). Density-dependent induction of discoidin-I synthesis in exponentially growing cells of *Dictyostelium discoideum*. *Differentiation* **34**, 79–87.

Corney, A. J., Richards, A. J., Phillpots, T., and Hames, B. D. (1990). Developmental regulation of cell-type-enriched mRNAs in *Dictyostelium discoideum*. *Mol. Microbiol.* **4**, 613–623.

Coukell, M. B., and Cameron, A. M. (1990). Calcium depletion of *Dictyostelium* cells selectively inhibits cyclic nucleotide phosphodiesterase synthesis at a post-transcriptional step. *J. Cell Sci.* **97**, 649–657.

Coukell, M. B., Lappano, S., and Cameron, A. M. (1983). Isolation and characterization of cAMP unresponsive (frigid) aggregation-deficient mutants of *Dictyostelium discoideum. Dev. Genet.* **3**, 283–297.

Coukell, M. B., Cameron, A. M., and Adames, N. R. (1992). Involvement of intracellular calcium in protein secretion in *Dictyostelium discoideum. J. Cell Sci.* **103**, 371–380.

Darmon, M., Brachet, P., and Pereira da Silva, L. H. (1975). Chemotactic signals induce cell differentiation in *Dictyostelium discoideum. Proc. Natl. Acad. Sci. U.S.A.* **72**, 3163–3166.

Datta, S., and Firtel, R. A. (1987). Identification of the sequences controlling cyclic AMP regulation and cell-type-specific expression of a prestalk-specific gene in *Dictyostelium discoideum. Mol. Cell. Biol.* **7**, 149–159.

Datta, S., and Firtel, R. A. (1988). An 80-bp *cis*-acting regulatory region controls cAMP and development regulation of a prestalk gene in *Dictyostelium. Genes Dev.* **2**, 294–304.

De Gunzburg, J., and Véron, M. (1982). A cAMP-dependent protein kinase is present in differentiating *Dictyostelium discoideum* cells. *EMBO J.* **1**, 1063–1068.

De Gunzburg, J., Part, D., Guiso, N., and Véron, M. (1984). An unusual adenosine 3′,5′-phosphate dependent protein kinase from *Dictyostelium discoideum. Biochemistry* **23**, 3805–3812.

De Wit, R. J. W., Arents, J. C., and Van Driel, R. (1982). Ligand binding properties of the cytoplasmic cAMP-binding protein of *Dictyostelium discoideum. FEBS Lett.* **145**, 150–154.

Dingermann, T., Reindl, N., Werner, H., Hildebrandt, M., Nellen, W., Harwood, A., Williams, J., and Nerke, K. (1989). Optimization and in situ detection of Escherichia coli β-galactosidase gene expression in *Dictyostelium discoideum. Gene* **85**, 353–362.

Dominov, J. A., and Town, C. D. (1986). Regulation of stalk and spore antigen expression in monolayer cultures of *Dictyostelium discoideum* by pH. *J. Embryol. Exp. Morphol.* **96**, 131–150.

Early, V. E., and Williams, J. G. (1988). A *Dictyostelium* prespore-specific gene is transcriptionally repressed by DIF in vitro. *Development* **103**, 519–524.

Esch, R. K., and Firtel, R. A. (1991). cAMP and cell sorting control the spatial expression of a developmentally essential cell-type-specific *ras* gene in *Dictyostelium. Genes Dev.* **5**, 9–21.

Europe-Finner, G. N., and Newell, P. C. (1986). Inositol 1,4,5 trisphosphate induces calcium release from a nonmitochondrial pool in amoebae of *Dictyostelium. Biochim. Biophys. Acta* **887**, 335–340.

Europe-Finner, G. N., and Newell, P. C. (1987). Cyclic AMP stimulates accumulation of inositol trisphosphate in *Dictyostelium. J. Cell Sci.* **87**, 221–229.

Evans, R. M. (1988). The steroid and thyroid hormone receptor superfamily. *Science* **240**, 889–895.

Firtel, R. A., and Chapman, A. L. (1990). A role for cAMP-dependent protein kinase in early *Dictyostelium* development. *Genes Dev.* **4**, 18–28.

Fishell, B. R., Manrow, R. E., and Dottin, R. P. (1982). Developmental regulation of multiple forms of UDPglucose pyrophosphorylase of *Dictyostelium. Dev. Biol.* **92**, 175–187.

Fishell, N. R., Ragheb, J. A., *et al.* (1985). Molecular cloning of a cDNA complementary to a UDP-glucose pyrophosphorylase mRNA of *Dictyostelium discoideum. Dev. Biol.* **110**, 369–381.

George, R. P. (1977). Disruption of multicellular organization in the cellular slime molds by cyclic AMP. *Cell Diff.* **5**, 293–300.

Gerisch, G., Fromm, H., Huesgen, A., and Wick, U. (1975). Control of cell-contact sites by cyclic AMP pulses in differentiating *Dictyostelium* cells. *Nature (London)* **255**, 547–549.

Gibson, F. P., and Hames, D. (1988). Characterization of a spore protein inducing factor from *Dictyostelium discoideum. J. Cell Sci.* **89**, 387–395.

Gibson, F. P., Schofield, T., and Hames, B. D. (1991). SPIF regulates spore coat protein synthesis by *Dictyostelium discoideum* at the translational level. *J. Cell Sci.* **100**, 357–364.
Ginsburg, G., and Kimmel, A. R. (1989). Inositol trisphosphate and diacylglycerol can differentially modulate gene expression in *Dictyostelium. Proc. Natl. Acad. Sci. U.S.A.* **86**, 9332–9336.
Gomer, R. H., Armstrong, D., Leichtling, B. H., and Firtel, R. A. (1986). cAMP induction of prespore and prestalk gene expression in *Dictyostelium* is mediated by the cell-surface cAMP receptor. *Proc. Natl. Acad. Sci. U.S.A.* **83**, 8624–8628.
Gomer, R. H., Yuen, I. S., and Firtel, R. A. (1991). A secreted 80×10^3 M_r protein mediates sensing of cell density and the onset of development in *Dictyostelium. Development* **112**, 269–278.
Grabel, L., and Loomis, W. F. (1978). Effector controlling accumulation of *N*-acetylglucosaminidase during development of *Dictyostelium discoideum. Dev. Biol.* **64**, 203–209.
Grant, C. E., and Tsang, A. (1990). Cloning and characterization of cDNAs encoding a novel cyclic AMP-binding protein in *Dictyostelium discoideum. Gene* **96**, 213–218.
Green, A. A., and Newell, P. C. (1975). Evidence for the existence of two types of cAMP binding sites in aggregating cells of *Dictyostelium discoideum. Cell* **6**, 129–136.
Gross, J. D., Bradbury, J., Kay, R. R., and Peacey, M. J. (1983). Intracellular pH and the control of cell differentiation in *Dictyostelium discoideum. Nature (London)* **303**, 244–245.
Haberstroh, L., and Firtel, R. (1990). A spatial gradient of expression of a cAMP-regulated prespore cell-type-specific gene in *Dictyostelium. Genes Dev.* **4**, 596–612.
Haberstroh, L., Galindo, J., and Firtel, R. (1991). Developmental and spatial regulation of a *Dictyostelium* prespore gene: *cis*-Acting elements and a cAMP-induced, developmentally regulated DNA binding activity. *Development* **113**, 947–958.
Hadwiger, J. A., Wilkie, T. M., Strathmann, M., and Firtel, R. A. (1991). Identification of *Dictyostelium* Gα genes expressed during multicellular development. *Proc. Natl. Acad. Sci. U.S.A.* **88**, 8213–8217.
Haribabu, B., and Dottin, R. P. (1986). Pharmacological characterization of cyclic AMP receptors mediating gene regulation in *Dictyostelium discoideum. Mol. Cell Biol.* **6**, 2402–2408.
Haribabu, B., and Dottin, R. P. (1991a). Identification of a protein kinase multigene family of *Dictyostelium discoideum*: Molecular cloning and expression of a cDNA encoding a developmentally regulated protein kinase. *Proc. Natl. Acad. Sci. U.S.A.* **88**, 1115–1119.
Haribabu, B., and Dottin, R. P. (1991b). Homology cloning of protein kinase and phosphoprotein phosphatase sequences of *Dictyostelium discoideum. Dev. Genet.* **12**, 45–49.
Haribabu, B., Rajkovic, A., and Dottin, R. P. (1986). Cell–cell contact and cAMP regulate the expression of a UDP glucose pyrophosphorylase gene of *Dictyostelium discoideum. Dev. Biol.* **113**, 436–442.
Haribabu, B., Pavlovic, J., and Dottin, R. P. (1989). Regulation of the rates of transcription of several specific genes in response to exogenous cAMP in *Dictyostelium discoideum. Mol. Genet.* **8**, 1–5.
Haribabu, B., Pavlovic, J., Bodduluri, S. R., Doody, J. F., Ortiz, B. D., Mullings, S., Moon, B., and Dottin, R. P. (1991). Signal transduction pathways involved in the expression of the uridine diphosphoglucose pyrophosphorylase gene of *Dictyostelium discoideum. Dev. Genet.* **12**, 35–44.
Harwood, A. J., Hopper, N. A., Simon, M.-N., Bouzid, S., Véron, M., and Williams, J. G. (1992). Multiple roles for cAMP-dependent protein kinase during *Dictyostelium* development. *Dev. Biol.* **149**, 90–99.
Henderson, E. J. (1975). The cyclic adenosine 3′,5′-monophosphate receptor of *Dictyostelium discoideum. J. Biol. Chem.* **250**, 4730–4736.
Hjorth, A. L., Khanna, N. C., and Firtel, R. A. (1989). A *trans*-acting factor required for cAMP-induced gene expression in *Dictyostelium* is regulated developmentally and induced by cAMP. *Genes Dev.* **3**, 747–759.

Hjorth, A. L., Pears, C., Williams, J. G., and Firtel, R. A. (1990). A developmentally regulated trans-acting factor recognizes dissimilar G/C-rich elements controlling a class of cAMP-inducible *Dictyostelium* genes. *Genes Dev.* **4**, 419–432.

Inouye, K. (1985). Measurements of intracellular pH and its relevance to cell differentiation in *Dictyostelium discoideum. J. Cell Sci.* **76**, 235–245.

Insall, R., and Kay, R. R. (1990). A specific DIF binding protein in *Dictyostelium. EMBO J.* **9**, 3323–3328.

Jermyn, K. A., and Williams, J. G. (1991). An analysis of culmination in *Dictyostelium* using prestalk and stalk-specific cell autonomous markers. *Development* **111**, 779–787.

Jermyn, K. A., Berks, M., Kay, R. R., and Williams, J. G. (1987). Two distinct classes of prestalk-enriched mRNA sequences in *Dictyostelium discoideum. Development* **100**, 745–755.

Jermyn, K. A., Duffy, K. T. I., and Williams, J. G. (1989). A new anatomy of the prestalk zone in *Dictyostelium. Nature (London)* **340**, 144–146.

Jiminéz, B., Pestaña, A., and Fernandez-Renart, M. (1989). A phospholipid-stimulated protein kinase from *Dictyostelium discoideum. Biochem. J.* **260**, 557–561.

Johnson, R. L., Van Haastert, P. J. M., Kimmel, A. R., Saxe, C. L., III, Jastorff, B., and Devreotes, P. N. (1992). The cyclic nucleotide specificity of three cAMP receptors in *Dictyostelium. J. Biol. Chem.* **267**, 4600–4607.

Kay, R. R. (1982). cAMP and spore differentiation in *Dictyostelium discoideum. Proc. Natl. Acad. Sci. U.S.A.* **79**, 3228–3231.

Kay, R. R. (1989). Evidence that elevated intracellular cyclic AMP triggers spore maturation in *Dictyostelium. Development* **105**, 753–759.

Kay, R. R., and Jermyn, K. A. (1983). A possible morphogen controlling differentiation in *Dictyostelium. Nature (London)* **303**, 242–244.

Kay, R. R., Gadian, D. G., and Williams, S. R. (1986). Intracellular pH in *Dictyostelium*: A ^{31}P nuclear magnetic resonance study of its regulation and possible role in controlling cell differentiation. *J. Cell. Sci.* **83**, 165–179.

Kesbeke, F., Snaar-Jagalska, B. E., and Van Haastert, P. J. M. (1988). Signal transduction in *Dictyostelium* fgd A mutants with a defective interaction between surface cAMP receptors and a GTP-binding regulatory protein. *J. Cell Biol.* **107**, 521–528.

Kimmel, A. R. (1987). Different molecular mechanisms for cAMP regulation of gene expression during *Dictyostelium* development. *Dev. Biol.* **122**, 163–171.

Klein, P. S., Sun, T. J., Saxe, C. L., III, Kimmel, A. R., Johnson, R. L., and Devreotes, P. N. (1988). A chemoattractant receptor controls development in *Dictyostelium discoideum. Science* **241**, 1467–1472.

Konijn, T. M., Van De Meene, J. G. C., Bonner, J. T., and Barkley, D. S. (1967). The acrasin activity of adenosine 3′,5′-cyclic phosphate. *Proc. Natl. Acad. Sci. U.S.A.* **58**, 1152–1154.

Kopachik, W., Bergen, L. G., and Barclay, S. L. (1985a). Genes selectively expressed in proliferating *Dictyostelium* amoebae. *Proc. Natl. Acad. Sci. U.S.A.* **82**, 8540–8544.

Kopachik, W. J., Dhokia, B., and Kay, R. R. (1985b). Selective induction of stalk cell specific proteins in *Dictyostelium. Differentiation* **28**, 209–216.

Kumagai, A., and Okamoto, K. (1986). Prespore-inducing factors in *Dictyostelium discoideum*. Developmental regulation and partial purification. *Differentiation* **31**, 79–84.

Kumagai, A., Pupillo, M., Gundersen, R., Miake-Lye, R., Devreotes, P. N., and Firtel, R. A. (1989). Regulation and function of Gα protein subunits in *Dictyostelium. Cell* **57**, 265–275.

Kumagai, A., Hadwiger, J. A., Pupillo, M., and Firtel, R. A. (1991). Molecular genetic analysis of two Gα protein subunits in *Dictyostelium. J. Biol. Chem.* **266**, 1220–1228.

Lacombe, M-L., Podgorski, G. J., Franke, J., and Kessin, R. H. (1986). Molecular cloning and developmental expression of the cyclic nucleotide phosphodiesterase gene of *Dictyostelium discoideum. J. Biol. Chem.* **261**, 16811–16817.

Leichtling, B. H., Majerfeld, I. H., Spitz, E., Schaller, K. L., Woffendin, C., Kakinuma, S., and Rickenberg, H. V. (1984). A cytosolic cAMP-dependent protein kinase in *Dictyostelium discoideum*. II. Developmental regulation. *J. Biol. Chem.* **259**, 662–668.

Lilly, P., Klein, P., Theibert, A., Vaughan, R., Pupillo, M., Saxe, C., Kimmel, A., and Devreotes, P. (1988). Receptor G protein interactions in the development of *Dictyostelium*. *Bot. Acta* **101**, 123–127.

Loomis, W. F. (1969). Developmental regulation of alkaline phosphatase in *Dictyostelium discoideum*. *J. Bacteriol.* **100**, 417–422.

Louvion, J-F., Scholder, J-C., Pinaud, S., and Reymond, C. D. (1991). Two independent promoters as well as 5′untranslated regions regulate *Ddras* expression in *Dictyostelium*. *Nucleic Acids. Res.* **19**, 6133–6138.

Ludérus, M. E. E., Van der Most, R. G., Otte, A. P., and Van Driel, R. (1989). A protein kinase C-related enzyme activity in *Dictyostelium discoideum*. *FEBS Lett.* **253**, 71–75.

Ma, G. C. L., and Firtel, R. A. (1978). Regulation of the synthesis of two carbohydrate-binding proteins in *Dictyostelium discoideum*. *J. Biol. Chem.* **253**, 3924–3932.

Ma, P. C-C., and Siu, C-H. (1990). A pharmacologically distinct cyclic AMP receptor is responsible for the regulation of gp80 expression in *Dictyostelium discoideum*. *Mol. Cell. Biol.* **10**, 3297–3306.

McRobbie, S. J., Jermyn, K. A., Duffy, K., Blight, K., and Williams, J. G. (1988). Two DIF-inducible, prestalk-specific mRNAs of *Dictyostelium* encode extracellular matrix proteins of the slug. *Development* **104**, 275–284.

Malchow, D., Nanjundiah, V., Wurster, B., Eckstein, F., and Gerisch, G. (1978). Cyclic AMP-induced pH changes in *Dictyostelium discoideum* and their control by calcium. *Biochim. Biophys. Acta* **538**, 473–480.

Mann, S. K. O., and Firtel, R. A. (1989). Two-phase regulatory pathway controls cAMP receptor-mediated expression of early genes in *Dictyostelium*. *Proc. Natl. Acad. Sci. U.S.A.* **86**, 1924–1928.

Mann, S. K. O., and Firtel, R. (1991). A developmentally regulated, putative serine/threonine protein kinase is essential for development in *Dictyostelium*. *Mech. Dev.* **35**, 89–101.

Mann, S. K. O., Pinko, C., and Firtel, R. A. (1988). cAMP regulation of early gene expression in signal transduction mutants of *Dictyostelium*. *Dev. Biol.* **130**, 294–303.

Margolskee, J. P., Froshauer, S., Skrinska, R., and Lodish, H. F. (1980). The effects of cell density and starvation on early developmental events in *Dictyostelium discoideum*. *Dev. Biol.* **74**, 409–421.

Marshak, D. R., Clarke, M., Roberts, D. M., and Watterson, D. M. (1984). Structural and functional properties of calmodulin from the eukaryotic microorganism *Dictyostelium discoideum*. *Biochemistry* **23**, 2891–2899.

Mato, J. M., Krens, F. A., Van Haastert, J. P. M., and Konijn, T. M. (1977). 3′,5′-Cyclic AMP-dependent 3′,5′-cyclic GMP accumulation in *Dictyostelium discoideum*. *Proc. Natl. Acad. Sci. U.S.A.* **74**, 2348–2351.

May, T., Kern, H., Müller-Taubenberger, A., and Nellen, W. (1989). Identification of a *cis*-acting element controlling induction of early gene expression in *Dictyostelium discoideum*. *Mol. Cell. Biol.* **9**, 4653–4659.

May, T., Blusch, J., Sachse, A., and Nellen, W. (1991). A *cis*-acting element responsible for early gene induction by extracellular cAMP in *Dictyostelium discoideum*. *Mech. Dev.* **33**, 147–156.

Mehdy, M. C., and Firtel, R. A. (1985). A secreted factor and cyclic AMP jointly regulate cell-type-specific gene expression in *Dictyostelium discoideum*. *Mol. Cell. Biol.* **5**, 705–713.

Mehdy, M. C., Ratner, D., and Firtel, R. A. (1983). Induction and modulation of cell-type-specific gene expression in *Dictyostelium*. *Cell* **32**, 763–771.

Milne, J. L., and Coukell, M. B. (1991). A Ca^{2+} transport system associated with the plasma membrane of *Dictyostelium discoideum* is activated by different chemoattractant receptors. *J. Cell Biol.* **112**, 103–110.

Morris, H. R., Taylor, G. W., Masento, M. S., Jermyn, K. A., and Kay, R. R. (1987). Chemical structure of the morphogen differentiation inducing factor from *Dictyostelium discoideum. Nature (London)* **328**, 811–814.

Morrissey, J. H., Devine, K. M., and Loomis, W. F. (1984). The timing of cell-type-specific differentiation in *Dictyostelium discoideum. Dev. Biol.* **103**, 414–424.

Mutzel, R., Lacombe, M-L., Simon, M-N., De Gunzburg, J., and Véron, M. (1987). Cloning and cDNA sequence of the regulatory subunit of cAMP-dependent protein kinase from *Dictyostelium discoideum. Proc. Natl. Acad. Sci. U.S.A.* **84**, 6–10.

Newell, P. C. (1982). Cell surface binding of adenosine to *Dictyostelium* and inhibition of pulsatile signalling. *FEMS Microbiol. Lett.* **13**, 417–421.

Newell, P. C., and Ross, F. M. (1982). Inhibition by adenosine of aggregation center initiation and cyclic AMP binding in *Dictyostelium. J. Gen. Microbiol.* **128**, 2715–2724.

Newell, P. C., Ellingson, J. S., and Sussman, M. (1969). Synchrony of enzyme accumulation in a population of differentiating slime mold cells. *Biochim. Biophys. Acta* **177**, 610–614.

Noegel, A., Gerisch, G., Stadler, J., and Westphal, M. (1986). Complete sequence and transcript regulation of a cell adhesion protein from aggregating *Dictyostelium* cells. *EMBO J.* **5**, 1473–1476.

Oyama, M., and Blumberg, D. D. (1986). Interaction of cAMP with the cell-surface receptor induces cell-type specific mRNA accumulation in *Dictyostelium discoideum. Proc. Natl. Acad. Sci. U.S.A.* **83**, 4819–4823.

Parissenti, A. M., and Coukell, M. B. (1990). Effects of DNA and synthetic oligodeoxyribonucleotides on the binding properties of a cGMP-binding protein from *Dictyostelium discoideum. Biochim. Biophys. Acta* **1040**, 294–300.

Pavlovic, J., Banz, E., and Parish, R. W. (1986). Hypersensitive sites in the 5′ and 3′ flanking regions of the cysteine proteinase I gene of *Dictyostelium discoideum. Nucleic Acids Res.* **14**, 8703–8722.

Pavlovic, J., Haribabu, B., and Dottin, R. P. (1988). Transmembrane signal transduction regulates gene expression in *Dictyostelium discoideum. Dev. Genet.* **9**, 371–382.

Pavlovic, J., Haribabu, B., and Dottin, R. P. (1989). Identification of a signal transduction response sequence element necessary for induction of a *Dictyostelium discoideum* gene by extracellular cyclic AMP. *Mol. Cell. Biol.* **9**, 4660–4669.

Pears, C. J., and Williams, J. G. (1987). Identification of a DNA sequence element required for efficient expression of a developmentally regulated and cAMP-inducible gene of *Dictyostelium discoideum. EMBO J.* **6**, 195–200.

Pears, C. J., and Williams, J. G. (1988). Multiple copies of a G-rich element upstream of a cAMP-inducible *Dictyostelium* gene are necessary but not sufficient for efficient gene expression. *Nucleic Acids Res.* **16**, 8467–8486.

Pears, C. J., Mahbubani, H., and Williams, J. G. (1985). Characterization of two highly diverged but developmentally co-regulated cysteine proteinase genes in *Dictyostelium discoideum. Nucleic Acids Res.* **13**, 8853–8866.

Peters, D. J. M., Van Lookeren Campagne, M. M., Van Haastert, P. J. M., Spek, W., and Schaap, P. (1989). Lithium ions induce prestalk-associated gene expression and inhibit prespore gene expression in *Dictyostelium discoideum. J. Cell Sci.* **93**, 205–210.

Peters, D. J. M., Bominaar, A. A., Snaar-Jagalska, B. E., Brandt, R., Van Haastert, P. J. M., Ceccarelli, A., Williams, J. G., and Schaap, P. (1991a). Selective induction of gene expression and second-messenger accumulation in *Dictyostelium discoideum* by the partial chemotactic antagonist 8-*p*-chlorophenylthioadenosine 3′,5′-cyclic monophosphate. *Proc. Natl. Acad. Sci. U.S.A.* **88**, 9219–9223.

Peters, D. J. M., Cammans, M., Smit, S., Spek, W., Van Lookeren Campagne, M. M., and Schaap, P. (1991b). Control of cAMP-induced gene expression by divergent signal transduction pathways. *Dev. Genet.* **12**, 25–34.

Pfeifer, K., Prezant, T., and Guarente, L. (1987). Yeast HAP1 activator binds to two upstream activation sites of different sequence. *Cell* **49**, 19–27.

Pupillo, M., Kumagai, A., Pitt, G. S., Firtel, R. A., and Devreotes, P. N. (1989). Multiple α subunits of guanine nucleotide-binding proteins in *Dictyostelium. Proc. Natl. Acad. Sci. U.S.A.* **86**, 4892–4896.

Ramji, D. P., Donovan, J. T. J., and Hames, B. D. (1991). Sequence and expression analysis of a cAMP-responsive gene regulated during late development of *Dictyostelium discoideum. Mol. Microbiol.* **5**, 427–432.

Raper, K. B. (1940). Pseudoplasmodium formation and organization in *Dictyostelium discoideum. J. Elisha Mitchell Sci. Soc.* **59**, 241–282.

Raper, K. B., and Fennell, D. I. (1952). Stalk formation in *Dictyostelium. Bull. Torrey Club* **79**, 25–51.

Reeves, R. (1984). Transcriptionally active chromatin. *Biochim. Biophys. Acta* **782**, 343–393.

Reymond, C. D., Gomer, R. H., Mehdy, M. C., and Firtel, R. A. (1984). Developmental regulation of a *Dictyostelium* gene encoding a protein homologous to mammalian *ras* protein. *Cell* **39**, 141–148.

Richards, A. J., Corney, A. J., and Hames, B. D. (1990). Cell-type-specific genes expressed late in *Dictyostelium* development show markedly different responses to 3′,5′ cyclic AMP. *Mol. Microbiol.* **4**, 1279–1291.

Richardson, D. L., Hong, C. B., and Loomis, W. F. (1991). A prespore gene, *Dd31*, expressed during culmination of *Dictyostelium discoideum. Dev. Biol.* **114**, 269–280.

Riley, B. B., and Barclay, S. L. (1990). Conditions that alter intracellular cAMP levels affect expression of the cAMP phosphodiesterase gene in *Dictyostelium. Proc. Natl. Acad. Sci. U.S.A.* **87**, 4746–4750.

Roesler, W. J., Vandenbark, G. R., and Hanson, R. W. (1988). Cyclic AMP and the induction of eukaryotic gene transcription. *J. Biol. Chem.* **265**, 9063–9066.

Roos, W., Nanjundiah, V., Malchow, D., and Gerisch, G. (1975). Amplification of cyclic-AMP signals in aggregating cells of *Dictyostelium discoideum. FEBS Lett.* **53**, 139–142.

Ross, F. M., and Newell, P. C. (1981). Streamers: Chemotactic mutants of *Dictyostelium discoideum* with altered cyclic GMP metabolism. *J. Gen. Microbiol.* **127**, 339–350.

Roth, R., and Sussman, M. (1968). Trehalose 6-phosphate synthetase (uridine diphosphate glucose: D-glucose 6-phosphate 1-glucosyltransferase) and its regulation during slime mold development. *J. Biol. Chem.* **243**, 5081–5087.

Rutherford, C. L., Taylor, R. D., Frame, L. T., and Auck, R. L. (1982). A cyclic AMP dependent protein kinase in *Dictyostelium discoideum. Biochem. Biophys. Res. Commun.* **108**, 1210–1220.

Saxe, C. L., III, Johnson, R. L., Devreotes, P. N., and Kimmel, A. R. (1991). Expression of a cAMP receptor gene of *Dictyostelium* and evidence for a multigene family. *Genes Dev.* **5**, 1–8.

Schaap, P., and Van Driel, R. (1985). Induction of post-aggregative differentiation in *Dictyostelium discoideum* by cAMP. Evidence of involvement of the cell surface cAMP receptor. *Exp. Cell Res.* **159**, 388–398.

Schaap, P., and Wang, M. (1986). Interactions between adenosine and oscillatory cAMP signaling regulate size and pattern in *Dictyostelium. Cell* **45**, 137–144.

Schaap, P., Van Lookeren-Campagne, M. M., Van Driel, R., Spek, W., Van Haastert, P. J. M., and Pinas, J. (1986). Postaggregative differentiation induction by cyclic AMP in *Dictyostelium*: Intracellular transduction pathway and requirement for additional stimuli. *Dev. Biol.* **118**, 52–63.

Schatzle, J., Rathi, A., Clarke, M., and Cardelli, J. A. (1991). Developmental regulation of the α-mannosidase gene in *Dictyostelium discoideum*: Control is at the level of transcription and is affected by cell density. *Mol. Cell. Biol.* **11**, 3339-3347.

Shaffer, B. M. (1975). Secretion of cyclic AMP induced by cyclic AMP in the cellular slime mould *Dictyostelium discoideum*. *Nature (London)* **255**, 549–552.

Simon, M-N., Driscoll, D., Mutzel, R., Part, D., Williams, J., and Véron, M. (1989). Overproduction of the regulatory subunit of the cAMP-dependent protein kinase blocks the differentiation of *Dictyostelium discoideum*. *EMBO J.* **8**, 2039–2043.

Singleton, C. K., Delude, R. L., and McPherson, C. E. (1987). Characterization of genes which are deactivated upon the onset of development in *Dictyostelium discoideum*. *Dev. Biol.* **119**, 433–441.

Singleton, C. K., Gregoli, P. A., Manning, S. S., and Northington, S. J. (1988). Characterization of genes which are transiently expressed during the preaggregative phase of development of *Dictyostelium discoideum*. *Dev. Biol.* **129**, 140–146.

Sussman, M., and Osborn, M. J. (1964). UDP-galactose polysaccharide transferase in the cellular slime mold, *Dictyostelium discoideum:* Appearance and disappearance of activity during cell differentiation. *Proc. Natl. Acad. Sci. U.S.A.* **52**, 81–87.

Takemoto, K., Takeuchi, I., and Tasaka, M. (1980). cAMP regulation of the expression of prespore-specific genes, *SP96* and *Dp87*, in disaggregated slug cells of *Dictyostelium discoideum*. *Cell Differ. Dev.* **31**, 89–96.

Takeuchi, I. (1963). Immunochemical and immunohistochemical studies on the development of the cellular slime mold *Dictyostelium mucoroides*. *Dev. Biol.* **8**, 1–26.

Tan, J. L., and Spudich, J. A. (1990). Developmentally regulated protein-tyrosine kinase genes in *Dictyostelium discoideum*. *Mol. Cell. Biol.* **10**, 3578–3583.

Telser, A., and Sussman, M. (1971). Uridine diphosphate galactose-4-epimerase, a developmentally regulated enzyme in the cellular slime mold *Dictyostelium discoideum*. *J. Biol. Chem.* **246**, 2252–2257.

Thaller, C., and Eichele, G. (1987). Identification and spatial distribution of retinoids in the developing chick limb bud. *Nature (London)* **327**, 625–628.

Theibert, A., and Devreotes, P. N. (1984). Adenosine and its derivatives inhibit the cAMP signaling response in *Dictyostelium discoideum*. *Dev. Biol.* **106**, 166–173.

Theibert, A., and Devreotes, P. N. (1986). Surface receptor-mediated activation of adenylate cyclase in *Dictyostelium*. Regulation by guanine nucleotides in wild-type cells and aggregation deficient mutants. *J. Biol. Chem.* **261**, 15121–15125.

Tomchik, K. J., and Devreotes, P. N. (1981). Adenosine 3′,5′-monophosphate waves in *Dictyostelium discoideum*: A demonstration by isotope dilution-fluorography. *Science* **212**, 443–446.

Town, C. D. (1984). Differentiation of *Dictyostelium discoideum* in monolayer cultures and its modification by ionic conditions. *Differentiation* **27**, 29–35.

Town, C., and Gross, J. (1978). The role of cyclic nucleotides and cell agglomeration in postaggregative enzyme synthesis in *Dictyostelium discoideum*. *Dev. Biol.* **63**, 412–420.

Town, C., Gross, J., and Kay, R. (1976). Cell differentiation without morphogenesis in *Dictyostelium discoideum*. *Nature (London)* **262**, 717–719.

Tsang, A. S., and Tasaka, M. (1986). Identification of multiple cyclic AMP-binding proteins in developing *Dictyostelium discoideum* cells. *J. Biol. Chem.* **261**, 10753–10759.

Van Haastert, P. J. M. (1984). Guanine nucleotides modulate cell surface cAMP-binding sites in membranes from *Dictyostelium discoideum*. *Biochem. Biophys. Res. Commun.* **124**, 597–604.

Van Haastert, P. J. M., and Kien, E. (1983). Binding of cAMP derivatives to *Dictyostelium discoideum* cells. Activation mechanism of the cell surface cAMP receptor. *J. Biol. Chem.* **258**, 9636–9642.

Van Haastert, P. J. M., Snaar-Jagalska, B. E., and Janssens, P. M. W. (1987). The regulation of adenylate cyclase by guanine nucleotides in *Dictyostelium discoideum* membranes. *Eur. J. Biochem.* **162**, 251–258.

Van Haastert, P. J. M., De Vries, M. J., Penning, L. C., Roovers, E., Van der Kaay, J., Erneux, C., and Van Lookeren Campagne, M. M. (1989). Chemoattractant and guanosine 5′-[α-thio]triphosphate induce the accumulation of inositol 1,4,5-trisphosphate in *Dictyostelium* cells that are labelled with [^{3}H]inositol by electroporation. *Biochem. J.* **258**, 577–586.

Van Lookeren Campagne, M. M., Schaap, P., and Van Haastert, P. J. M. (1986). Specificity of adenosine inhibition of cAMP-induced responses in *Dictyostelium* resembles that of the P site of higher organisms. *Dev. Biol.* **117**, 245–251.

Van Lookeren Campagne, M. M., Aerts, R. J., Spek, W., Firtel, R. A., and Schaap, P. (1989). Cyclic-AMP-induced elevation of intracellular pH precedes, but does not mediate, the induction of prespore differentiation in *Dictyostelium discoideum*. *Development* **105**, 401–406.

Van Ments-Cohen, M., and Van Haastert, P. J. M. (1989). The cyclic nucleotide specificity of eight cAMP-binding proteins in *Dictyostelium discoideum* is correlated into three groups. *J. Biol. Chem.* **264**, 8717–8722.

Vauti, F., Morandini, P., Blusch, J., Sachse, A., and Nellen, W. (1990). Regulation of the discoidin Iα gene in *Dictyostelium discoideum*: Identification of individual promoter elements mediating induction of transcription and repression by cyclic AMP. *Mol. Cell. Biol.* **10**, 4080–4088.

Wang, M., and Schaap, P. (1989). Ammonia depletion and DIF trigger stalk cell differentiation in intact *Dictyostelium discoideum* slugs. *Development* **105**, 569–574.

Wang, M., Van Haastert, P. J. M., and Schaap, P. (1986). Multiple effects of differentiation-inducing factor on prespore differentiation and cyclic-AMP signal transduction in *Dictyostelium*. *Differentiation* **33**, 24–28.

Wang, M., Roelfsema, J. H., Williams, J. G., and Schaap, P. (1990). Cytoplasmic acidification facilitates but does not mediate DIF-induced prestalk gene expression in *Dictyostelium discoideum*. *Dev. Biol.* **140**, 182–188.

Wanner, R., and Wurster, B. (1990). Cyclic GMP-activated protein kinase from *Dictyostelium discoideum*. *Biochim. Biophys. Acta* **1053**, 179–184.

Weijer, C. J., and Durston, A. J. (1985). Influence of cAMP and hydrolysis products on cell type regulation in *Dictyostelium discoideum*. *J. Embryol. Exp. Morphol.* **86**, 19–37.

Wick, U., Malchow, D., and Gerisch, G. (1978). Cyclic-AMP stimulated calcium influx into aggregating cells of *Dictyostelium discoideum*. *Cell Biol. Intern. Rep.* **2**, 71–79.

Wilkinson, D. G., Wilson, J., and Hames, B. D. (1985). Spore coat protein synthesis during development of *Dictyostelium discoideum* requires a low-molecular-weight inducer and continued multicellularity. *Dev. Biol.* **107**, 38–46.

Williams, J. G., Ceccarelli, A., McRobbie, S., Mahbubani, H., Kay, R. R., Earley, A., Berks, M., and Jermyn, K. A. (1987). Direct induction of *Dictyostelium* prestalk induction by DIF provides evidence that DIF is a morphogen. *Cell* **49**, 185–192.

Wu, L., and Devreotes, P. N. (1991). *Dictyostelium* transiently expresses eight distinct G protein α subunits during its developmental program. *Biochem. Biophys. Res. Commun.* **179**, 1141–1147.

Wu, L., and Franke, J. (1990). A developmentally regulated and cAMP-repressible gene of *Dictyostelium discoideum*: Cloning and expression of the gene encoding cyclic nucleotide phosphodiesterase inhibitor. *Gene* **91**, 51–56.

Wurster, B., Schubiger, K., Wick, U., and Gerisch, G. (1977). Cyclic GMP in *Dictyostelium discoideum*. Oscillations and pulses in response to folic acid and cyclic AMP signals. *FEBS Lett.* **76**, 141–144.

Chapter 15

G Proteins of *Drosophila melanogaster*

JAMES B. HURLEY
Department of Biochemistry and Howard Hughes Medical Institute
University of Washington
Seattle, Washington 98195

I. Introduction

G proteins constitute a diverse family of proteins whose members participate in a variety of biochemical signal transduction mechanisms. These proteins mediate such physiological processes as the yeast mating pheromone response, many types of hormonal responses, and the responses of photoreceptors to light (Simon *et al.*, 1991). In *Drosophila*, evidence suggests that G proteins participate in several well-characterized physiological processes. The roles of G proteins in this system may be more diverse than in other genetically tractable organisms.

SIGNAL TRANSDUCTION
Prokaryotic and Simple Eukaryotic Systems

Much of the information on G proteins in *Drosophila* has been acquired through the isolation of genes that encode G protein subunits. Biochemical approaches, including toxin labeling, immunoreactivity, and assays of guanyl nucleotide binding and hydrolysis, have provided additional information on some G proteins. Genetic analysis has been disappointing with respect to the identification and determination of the role of G proteins; only the *concertina* mutant has identified a G protein subunit. However, other components of some G protein-mediated systems have been identified by genetic analysis. This chapter summarizes progress toward identifying *Drosophila* G proteins and determining their physiological roles.

II. Characteristics of the Known *Drosophila* G Proteins

Drosophila G protein genes have been identified by a variety of approaches, as described in this section. In most cases, no genetic results indicate the role of the particular G protein subunits. However, the temporal and spatial expression patterns provide some tentative information on possible roles of the proteins.

A. α Subunits

1. α_s Quan and colleagues (Quan *et al.*, 1989) screened a *Drosophila* genomic DNA library with a mammalian α_s cDNA probe at reduced stringency and identified a single gene encoding a protein 70% identical to mammalian α_s. The approximately 4.5-kb gene can be spliced two different ways to encode either a long or a short form of α_s (Quan and Forte, 1990). The two proteins differ by three amino acids inserted after residue 309. The gene is located at 99C near the distal end of the third chromosome. No mutants of any phenotype have been identified at this locus. *Drosophila* α_s mRNA has been detected throughout the *Drosophila* nervous system (Quan *et al.*, 1989). α_s protein is present at highest levels in cell membranes of neuronal cell bodies in the brain and the thoracic ganglion (Wolfgang *et al.*, 1990). An arginine at position 189 in *Drosophila* α_s is a likely target for cholera toxin-catalyzed ADP-ribosylation. No cysteine occurs near the C terminus, the usual ADP-ribosylation site recognized by pertussis toxin.

2. α_o A single *Drosophila* α_o gene was isolated by screening a genomic DNA library at reduced stingency with a mammalian α_o cDNA as a probe (de Sousa *et al.*, 1989; Thambi *et al.*, 1989; Yoon *et al.*, 1989). The gene is located at 47A on the second chromosome. No mutants of any phenotype at this locus have been reported. Two different α_o mRNAs are produced depending on which of two alternative promoters is used (Yoon *et al.*, 1989), but the spatial and developmental expression from both promoters is similar. *In situ* hybridization detected only a slight difference in the expression of maternal transcripts from the two promoters in nurse cells and oocytes (de Sousa *et al.*, 1989.) The transcripts encode proteins with slightly different amino termini, both of which include a consensus sequence for N-terminal myristoylation (Gordon *et al.*, 1991). The amino acid sequences of the two *Drosophila* α_os are 83% identical to that of mammalian α_o. A cysteine near the C terminus is a potential target for ADP-ribosylation by the pertussis toxin S1 subunit. Pertussis toxin treatment combined with immunoblotting (Thambi *et al.*, 1989; Guillen *et al.*, 1990) revealed that the *Drosophila* α_o subunit is a pertussis toxin substrate.

3. α_i A *Drosophila* α_i gene was identified by reduced stringency screening of a *Drosophila* genomic DNA library (Provost *et al.*, 1988). This gene is located at 65C on the left arm of the *Drosophila* third chromosome. No mutants of any phenotype have been identified at this locus. An α_i transcript of 2.3 kb has been detected that is most abundant in very early embryos (Provst *et al.*, 1988). *In situ* hybridization of adult female *Drosophila* demonstrates that the most abundant α_i mRNA transcripts are maternal, that is, they are produced in ovarian nurse cells and then released into the developing oocyte (Hurley *et al.*, 1990). α_i mRNA transcripts are present only at barely detectable levels in other tissues but α_i cDNA clones have been isolated from head mRNA. α_i protein has been reported to be present in photoreceptor cell terminations, glomerulae of the antennal lobes, and the ocellus (Wolfgang *et al.*, 1990). Although, the amino acid sequences of *Drosophila* and mammalian α_is are 80% identical, the pertussis toxin modification site present in all mammalian α_is is missing in *Drosophila* α_i. The N terminus includes a consensus sequence for N-terminal myristoylation (Gordon *et al.*, 1991).

4. α_q The *Drosophila* α_q gene was identified as one of several genes expressed only in the *Drosophila* compound eye (Lee *et al.*, 1990). The gene is located at 49B on the second chromosome. No *Drosophila* mutations are known to map to that locus. An α_q transcript has been detected by Northern blots. *In situ* hybridization revealed α_q gene expression only in *Drosophila* photoreceptors, where it is proposed to play a role in phototransduction. The amino acid sequence of *Drosophila* α_q is 76% identical to mammalian α_q (Strathman and Simon, 1990). Members of the α_q subfamily of mammalian G proteins—α_q, α_{11}, and α_{13}—activate phospholipase C *in vitro* (Taylor *et al.*, 1990; Smrcka *et al.*, 1991). The presence of a specific phospholipase C in photoreceptors and its

GTP sensitive activation by light suggest that α_q may mediate the action of light on phospholipase C. No pertussis toxin modification site is found in *Drosophila* α_q, consistent with the observation that no pertussis toxin substrates exist in the *Drosophila* compound eye (Hopkins *et al.*, 1988).

5. *concertina* The *concertina* gene encodes a 53-kDa protein 56% identical to mammalian α_{12} and α_{13} (Strathmann and Simon, 1991). This gene includes all the regions of G protein α subunits thought to interact with GTP. Of 82 amino acids that are nearly invariant in G protein α subunits, 81 are conserved in *concertina*-encoded proteins. Unlike other α subunits, the concertina protein does not include the consensus sequence for N-terminal myristoylation.

B. β Subunits

1. *Gbb* The *Drosophila GBB* gene was identified by probing a *Drosophila* genomic DNA library at reduced stringency with a mammalian G protein β subunit cDNA (Yarfitz *et al.*, 1988). The *GBB* gene is located at 13F on the *Drosophila* X chromosome. No mutants of any phenotype have been identified at this locus. Although no introns occur within the coding region of the *GBB* gene, transcripts of at least five different sizes have been detected. These transcripts probably are derived from alternative splicing or alternative promoters. The significance of these different transcripts is not clear since each one includes the same coding region of *GBB*. Also little difference is seen in the developmental pattern of expression of these transcripts, except in very early embryos in which only two small maternal transcripts are present. *GBB* mRNA transcripts are not present in photoreceptors but are present throughout the adult *Drosophila* central nervous system, including the thoracic ganglion. The Gbb amino acid sequence is 82–84% identical to mammalian G protein β subunit sequences, and includes a 43-amino-acid repetitive sequence found in all G protein β subunits. The function of this repeat is unknown, but it is also present in other unrelated proteins such as *Drosophila* enhancer of split and yeast Cdc4 (Yarfitz *et al.*, 1991). These proteins do not appear to have any common functions.

2. *Gbe* Antibodies raised against a TrpE/Gbb fusion protein reacted with the entire adult *Drosophila* nervous system, including the eye (Yarfitz *et al.*, 1991). This result was surprising because mRNA transcripts derived from the *GBB* gene were absent from the eye. The fusion protein antibody detected two polypeptides, 37 kDa and 39 kDa, on immunoblots of SDS–PAGE analyses of *Drosophila* head homogenates. Only the 37-kDa polypeptide was

detected in *Drosophila* body homogenates. The 39-kDa polypeptide present in wild-type head homogenates was absent from heads of mutant *Drosophila* that lack eyes. These results prompted Yarfitz and colleagues (1991) to screen a *Drosophila* head cDNA expression library with the fusion protein antibody. Positive phage then were screened using the *GBB* cDNA as a probe. Clones that reacted with the antibody but did not hybridize to *GBB* cDNA were analyzed further. These clones identified a G protein β subunit expressed only in photoreceptors. The *GBE* gene maps to the 76C locus on the *Drosophila* third chromosome. Two mutants have been obtained that express significantly reduced levels of Gbe protein (P. Dolph, S. Yarfitz, J. Hurley, and C. Zuker, unpublished results). The visual phenotypes of these mutants currently are being investigated. A single transcript has been detected from the *GBE* gene. *In situ* hybridization reveals that this mRNA is most abundant in adult *Drosophila* photoreceptors and ocelli (Yarfitz *et al.*, 1991; M. Spencer, unpublished observations). Very low levels were also detected in ovaries (M. Spencer, unpublished observations). *GBE* transcripts were not detected in any other tissues or at any other developmental stage. Gbe appears to define a new class of G protein β subunits. The amino acid sequence of Gbe is distantly related to Gbb and only 45% identical to a β subunit consensus sequence derived from all β subunit sequences published to date (Yarfitz *et al.*, 1991). The yeast G protein β subunit (Whiteway *et al.*, 1989), which bears even less resemblance to the β consensus, is known by genetic methods to interact with an α subunit. Gbe may interact with a G protein α subunit such as the eye-specific α_q (Lee *et al.*, 1990). However, this sort of interaction needs to be confirmed by biochemical methods.

C. γ Subunit

1. γ_1 A single γ subunit gene has been identified among clones that are expressed predominantly in the head (Ray and Ganguly, 1992). The predicted gene product shows 46%, 43%, and 28% amino acid identity to the bovine γ_2, γ_3, and γ_t sequences, respectively. The gene is located at 44CD on the second chromosome. No known mutants exist at this locus that could be candidates for mutations in this gene. Transcripts of several sizes have been detected. A 2.6-kb transcript is expressed only in the head, a 1.3-kb transcript is expressed only in the body, and the most prevalent transcript of 1.1 kb is expressed in both the head and the body, although expression is higher in the head. Expression of the 1.1-kb transcript peaks between 9 and 12 hr of embryogenesis and is expressed predominantly in the nervous system of adult flies. The RNAs are present in *eym* (*eye missing*) mutants. This expression pattern suggests that γ_1 may play roles in neuronal signal transduction and differentiation.

III. Biochemical Evidence for G Proteins in *Drosophila*

Biochemical approaches that have been used for the identification of G protein α subunits in mammalian cells include labeling with cholera and pertussis toxins and analysis of guanine nucleotide binding and hydrolysis. These approaches also have provided some information on potential α subunits in *Drosophila*.

A. Toxin Labeling

One of the most effective ways to identify G proteins in a tissue is by incubation with [^{32}P]NAD and cholera toxin or pertussis toxin. The toxin transfers the [^{32}P]ADP–ribose portion of NAD to the G protein substrate, which then can be visualized by SDS–PAGE and autoradiography (Hopkins *et al.*, 1988). The only major membrane-associated pertussis toxin substrate in *Drosophila* is a 39-kDa protein that appears to be α_o on the basis of its immunoreactivity with an α_o-specific antibody (Thambi *et al.*, 1989; Guillen *et al.*, 1990).

On treatment with trypsin, ADP-ribosylated *Drosophila* α_o yields a soluble 15-kDa fragment containing the ADP–ribose group (Hopkins *et al.*, 1988). However, the presence of 5′-γthio-GTP (GTPγS), a nonhydrolyzable form of GTP, protects one of the proteolytic sites in α_o and stabilizes a 37-kDa ADP-ribosylated fragment. Similar effects of trypsin and guanyl nucleotides on mammalian G protein α subunits have been observed (Hurley *et al.*, 1984).

Cholera toxin specifically modifies an arginine near the middle of the primary sequence of certain mammalian G protein α subunits such as mammalian α_s and α_{olf}. This arginine is highly conserved and essential for GTP hydrolysis (see Bourne *et al.*, 1991). Cholera toxin substrates are also present in *Drosophila* homogenates (Hopkins *et al.*, 1988). One is approximately 45 kDa and may correspond to the *Drosophila* α_s subunit (Quan *et al.*, 1989). The other is a soluble 37-kDa protein that also may be a substrate for an endogenous ADP-ribosyl transferase (Hopkins *et al.*, 1988).

Even without added pertussis toxin, a soluble 37-kDa protein can be modified reproducibly simply by incubating *Drosophila* homogenates with [^{32}P]NAD. This modification has been shown to include ADP–ribose by observing the release of 5′[^{32}P]AMP on treatment with either NaOH or snake venom phosphodiesterase (Hopkins *et al.*, 1988). The identity of this 37-kDa protein, the site at which it is modified, and the endogenous mechanism behind its modification are unknown. This protein differs from the 39-kDa *Drosophila* α_o because it does not associate with membranes. Also, the protein does not appear to be a C-terminal proteolytic fragment of α_o because it fails to react with an antibody against the α_o C terminus. This same antibody reacts with a 37-kDa C-terminal tryptic fragment of *Drosophila* α_o (T. Neubert and J. Hurley, unpublished results).

B. Guanyl Nucleotide Binding and Hydrolysis

Blumenfeld and co-workers (Blumenfeld *et al.*, 1985) demonstrated that light stimulates GTP hydrolysis in homogenates of *Musca domestica* eyes. This activity has a submicromolar K_m, suggesting that it reflects the activity of a rhodopsin-stimulated G protein. Devary and colleagues (Devary *et al.*, 1987) then showed that light also stimulates binding of GTPγS to these membranes and that the light-induced binding of guanyl nucleotides stimulates phospholipase C activity in these homogenates. The light-stimulated binding of GTPγS also was demonstrated in homogenates of *Drosophila* heads. These experiments suggest that a G protein exists in fly photoreceptors that couples light stimulation of rhodopsin to activation of phospholipase C, as described subsequently.

IV. Genetic Approaches to Studying *Drosophila* G Proteins

The *concertina* gene is the only G protein gene that has been identified on the basis of a mutant phenotype (Parks and Wieschaus, 1991). No other *Drosophila* G proteins have been associated with a physiological mutant phenotype. Perhaps some of these G proteins have multiple functions, including some required for viability, which could obscure phenotypes that would indicate a role in a signal transduction process. The absence of α_q and *GBE* mutants from screens for vision mutants could be the result of bias of those screens for mutations on the X chromosome.

An attraction of studying G proteins in *Drosophila* is that genetic approaches can be taken to study the physiological roles of these proteins. Mutant genes might be selected by screening mutagenized flies with a chromosome with a large deficiency at the G protein subunit locus. The disadvantage of this approach is that no reliably predictable phenotype exists for which to screen. A more successful approach may be to screen individual flies for the absence of mRNA or protein corresponding to a particular G protein subunit.

An interesting alternative approach has been used (Fitch *et al.*, 1993). *Drosophila* α_o was inactivated conditionally *in vivo* by expressing pertussis toxin under the control of the hsp70 heat-shock promoter. Pertussis toxin was expressed primarily in the eye and the gut. This expression altered the visual response and prevented flies from eating. In addition, pertussis toxin expression in embryos at certain stages halted development. Because some of the effects of pertussis toxin expression are lethal, suppressor mutations in genes involved in α_o-mediated pathways may be identifiable.

V. Physiological Processes Mediated by G Proteins

A. Phototransduction

Transducin, the G protein of vertebrate rod and cone photoreceptors, has been studied extensively (Hurley, 1992). On stimulation by photolyzed rhodopsin, transducin activates a cGMP phosphodiesterase to hydrolyze cyclic GMP. The resulting depletion of intracellular cGMP shuts down cation channels and hyperpolarizes the cell.

Invertebrate photoreceptors respond in a similar manner. Rhabdomeres are the photosensitive segments of *Drosophila* photoreceptors. Light depolarizes them by stimulating plasma membrane cation channels (Smith *et al.*, 1991). Molecular and genetic evidence indicates that this process requires an eye-specific phospholipase C (Bloomquist *et al.*, 1988). The gene encoding this enzyme is defective in *norpA* mutants whose photoreceptors fail to respond to light.

Biochemical evidence that *Drosophila* phospholipase C is regulated by light has not been reported, but light stimulation of phospholipase C from *Musca* eyes has been observed (Devary *et al.*, 1987). GTP and GTP analogs enhance this effect, suggesting that a GTP-binding protein participates in the invertebrate photoresponse. This notion is supported by the observation that GTP or GTP analogs injected into *Limulus* photoreceptors affect the photoresponse (Fein, 1986).

The defective phospholipase C in *norpA* mutants is of the type regulated by the G_q class of heterotrimeric G protein α subunits (Taylor *et al.*, 1990; Smrcka *et al.*, 1991). This subfamily of G proteins includes α_q, α_{11}, and α_{13} (Simon *et al.*, 1991). A gene that encodes a *Drosophila* homologue of α_q has been identified that is expressed only in *Drosophila* eyes (Lee *et al.*, 1991). Therefore rhodopsin stimulation is proposed to be coupled to phospholipase C activation by this α_q subunit.

G protein α subunits generally interact with a $\beta\gamma$ complex and require this interaction to couple with receptors (Gilman, 1987). The *Drosophila GBB* gene is expressed throughout the adult *Drosophila* nervous system except in eyes (Yarfitz *et al.*, 1988). Therefore its product is not a candidate for a β subunit involved in phototransduction. The *GBE* gene is expressed almost entirely in *Drosophila* eyes and ocelli (Yarfitz *et al.*, 1991). The Gbe protein, with the eye-specific α_q and an as yet unidentified γ subunit, may participate in *Drosophila* phototransduction by transducing light activation of rhodopsin into stimulation of phospholipase C. Genetic and biochemical experiments are underway to test this idea.

Mutant *Drosophila* defective in their visual response have been isolated and many of the mutated genes have been identified (reviewed by Smith *et al.*, 1991). None have been reported to encode G protein subunits. One possible explanation for this result is that G proteins are not essential for *Drosophila* phototransduction. However, although the *Drosophila* X chromosome has been screened exhaustively for mutations affecting the photoresponse, the autosomes have not; therefore, an alternative explanation is that the relevant G protein genes are on autosomes. The findings that the gene for α_q maps to the second chromosome and the gene for Gbe maps to the third chromosome support this explanation. Other possible explanations are that phototransduction G proteins have other functions that are essential or that redundant phototransduction G protein genes exist.

B. Olfaction

Like vertebrate phototransduction, vertebrate olfactory transduction may be mediated by a G protein (Reed, 1992). Mammalian olfactory epithelia express a large family of transmembrane proteins that are proposed to be odorant receptors (Buck and Axel, 1991). A G protein α subunit, α_{olf}, is present specifically in rat olfactory epithelial cells where it may be stimulated by these receptors. α_{olf} activates adenylyl cyclase (Reed, 1992), so cAMP may be a second messenger required for olfactory transduction.

Is *Drosophila* olfactory transduction also mediated by a G protein? *Drosophila* olfaction mutants have been identified and the genes responsible for several of them have been characterized (for examples, see McKenna *et al.*, 1989). None of these gene products resembles components of a G protein-mediated signal transduction mechanism. However, the number of mutants that have been characterized is small, so G proteins involved in *Drosophila* olfaction easily could have been missed.

Other lines of evidence suggest that G proteins may be present in *Drosophila* olfactory receptors. An antimammalian G_o antiserum reacts with a protein in *Drosophila* antennae (Schmidt *et al.*, 1989). The identity of the G protein is uncertain because the antiserum used in these experiments may react with more than one *Drosophila* protein. The G_o antiserum used in these experiments also reacts with the *Drosophila* eye, a tissue in which *Drosophila* α_o mRNA is not present.

A G protein has been detected in antennae of another insect, *Manduca sexta,* using pertussis toxin to [^{32}P]ADP-ribosylate G proteins specifically (Hopkins *et al.*, 1988). However, G proteins were not detected in *Drosophila* antennae using either pertussis toxin or G protein specific antibodies (T. Neubert and

J. Hurley, unpublished observations). Because these reagents do not necessarily identify all G proteins, the question of G protein involvement in *Drosophila* olfaction requires further investigation.

C. Memory

Wild-type *Drosophila* demonstrate associative learning and memory. For example, they can be trained to associate an odor with a mild electric shock (Quinn *et al.*, 1974). Mutants such as *dunce* and *rutabaga* have been identified that are deficient in the ability to acquire and store memories (Dudai *et al.*, 1988).

A factor that influences *Drosophila* memory is cAMP. The *rutabaga* and *dunce* genes encode products with sequence similarity to an adenylyl cyclase (Levin *et al.*, 1992) and a cAMP phosphodiesterase (Chen *et al.*, 1986), enzymes that synthesize and degrade cAMP. Eukaryotic adenylyl cyclases are regulated by G proteins, suggesting that G proteins could influence memory. However, no memory mutants that have been characterized have genetic deficiencies in G protein subunits. Both the *rutabaga* and *dunce* genes appear to be expressed primarily in mushroom bodies of the *Drosophila* brain, an area thought to function in olfactory learning and memory (Levin *et al.*, 1992). Perhaps the role of these genes is specific to olfaction, whereas the G proteins that regulate adenylyl cyclase activity needed for memory are essential for other vital functions, leading to a lethal or pleiotropic phenotype when mutated.

Although G protein genes are not among the genetic loci identified as responsible for the memory deficiency phenotype, their levels are elevated in some learning mutants (Guillen *et al.*, 1990). Presumably, this change reflects an attempt of the cells to compensate for the primary deficiencies that affect cAMP.

D. Development

The *concertina* gene is required for normal gastrulation during *Drosophila* embryonic development. The products of this gene show sequence similarity to G protein α subunits (Parks and Wieschaus, 1991). *Concertina* mutants fail to form a normal ventral furrow during gastrulation. mRNA transcripts from α_s, α_i, α_o, *GBB*, and γ_1 are present during early embryonic development. Whether this expression pattern reflects a role in development is unknown. However, expression of pertussis toxin in stage 6 to stage 9 *Drosophila* embryos is lethal (Fitch *et al.*, 1993). Because pertussis toxin ADP-ribosylates and inactivates *Drosophila* α_o, its lethality in these embryos suggests that α_o may participate in some aspect of signal transduction necessary for development or in some essential homeostatic process. This result underscores the difficulty of analyzing mutations in certain G proteins in *Drosophila*. If a particular G protein has multiple functions, gene disruption experiments are likely to be difficult. A more promising approach may be to inactivate particular G protein subunits conditionally or to express mutant G proteins in specific cell types.

VI. Summary

The number of G protein subunits identified in *Drosophila* is small compared with the number found in mammals. This finding might indicate a smaller pool of G protein genes. Repeated screens of *Drosophila* DNA and mRNA have identified the same genes. Now that *Drosophila* G protein genes have been identified, thorough biochemical and genetic analyses of their functions and the pathways that they mediate must begin.

Acknowledgments

I wish to thank Stuart Yarfitz for his suggestions and Janet Kurjan for patience, helpful suggestions, and editing.

References

Bloomquist, B. T., Shortridge, R. D., Schneuwly, S., Perdew, M., Montell, C., Steller, H., Rubin, G., and Pak, W. L. (1988). Isolation of a putative phospholipase C gene of *Drosophila*, norpA and its role in phototransduction. *Cell* **54**, 723–733.

Blumenfeld, A., Erusalimsky, J., Heichal, O., Selinger, Z., and Minke, B. (1985). Light-activated guanosinetriphosphatase in *Musca* eye membranes resembles the prolonged depolarizing afterpotential in photoreceptor cells. *Proc. Natl. Acad. Sci. U.S.A.* **82**, 7116–7120.

Bourne, H. R., Sanders, D. A., and McCormick, F. (1991). The GTPase superfamily: Conserved structure and molecular mechanism. *Nature (London)* **349**, 117–127.

Buck, L., and Axel, R. (1991). A novel multigene family may encode odorant receptors: A molecular basis for odor recognition. *Cell* **65**, 175–187.

Chen, C. N., Denome, S., and Davis, R. L. (1986). Molecular analysis of cDNA clones and the corresponding genomic coding sequences of the *Drosophila* dunce gene, the structural gene for cAMP phosphodiesterase. *Proc. Natl. Acad. Sci. U.S.A.* **289**, 9313–9317.

de Sousa, S. M., Hoveland, L. L., Yarfitz, S., and Hurley, J. B. (1989). The *Drosophila* α_o-like G protein gene produces multiple transcripts and is expressed in the nervous system and in ovaries. *J. Biol. Chem.* **264**, 18544–18551.

Devary, O., Heichal, O., Blumenfeld, A., Cassel, D., Suss, E., Barash, S., Rubinstein, C. T., Minke, B., and Selinger, A. (1987). Coupling of photoexcited rhodopsin to inositol phospholipid hydolysis in fly receptors. *Proc. Natl. Acad. Sci. U.S.A.* **84**, 6939–6943.

Dudai, Y. (1988). Neurogenetic dissection of learning and short-term memory in *Drosophila*. *Ann. Rev. Neurosci.* **11**, 537–563.

Fein, A. (1986). Blockade of visual excitation and adaptation in *Limulus* photoreceptor by GDPβ-S. *Science* **232**, 1543–1545.

Fitch, C. L., de Sousa, S. M., O'Day, P., Neubert, T., Plantilla, C., Spencer, M., Yarfitz, S., Apte, D., and Hurley, J. B. (1993). Pertussis toxin expression in *Drosophila* alters the visual response and blocks eating behavior. *Cell Signalling* **5**, 187–207.

Gilman, A. G. (1987). G proteins: Transducers of receptor-generated signals. *Ann. Rev. Biochem.* **56**, 615–649.

Gordon, J. I., Duronio, R. J., Rudnick, D. A., Adams, S. P., and Goke, G. W. (1991). Protein N-myristoylation. *J. Biol. Chem.* **266**, 8647–8650.

Guillen, A., Jallon, J. M., Fehrentz, J. A., Pantaloni C., Bockaert, J., and Homborger, V. (1990). A G_o-like protein in *Drosophila* melanogaster and its expression in memory mutants. *EMBO J.* **9**, 1449–55.

Hopkins, R. S., Stamnes, M. A., Simon, M. I., and Hurley, J. B. (1988). Cholera toxin and pertussis toxin substrates and endogenous ADP-ribosyltransferase activity in *Drosophila* melanogaster. *Biochim. Biophys. Acta* **970**, 355–362.

Hurley, J. B. (1990). A likely history of G protein genes. *Biochem. Soc. Symp.* **56**, 81–84.

Hurley, J. B. (1992). Signal transduction enzymes of vertebrate photoreceptors, *J. Bioenerg. Biomem.* **24**, 219–225.

Hurley, J. B., Yarfitz, S., Provost, N., and de Sousa, S. (1990). *Drosophila* G proteins. *In* "G Proteins and Signal Transduction," pp 158–167. Rockefeller University Press, New York.

Hurley, J. B., Simon, M. I., Teplow, D. B., Robishaw, J. D., and Gilman, A. G. (1984). Homologies between signal transducing G proteins and ras gene products. *Science* **226**, 860–862.

Lee Y.-J., Dobbs, M. B., Veradi, M. L., and Hyde, D. R. (1990). *dgq*: A *Drosophila* gene encoding a visual system specific Gα molecule. *Neuron* **5**, 889–898.

Levin, L. R., Han, P.-L., Hwang, P. M., Feinstein, P. G., Davis, R. L., and Reed, R. R. (1992). The *Drosophila* learning and memory gene *rutabaga* encodes a Ca^{2+}/calmodulin-responsive adenylyl cyclase. *Cell* **68**, 479–489.

McKenna, M., Monter, P., Hlefland, S. L., Woodard, C., and Carlson, J. (1989). A simple chemosensory response in *Drosophila* and the isolation of acj mutants in which it is affected. *Proc. Natl. Acad. Sci. U.S.A.* **86**, 8118–8122.

Parks, S., and Wieschaus, E. (1991). The *Drosophila* gastrulation gene *concertina* encodes a Gα-like protein. *Cell* **64**, 447–458.

Provost, N. M., Somers, D. E., and Hurley, J. B. (1988). A *Drosophila* melanogaster G protein α subunit gene is expressed primarily in embryos and pupae. *J. Biol. Chem.* **263**, 12070–12076.

Quan, F., and Forte, M. A. (1990). Two forms of *Drosophila* melanogaster G_s α are produced by alternate splicing involving an unusual splice site. *Mol. Cell. Biol.* **10**, 910–917.

Quan, F., Wolfgang, W. J., and Forte, M. A. (1989). The *Drosophila* gene coding for the a subunit of a stimulatory G protein is preferentially expressed in the nervous system. *Proc. Natl. Acad. Sci. U.S.A.* **86**, 4321–4325.

Quinn, W. G., Harris, W. A., and Benzer, S. (1974). Conditioned behavior in *Drosophila melanogaster*. *Proc. Natl. Acad. Sci. U.S.A.* **71**, 708–712.

Ray, K., and Ganguly, R. (1992). The *Drosophila* G protein γ subunit gene (*D-Gγ1*) produces three developmentally regulated transcripts and is predominantly expressed in the central nervous system. *J. Biol. Chem.* **267**, 6086–6092.

Reed, R. (1992). Signalling pathways in odorant detection. *Neuron* **8**, 205–209.

Schmidt, C. J., Garen-Fazio, S., Chow, Y., and Neer, E. J. (1989). Neuronal expression of a newly identified *Drosophila* melanogaster G protein α_o subunit. *Cell Regul.* **1**, 125–134.

Simon, M. I., Strathman, M. P., and Gautam, N. (1991). Diversity of G proteins in signal transduction. *Science* **252**, 802–808.

Smith, D. P., Stamnes, M. A., and Zuker, C. S. (1991). Signal transduction in the visual system of *Drosophila*. *Ann. Rev. Cell Biol.* **7**, 161.

Smrcka, A. V., Hepler, J. R., Brown, K. O., and Sternweis, P. C. (1991). Regulation of phosphoinositide-specific phospholipase C activity by purified α_q. *Science* **251**, 804–807.

Strathmann, M., and Simon, M. I. (1990). G protein diversity: A distinct class of α subunits is present in vertebrates and invertebrates. *Proc. Natl. Acad. Sci. U.S.A.* **87**, 9113–9117.

Strathmann, M. P., and Simon, M. I. (1991). $G\alpha_{12}$ and $G\alpha_{13}$ subunits define a fourth class of G protein α subunits. *Proc. Natl. Acad. Sci. U.S.A.* **88**, 5582–5586.

Taylor, S. J., Smith, J. A., and Exton, J. H. (1990). Purification from bovine liver membranes of a guanine mnucleotide-dependent activator of phosphoinositide-specific phospholipase c. *J. Biol. Chem.* **265**, 17150–17156.

Thambi, M. C., Quan, R., Wolfgang, W. J., Spiegel, A., and Forte, M. (1989). Immunological and molecular characterization of G_o α-like proteins in the *Drosophila* central nervous system. *J. Biol. Chem.* **264**, 18552–18560.

Whiteway, M., Hougan, L., Dignard, D., Thomas, D. Y., Bell, L., Saari, G. C., Grant, F. J., O'Hara, P., and MacKay, V. L. (1989). The STE4 and STE18 genes of yeast encode potential β and γ subunits of the mating factor receptor-coupled G protein. *Cell* **56**, 467–477.

Wolfgang, W. J., Quan, F., Goldsmith, P., Unson, C., Spiegel, A., and Forte, M. (1990). Immunolocalization of G protein a subunits in the *Drosophila* CNS. *J. Neurosci.* **10**, 1014–1024.

Yarfitz, S., Provost, N., and Hurley, J. B. (1988). Cloning a *Drosophila melanogaster* guanine nucleotide regulatory protein β-subunit gene and characterization of its expression during development. *Proc. Natl. Acad. Sci. U.S.A.* **85**, 7134–7138.

Yarfitz, S., Niemi, G. A., McConnell, J. L., Fitch, C. L., and Hurley, J. B. (1991). A Gβ protein in the *Drosophila* compound eye is different from that in the brain. *Neuron* **7**, 429–438.

Yoon, J., Shortridge, F. D., Bloomquist, B. T., Schneuwly, S., Perdew, M. H., and Pak, W. L. (1989). Molecular characterization of a *Drosophila* gene encoding a G_o α subunit homolog. *J. Biol. Chem.* **264**, 18536–18543.

Chapter 16

Signal Transduction during *Caenorhabditis elegans* Vulval Determination

RAFFI V. AROIAN and PAUL W. STERNBERG

Division of Biology
California Institute of Technology
Pasadena, California 91125

I. General Aspects of Signal Transduction and *Caenorhabditis elegans* Development

Development of multicellular organisms involves the proliferation of the single-celled egg into numerous cells that subsequently can differentiate into discrete cell types, for example, liver, muscle, or neuronal, as the organism matures. Two primary mechanisms are thought to be responsible for this development. The identity of cells can be determined solely by factors inherited by their progenitor. This type of development is cell autonomous, because cells take on identity without informational input from neighboring cells or the environment. Conversely, the identity of cells can be specified by their location, that is, where they are and what other cells they neighbor. This type of development is cell nonautonomous, because the developing cells require information from the outside to determine the changes that occur inside. The latter type of development requires signal transduction, the mechanism by which cells take in information from their surroundings. [For an in-depth review of the processes involved in animal development, see Davidson (1990).]

The development of the free-living nematode *Caenorhabditis elegans* provides a unique opportunity to study signal transduction by means of three methods: cell biology, genetics, and molecular biology. Each adult *C. elegans* hermaphrodite contains 959 somatic nuclei, the identity and relative geometry of which are known and essentially identical. Each individual hermaphrodite therefore can be thought of as an identical "tissue-culture dish" in which a complex arrangement of known cells and tissues is reproduced faithfully, providing an excellent opportunity for *in vivo* cell biology. Using a laser microbeam and various mutations, the local environment around a cell or group of cells can be perturbed and the effects on those cells assessed. This analysis can yield insights into how various cells and tissues interact. Genetics can be used to identify and characterize the genes involved in a given developmental pathway because small perturbations in the molecules involved in development can have major consequences for the final outcome (e.g., hermaphrodites that cannot lay eggs). The interactions of the genes also can be ascertained by examining phenotypes of double mutants (epistasis analysis). Chromosomal deficiencies and duplications permit variation of dose in a controlled way. Mosaic analysis can be used to determine in which tissues a specific gene acts. Finally, molecular biology can be used to identify the proteins involved in development and molecular genetic techniques, for example, transgenic nematodes, can be used to probe the functions of these molecules. As discussed in this chapter, the study of the development of the hermaphrodite vulva makes use of all these tools to unravel a complex and fascinating signal transduction pathway.

Because the power of vulval development as a system in which to study signal transduction relies on cell biology, genetics, and molecular biology, this chapter offers a comprehensive look at each method. Vulval development is introduced through cell biology and cell biological experiments. These experiments were the first to suggest that vulval induction involves signal transduction. Next, the isolation of mutations in genes responsible for vulva development is discussed, followed by the molecular identification of some of these genes. Some of the identified genes are similar to genes encoding signal transduction components of other systems, for example, the epidermal growth factor (EGF) receptor gene, *ras*, and *Notch*. Genetic experiments that order the genes in the vulval pathway and that elucidate the interactions among these genes are reviewed. This information is combined with cell biology and molecular biology data to model the signal transduction events responsible for vulval development and to model how three different outcomes from those signal transduction events are produced.

In addition to providing a summary of data and interpretation, this chapter also discusses how the data were obtained, on what basis a given interpretation is made, and caveats in making that interpretation. This attention to critical thinking as well as data serves several purposes: (1) it provides a deeper understanding of the field; (2) it permits many different aspects of vulval determination to be brought together in a single chapter; and (3) it provides a foundation for understanding future results—not only their meaning but also how they were obtained. For a briefer review on vulval determination, see Horvitz and Sternberg (1991).

II. Cell Biology of the *Caenorhabditis elegans* Hermaphrodite Vulva—Development and Signal Transduction

A. Development of the Vulva

Caenorhabditis elegans comes in two sexes, hermaphrodite and male. The two sexes share many similarities—for example, pharynx, digestive system, overall musculature, and nervous system—but exhibit major differences with respect to the reproductive system. Hermaphrodites can be thought of as females with a limited amount of sperm. They produce both oocytes and sperm, and self-fertilization can occur in the gonad (Fig. 16.1). Sperm from a hermaphrodite can fertilize its own oocytes in the spermatheca, producing self-progeny. After a few rounds of division, the fertilized egg is ejected through an opening in the ventral side of the hermaphrodite, the vulva, and the zygote hatches about 12 hr later. Males, on the other hand, produce only sperm and can introduce these sperm into hermaphrodites to produce cross-progeny, half of which are male and half of which are hermaphrodites.

The wild-type vulva is composed of 22 nuclei and develops identically in each wild-type hermaphrodite (Fig. 16.2A). These 22 nuclei are derived from three cells, located in the ventral hypodermis, that are called P5.p, P6.p, and P7.p (Sulston and Horvitz, 1977). From 29 hr after hatching until 34 hr after hatching, these three cells begin to undergo three rounds of division. In comparison, the three ventral cells that flank P5.p and P7.p, namely P3.p, P4.p, and P8.p, divide only once and fuse with hyp7, a large multinucleated hypodermal cell that extends most of the length of the body. The cells P3.p, P4.p, and P8.p therefore normally do not contribute to the hermaphrodite vulva, although they have the potential to do so (see subsequent discussion).

The term lineage, in the case of the vulva, is used to describe the set of divisions that the cells P3.p–P8.p and their progeny undergo. Thus, the cell

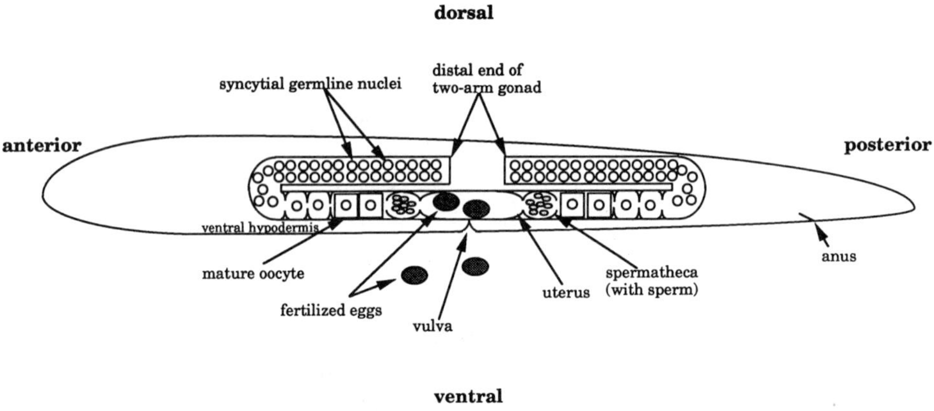

Figure 16.1 Reproduction in an adult *C. elegans* hermaphrodite. Syncytial germline nuclei are located at the distal end of each gonad arm. As they move proximally toward the vulva, they mature into oocytes, which are fertilized by sperm as they pass through the spermatheca. The hermaphrodite stores its own sperm in the spermatheca, but also can store sperm from a male. The fertilized eggs end up in the uterus where, after several rounds of division, they are ejected through the vulva. The adult hermaphrodite length is about 1 mm.

P6.p executes the following lineage (Fig. 16.2B): it divides along the longitudinal (anterior–posterior) axis of the animal to produce two daughters, both of which divide longitudinally to produce four granddaughters, all of which then divide along the transverse (left–right) axis to form eight vulval cells. This final round of division from four to eight cells is abbreviated TTTT to indicate that all four cells divided transversely. Because the decision of which lineage to execute is thought to occur primarily in the cells P3.p–P8.p (see subsequent discussion), a lineage often is abbreviated to include only a given cell and the final division in its lineage. Thus, P6.p is said to execute a TTTT lineage, although in reality the granddaughters of P6.p divide TTTT to produce eight vulval nuclei.

B. Vulval Development Requires a Signal from the Gonad

Although the timing and pattern of the cell divisions that produce the vulva are invariant from hermaphrodite to hermaphrodite, Sulston and Horvitz (1977) hypothesized that vulval development requires an external influence. This hypothesis was based on the fact that, although P5.p, P6.p, and P7.p always divide in the fixed pattern just described, the cells that give rise to P5.p, P6.p, and P7.p are determined variably. For example, P5.p is derived from the cell P5 (P5.p is its posterior daughter) and P6.p is derived from P6, but the cells P5 and P6 themselves come from a pool of two cells, one of which randomly becomes P5 and the other P6 (Fig. 16.3). Thus, although P5.p and P6.p give rise to different division patterns (Fig. 16.2), this difference cannot be entirely the result of ancestry, suggesting that cell nonautonomous development and signal transduction are involved. Similarly, P7.p and P8.p are derived randomly from a pool of two cells.

Based on the symmetry of the vulva relative to the somatic gonad, that is, the center of the vulva is located near the center of the gonad, Sulston and Horvitz, hypothesized that the gonad was responsible for signaling the development of the vulva (Sulston and Horvitz, 1977). The gonad is distinct from the ventral hypodermis (Fig. 16.1) and by ancestry is unrelated to the P3.p–P8.p cells. Removal of the two precursor cells that make the somatic gonad using a laser microbeam resulted in the absence of vulva formation (Fig. 16.4B; Sulston and White, 1980). Instead P5.p, P6.p, and P7.p behaved the way P3.p, P4.p, and P8.p normally do, that is, all six cells divided once and fused with the hypodermis.

This observation was extended to show that one cell in the gonad, the anchor cell, which normally is located directly above (dorsal to) P6.p, is both necessary and sufficient for P5.p, P6.p, and P7.p progeny to form vulva (Fig. 16.4C,D; Kimble, 1981). In addition, the stimulation of these three cells by the anchor cell to form vulva occurs before the first round of division, since

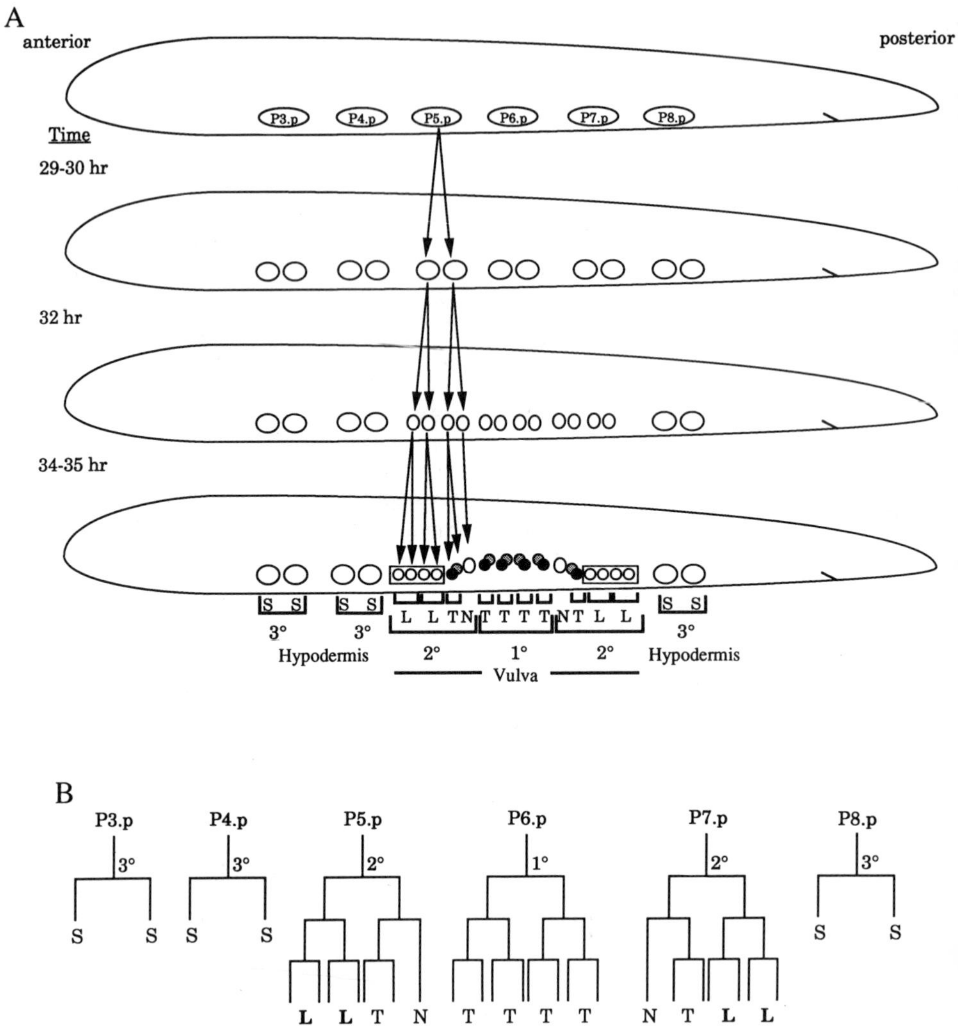

Figure 16.2 Cell divisions responsible for producing the vulva. (A) A *C. elegans* hermaphrodite is shown schematically in the same orientation as in Fig. 16.1. The divisions of P5.p, P6.p, and P7.p to generate the 22 vulval nuclei are shown. The divisions of P5.p and its progeny are connected by arrows to show the relationship between the cells of each generation. Numbers to the left show the approximate time in hours from hatching at which a given round of division occurs. The first two rounds of division for P5.p, P6.p, and P7.p and their progeny are similar. They and their daughters divide longitudinally along the ventral hypodermis of the hermaphrodite. The division patterns diverge at the third and final round. All four granddaughters of the P6.p cell divide transversely (T)—perpendicular to the long axis of the nematode along the left–right axis. (Shading indicates that cells are not located in the plane of the page.) We refer to this pattern of division as TTTT. The four granddaughters of P7.p execute the mirror image division patterns of

ablation of the anchor cell during or after the first round of division leads to generally normal production of vulval cells (Fig. 16.4E; Kimble and White, 1981; Sternberg and Horvitz, 1986). Based on these results and on other observations described subsequently, the primary determination of vulva versus hypodermis is thought to occur mostly in the P5.p, P6.p, and P7.p cells themselves, although their great-granddaughters eventually form the vulva. The action of the anchor cell on the ventral hypodermal cells P5.p–P7.p to produce vulva is called induction since the development of the vulva is dependent on a cell that itself does not contribute to that structure (Sulston and White, 1980)

C. Regulation of Vulval versus Hypodermal Fate among the Vulval Precursor Cells

Although these experiments show that the anchor cell signals the formation of the vulva, they do not indicate how P6.p is instructed to take on a different fate than P5.p and how P7.p is instructed to take on a different fate than P8.p (recall that the selection of P5 or P6 is random, as is P7 or P8). This question was addressed by ablation of individual cells in the ventral hypodermis (Fig. 16.5; Sulston and White, 1980; Sternberg and Horvitz, 1986). When P6.p was ablated in an early larval stage, either P5.p or P7.p could move into the position normally occupied by P6.p and behave the way P6.p normally would. Similarly, when P5.p was ablated or moved to replace P6.p, P4.p was able to replace P5.p and generate seven vulval nuclei instead of the normal two hypodermal nuclei. Also, P8.p could replace P7.p. In the most extreme case, P5.p, P6.p, and P7.p were ablated and P3.p, P4.p, and P8.p, which normally produce hypodermis, produced vulva. However, the other P cell daughters, P1.p, P2.p, and P9.p–P11.p, were never found to take part in the formation of vulva.

The cells P3.p, P4.p, P5.p, P6.p, P7.p, and P8.p therefore are thought to form an equivalence group in which all are equally competent to receive the anchor cell signal and differentiate into vulva. These six cells are referred to collectively as the vulval precursor cells (VPCs). The presence of P6.p normally

---

the four P5.p granddaughters, which include longitudinal (L), transverse (T), and no (N) divisions. From anterior to posterior, the pattern of division for P5.p is LLTN and for P7.p NTLL. The LL portions of P5.p and P7.p are boxed; the other vulval nuclei are shown pulled away from the ventral cuticle to emphasize that the vulval nuclei resulting from the LL divisions adhere to the cuticle. The final round of division is shown schematically; although the orientation of the divisions (L,T,N) is correct, the final relative placement of the cells is not exact. The surrounding cells—P3.p, P4.p, and P8,p—divide once at about the same time and in the same orientation as P5.p, P6.p, and P7.p cells but their daughters do not divide. These daughters do not contribute to the vulva but fuse with the hypodermal syncytium and become hypodermis (lineage abbreviated S S for syncytial). The designations 1°,2°, and 3° are discussed in the text. (B) This schematic is often simplified as a lineage tree. The L divisions of P5.p and P7.p are boldfaced to emphasize adherence to the cuticle. Data from Sulston and Horvitz (1977).

prevents the progeny of P5.p or P7.p from exhibiting the TTTT pattern, whereas the presence of P5.p and P7.p, respectively, prevents the progeny of P4.p and P8.p from forming vulva and exhibiting the LLTN or NTLL pattern (see Fig. 16.2 for a description of division patterns). At least two cells must be removed for P3.p to form vulva.

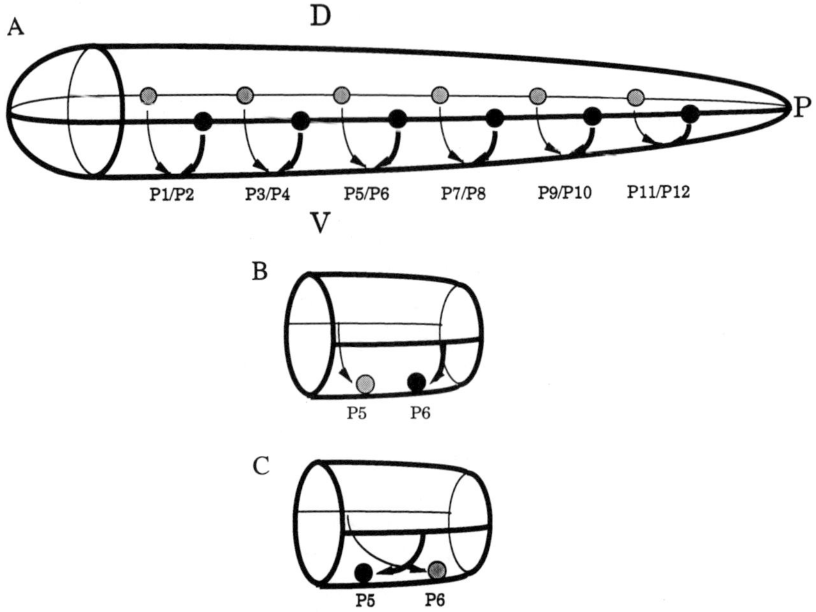

Figure 16.3 Random intercalation of the cells that produce the vulva. A, anterior; P, posterior; D, dorsal; V, ventral. Cells on the right side of the nematode are shaded lightly; cells on the left side are shaded darkly. (A) At hatching, 12 cells are located in 6 left–right pairs along the length of a hermaphrodite. Soon after hatching, these cells migrate circumferentially down into the ventral cord. (B) As shown for the third left–right pair, half the time the cell located on the right side will intercalate anterior to the cell on the left. (C) The other half of the time, the cell located on the right will intercalate posterior to the cell on the left. Because the cell that ends up anterior becomes P5, the cell that becomes P5 is not fixed but can be either the left or the right cell of the third left–right pair; the other cell becomes P6. P5 will divide soon after migration and give rise to P5.p (the posterior daughter of P5); P6 will give rise to P6.p. The granddaughters of P5.p always divide in the LLTN pattern, whereas the granddaughters of P6.p divide in the TTTT pattern (Fig. 16.2). Thus, the determination of which progeny will execute the TTTT pattern (P6.p granddaughters) and which progeny will execute the LLTN pattern (P5.p granddaughters) is not fixed entirely by ancestry (i.e., cell-autonomous development) but must be determined interactively (i.e., cell-nonautonomous development). This determination must occur after P5 and P6 migrate but before the granddaughters of P5.p and P6.p divide. Because P7 and P8 also form a variable left–right pair, the decision to produce vulval progeny (P7.p) rather than hypodermal progeny (P8.p) also must be cell nonautonomous. Data from Sulston and Horvitz (1977).

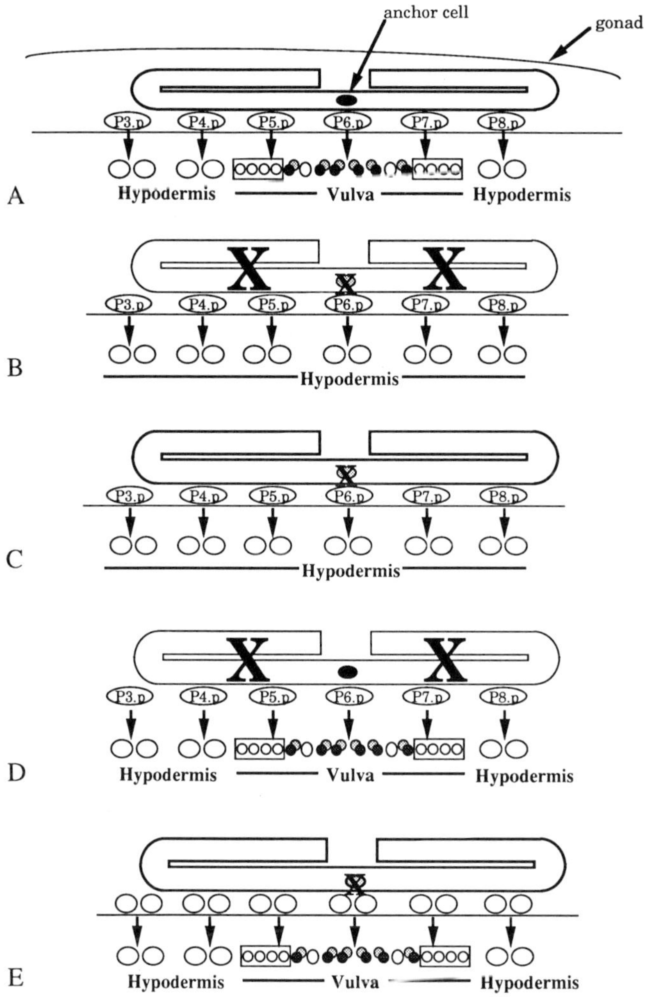

Figure 16.4 The anchor cell of the somatic gonad is necessary and sufficient to induce a vulva. (A) Abbreviated vulval development is shown from the cells P3.p–P8.p to the final set of cells derived from these cells. The gonad and the anchor cell (which is in the gonad) are shown also. The orientation is the same as in Fig. 16.1. (B) If the entire gonad is eliminated before the cells P3.p–P8.p divide (designated by X), P3.p–P8.p all divide once and fuse with the hypodermis and no vulva is formed. (C) The elimination of one of the somatic gonad cells, the anchor cell, produces the same result, indicating that the anchor cell is necessary for differentiation of the vulva. (D) Elimination of all the gonad cells except the anchor cell results in normal production of the vulva, indicating that the anchor cell is sufficient for vulval differentiation. (E) However, if the anchor cell is ablated after cells P3.p–P8.p have divided, subsequent rounds of division occur normally, indicating that the anchor cell plays its role before the first round of division. Data from Sulston and White (1980); Kimble (1981); Sternberg and Horvitz (1986).

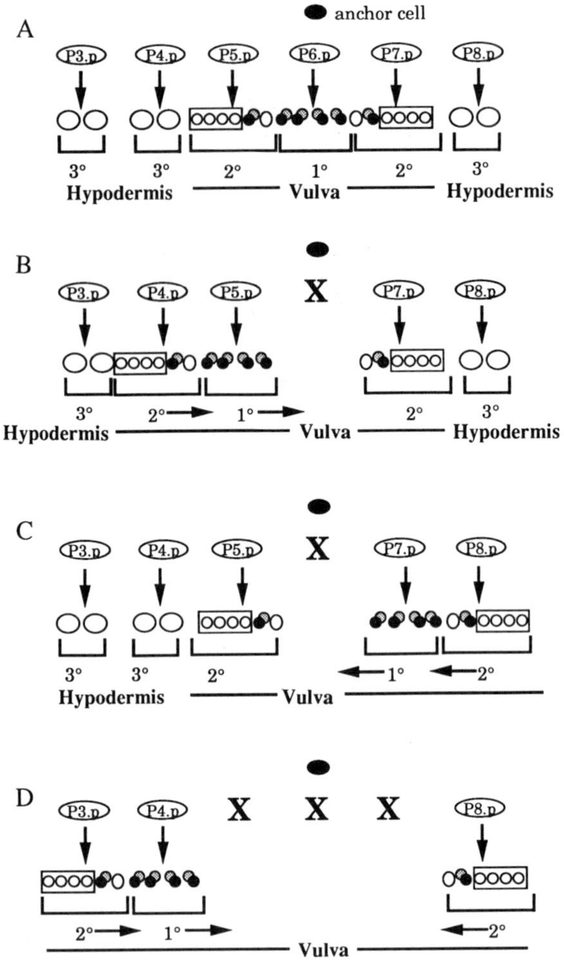

Figure 16.5 Regulation among the cells P3.p–P8.p. (A) Wild-type induction. Cells P5.p–P7.p are induced to form vulva. (B) If P6.p is ablated, either genetically or by laser, then P5.p can replace P6.p and produce a TTTT lineage instead of the usual LLTN. P4.p can fill the space vacated by P5.p and generate the vulval LLTN lineage instead of the usual hypodermal S S lineage. (C) Sometimes P7.p rather than P5.p will replace P6.p. In this case, P8.p replaces P7.p. (D) If all three cells that normally give rise to vulva, that is, P5.p–P7.p, are ablated, then the three cells that normally produce hypodermis—P3.p, P4.p, and P8.p—can replace them. The other P*n*.p cells—P1.p, P2.p, P9.p, P10.p, and P11.p—were never found to take part in vulval replacement; although these five cells produce hypodermis as do P3.p, P4.p, and P8.p, they differ from these latter cells because they do not divide before fusing (lineage S). The S S hypodermal lineage is, therefore, thought to be indicative of a cell in the vulva equivalence group, whereas the S lineage indicates a cell that is not. See Fig. 16.2 for a description of lineage designations. Data from Sulston and White (1980); Sternberg and Horvitz (1986).

The lineage adopted by P6.p is referred to as 1° because it is the "highest" in a hierarchy of lineages, that is, a VPC that normally executes a 1° lineage can be replaced by a VPC that normally executes an LLTN or an S S (hypodermal) lineage. Likewise, the LLTN or NTLL lineage is designated 2° since it can be replaced by a VPC that normally executes an S S lineage but not by a cell that executes a 1° lineage. Finally, the hypodermal S S lineage is designated as 3°. Ablation and other experiments have shown that the VPCs are completely equipotential, that is, P3.p, P4.p, P5.p, P6.p, P7.p, and P8.p each can execute 1°, 2°, and 3° lineages (Sulston and Horvitz, 1977; Sulston and White, 1980; Sternberg and Horvitz, 1986; Thomas *et al.*, 1990).

The experiments described suggest that the location of a VPC is important in determining which lineage it executes. Location of a VPC could be an important factor because of its: (1) distance to the inductive anchor-cell signal or (2) position with respect to other VPCs (i.e., VPC–VPC interactions could be important). To determine whether distance from the anchor cell matters, all VPCs but one were ablated in the presence of the anchor cell (Sternberg and Horvitz, 1986). On isolation, a VPC often would move toward the anchor cell and end up at a variable distance from it. The closer a VPC was to the signal, the higher the fate it adopted. All three fates—1°, 2°, and 3°—were possible for an isolated VPC (Fig. 16.6A). This result suggests that the anchor-cell signal alone is sufficient to generate the three different cell types and that VPC–VPC interactions are not absolutely required, although such interactions cannot be ruled out by these experiments (in fact, VPC–VPC interactions probably do occur; see Section IV). In the simplest interpretation, the anchor cell emits a graded signal that is received and transduced by the VPCs–the closer a VPC is to that signal, the greater the amount of signal it receives, and the "higher" the fate that VPC executes (Fig. 16.6B). Alternatively, the anchor cell may emit two signals, one that acts over a very short range to specify 1° fate and another that acts over a somewhat larger range to specify 2° fate (Fig. 16.6C).

Other experiments also suggest that the anchor-cell signal can act at a distance (Thomas *et al.*, 1990). Hermaphrodites in which the anchor cell was displaced dorsally from the VPCs because of the *dig-1* mutation were examined. Induction of the VPCs still occurred, albeit often abnormally, although the anchor cell was presumably no longer touching the VPCs. This result suggests that the action of the anchor-cell signal does not require contact between the anchor cell and the VPCs. Hence, the signal is probably diffusible.

D. Determination of Vulval Precursor Cell Fate Seems to Occur in the Vulval Precursor Cells

The inductive action of the anchor cell most likely acts on the VPCs and not on their progeny since ablation of the anchor cell just before or during VPC division does not affect the lineages (Kimble and White, 1981; Sternberg and

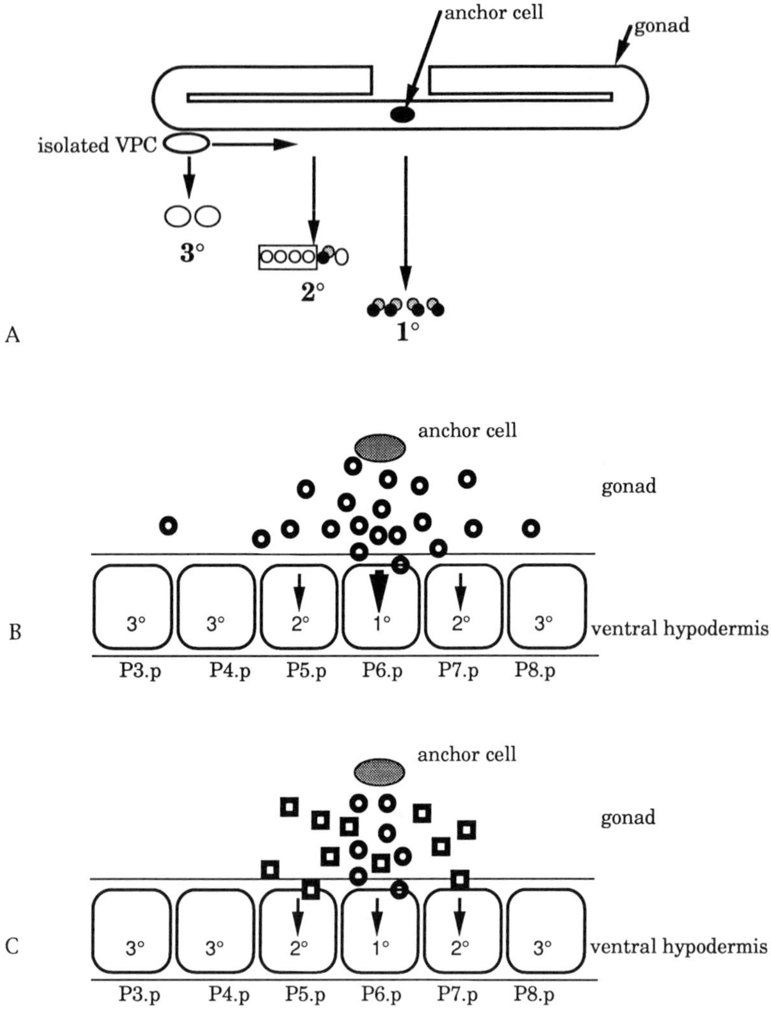

Figure 16.6 Fates of an isolated VPC suggest that the anchor cell signal is graded. Five of the six VPC cells were ablated. (A) The remaining cell sometimes migrates toward the anchor cell; the lineage executed by that cell correlates with its final distance from the anchor cell. If its final position is close to the anchor cell, it executes a 1° fate. If it ends up further away, it executes a 2° fate. If its ultimate location is yet further, it executes a 3° fate. (B) One simple model is that the strength of the anchor cell signal is interpreted by each cell, resulting in 1°, 2°, or 3° fate depending on how strongly or at what time the signal transduction apparatus is activated in a VPC. The inductive signal is represented by circles. (C) Alternatively, two signals, one that acts over a very short range to promote 1° fate (○) and another that acts over a somewhat larger range to promote 2° fate (□), may exist. Data from Sternberg and Horvitz (1986); P. W. Sternberg, unpublished results; M. Herman and R. Horvitz, unpublished results.

Horvitz, 1986). When interpreting all ablation experiments, caution should be exercised because the influence of a cell does not necessarily stop at the time of ablation. Debris from the ablated anchor cell may affect the VPCs, or the anchor-cell signal itself could persist if, for example, the signal is stored in the extracellular matrix. Nonetheless, the idea that the anchor cell acts on the VPCs was strengthened by studying vulval induction in *lon-1* and *dig-1* animals (Sternberg and Horvitz, 1986; Thomas *et al.*, 1990). The *lon-1* mutation results in hermaphrodites that are about 50% longer than wild-type; the anchor cell in these hermaphrodites occasionally is positioned between P6.p and P5.p rather than directly over P6.p. If the anchor-cell signal acts on the daughters of the VPCs and not on the VPCs themselves, it should be possible to generate an intact 1° lineage from the posterior daughter of P5.p and the anterior daughter of P6.p (i.e., from P3.p to P8.p, one should see the following division patterns: S S, S LL, TN <u>TT</u>, <u>TT</u> NT, LL S, S S; 1° lineage underlined). In fact, this pattern was never found; either P5.p or P6.p, but never a combination of P5.p and P6.p daughters, gave rise to intact 1° lineages. Similarly, in *dig-1* hermaphrodites the anchor cell sometimes is located initially between two VPCs. As in *lon-1* mutants, the daughters of two different VPCs were never found to combine to generate an intact 1° or 2° lineage.

These data do not exclude the possibility that other signaling occurs after the anchor cell has ceased to play its role. For example, communication among the VPC daughters could be necessary to execute the pattern established in the VPCs by the anchor cell. One way to address this possibility is to ablate VPC daughters and look for perturbations in lineage. For example, when the VPC daughter cell that gives rise to the TN part of a 2° lineage was ablated in several animals, the other daughter still gave rise to an L L lineage (Sternberg and Horvitz, 1986). Based on these negative results, a good working hypothesis is that the entire vulval lineage from P5.p, P6.p, and P7.p to their great-granddaughters is determined by inductive events that occur before P5.p, P6.p, and P7.p divide.

The 1°, 2°, and 3° lineages executed by VPCs seem to be an intrinsic property of the VPCs. The VPCs are preprogrammed to execute a fixed repertoire of lineages called sublineages—either TTTT, LLTN, or S S (Sternberg and Horvitz, 1986)—the selection of which depends on the outcome of signal transduction events that occur in the VPCs. This selection is called vulval determination. The sublineage selected by a VPC often is referred to as its fate. Thus, for example, P6.p can be said to select a 1° fate. Consistent with this notion of sublineage, no physical or genetic alteration has generated more than the three rounds of division associated with 1° and 2° lineages. With the exception of hybrid lineages (see subsequent discussion), genetic and physical perturbations of VPCs result in lineages generally identifiable as 1°, 2°, or 3° (Sternberg and Horvitz, 1986; Sternberg, 1988b). The cell biology data are summarized in Fig. 16.7.

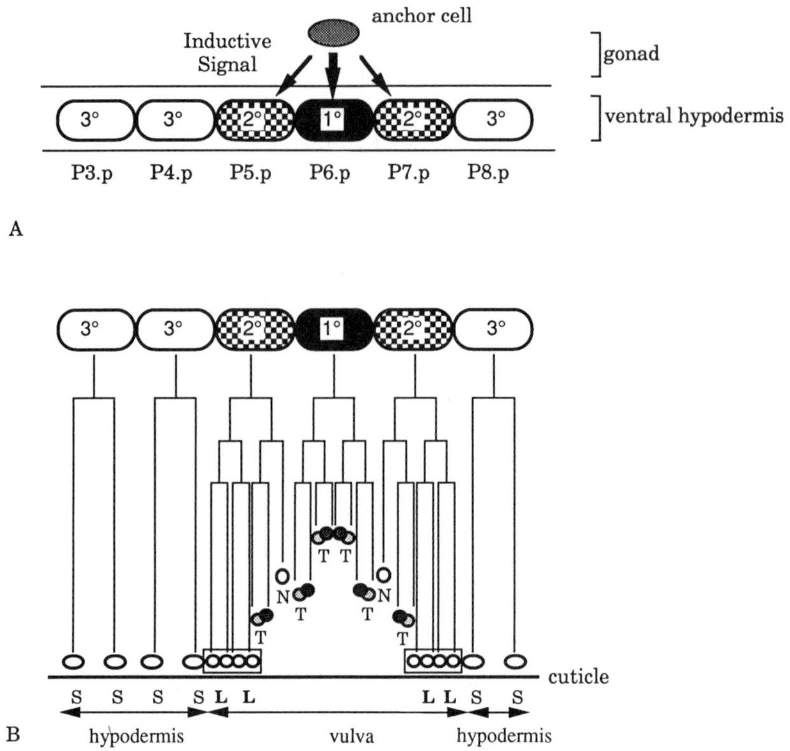

Figure 16.7. Model of vulval development that is suggested by cell biological data, as described in Section II. (A) The anchor cell in the gonad releases a diffusible graded signal that is transduced by the six vulval precursor cells (VPCs). These cells all have the potential to select any one of three possible fates; the anchor cell signal plays an important role in determining which fate each VPC selects. The cell closest to the anchor cell, P6.p, receives the most signal and selects 1° vulval fate; the next two cells, P5.p and P7.p, receive less signal and select 2° vulval fate; the outer three cells select 3° nonvulval fate. In the absence of the anchor cell, all six cells select 3° fate. (B) Once the VPCs have selected their fate, no further inputs appear to be required to execute vulval fate; development proceeds cell autonomously. The outer three cells divide once and fuse with the hypodermis. P5.p and P7.p each give rise, through three rounds of division, to seven vulval nuclei, four of which adhere to the cuticle (2° lineage). P6.p gives rise to eight vulval nuclei, none of which adhere to the cuticle but the central four of which adhere to the anchor cell (1° lineage). The 22 nuclei from P5.p, P6.p, and P7.p make up the full complement of cells required to make a wild-type vulva. The relative geometry of the 22 nuclei shown in this figure approximates the *in vivo* geometry immediately after the nuclei all have formed. See Fig. 16.2 for orientation and definition of T, N, L, and S.

E. Cell Types and Morphogenesis

Vulval development is more than the signal transduction events associated with induction of the VPCs. The processes that occur subsequent to vulval determination offer the opportunity to study processes downstream of signal transduction.

What are the characteristics of the differentiated vulval cells? Wild-type axes of division establish three vulval cell types—T, N, and L—that can be classified further. For example, of the 12 nuclei derived from T divisions (i.e., LLTN T-TT-T NTLL), only those derived from the central two (underlined) are capable of adherence to the anchor cell (Sternberg and Horvitz, 1989). [The components of the vulva are referred to as nuclei and not cells because some of the nuclei are syncytial (J. White, personal communication).] An important characteristic of 2° lineages is the adherence to the cuticle of the progeny of the L divisions. In some mutant backgrounds, L divisions are seen whose progeny do not adhere; these cases often are interpreted as part of abnormal 1° lineages (Sternberg, 1988b). Another issue is what generates the mirror image symmetry in the vulva, that is, the polarity of the 2° lineages adopted by P5.p (LLTN) and P7.p (NTLL), for example.

Following the final set of divisions, the vulva undergoes a process of morphogenesis in which the 22 nuclei move in a coordinated and defined fashion (Sulston and Horvitz, 1977). This movement results in the proper positioning of the vulval cells and attachment to the uterus to form a functional vulva. Ablation of the anchor cell just prior to VPC division results in abnormal morphogenesis, although the proper complement of cells is generated (Kimble, 1981). Thus, the anchor cell is needed for both induction of the VPCs and proper movement of their great-granddaughters. Ablation of the anchor cell at or soon after the first division of the VPCs results in normal lineages and morphogenesis, but abnormal attachment of the vulva to the uterus, suggesting that the anchor cell is needed for this process also. Finally, during the final stages of morphogenesis the anchor cell disappears, thereby allowing a passage between the uterus and the vulva.

An oddity of vulval cell biology involves P3.p. P3.p executes a 3° vulval lineage like P4.p and P8.p (lineage S S) 50% of the time. However, the other 50% of the time, P3.p behaves like P1.p, P2.p, P9.p, P10.p, and P11.p, which do not divide but fuse directly with the hypodermis (lineage S; Sulston and White, 1980; Sternberg and Horvitz, 1986). In the latter case, P3.p is thought to be incapable of being a part of the vulval equivalence group (Sulston and White, 1980). This decision may be pertinent to understanding the generation of the entire vulval equivalence group (see discussion of generation mutants Section III,E). For the remainder of this chapter, we will ignore the nonvulval fate of P3.p.

III. Identification of Mutations that Affect *Caenorhabditis elegans* Vulval Development

A. Overview of Genetics as a Tool to Study Vulval Development

Several characteristics make genetics a powerful approach to study *C. elegans* vulval determination. All the genes presently known to participate in *C. elegans* vulval determination were identified first via genetic techniques. First, the vulva is not essential for propagation of *C. elegans*, thus making it possible to select for mutations that perturb vulval development. Hermaphrodites that lack a vulva, for example, via anchor-cell ablation, cannot lay eggs but still can produce self-progeny (Horvitz and Sulston, 1980). Second, mutations that eliminate or alter the vulva are easily detectable under a dissecting microscope, making large-scale morphological screens for mutants in the inductive process feasible. (For reference, a dissecting microscope generally is used at a magnification of 6×–50× and allows for visualization of up to hundreds of nematodes at a time. To visualize VPC fate (see Section II), however, a dissecting microscope is inadequate and Nomarski optics at 1000–1250× magnification are necessary. At such magnification, only one hermaphrodite at a time can be studied.) Third, *C. elegans* itself is a genetically tractable organism. The generation time is fast (an embryo can develop into a sexually mature hermaphrodite in about 3.5 days). A good genetic map exists, as do genetic tools such as balancers, rearrangements, deficiencies, and duplications. The hermaphroditic nature of *C. elegans* facilitates screens for random mutations that affect the vulva and screens for suppressors of known mutations. The availability of fertile males facilitates the ability to isolate more alleles of a given mutation (see subsequent discussion). Fourth, the effects of mutations can be interpreted at the cellular level, and ablations can be performed to gain greater insights into the signal transduction process.

B. Isolation of Vulvaless and Multivulva Mutations and Associated Cellular Defects

Random mutagenesis screens have been fruitful in identifying important components of vulval determination (Horvitz and Sulston, 1980; Sulston and Horvitz, 1981; Ferguson and Horvitz, 1985). Without prior knowledge of what to expect, two classes of easily detectable abnormalities associated with the production of mutant vulvas were identified (Horvitz and Sulston, 1980; Sulston and Horvitz, 1981): the egg-laying defective (Egl) phenotype, detectable by bloated worms or the "bag-of-worms" appearance, and the multivulva (Muv) phenotype, detectable by up to five protrusions on the ventral surface (Fig. 16.8). In a random screen, wild-type hermaphrodites are mutagenized, generally with ethyl methane sulfonate (EMS; Brenner, 1974), and allowed to

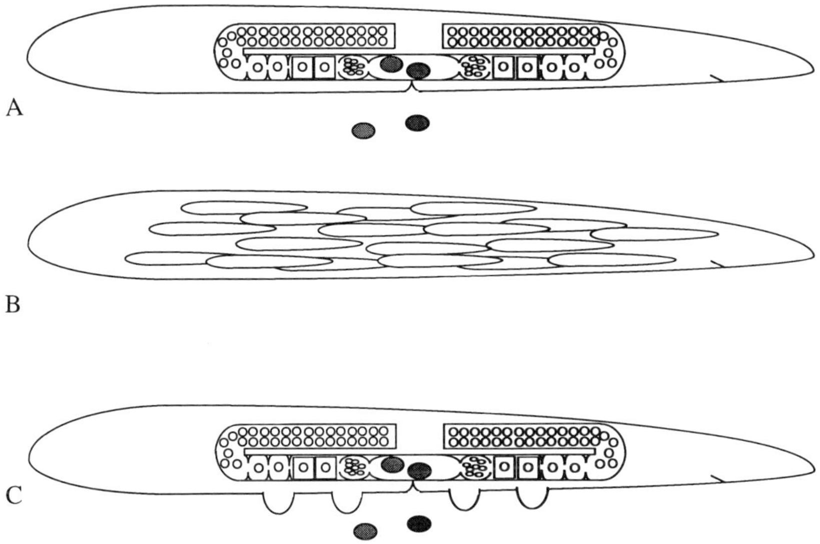

Figure 16.8 Two broad phenotypes of vulval mutants as seen with a dissecting microscope (see Fig. 16.1 for orientation and structures). (A) Wild-type hermaphrodites generally have zero or one small protrusion at the vulva and are able to lay eggs. (B) Mutations that prevent vulval formation have an egg-laying defective or Egl phenotype. In the absence of a vulval opening, oocytes are fertilized normally but subsequently cannot be ejected. Fertilized eggs hatch internally, eventually resulting in a "bag-of-worms" appearance. The hatched larvae feed off the mother and eventually eat their way out, killing the mother. (C) Mutations that increase the amount of vulval tissue show a Muv phenotype. The extra vulval tissue generated results in ventral protrusions that are easily discernible under a dissecting microscope.

produce self-progeny (Fig. 16.9). Both the first generation progeny (F_1) and the second generation progeny (F_2) are screened for mutant phenotypes, allowing the isolation of dominant and recessive Egl and Muv mutations.

The phenotype associated with many Egl mutants is caused by a lack of or reduction in vulval induction, that is, less than three and often no VPCs adopt 1° or 2° vulval fates (Fig. 16.10C,D). In the most severe case, all VPCs execute a 3° hypodermal fate. The latter class of Egl mutants is called specifically the vulvaless (or Vul) mutants, so named because no vulval tissue forms (Horvitz and Sulston, 1980). (Note that although all Vul hermaphrodites are Egl, not all Egl hermaphrodites are Vul.) Mutations in any of the following seven genes can result in a Vul phenotype: *lin-2, lin-3, lin-7, lin-10, lin-45, let-23,* and *let-60* (Horvitz and Sulston, 1980; Sulston and Horvitz, 1981; Ferguson and Horvitz, 1985; Ferguson *et al.*, 1987; Sternberg and Horvitz, 1989; Beitel *et al.*, 1990; Han *et al.*, 1990; Aroian and Sternberg, 1991). The genes *lin-12* and *lin-25*, which also can mutate to an Egl phenotype, are discussed later in the chapter.

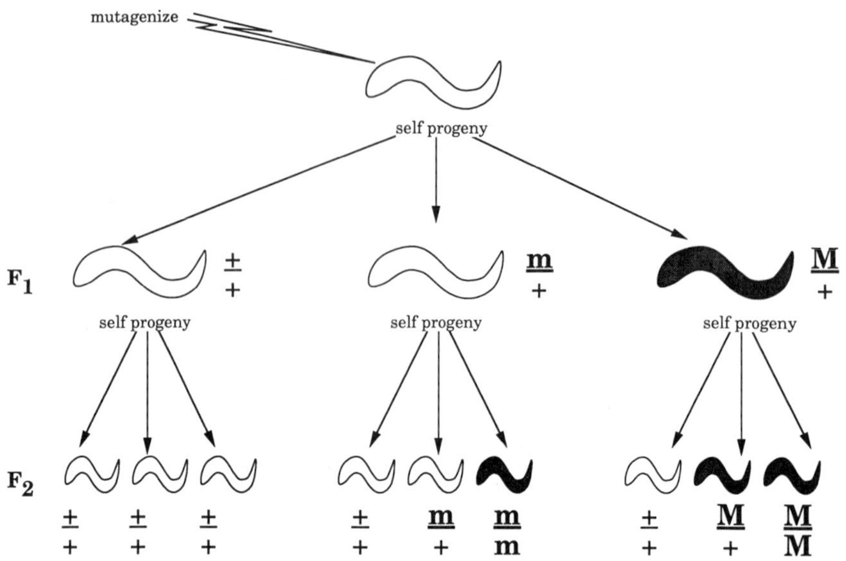

Figure 16.9 Random screen for vulval mutants. Fourth stage wild-type (+/+) larval hermaphrodites are incubated with EMS to mutagenize their germline. Random screens for vulval mutations are technically easy and can be completed in about 1 week—hermaphrodites are mutagenized, allowed to produce self-progeny for two generations, and screened for Egl or Muv worms. First generation (F_1) self-progeny are often wild-type (+/+) like their mother, but occasionally a mutation is introduced in a gene involved in vulval development. If this mutation is dominant (Mutant or M), Egl or Muv worms may develop in the F_1 generation. If the mutations is recessive (mutant or m), Egl or Muv worms may be detected in the self-progeny of the F_1, that is, the F_2 generation.

Conversely, VPCs in Muv mutants lack the 3° nonvulval fate (Fig. 16.10E). Mutations in the *lin-1, lin-13, lin-15, let-60 ras* (formerly called *lin-34*), or synthetic multivulva genes (see subsequent text) can result in a Muv phenotype (Horvitz and Sulston, 1980; Sulston and Horvitz, 1981; Ferguson and Horvitz, 1985; Beitel *et al.*, 1990; Han *et al.*, 1990). The most severe Muv phenotype results in all six VPCs executing only vulval fates, 1° and 2° (Sulston and Horvitz, 1981; Ferguson *et al.*, 1987; Sternberg, 1988b; Sternberg and Horvitz, 1989). The extra vulval tissue generated in a Muv hermaphrodite is responsible for the prominent ventral protrusions, also called pseudovulvae, discernible under the dissecting microscope (Fig. 16.8; Sulston and Horvitz, 1981). The number of protrusions, from one to five, depends on the severity of the Muv phenotype and the exact vulval lineages executed.

Two types of screens in addition to the random screens have been used to isolate vulval mutations: F_1 noncomplementation and suppressor isolation (Ferguson and Horvitz, 1985; Han *et al.*, 1990; Kim and Horvitz, 1990; Beitel *et al.*, 1990; Aroian and Sternberg, 1991). An F_1 noncomplementation screen is a powerful way to generate more mutations in a particular gene (Fig. 16.11). For example, this scheme was used to generate 14 new *let-23 RTK* alleles (receptor tyrosine kinase; Aroian and Sternberg, 1991). In a suppressor screen, mutant hermaphrodites that are either Muv or Vul are mutagenized, and the F_1 or F_2 generation is screened for suppression of that phenotype (Fig. 16.12). For instance, dominant negative *let-60 ras* Vul mutations were isolated as suppressors of the *lin-15* Muv gene (Beitel *et al.*, 1990; Han *et al.*, 1990). The hermaphroditic nature of *C. elegans* makes such a screen readily feasible. The screen can be set up so suppressors are easy to identify in the original mutant background. Such a screen is restrictive because it only selects for mutations that can suppress the original mutant. Further, if the original mutant results in a complete loss of gene function, the screen selects for genes that act only downstream of the original mutation.

C. Methods for Cloning *Caenorhabditis elegans* Genes

The availability of strains with active transposition, the presence of an extensive physical map, and the ability to perform germ-line transformation experiments all combine to allow the facile cloning of *C. elegans* genes. To date, the molecular identities of the genes *lin-10, lin-11, lin-12, let-23 RTK*, and *let-60 ras* have been described (Greenwald, 1985; Yochem *et al.*, 1988; Aroian *et al.*, 1990; Freyd *et al.*, 1990; Han and Sternberg, 1990; Kim and Horvitz, 1990). Cloning strategies have made use of transposon tagging (*lin-10, lin-11*, and *lin-12*) and physical mapping or germ-line transformation (*let-23 RTK, let-60 ras*).

Although transposons are found inserted in the genomes of all *C. elegans* strains, germ-line transposition only occurs in some of them. For example, germ-line transposition of the Tc1 family of transposable elements does not occur in the standard laboratory *C. elegans* strain, called variety Bristol or N2, but does occur in the *C. elegans* strain TR679 (Collins *et al.*, 1987). Therefore, a Tc1 transposon in the strain TR679 can insert into a gene of interest, causing a heritable mutant phenotype. This transposition event can be identified by performing either an F_2 random screen or an F_1 noncomplementation screen similar to those performed with EMS as a mutagen. Once a transposon mutation has been isolated in a gene of interest, genetic mapping and Southern analysis probing with cloned transposons are combined to identify the single transposition event responsible for the mutant phenotype. Inducing the transposon to excise by reverting the mutant phenotype is also useful to identify the mutagenic transposon. Once the responsible transposon has been identified, flanking DNA can be isolated, which presumably contains or is tightly linked to the gene of interest. One limitation of this technique is that not all genes are good targets for transposon insertion.

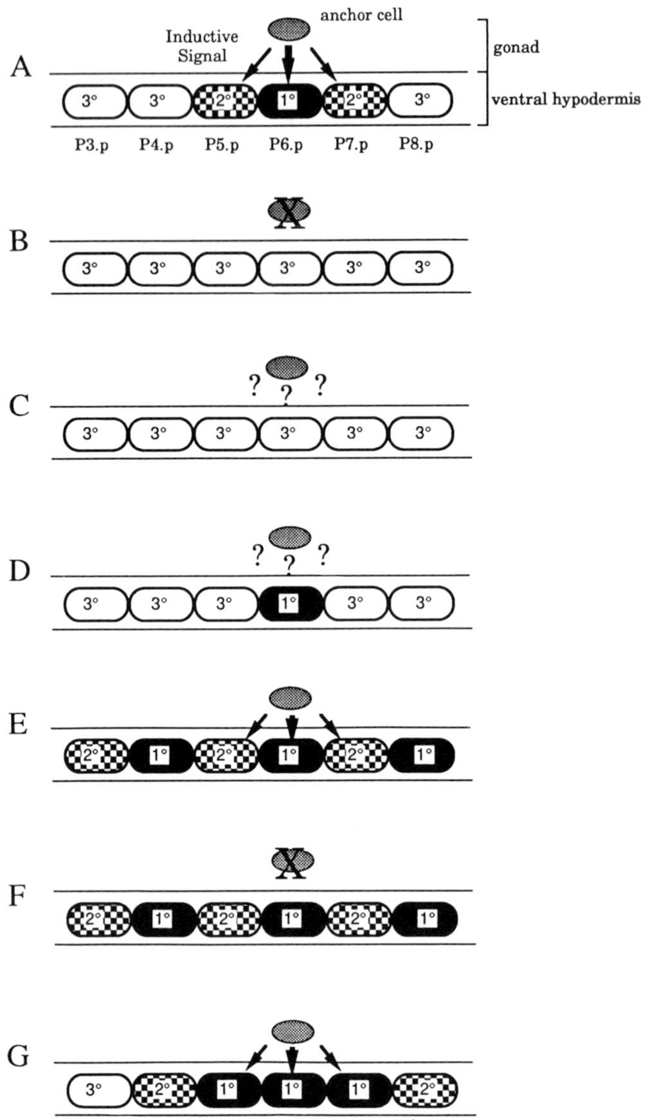

Fig 16.10 The effects of *vulvaless* and *multivulva* mutations on vulval determination. For reference, (A) wild-type induction and (B) induction in a hermaphrodite lacking an anchor cell are shown. (C) The anchor cell minus phenotype is mimicked in a fully penetrant *vulvaless* (Vul) mutant; all VPCs divide and fuse with the hypodermis (3° fate), although the anchor cell is intact. No vulval tissue forms. The anchor cell signal may or may not be normal in a given Vul mutant (?) since the Vul phenotype mimics the anchor cell ablation. (D) Although the fully penetrant phenotype occurs often with Vul mutations, most Vul alleles occasionally show partial induction. This phenotype may reflect the fact that the mutations are not nulls or that the genes involved can be

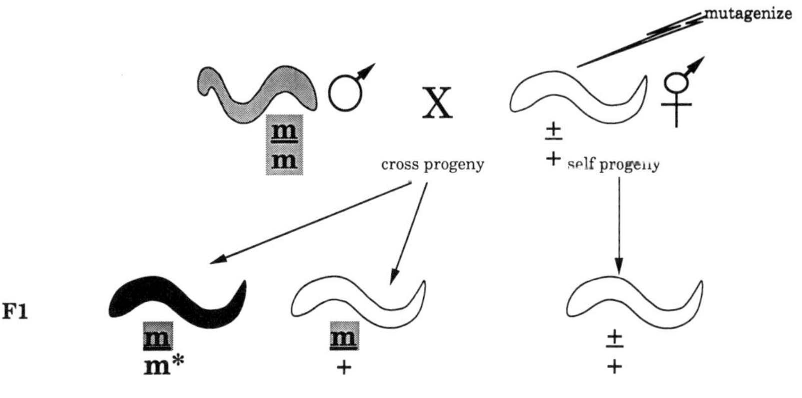

Figure 16.11 F_1 noncomplementation mutagenesis screen. The germline of a wild-type hermaphrodite (+/+) is mutagenized and mated with homozygous mutant males (m/m). (The male in the figure is shaded lightly, as are the mutant m copies contributed by the male.) Self-progeny from the hermaphrodite will be phenotypically wild type, unless a new dominant mutation has been introduced (which is sorted out later). Most cross-progeny from the mating between the male and the hermaphrodite will be m/+. Because m is recessive, these F_1 progeny will show a wild-type phenotype. Occasionally, however, a new mutation in the same gene (m*) will be generated in an oocyte of the original mutagenized hermaphrodite. When this gamete is fertilized by the mutant male, the resulting F_1 progeny (m/m*) will show a mutant vulval phenotype (dark shading) because the new allele fails to complement the allele introduced by the mutant male. The screen also can be performed in reverse, so wild-type males are mutagenized and the original mutant chromosome m is introduced via hermaphrodites. A recessive unrelated marker often is included with either screen to allow cross-progeny to be distinguished from self-progeny.

An alternative approach utilizes correlation of genetic and physical maps, followed by germ-line rescue experiments. DNA spanning most of the *C. elegans* genome has been collected into large stretches of overlapping cosmid and yeast artificial chromosome (YAC) clones called contigs (Coulson *et al.*, 1986,1988). This physical DNA map has been correlated with the genetic map in many areas of the genome. Most genes can be mapped readily relative to

---

bypassed to some extent (see text). (E) Conversely, in a *multivulva* (Muv) hermaphrodite, no 3° fates are seen; all VPCs execute 1° or 2° vulval fate. (F) This induction of vulval fate in Muv mutants occurs even if the anchor cell is ablated, suggesting that the Muv mutations do not affect the anchor cell signal. Weaker Muv mutations, such as *lin-13* or weak alleles of *lin-15*, result in an intermediate phenotype in which more than three but fewer than six of the VPCs execute vulval fates (Ferguson *et al.*, 1987; P. W. Sternberg, unpublished observations). (G) The hyperinduced (Hin) phenotype is discussed in the text but is shown here for comparison. Like the Muv phenotype, extra 1° and 2° fates are seen. Unlike the Muv phenotype, the VPCs in a Hin mutant still behave as though they are responding to the anchor cell signal: (1) a gradation of VPC fate occurs around the anchor cell and (2) VPCs in a Hin hermaphrodite will select 3° fate if the anchor cell is ablated.

genetic or physical (e.g., restriction fragment length polymorphisms, RFLP) markers and, if the physical map is well characterized in that area, narrowing down the region of DNA containing the gene of interest is possible. For example, in the case of *let-23 RTK*, mapping of the gene between a RFLP and a deficiency breakpoint narrowed the region containing the gene to about 200 kb (Aroian *et al.*, 1990). The exact location of the gene in this 200-kb region was then determined by germ-line transformation experiments (Fig. 16.13).

D. Molecular Identity of *lin-10, lin-11, lin-12, let-23 RTK,* and *let-60* ras

In this section, the molecular identities of the vulval genes published to date are reviewed. The relevance of these identities to the vulval determination pathway is discussed in later sections.

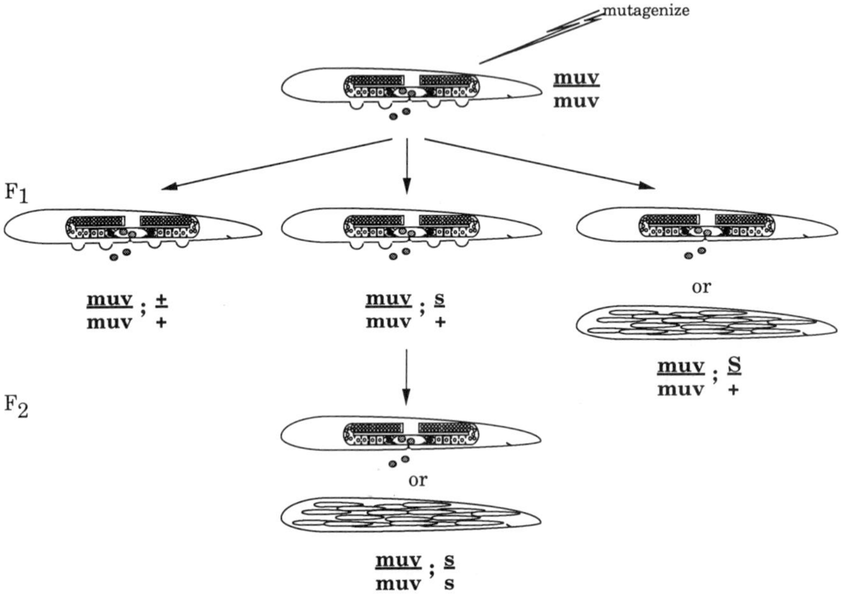

Figure 16.12 Suppressor screening. New mutations in the vulval determination pathway can be isolated by mutagenizing mutant Muv or Vul hermaphrodites and screening for suppression of that phenotype. In the example shown, Muv hermaphrodites with multiple ventral protrusions are mutagenized and allowed to produce self-progeny. A dominant suppressor (S) will result in an F_1 progeny with no ventral protrusions (muv/muv; S/+); the progeny may or may not be Egl. A recessive suppressor (s) will result in some F_2 progeny with no ventral protrusions (muv/muv; s/s). The suppressor gene can be isolated from the Muv mutation and analyzed. In the case of a Vul mutation, suppressors can be identified by the presence of eggs on a plate or by hermaphrodites that no longer form a bag of worms.

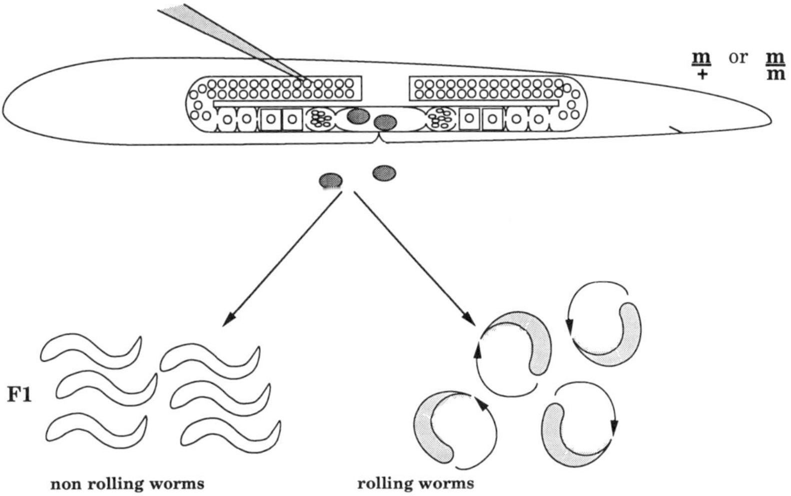

Figure 16.13 Germline transformation in *C. elegans.* The distal gonad syncytium of a hermaphrodite mutant for a *vul* or *muv* gene (either heterozygous m/+ or homozygous m/m) is injected simultaneously with two DNAs: (1) candidate DNA that may encode the wild-type *vul* or *muv* gene and (2) marker DNA that tests for expression of injected DNAs. In this example, the marker DNA is a plasmid that contains the dominant cuticle mutation *rol-6(d)*. The injected DNAs sometimes are expressed in the F_1 progeny of the injected mother. Such progeny will have a twisted cuticle and will gyrate on the plate because of expression of the dominant cuticle mutation. These progeny also are likely to express the candidate *vul/muv* DNA because the *rol-6(d)* and candidate DNAs concatamerize into long linear arrays. Progeny that roll therefore are tested for rescue of the mutant Vul or Muv phenotype. Rescue of the mutant phenotype indicates that the candidate DNA contains a functional wild-type *vul* or *muv* gene. Nonrescue indicates that the candidate DNA does not contain a functional gene. Since the candidate DNA can be cosmid or plasmid, one can test ≥40 kb genomic sequence at a time and quickly narrow down the region containing the *vul* or *muv* gene. Germline transformation is also useful for assessing *in vivo* effects of mutations made *in vitro*. For transformation protocols, see Fire (1986) and Mello *et al.* (1991). For actual instances of germline rescue of vulval mutations, see Kim and Horvitz (1990), Freyd *et al.* (1990), Han and Sternberg (1990), and Aroian *et al.* (1990). For using germline transformation to study mutations made *in vitro*, see Han and Sternberg (1991).

The predicted protein sequence of *lin-10* shows no similarity to known proteins (Kim and Horvitz, 1990). Although *lin-10* mutants show only defects in the vulva, expression data indicate that the gene is transcribed in many other cells and throughout development. For example, *lin-10* mRNA levels are not substantially altered in *lin-26* mutant hermaphrodites that completely lack VPCs (Kim and Horvitz, 1990).

The *lin-11* gene, which is necessary for execution of the NT portion of 2° lineages, encodes a protein with a homeodomain and two tandem copies of a putative metal-binding domain found in two other homeodomain-containing genes (Freyd *et al.*, 1990). The *lin-11* gene is likely to encode a transcription factor.

The *lin-12* gene encodes a putative membrane-spanning protein with striking similarity in overall architecture to the *C. elegans glp-1*, the *Drosophila* Notch, *Xenopus* Xotch, and human TAN-1 proteins (Greenwald, 1985; Yochem *et al.*, 1988; Austin *et al.*, 1989; Yochem and Greenwald, 1989; Coffman *et al.*, 1990; Ellisen *et al.*, 1991). In its extracellular domain, *lin-12* contains 13 EGF-like repeats and three tandem copies of a cysteine-rich motif. In its intracellular domain, *lin-12* contains six copies of a motif identified in the yeast genes *cdc10/swi6*, which may provide interaction surfaces for binding with other proteins (Thompson *et al.*, 1991). All members of this family contain these motifs, although the number of EGF-like repeats varies. The *C. elegans glp-1* protein is involved in cell–cell interactions in the hermaphrodite germ line and anterior pharynx and may be a receptor for close-range interactions (Austin and Kimble, 1987; Priess and Thomson, 1987). The *Drosophila* Notch protein also appears to act as a receptor for close-range signals and is involved in the determination of epidermal or neuronal fates (reviewed in Simpson, 1990).

The *let-23 RTK* protein belongs to the EGF receptor family of tyrosine kinases because it contains (1) a tyrosine kinase domain with highest similarity to human EGF receptor (44% identity), (2) an appropriately positioned putative membrane-spanning domain, (3) two cysteine-rich motifs in the extracellular domain with intercysteine spacing characteristic of the EGF receptor family, and (4) limited similarity in the ligand-binding region with other members of the EGF receptor family (Aroian *et al.*, 1990). EGF receptor and its ligands can elicit diverse cellular responses including proliferation, inhibition of proliferation, and fate determination in mammals and *Drosophila*. (For reviews in mammalian systems, see Ullrich and Schlessinger, 1990; Sporn and Roberts, 1988; for *Drosophila* EGF receptor, see Price *et al.*, 1989; Schejter and Shilo, 1989.) EGF receptor responses are elicited on binding of ligand to the receptor extracellular domain and subsequent activation of tyrosine kinase activity in the receptor cytoplasmic domain.

The *let-60 ras*-encoded protein is highly similar to *ras*-encoded proteins of other organisms (Han and Sternberg, 1990). For example, the predicted *let-60 ras*-encoded protein is 83% identical to human N-*ras*-encoded protein in its first 164 amino acids, a region that contains the guanine nucleotide binding domains, intrinsic GTPase activity, and the so-called effector domain. Like the EGF receptor family, the Ras family of proteins contains important components of signal transduction involved in cell growth and differentiation; these proteins are thought to be downstream targets of growth factor receptors (Weissman and Aaronson, 1983; Mulcahy *et al.*, 1985). The Ras family of proteins contains small (~21 kDa) guanine nucleotide binding proteins that can switch from an inactive GDP-bound form to an active GTP-bound form (reviewed by Bourne *et al.*, 1991). The function of the *let-60 ras*-encoded protein is likely to be similar to that of other Ras proteins (see subsequent text).

E. Vulval Development beyond Vulval Induction

Although the remainder of this chapter primarily details the signal transduction pathway involved in vulval determination (i.e., selection of 3°, 2°, or 1° fate), vulval development involves more than just vulval determination. The Vul and Muv mutants discussed earlier are called determination mutants because they alter the determination of vulval fate. Screens for vulval mutations reveal three other groupings of mutants (Ferguson *et al.*, 1987). Like Vul mutants, generation mutants cannot lay eggs. However, these mutants are Egl not because of defects in vulval determination, but because the cells that make up the vulval equivalence group are never properly generated. For example, in *n300* mutant hermaphrodites, the cells P3.p–P8.p behave the way P1.p, P2.p, and P9.p–P11.p normally do; they do not divide but fuse directly with the hypodermis (Ferguson *et al.*, 1987). These cells do not become vulva precursor cells and therefore are incapable of responding to the anchor-cell signal and executing 1°, 2°, or 3° fates. The expression or execution genes are good candidates for targets of the determination genes because mutations in these genes appear to act downstream of the determination genes. These mutations do not affect which fate the VPCs select, but affect the execution of specific lineages. All execution mutants known to date affect aspects of 2° fate execution (see subsequent discussion). The last category consists of mutations not easily amenable to classification. For example, mutations in *lin-24* and *lin-33* genes can result in the death of some of the P3.p–P8.p cells. These mutations are thought to encode abnormal products that are toxic to these cells (Ferguson and Horvitz, 1985).

These different mutant categories, all of which result in abnormal vulval lineages, underscore the importance of vulval cell biology in categorizing mutants. Further, not all mutations that result in an Egl phenotype or in ventral protrusions overtly affect vulval lineages. For example, mutations that affect the egg-laying musculature or neuronal circuitry also can lead to an Egl phenotype (Waterston *et al.*, 1980; Trent *et al.*, 1983; Desai and Horvitz, 1989). In these mutants, the VPCs and their lineages are normal. The well-characterized vulval cell biology therefore allows any new mutation to be examined at the cellular level to ascertain whether or not it affects vulval lineage and, if so, in what specific way.

Interestingly, the positioning and development of some other components of the egg-laying system noted here, for example, musculature and neuronal circuitry, also depend on signal transduction events that involve both the vulva and the somatic gonad. Vulval cells from the 1° and 2° lineages are important for branching of the ventral type C motor neurons, for positioning of the vulval muscle cells, and for the induction of neuropeptide in certain gonadal cells (Li and Chalfie, 1990). Further, the somatic gonad, which is responsible for vulval induction via the anchor cell, is also important for proper migration of an important component of the egg-laying system—the sex myoblasts (Thomas *et al.*, 1990).

F. Other Effects of the Vul/Mux Genes

Mutations in many of the genes mentioned in this chapter have effects in tissues other than the vulva. The *lin-12* gene affects at least 10 developmental decisions in the hermaphrodite and male, including the production of the anchor cell (Greenwald *et al.*, 1983). Mutations in the Vul genes *lin-3*, *let-23 RTK*, and *let-60 ras* share at least three phenotypes apart from Vul, including larval lethality, sterility, and male tail defects (Ferguson and Horvitz, 1985; Han *et al.*, 1990; Aroian and Sternberg, 1991; H. M. Chamberlin, M. Han, R. J. Hill, and P. W. Sternberg, unpublished observations). Muv mutants can affect the male hypodermis, the hermaphrodite head region, and hermaphrodite fertility (Sulston and Horvitz, 1981; Ferguson and Horvitz, 1985, 1989). Mutations in *lin-11* show defects in the male tail and in movement (Ferguson and Horvitz, 1985). However, some genes such as *lin-2*, *lin-7*, and *lin-10* seem to operate primarily in the development of the vulva (Ferguson and Horvitz, 1985; Kim and Horvitz, 1990). The vulval determination pathway therefore utilizes genes that are generally involved in *C. elegans* development and also specifically tailored for the development of the vulva.

IV. Building a Signal Transduction Pathway for Determination of Vulval or Hypodermal Fate

A. Genes associated with Vul Mutations Normally Stimulate Vulval Fate

Genetic evidence suggests that the vulvaless phenotype associated with mutations in *lin-2, lin-3, lin-7, lin-10, let-23 RTK,* or *let-60 ras* is the result of loss of gene function (Table I). Eliminating the activity of any one of these genes prevents the VPCs from selecting 1° and 2° vulval fate and results in the VPCs adopting 3° nonvulval fate. These genes, therefore, normally stimulate selection of vulval rather than hypodermal fate and could be required for producing the anchor-cell signal, for reception of the signal at the VPCs, or for transduction of that signal in the VPCs.

For *lin-3*, *let-23 RTK*, and *let-60 ras*, the evidence suggests that complete loss of gene function results in a fully penetrant Vul phenotype, that is, all six VPCs always execute 3° fates (Ferguson and Horvitz, 1985; Sternberg and Horvitz, 1989; Han *et al.*, 1990; Aroian and Sternberg, 1991; R. Hill and P. W. Sternberg, unpublished results). The previous statement is actually a simplification. Directly determining the fate of the VPCs in complete loss-of-function

TABLE I
Vulval Phenotypes Resulting from Loss of Function in Various Genes[a]

Gene	Vulval phenotype	Criteria[b]	References
lin-2	Vul-incomplete	1,2	Ferguson and Horvitz (1985); Ferguson *et al.* (1987); Sternberg and Horvitz (1989)
lin-3	Vul-complete	1,3	Ferguson and Horvitz (1985); Ferguson *et al.* (1987); Sternberg and Horvitz (1989); R. Hill and P. W. Sternberg (unpublished observations)
lin-7	Vul-incomplete	1,2,4	Ferguson and Horvitz (1985); Ferguson *et al.* (1987); Sternberg and Horvitz (1989)
lin-10	Vul-incomplete	1,2,4,5,6	Ferguson and Horvitz (1985); Ferguson *et al.* (1987); Sternberg and Horvitz (1989); Kim and Horvitz (1990)
let-23	Vul-complete	1,3,5	Ferguson and Horvitz (1985); Ferguson *et al.* (1987); Sternberg and Horvitz (1989); Aroian and Sternberg (1991)
let-60	Vul-complete	1,6	Han *et al.* (1990); Han and Sternberg (1991)
lin-1	Muv	1,2,4	Ferguson and Horvitz (1985); Ferguson *et al.* (1987)
lin-13	Muv	1,5	Ferguson and Horvitz (1985); Ferguson *et al.* (1987)
lin-15	Muv	1,5	Ferguson and Horvitz (1985); Ferguson *et al.* (1987); L. Huang and P. W. Sternberg (unpublished observations)

[a]See text for discussion of complete/incomplete Vul phenotypes. Note that mutations in some of these genes can affect more than just vulval determination (e.g., *lin-13, let-23 RTK*, and *let-60 ras* can be mutated to larval lethality).

[b](1) Recessive phenotype. Consistent with loss of function; mutations in these genes are recessive (*let-60* also has dominant alleles, see text). (2) Number of alleles. Isolation of many alleles with the same phenotype suggests that such alleles are complete loss of function or null because such alleles are apt to be the most common. (3) Allelic series. In some cases, alleles with more severe phenotypes (e.g., lethality) enhance the penetrance of the vulval phenotype of less severe alleles. Such an effect suggests that the vulval phenotype is due to loss of function. For example, the reduction-of-function *let-23* allele *let-23(sy1)* shows little vulval induction but does not show larval lethality. The *let-23* allele *let-23(sy15)* has a more severe phenotype (i.e., larval lethality) and presumably has less *let-23* activity than *sy1*. The *trans* heterozygote *let-23(sy1)/let-23(sy15)* shows no vulval induction suggesting that, as *let-23* activity is lowered from *let-23(sy1)/let-23(sy1)* to *let-23(sy1)/let-23(sy15)*, vulval induction is eliminated completely and that loss of *let-23* function results in no vulval induction. (4) Amber alleles. Amber alleles often, but not always, decrease or eliminate gene activity because they often result in a truncated protein. (5) Deficiency enhanced phenotype. Alleles in *trans* to a deficiency (i.e., a deletion) that show a more penetrant phenotype suggest a loss of function. In the case of *lin-10*, a deficiency did not enhance the phenotype, suggesting that the phenotype was already the null phenotype. (6) Other. For *lin-10*, alleles generated by transposon insertion are Vul. For *lin-10*, see Kim and Horvitz (1990).

lin-3, *let-23 RTK*, or *let-60 ras* alleles is not possible because the complete loss-of-function phenotype for all three genes is early larval lethality. However, complete loss of function for any one of these genes can be inferred to result in no vulval induction based on strong reduction of function alleles, deficiency arguments, and allelic series arguments (Han *et al.*, 1990; Aroian and Sternberg, 1991; R. Hill and P. W. Sternberg, unpublished results). The lethality associated with complete loss-of-function *let-60 ras* alleles can be rescued maternally using activated *let-60 ras* mutations. Thus, induction in complete loss-of-function *let-60 ras* animals can be measured. Since complete loss of function in these genes reproduces the effect of removing the anchor cell, assuming that these genes act directly in the pathway between the anchor cell and the selection of vulval fate is the simplest conclusion (Fig. 16.14). For *let-23 RTK*, the genetic evidence in addition to its molecular structure suggests that the protein is the receptor for the anchor-cell signal in the VPCs (Aroian *et al.*, 1990). For *let-60 ras*, the genetic and molecular evidence suggests that this protein is a necessary signal transduction component in the VPCs downstream of *let-23 RTK* (subsequent text).

For *lin-2*, *lin-7*, and *lin-10*, complete loss of gene activity does not result in a fully penetrant Vul phenotype (Horvitz and Sulston, 1980; Sulston and Horvitz, 1981; Ferguson and Horvitz, 1985; Sternberg and Horvitz, 1989; Kim and Horvitz, 1990). Even in animals carrying putative complete loss-of-function or null alleles, all VPCs do not always execute 3° hypodermal fates, although the number of VPCs that execute 1° and 2° vulval fates is reduced

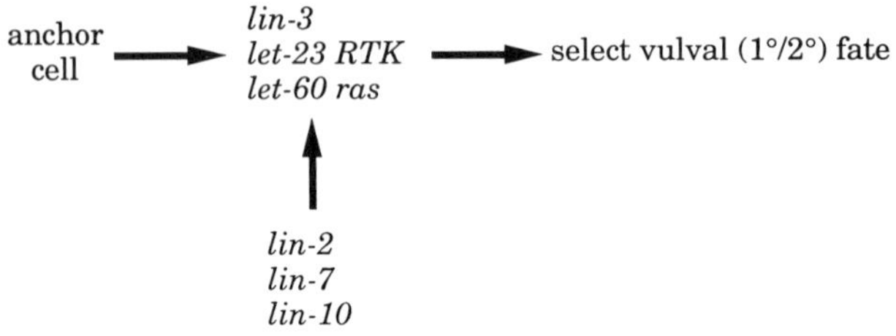

Figure 16.14 Assigning *vul* genes in a simple pathway. The genes *lin-2*, *lin-3*, *lin-7*, *lin-10*, *let-23 RTK*, and *let-60 ras* all are required for stimulation (→) of vulval fate because eliminating any of these genes results in no or little selection of vulval fate. Anchor cell influence on the selection of vulval fate is hypothesized to act through *lin-3*, *let-23 RTK*, and *let-60 ras* because elimination of any one of these genes mimics ablation of the anchor cell. The genes *lin-2*, *lin-7*, and *lin-10*, however, are shown to act in a pathway that is not in line with the anchor cell because elimination of any one of these genes does not fully mimic ablation of the anchor cell. For simplicity, these genes are assumed to act through *lin-3*, *let-23 RTK*, and/or *let-60 ras*, but could act through other genes. An alternative explanation is that these three genes are indeed in line with the anchor cell signal, but a different anchor cell signal is involved or another slightly redundant component acts with these three genes that can compensate weakly for their loss. A more complete model is given later in Figure 16.18.

markedly. These data suggest that these genes do not act directly in the pathway from anchor cell to selection of vulval fate because if they did, elimination of their activities should fully reproduce elimination of the anchor cell (Fig. 16.14). This conclusion is supported by the finding that the partial induction seen in *lin-7* and *lin-10* mutant hermaphrodites is eliminated if the anchor cell is ablated (Sternberg and Horvitz, 1989). Rather, these genes may act indirectly by modifying or stabilizing components in the direct pathway. For example, one of these genes may stabilize *let-23 RTK*. In the absence of that gene, *let-23 RTK* is unstable and is present in reduced quantity so only a small amount of induction can occur. (See Fig. 16.14 for alternative interpretations of the *lin-2*, *lin-7*, and *lin-10* data, for example, functional redundancy.)

B. Genes Associated with Muv Mutants Normally Inhibit Vulval Fate

The Muv phenotype associated with mutations in *lin-1*, *lin-13*, and *lin-15* is caused by a decrease of gene function (Table I). In the absence of any of these genes, all VPCs execute 1° or 2° vulval fates. Thus, these genes are required for the VPCs to inhibit 1° and 2° vulval fates and to adopt 3° nonvulval fates (Fig. 16.15A).

In some instances, the manifestation of the Muv phenotype requires mutations in two genes (Horvitz and Sulston, 1980; Ferguson *et al.*, 1987; Ferguson and Horvitz, 1989). This effect is known as the synthetic Muv or syn-Muv phenotype. Syn Muv mutations fall into two classes: A (e.g., *lin-8*, *lin-38*) and B (e.g., *lin-9*, *lin-35*, *lin-36*, *lin-37*). Any combination of A-only mutants (e.g., *lin-8*, or *lin-8*,*lin-38* double mutant) or B-only mutants (e.g., *lin-9* or *lin-9*; *lin-36* double mutant) results in a wild-type vulva, but any combination of an A and a B mutant (e.g., *lin-8*; *lin-9* or *lin-8*; *lin-35* double mutants) results in a Muv phenotype (Ferguson and Horvitz, 1989). These results suggest that the A and B classes of genes define two separate pathways that are functionally redundant with respect to inhibition of 1° or 2° vulval fates; eliminating any component of just one pathway is not sufficient to result in a phenotype change but eliminating components from both pathways can result in a change in phenotype (Fig. 16.15B; Ferguson and Horvitz, 1989). Aside from its synthetic nature, the syn-Muv phenotype is the same as the Muv phenotype.

Although many alleles of *lin-7*, *lin-13*, or *lin-15* alone are sufficient for the Muv phenotype, some alleles of *lin-13* and *lin-15* also exhibit syn-Muv behavior. For example, the *lin-15* allele *n767* behaves like a class A syn-Muv and the *lin-15* allele *n744* behaves like a class B syn-Muv. Nonetheless, the existence of alleles of each of these two genes that give rise to a Muv phenotype alone suggests that these genes operate in both A and B pathways (Ferguson and Horvitz, 1989).

The Muv phenotype is independent of the anchor-cell signal—if the anchor cell is eliminated in the background of Muv mutants, the VPCs still

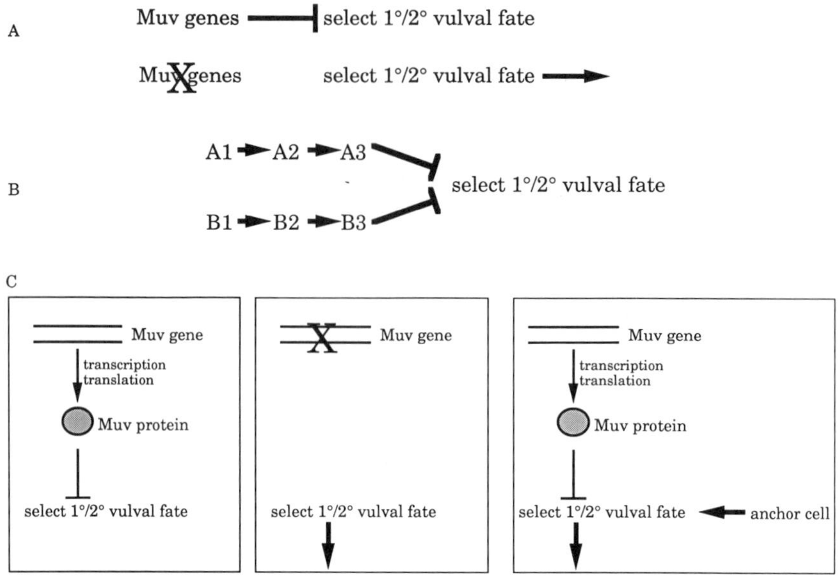

Figure 16.15 Action of the *muv* genes. (A) The *muv* genes are necessary to turn off vulval fate because, in their absence, vulval fate proceeds. (B) The *muv* genes also appear to act in two redundant pathways. Elimination of one or more components in either the A or the B pathway still results in normal vulval differentiation. However, simultaneously eliminating components from both pathways does result in a lack of vulval fate inhibition (i.e., a Muv phenotype). Some genes, such as *lin-15*, can be mutated to an A^-, B^-, or A^-B^- phenotype, suggesting that this gene operates in both pathways. (C) *muv* gene activity can be eliminated in two ways. A wild-type *muv* gene produces a protein that inhibits vulval fate selection (*left*). If this gene is eliminated by mutation, this inhibitory activity is eliminated (*center*). In a wild-type hermaphrodite, the activity of the *muv* gene presumably is eliminated in P5.p–P7.p by the anchor cell signal, which overrides the *muv* gene (*right*). The anchor cell signal probably does not act through the *muv* genes (see text and Fig. 16.16). See Ferguson and Horvitz (1989) and Herman and Hedgecock (1990) for more details.

execute 1° and 2° vulval fates (Fig. 16.10F; Ferguson *et al.*, 1987; Sternberg, 1988b; Beitel *et al.*, 1990; Han *et al.*, 1990). Thus, in the absence of Muv gene activity, a VPC automatically or intrinsically adopts 1° or 2° vulval fate with or without the anchor-cell signal. In other words, all that is required for a VPC to execute vulval fate is for the Muv genes to be turned off. This elimination is achieved genetically in the case of Muv mutants. In the wild-type hermaphrodite, however, the simplest model would be that the anchor-cell signal overrides the effect of the Muv genes in P5.p–P7.p (Fig. 16.15C; Herman and Hedgecock, 1990).

Just as *lin-2*, *lin-7*, and *lin-10* appear not to act directly in the pathway from the anchor-cell signal to the selection of vulval fate, three observations suggest that at least one of the Muv genes, *lin-15*, also does not act directly. First, although all VPCs select 1° or 2° fates in a *lin-15(lf)* (i.e., *lin-15* complete loss-of-function) mutant irrespective of the presence of the anchor cell, the pattern of 1° and 2° fate seen is affected by the anchor cell (Sternberg, 1988b; Fig. 16.16A,B). Second, the anchor cell influences the patterns of induction seen in *lin-15(lf)* hermaphrodites in which all but two VPCs have been ablated (Fig. 16.16C,D). Third, *lin-15(lf)* hermaphrodites that also carry a *let-60(dn)*

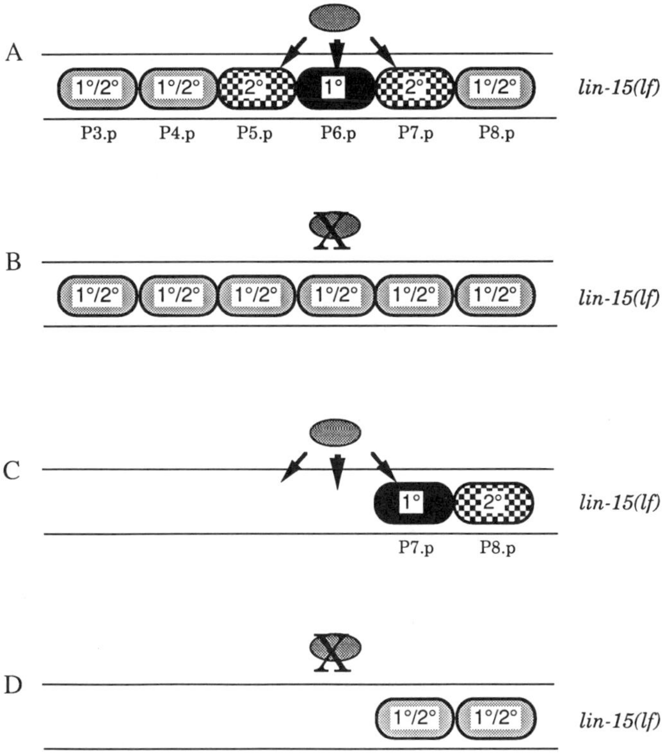

Figure 16.16 The *muv* gene *lin-15* is likely to act in a pathway that is independent of the anchor cell signal. In a hermaphrodite completely lacking *lin-15* activity [i.e., *lin-15(lf)*], the VPCs select only 1° or 2° vulval fates. (A) If the anchor cell is present in these mutant animals, P6.p always selects 1° fate, P5.p and P7.p select 2° fate, and the other VPCs select either 1° or 2° fate. (B) However, if the anchor cell is ablated in these mutant animals, the fates of all VPCs are variable. For example, P6.p has an equal probability of being 1° or 2°. Thus, the anchor cell can affect the VPCs although *lin-15* is absent, suggesting that the anchor cell signal does not act through *lin-15*. (C) Similarly, if all but two VPCs are ablated in a *lin-15(lf)* mutant hermaphrodite, the VPCs closest to the anchor cell always adopts a 1° fate. (D) However, if the anchor cell is ablated, the fate of both VPCs is randomized. Data from Sternberg (1988b).

mutant (see subsequent text) often have wild-type vulvas. This induction is anchor-cell dependent (Han *et al.*, 1990). In all three instances, the anchor cell is able to exert an influence although *lin-15* activity has been eliminated. Whether other Muv mutations also act independently of the anchor-cell signaling pathway is unknown.

C. Vul/Muv Interactions

The Muv and Vul genes act antagonistically on the VPCs: whereas Muv genes inhibit 1° and 2° vulval fates and promote nonvulval 3° fate (loss of a Muv gene leads to no 3° fate in all VPCs), Vul genes promote vulval 1° and 2° fates (loss of a Vul gene leads to no 1° and 2° fates in all VPCs). Therefore, Muv genes must be activated to select 3° fate and turned off to select 1° or 2° fate; the converse must be true for Vul genes (Fig. 16.17A). A gene that negatively regulates a Muv gene would be expected to mutate to a Vul phenotype because, if a Muv gene could not be turned off, 3° fate would be promoted. For example, assume *lin-2* negatively regulates *lin-15* (Fig. 16.17B). A *lin-2(lf)* mutant therefore would result in a Vul phenotype because *lin-15* cannot be turned off, thereby promoting all 3° fates. Alternatively, a gene that turns off a Vul gene would be expected to mutate to a Muv phenotype because, if a Vul gene could not be turned off, 1° or 2° fate would be promoted. For example, assume *lin-15* negatively regulates *lin-2* (Fig. 16.17B). A *lin-15(lf)* mutant therefore would result in a Muv phenotype because *lin-2* cannot be turned off, thereby promoting all 1° and 2° fates. The difference between these two cases is the order in which the genes act. The Vul gene regulates the Muv gene in the first case, whereas the Muv gene regulates the Vul gene in the second case.

For a given Muv/Vul pair, determining their relative order is possible by making the double Muv/Vul mutant (Fig. 16.17). If a given Muv/Vul double mutant is Muv, the Vul gene normally regulates the Muv gene (Fig. 16.17C). If the Muv gene activity is eliminated by the mutation, then the presence or absence of the Vul gene is superfluous because the normal action of the Vul gene is to eliminate Muv gene activity. Conversely, if a given Muv/Vul double mutant is Vul, the Muv gene normally regulates the Vul gene. A Muv/Vul double mutant may be neither Muv nor Vul; the two mutations could be coexpressed. This result would suggest that the two genes do not act in a linear order, but that their activities converge antagonistically on a third gene (Fig. 16.17B). Interpretation of these experiments relies heavily on the use of mutations in each gene that eliminate activity, that is, complete loss-of-function alleles of the Muv and Vul genes must be used.

Double mutants between many of the Muv and Vul genes have been made, although for some combinations incomplete loss-of-function alleles were used (Ferguson *et al.*, 1987; Han *et al.*, 1990; Aroian and Sternberg,

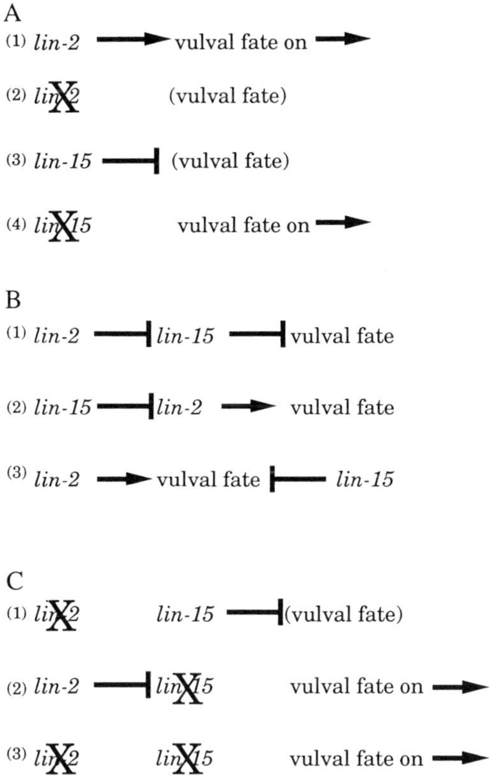

Figure 16.17 Ordering genes in a regulative pathway. (A) The gene *lin-2* is required for selection of vulval fate (1) because if *lin-2* is eliminated (2), vulval fate is not selected. Similarly, *lin-15* is required for inhibition of vulval fate (3) because in its absence (4), vulval fate is selected. (B) These two genes can act in three simple ways: (1) *lin-2* negatively regulates *lin-15*, which negatively regulates vulval fate; (2) *lin-15* negatively regulates *lin-2*, which stimulates vulval fate; and (3) *lin-2* and *lin-15* exert their influences independently of each other. (C) If the first combination were true, then elimination of *lin-2* would result in inhibition of vulval fate (i.e., Vul) because *lin-15* could not be inactivated (1). Elimination of *lin-15* would result in automatic selection of vulval fate (i.e., a Muv phenotype) because *lin-15* is required for inhibition of vulval fate (2). Because *lin-2* requires the presence of *lin-15* to affect vulval fate, the double mutant *lin-2(lf) lin-15(lf)* and the single mutant *lin-15(lf)* would exhibit the same phenotype—Muv (3).

Similarly, if the order of the genes is reversed (i.e., *lin-15* inhibits *lin-2*), the double mutant *lin-2(lf) lin-15(lf)* would exhibit the same phenotype as *lin-2(lf)* alone—Vul. Thus, if the two genes act together in the same pathway, the relative order of the two genes can be determined by examining the phenotype of the double mutant (not shown). The third combination, that is, that *lin-2* and *lin-15* affect vulval fate independently, is more difficult to prove. The double mutant *lin-2(lf) lin-15(lf)* could be Muv or Vul depending on the default fate selection in the absence of both genes. However, an unambiguous result would be obtained if the double mutant were neither Muv nor Vul but showed an intermediate phenotype (e.g., wild-type). This result is referred to as coexpression of phenotypes and indicates either that the genes are acting in independent pathways or that the mutations used do not eliminate function completely (also not shown).

1991). The simplest pathway inferred from the double mutant data in which complete or near loss-of-function mutations were used is shown in Fig. 16.18. Genes not shown in the pathway are discussed in the notes.

Of particular note is the relationship of *let-23 RTK* to *let-60 ras*. Both genetic and molecular data suggest that *let-60 ras* operates downstream of *let-23*

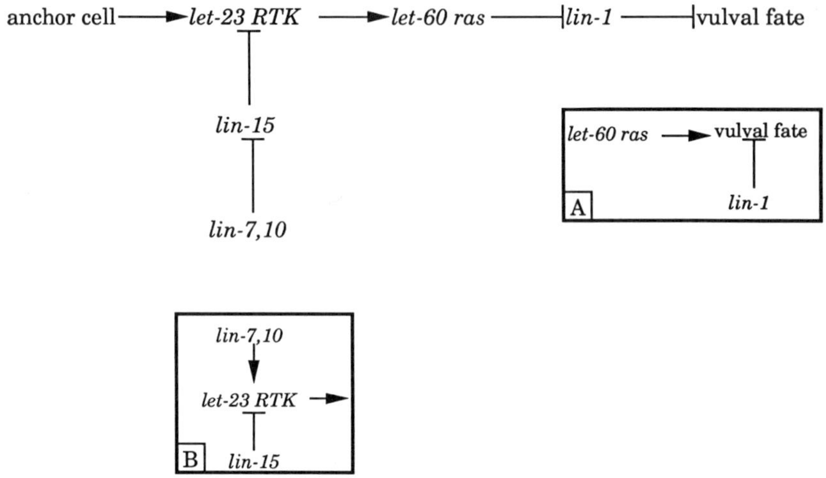

Figure 16.18 Pathway of vulval determination (1°/2° vs. 3° fate) as determined by double mutant phenotypes. The anchor cell signal is received by *let-23 RTK* which allows *let-23 RTK* to overcome inhibition by *lin-15*. This step leads to activation of *let-60 ras*. The effect of activated *let-60 ras* is to inhibit *lin-1*, which allows vulval induction to occur. Other genes are likely to exist between *let-60 ras* and *lin-1* and possibly between *let-23 RTK* and *let-60 ras*. This pathway was put together using the logic detailed in Fig. 16.17 and incorporates published data in which complete-loss-of-function alleles were used to make the double mutants or in which the epistatic relationship was clear from the double mutant phenotype (Ferguson *et al.*, 1987; Han *et al.*, 1990; Aroian and Sternberg, 1991). The following additional assumptions were used. First, *lin-7*, *lin-10*, and *lin-15* are assumed to operate in a pathway independent of the anchor cell. Second, if no coexpression of phenotype occurs, the genes are arranged in dependent pathways, although independent pathways are also possible (see notes to *lin-1*, *lin-7*, and *lin-10*). *lin-1* is shown to act in line with the anchor cell signal because it is fully epistatic to all *vul* genes shown. However, analogous to *lin-15*, *lin-1* may act in a pathway independent of the anchor cell (see Box A). *lin-2* probably acts at the same step as *lin-7* and *lin-10*. *lin-3* may act upstream in the pathway, but a nonnull allele was used in making double mutants, making interpretation uncertain. *lin-7* and *lin-10* are shown in the same pathway as and upstream from *lin-15* because the double mutant phenotypes were Muv. However, other considerations (e.g., hyperinduced phenotypes; see text) suggest that *lin-7* and *lin-10* may act with *let-23 RTK* in a *lin-15*-independent pathway (see Box B). The *lin-8* and *lin-9* synthetic Muv phenotype appears to be coexpressed with *lin-7* and *lin-10*. If so, these syn-*muv* genes would act at the same step as *lin-7* and *lin-10*. However, nonnull alleles of both *lin-8* and *lin-9* were used, so their relationship to *lin-7* and *lin-10* is unclear. The *lin-13* Muv phenotype was suppressed by all Vul mutations except *lin-3*. However, only weak nonnull alleles of this gene exist.

RTK. First, hyperactive *let-60 ras* can bypass the need for *let-23 RTK*—gain-of-function *let-60 ras* mutations are able to overcome fully the vulval defects in a *let-23 RTK* complete loss-of-function mutant (Han *et al.*, 1990). Second, the Vul phenotype of a strongly defective *let-23* allele is suppressed by over-expression of *let-60 ras* (Han and Sternberg, 1990). Thus, *let-23 RTK* is no longer needed if *let-60 ras* is activated or overexpressed, suggesting that *let-23 RTK* normally acts to activate *let-60 ras* (directly or through other genes). These data provide dramatic *in vivo* support for mammalian tissue culture data that suggest that Ras operates downstream of the EGF receptor, and also indicate that molecules similar to EGF receptor and Ras can act in a linear pathway in the determination of cell fate.

The Muv gene *lin-1* is unusual. Unlike other Muv genes, the double mutant between *lin-1* and any Vul gene (except *lin-25*; see subsequent discussion), is Muv, suggesting that *lin-1* acts downstream of the Vul genes. Further, unlike all the other Muv genes, no syn-Muv *lin-1* alleles exist. Finally, the 1° and 2° lineages in a *lin-1* mutant are not always wild-type and can be difficult to interpret, unlike other Muv mutants (Ferguson *et al.*, 1987; P. W. Sternberg, unpublished observations). This gene may, therefore, act differently than the other Muv genes.

D. Place and Time of Action

The previous discussion has considered the role and functional relationship of various components but not where and when they act. Where a gene acts can be determined in two ways—molecularly and genetically. Fusion of the regulatory regions of a gene with *lacZ* can be used to determine where a gene is transcribed (Way and Chalfie, 1989; Fire *et al.*, 1990). Antibody staining can be used to determine in which cells protein product is present (Ruvkun and Giusto, 1989). mRNA expression levels also can suggest where a gene is expressed. For example, although *lin-10* mutations are known to affect only the vulva, *lin-10* transcript levels in hermaphrodites lacking an anchor cell or lacking VPCs is not altered substantially from expression in wild-type (Kim and Horvitz, 1990). This result suggests that, whether or not *lin-10* is expressed in the VPCs, the gene is expressed in other cells.

Unlike molecular experiments, which can indicate where a gene is expressed or whether protein product is present, genetic mosaic experiments can suggest in which cells a gene actually performs its function (reviewed in Herman, 1989). This type of experiment makes use of a free duplication, which is essentially an unstable minichromosome. When the cell containing this minichromosome divides, the normal chromosome pairs segregate properly, but this unstable minichromosome has a low probability of segregating improperly. Thus, the minichromosome may be present in some cells and not others. If the experiment is carried out in hermaphrodites mutant for a gene of interest [e.g., *lin-15(lf)*] but containing a minichromosome with a wild-type copy of the gene (e.g., *lin-15*$^{+}$), generating mosaic hermaphrodites in which some cells are *lin-15(lf)* and some cells are *lin-15*$^{+}$ is possible. Using other genetic markers to determine which cells are actually *lin-15(lf)* and which are *lin-15*$^{+}$ and noting the vulval phenotype of different mosaic animals, one can determine in which cells a wild-type copy of *lin-15* is needed to generate a wild-type hermaphrodite and in which cells *lin-15* must be absent to generate a mutant hermaphrodite.

With respect to vulval development, such an experiment has been performed only for *lin-15* (Herman and Hedgecock, 1990). An individual hermaphrodite in which all six VPCs carried a wild-type *lin-15* gene still could express a *lin-15* Muv phenotype at times if other cells in the hermaphrodite were mutant for *lin-15*. Thus, wild-type *lin-15* in the VPCs is not sufficient for wild-type vulva, suggesting that *lin-15* acts outside the VPCs (although the data do not exclude that *lin-15* also could act in the vulva). Further, animals that lacked an anchor cell, in which all six VPCs were wild-type for *lin-15*, also could express the mutant phenotype, arguing that the mutant phenotype seen in these VPCs was not caused by the anchor cell. In addition, the VPCs P3.p, P4.p, and P8.p sometimes could express a wild-type 3° phenotype when mutant for *lin-15*, indicating that induction in these VPCs was being repressed by some other cells.

The action of *lin-15* seems to be spread out among different lineages that share the characteristic of contributing to hyp7, the large multinucleated hypodermal cell that surrounds the VPCs. Combining all these observations, Herman and Hedgecock (1990) hypothesized that *lin-15* exerts an inhibitory effect on vulval induction from hyp7. Thus, three tissues appear to be involved in vulval determination: (1) the large hypodermal syncytium hyp7, which promotes 3° hypodermal fate in the VPCs; (2) the gonadal anchor cell, which induces some of the VPCs to overcome the effects of hyp7; and (3) the VPCs themselves. An updated molecular genetic model for vulval determination is shown in Fig. 16.19.

The question of when Muv and Vul genes act can be addressed using temperature-sensitive alleles and temperature-shift experiments. Such experiments have been performed for the Muv genes *lin-13* and *lin-15*, the Vul genes *lin-10* and *let-23 RTK*, the 2°-determining gene *lin-12*, and *lin-25* (Greenwald *et al.*, 1983; Ferguson *et al.*, 1987; Kim and Horvitz, 1990). All these genes except *lin-25* (discussed subsequently) appear to act before the VPCs undergo their first round of division, consistent with cell ablation studies that suggest that determination occurs before the VPCs divide. In addition, the Muv genes appear to act over a broad time interval that may precede the action of the Vul genes, which in turn may or may not precede the action of *lin-12*. The action of *lin-15* in the hypodermal syncytium may explain the long,

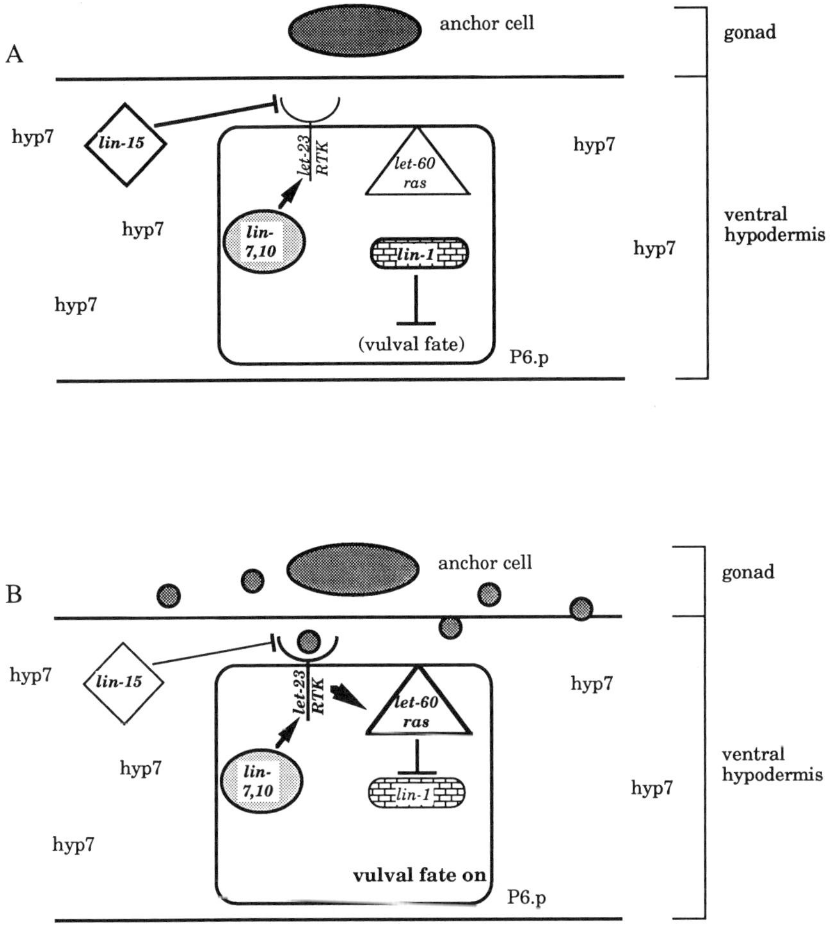

Figure 16.19 Molecular genetic model for determination of vulval versus hypodermal fates. For simplicity, only P6.p is shown. The pathway is from Fig. 16.18 except the alternative given in Box B has been chosen (because of hyperinduction data). (A) In a wild-type hermaphrodite without an anchor cell signal, *lin-15* acting from hyp7 (shown surrounding P6.p) inhibits *let-23 RTK* kinase activity in the VPCs. As a result, *let-60 ras* is not activated and vulval fate is not turned on because *lin-1* prevents selection of vulval fate. A *lin-15(lf)* mutant displays an anchor cell-independent Muv phenotype because (we hypothesize) *let-23 RTK* has a basal kinase activity that can activate *let-60 ras* even in the absence of ligand. (B) In a *lin-15*$^+$ animal, the anchor cell ligand reaches P6.p, is bound by *let-23 RTK*, and allows *let-23 RTK* to overcome inhibition by hyp7. Kinase activity from *let-23 RTK* activates *let-60 ras*, which activates vulval fate either by inhibiting *lin-1* or by overcoming *lin-1* inhibition. The specific mode of action of *lin-15* shown, that is, inhibit the extracellular domain of *let-23 RTK*, is arbitrary. The genes *lin-7* and *lin-10* are shown to act in the VPCs to stabilize *let-23 RTK*. Alternatively, *lin-7* and *lin-10* may act in hyp7 or in the VPCs to inhibit *lin-15* (see Fig. 16.18).

gradual, temperature-sensitive period seen with *lin-15* because hyp7 cumulatively recruits nuclei during this time. These data can be interpreted to support and extend the model presented earlier. The Muv genes establish their effects on the VPCs early, promoting 3° fate. Subsequently, the anchor-cell signal allows the VPCs to overcome these effects. The data are also consistent with the notion that these genes all act at or around the same time. Caveats for temperature-shift experiments include (1) the assumption that the temperature shift affects protein activity and not synthesis or stability and (2) error due to the amount of time it takes to inactivate or reactivate the protein after the temperature shift.

V. Vulval Precursor Cell–Cell Interactions and Distinguishing between Primary and Secondary Fates

A. Primary Fate Is Vulval Precursor Cell Default or Intrinsic Fate

The discussion has covered the molecular genetics of how 1° or 2° vulval fate versus 3° nonvulval fate is determined. What distinguishes the two vulval fates, 1° and 2°, from each other has not yet been addressed. As discussed here, VPC–VPC communication and the gene *lin-12* have prominent roles in this decision.

In the absence of outside influences, current data suggest that a VPC will adopt 1° fate. This conclusion is made because, in a *lin-15(lf)* mutant hermaphrodite (hence, no hyp7 influence) with no anchor cell (hence, no anchor-cell signal) and in which all but one VPC has been laser ablated (hence, no VPC–VPC influences), an isolated VPC will select 1° fate (Fig. 16.20A,B; Sternberg, 1988b). Presumably, vulval fate is selected because of basal *let-23 RTK* kinase activity in the absence of all outside influences. This basal activity then activates the *let-60 ras* pathway which results in a default of 1° fate. We call this pathway leading to selection of 1° fate the intrinsic pathway because it appears to be an intrinsic property of a VPC. A prediction of this model is that, if a VPC is placed alone in tissue culture, that VPC will select 1° fate, although we note that such an experiment is not currently possible.

B. In the Absence of *lin-15*, Vulval Precursor Cell–Cell Communication Is Responsible for Selection of Secondary Fate

If a VPC by default selects 1° fate, what causes selection of 2° and 3° fates? As modeled earlier, the nonvulval 3° fate is selected when *lin-15*, and perhaps other Muv genes acting from hyp7, inhibits *let-23 RTK* and activation of the intrinsic pathway. On the other hand, the selection of 2° fate seems to involve communication among the VPCs themselves.

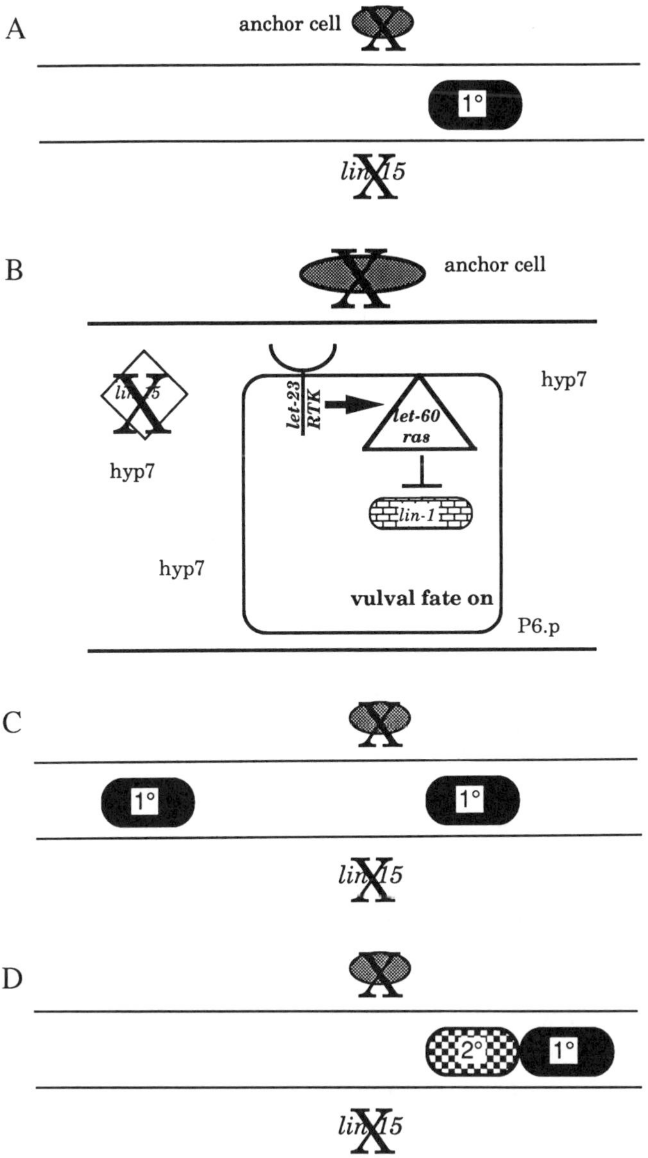

Figure 16.20 The intrinsic pathway and VPC–VPC interactions. (A) In the absence of the anchor cell signal, *lin-15*, and neighboring VPCs, a VPC will select 1° fate. (B) One possible interpretation is that *let-23 RTK* has basal kinase activity in the absence of ligand when *lin-15* has been turned off, and that this "default" or "intrinsic" mode results in selection of 1° fate. (C) In an anchor cell-minus *lin-15(lf)* hermaphrodite, two nonadjacent VPCs both will select 1° fates. (D) However, if the two VPCs are adjacent, then one will select 1° fate and the other 2° fate, suggesting that VPC–VPC communication results in lateral inhibition of the intrinsic pathway in one of the VPCs, leading to the selection of 2° fate. Data from Sternberg (1988b).

Unlike fully penetrant Vul mutants in which the pattern of VPC fate is fixed, that is, all 3°, the pattern of VPC fates in fully penetrant Muv mutants is somewhat variable and differs with and without the presence of an anchor cell (Ferguson *et al.*, 1987; Sternberg, 1988b; Sternberg and Horvitz, 1989). In a *lin-15(lf)* Muv mutant in which the anchor cell is present, P6.p adopts a 1° fate, P5.p and P7.p adopt 2° fate, but the fates adopted by P3.p, P4.p, and P8.p are variable, that is, each can be 1° or 2° (Fig. 16.16). In the absence of an anchor cell, the fates of all six VPCs are variable, 1° or 2°. Despite this variability, the presence of adjacent 1° fates in the *lin-15(lf)* mutant, with or without an anchor cell, is rare, whereas the presence of adjacent 2° fates is not (Sternberg, 1988). In no case are three consecutive 2° fates seen. Most often, 1° and 2° cells alternate with one another. This pattern suggests that some active mechanism prevents the formation of adjacent 1° cells.

In experiments with *lin-15(lf)* animals in which all but two VPCs are ablated, both VPCs select intrinsic 1° fate if not in contact with one another, but one VPC selects 1° fate and the other 2° fate if the VPCs are in contact (Fig. 16.20C,D). This result suggests that adjacent VPCs communicate with one another to prevent the formation of adjacent 1° cells. This communication has been called lateral inhibition or lateral signaling because it occurs laterally between VPCs.

Although this conclusion does not depend on the presence or absence of the anchor cell, the anchor cell does bias which cell executes the 1° fate (see Fig. 16.16C,D). In the absence of the anchor cell, two adjacent VPCs battle over which will adopt the 1° fate. At random, one VPC wins, so the other VPC is forced to adopt a 2° fate. In the presence of the anchor cell, the VPC closer to the inducing signal always selects the 1° fate, forcing its adjacent VPC neighbor to adopt a 2° fate. This result helps explain the patterns seen when all six VPCs are present in a *lin-15* Muv mutant (Fig. 16.16A,B). If the anchor cell is present, the VPC closest to the anchor cell, P6.p, adopts a 1° fate, forcing P5.p and P7.p into 2° fate via lateral signaling. In the absence of the anchor cell, P6.p is no longer biased to execute the 1° fate, resulting in a more random pattern of 1° and 2° fates.

C. *lin-12* Is Necessary and Sufficient to Specify Secondary Fate

The activation of 2° fate requires the *lin-12* gene. Mutations that eliminate *lin-12*, designated *lin-12(∅)*, result in VPCs executing only 1° and 3° fates in an otherwise wild-type hermaphrodite (Fig. 16.21B; Greenwald *et al.*, 1983; Sternberg and Horvitz, 1989). Thus, *lin-12* is necessary for selection of 2° fate in wild-type. (Phenotypically, *lin-12(∅)* animals often display one large protrusion at their vulva.) Further, in a *lin-12(∅), lin-15(lf)* double mutant, all six VPCs execute 1° fates (Fig. 16.21C; Sternberg and Horvitz, 1989). This result

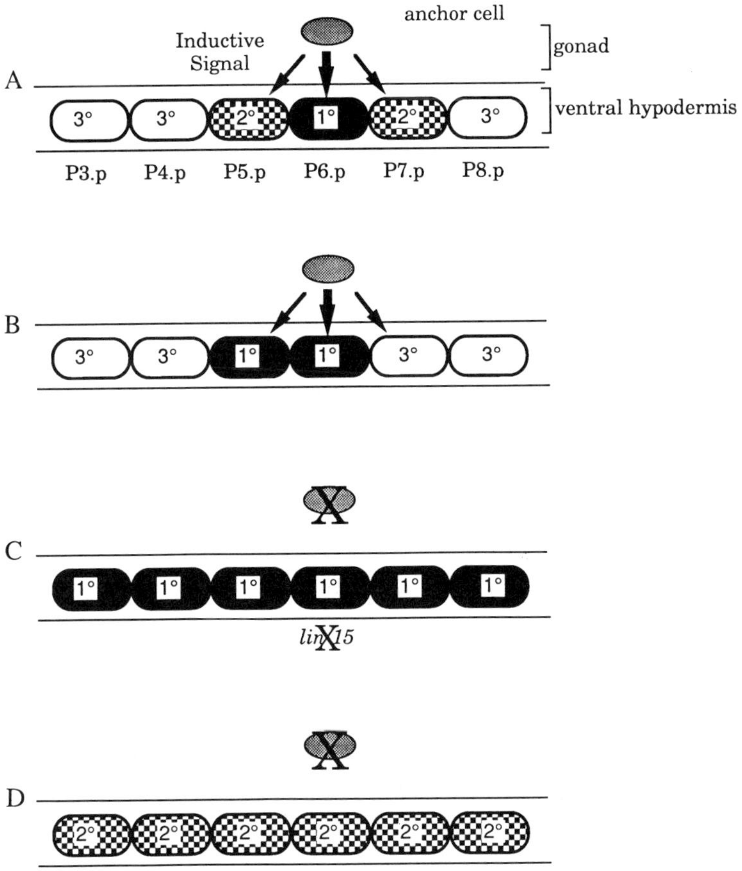

Figure 16.21 *lin-12* is necessary and sufficient for 2° fate selection. (A) Wild-type vulval induction. (B) When *lin-12* is eliminated, only 1° and 3° fates are selected. Note that (1) although two VPCs are shown selecting 1° fate, the actual number can vary and (2) although only one anchor cell is shown, *lin-12(∅)* hermaphrodites can have multiple anchor cells. (C) If, in a *lin-12(∅)* hermaphrodite, the intrinsic pathway is activated by the elimination of *lin-15*, then all VPCs select 1° fate. Thus, the selection of 2° fate in a *lin-15(lf)* animal requires *lin-12*. See text for interpretation. (D) Gain-of-function *lin-12(d)* (dominant) mutations result in all VPCs executing 2° fate. The anchor cell is missing because *lin-12(d)* mutations also result in absence of the anchor cell. Data from Greenwald *et al.* (1983); Sternberg and Horvitz (1989); W. Katz and P. W. Sternberg, unpublished results.

suggests that *lin-12* is required for lateral signaling, that is, selection of 2° fate in a *lin-15(lf)* background.

In addition to *lin-12(∅)* mutants, semidominant gain-of-function *lin-12* mutants [designated *lin-12(d)*] exist that behave as though they increase *lin-12* activity above wild-type levels (Greenwald *et al.*, 1983). All six VPCs in mutant *lin-12(d)* hermaphrodites that lack an anchor cell execute 2° fates (Fig. 16.21D; Greenwald *et al.*, 1983; Sternberg and Horvitz, 1989). Thus, activation of *lin-12* is sufficient for selection of 2° fate. Consistent with this result, all six VPCs in a *lin-12(d)* hermaphrodite select 2° fates when the Vul genes are eliminated (Sternberg and Horvitz, 1989; Han *et al.*, 1990). Phenotypically, *lin-12(d)* hermaphrodites show multiple ventral protrusions because of the presence of the extra 2° lineages. In this respect, these hermaphrodites share some similarity with Muv hermaphrodites. However, we will not refer to the *lin-12(d)* phenotype as Muv since the cellular base of the two phenotypes are quite distinct.

These data are summarized in Fig. 16.22; the conclusions from these data are shown in Fig. 16.23. In the absence of all known external influences, a VPC will execute 1° fate intrinsically via the intrinsic pathway. The selection of 2° fate ensues when such an isolated VPC is given a neighbor; one VPC signals the other to execute 2° fate. Selection of 2° but not 1° fate requires *lin-12* because, if *lin-12* activity is eliminated, VPCs do not select 2° fate. The intrinsic pathway normally is needed for both 1° and 2° fate because Vul hermaphrodites express neither 1° nor 2° fates. However, activated *lin-12* can abrogate the need for the intrinsic pathway in the selection of 2° but not 1° fate because a *lin-12(d)* mutation results in all VPCs executing 2° fate in a hermaphrodite lacking an anchor cell or the Vul genes. In addition, when *lin-12* is activated, 2° fate selection is not affected by the inhibitory signal from hyp7 because, in a *lin-12(d), lin-15*$^+$ hermaphrodite, all VPCs are 2°.

The molecular nature of *lin-12* provides a compelling link between the lateral signal inferred and the *lin-12* phenotypes. Similar to the hypothesis for the related *Drosophila* protein Notch, *lin-12* may be a receptor for short-range signal, in particular the lateral signal from a 1° VPC to its neighbor (Fig. 16.22B; Sternberg and Horvitz, 1989). When this receptor is lacking, the lateral signal to be 2° cannot be received by a VPC. When this receptor is activated, a VPC is biased to select a 2° fate. Consistent with the notion of *lin-12* as a receptor, mosaic analysis has shown that *lin-12* acts cell autonomously in the determination of anchor cell and ventral uterine cell fate (Seydoux and Greenwald, 1989). Alternatively, *lin-12* may not be a receptor for the lateral signal. In a

---

Figure 16.22 Summary of data on how VPCs select the different fates. (A) A VPC isolated from the influence of neighboring VPCs, the anchor cell signal, and presumably hyp7 [via *lin-15(lf)*] will select 1° fate via an intrinsic pathway (I.P., which includes *let-23* RTK and *let-60 ras*; see Fig. 16.20B). (B,C) This I.P. is necessary because if *let-23 RTK* or *let-60 ras* is reduced severely in a *lin-15(lf)* or *lin-15*$^+$ background, all 3° fates are selected. (D) However *lin-15* is sufficient to inhibit the I.P. as long as the anchor cell signal is off. (E,F) If the anchor-cell signal is on, the inhibitory effects of *lin-15* can be overcome. (E) If the anchor cell signal is close, an isolated VPC will select 1° fate via the I.P. (F) If the anchor cell signal is further away, an isolated VPC will select 2° fate. Presumably, this selection of 2° fate also requires the I.P. because if the I.P. is elimi-

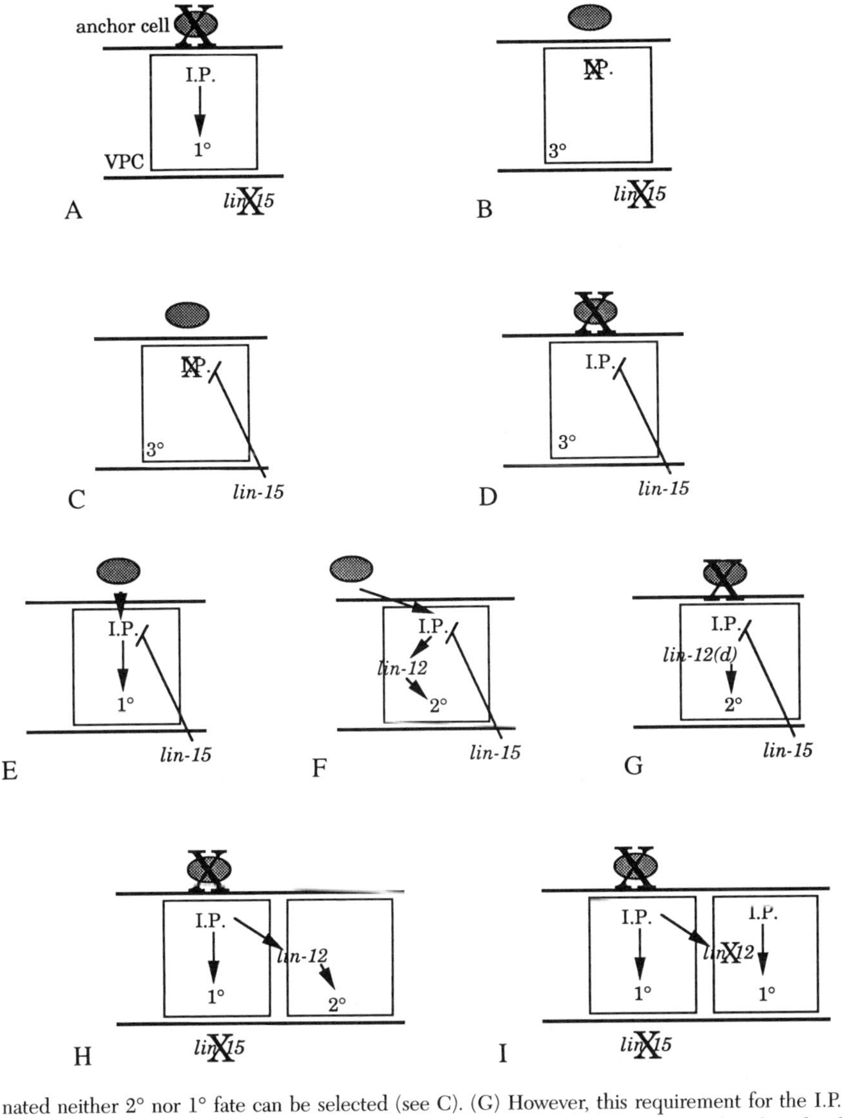

nated neither 2° nor 1° fate can be selected (see C). (G) However, this requirement for the I.P. can be bypassed if *lin-12* is activated via a *lin-12(d)* mutation. This result suggests that the role of the I.P. in selection of 2° fate may be to activate *lin-12* initially. Once activated, *lin-12* no longer needs the I.P. (H) Another way to activate the 2° pathway is via lateral signaling. Although the default fate of an isolated VPC is 1° (see A), this default can be overridden if a VPC has a neighbor. Thus, *lin-12* can be activated intracellularly and intercellularly. (I). If *lin-12* is eliminated, 1° fate is selected by all VPCs in a *lin-15(lf)* background. Perhaps *lin-12* is a receptor for the lateral signal. Data for A, D, E, and F are from isolated VPC experiments. Data for B, C, G, H, and I are from intact VPC experiments. Data for H and I also include experiments with two isolated VPCs.

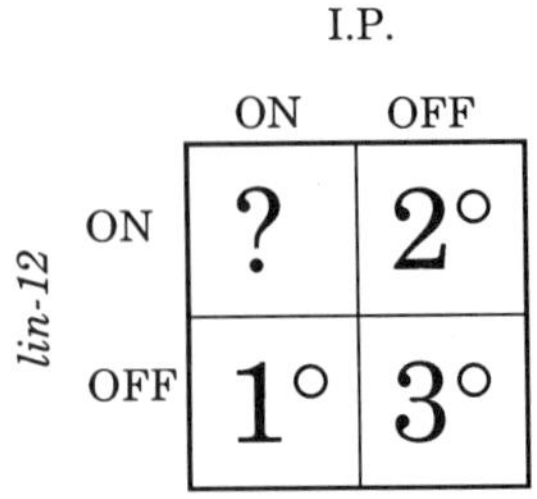

Figure 16.23 If the intrinsic pathway (I.P.) is on and the *lin-12* pathway is off, a VPC selects 1° fate. If the I.P. is off and *lin-12* is on, 2° fate ensues. If both pathways are off, 3° fate ensues. Since the pathways are mutually antagonistic, then what happens when both pathways are on depends on how strongly each is activated (see text).

lin-12(∅), lin-15(lf) double mutant, all 1° fates are selected because VPCs are unable to select 2° fate. The lateral signal still operates but the 2° pathway is blocked by the *lin-12(∅)* mutation. This confusing state may explain why the lineages executed by the VPCs, interpretable as 1°, are nonetheless unusual (see Sternberg and Horvitz, 1989).

D. Modeling the Selection of the Three Vulval Precursor Cell Fates in Wild Type

As summarized in Fig. 16.23, in the absence of the Vul genes 3° fate is selected unless *lin-12* is activated via *lin-12(d)*, which then leads to 2° fate. If the Vul genes are present and *lin-12* is off, a 1° fate ensues. Thus, the combination of two binary switches, I.P. (intrinsic pathway) ON/OFF and *lin-12* ON/OFF, allows determination of the three vulval fates (Sternberg and Horvitz, 1989). Muv genes such as *lin-15* affect determination by influencing the first switch, I.P. ON/OFF. However, this influence leaves unanswered what happens when both switches are on. This situation is important to consider because it is likely to occur in wild-type since: (1) *lin-12* is required for 2° fate and (2) the I.P. is required for 2° fate (in addition to 1° fate). Although Fig. 16.23 indicates that 2° fate ensues from *lin-12* ON/I.P. OFF, this conclusion is based on data from *lin-12(d)*, not wild-type, animals.

The data suggest that the situation when both switches are on is unstable since 1° fate and 2° fates are mutually antagonistic. First, whereas VPCs in a *lin-12(d)* hermaphrodite lacking an anchor cell all select 2° fates, P6.p in a *lin-12(d)* hermaphrodite with an anchor cell selects 1° fate (the others select 2° fates). Therefore, the anchor-cell signal and high activation of the intrinsic 1° pathway can override selection of 2° fate. Conversely, whereas an isolated VPC in a *lin-15(lf)* hermaphrodite always selects 1° fate, this selection is overridden in one of them when two adjacent VPCs are present. Therefore, 2° fate can inhibit 1° fate. The mutual antagonism is perhaps best represented in a *lin-12(d), lin-15(lf)* mutant with all six VPCs present—1° fate is present, so some of the 2° fates from *lin-12(d)* are suppressed, but fewer 1° fates occur than in a *lin-15(lf)* mutant alone, so the intrinsic 1° pathway also is suppressed in some VPCs.

This discussion also has ignored an important result from cell biology experiments in wild type, that is, that an isolated VPC at intermediate distance from the anchor cell can select 2° fate. One interpretation is that *lin-12* can be activated in two ways: (1) via the lateral signal from a neighboring VPC and (2) via intermediate activation of the intrinsic 1° pathway by the anchor cell. Further, activated *lin-12* serves two functions: (1) to inhibit the intrinsic 1° pathway within the same VPC and (2) to activate the subprogram required to execute 2° fate. A model summarizing how 1° vs 2° vs 3° fate might be determined is shown in Fig. 16.24.

E. Hyperinduction—Stimulatory and Inhibitory Pathways in Vulval Precursor Cells

Certain mutations in *lin-2, lin-7*, and *let-23 RTK* result in the hyperinduced or Hin phenotype (Ferguson and Horvitz, 1985; Aroian and Sternberg, 1991; G. Jongeward and P. W. Sternberg, unpublished results). The Hin phenotype is characterized by levels of vulval induction that are greater than wild-type, and is distinct from the Muv phenotype (Fig. 16.10G; Aroian and Sternberg, 1991). More 1° and 2° vulval fates occur in a Hin animal than in a wild-type animal but, unlike Muv induction patterns, VPCs closest to the anchor cell select higher fates than do cells further from the anchor cell. In addition, induction in Hin mutants is dependent on the anchor cell. If the anchor cell is ablated, all VPCs execute 3° fate. Therefore, unlike the VPCs in a Muv hermaphrodite, which are induced independently of the anchor cell, the VPCs in a Hin hermaphrodite still require the anchor-cell signal and are hypersensitive to it. Also, adjacent 1° fates can occur in Hin mutants, suggesting that lateral signaling has broken down or that higher activation of the intrinsic pathway has overridden the 2° pathway.

Surprisingly, mutations in *lin-2, lin-7*, and *let-23 RTK* that result in a Hin phenotype appear to reduce but not eliminate, gene function (Aroian and Sternberg, 1991; G. Jongeward and P. W. Sternberg, unpublished results). For example, the *let-23 RTK* allele *n1045* at 25°C results in the Vul phenotype when present in one copy (i.e., in *trans* to a deletion for *let-23 RTK*) but results in the Hin phenotype if present in two copies (i.e., when homozygous; Aroian and Sternberg, 1991). The first result suggests that *let-23 n1045* reduces *let-23 RTK* function. The second result suggests that two copies of this reduction

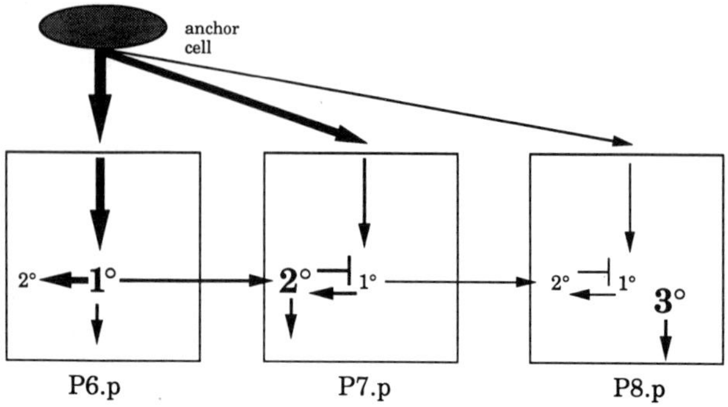

Figure 16.24 Model for determination of 1°, 2°, or 3° fate. In P6.p, the proximity of the anchor cell signal strongly activates the 1° intrinsic pathway (I.P.). This effect has three consequences. First, *lin-12* is activated in the same cell, although not enough to inhibit the strongly activated intrinsic pathway. Second, *lin-12* is activated in the neighboring cell P7.p via lateral signaling. Third, the 1° sublineage is selected. The net state is I.P. on and *lin-12* on. In P7.p, an intermediate level of anchor cell signal is received, moderately activating the intrinsic pathway which then activates *lin-12*. In addition, *lin-12* is activated by lateral signaling from P6.p. As a result, *lin-12* is activated to a high enough level and the intrinsic pathway to a low enough level that 1° fate is inhibited successfully by *lin-12*; 2° fate ensues. Possibly, P7.p emits a lateral signal, but this signal is apt to be weak since the intrinsic pathway is turned off. The net state is I.P. off and *lin-12* on. In P8.p, only a small amount of anchor cell signal is received, weakly activating the intrinsic pathway and *lin-12*. Further, *lin-12* is weakly activated by a P7.p lateral signal. As a result, the intrinsic pathway is inhibited and *lin-12* is never on to a high enough level to promote 2° fate. As a result, 3° fate ensues. The net state is I.P. off and *lin-12* off. The essential features of this model follow. (1) The level of anchor cell signal is important for how strongly the I.P. is turned on. (2) The intrinsic pathway turns on 2° fate and the lateral signal. (3) 2° fate is activated in two ways—by the intrinsic pathway and by the lateral signal coming from a neighboring cell. (4) The effect of 2° fate is to inhibit the intrinsic pathway. In a 2° VPC, this insures that 1° fate is turned off and that 2° fate ensues (because *lin-12* is on). In a 3° cell, this effect reinforces the low level of anchor cell signal. Presumably, the *lin-12* pathway is never highly activated in these 3° cells.

of function mutation somehow result in too much induction. Thus, if *let-23 RTK* is eliminated, no vulval indiction occurs, but if *let-23 RTK* is only partly reduced then more than wild-type induction occurs. This striking conclusion is supported by another *let-23 RTK* allele *mn224* (Aroian and Sternberg, 1991).

These data suggest that *let-23 RTK*, *lin-2*, and *lin-7* are involved in two different pathways, one that stimulates vulval fate and another that inhibits vulval fate (Fig. 16.25A; Aroian and Sternberg, 1991). Hin mutations such as *let-23 n1045* might cripple preferentially the ability of these proteins to operate in the inhibitory pathway over the stimulatory pathway, resulting in too much induction because inhibition is lacking. One simple model would be that the *let-23 RTK* activated-inhibitory process is more sensitive to a decrease in *let-23 RTK* activity levels than the *let-23 RTK* activated-stimulatory process. An initial reduction in *let-23 RTK* activity would, therefore, result in activation of stimulation but not inhibition (Fig. 16.25B). A further reduction in *let-23 RTK* activity, however, would affect both processes, leading to less than wild-type induction (Fig. 16.25C). An alternative model is that the Hin mutations change quantitative aspects of these proteins. For example, the C-terminal region of EGF receptor negatively regulates receptor activity because it contains autophosphorylation and receptor down-regulation sites. Sequencing indicates that the *let-23 n1045* mutation results in some protein truncated for the C terminus (R. V. Aroian and P. W. Sternberg, unpublished results). The *let-23 n1045* allele might, therefore, be Hin because the receptor is not properly regulated. Although this model could explain the *let-23* data, the *lin-2* and *lin-7* data are easier to incorporate with the level model presented earlier. The fact that *lin-2* and *lin-7* behave similarly to *let-23 RTK* (G. Jongeward and P. W. Sternberg, unpublished results) suggests that they may exert their influence on *let-23 RTK* independent of the Muv genes, because no evidence suggests that the Muv genes act in both pathways.

This inhibitory process associated with *let-23 RTK* could operate intracellularly or intercellularly (Fig. 16.25A). In an intracellular model, the *let-23 RTK* stimulatory and inhibitory pathways both would operate inside the same cell, consistent with, for example, the C-terminal truncation model. On the other hand, *let-23 RTK* could be involved in intracellular stimulation and intercellular inhibition. With this notion, combining the Hin phenotype with lateral signaling is possible. For example, *let-23 RTK* may be involved in activating the lateral signal between cells. A Hin phenotype then might result from a weakened lateral signal. Because the proposed role of the lateral signal is to activate *lin-12* in a neighboring VPC and thereby inhibit that intrinsic pathway (Fig. 16.25B), a consequence of a weaker lateral signal might be to raise the overall level of induction. Alternatively, *let-23* may be involved in an intercellular signal that is unrelated to the lateral signal. (See Fig. 16.25 for more discussion.)

VI. Dissecting Gene Functions and Downstream Elements

A. Sequencing Point Mutations

The availability of cellular techniques, genetic techniques, and molecular germ-line transformation techniques in the study of vulval determination promises to verify present knowledge of Ras- and EGF receptor-like genes and to yield new insights into their functional properties. The *let-23 n1045* hyper-induced allele is one potential example. Mutations associated with other *let-23*

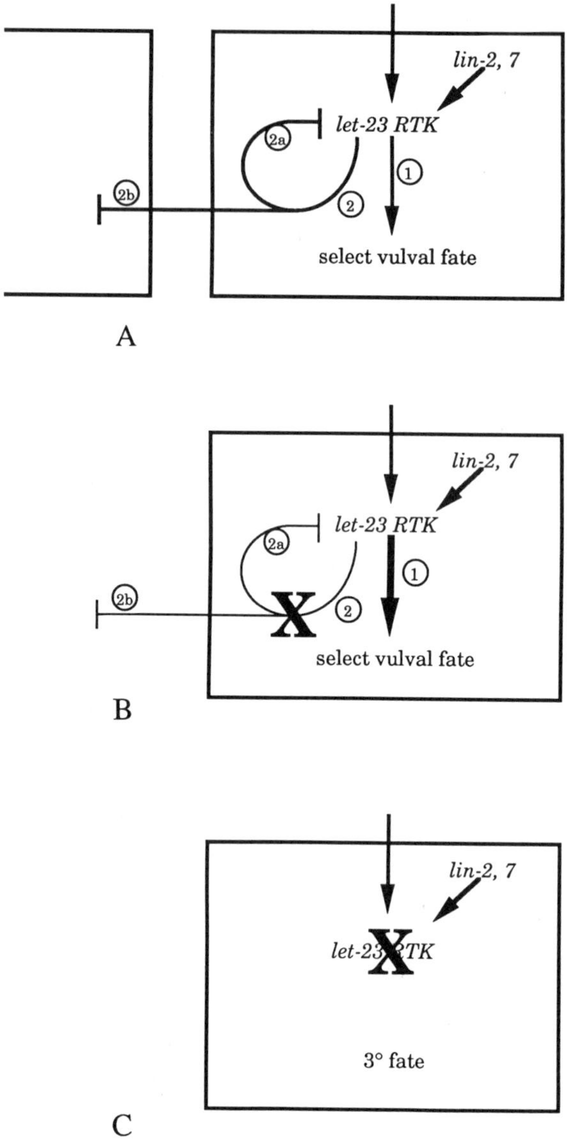

Figure 16.25 Modeling hyperinduction. (A) To explain the hyperinduction data, *let-23 RTK*, *lin-2*, and *lin-7* are likely to direct two pathways in the VPCs: (1) a stimulatory pathway required for selection of vulval fate and (2) an inhibitory pathway that normally acts to inhibit the stimulatory pathway either in the same cell (2a) or in a neighboring cell (2b). For example, as in intracellular inhibition (2a) EGF receptor is thought to activate phospholipase C-$_{\gamma}$, which, among other things, results in activation of protein kinase C, which inactivates receptor (reviewed by Ullrich

RTK alleles also have been sequenced. For example, two loss-of-function alleles result from missense mutations in conserved residues in the tyrosine kinase domain (R. V. Aroian and P. W. Sternberg, unpublished results). This result suggests that *let-23 RTK* has tyrosine kinase activity and that these conserved residues are functionally important.

Mutations in *let-60 ras* provide a powerful example of dissecting gene function with point mutation studies. Three types of *let-60 ras* mutations exist. Complete loss-of-function and reduction-of-function *let-60 ras* alleles result in a recessive Vul phenotype (as discussed earlier). In addition, dominant *let-60 ras* Muv mutations exist that behave like gain-of-function mutations in which the *let-60 ras* gene is constitutively on. These alleles are designated *let-60(gf)* (Beitel *et al.*, 1990; Han *et al.*, 1990). These mutations result in an anchor-cell independent Muv phenotype because the requirement for *let-60 ras* activation by *let-23 RTK* presumably is bypassed and the intrinsic pathway is constitutively on. Moreover, dominant Vul *let-60 ras* alleles exist that are designated *let-60(dn)* (dominant negative; Beitel *et al.*, 1990; Han *et al.*, 1990). A *let-60(dn)* allele in *trans* to a wild-type *let-60 ras* chromosome, that is, *let-60(dn)/+*, results in a Vul hermaphrodite. The data suggest that these *let-60(dn)* alleles reduce *let-60 ras* activity and also can poison wild-type *let-60 ras* activity, resulting in a dominant Vul phenotype (Han *et al.*, 1990; Han and Sternberg, 1991). Thus, the *let-60 ras* gene can be mutated three ways: (1) loss-of-function alleles display a recessive Vul phenotype, (2) dominant gain-of-function alleles display a dominant Muv phenotype, and (3) dominant negative alleles display a dominant Vul phenotype.

The sequences associated with these alleles indicate residues that, when altered (1) eliminate or reduce *ras* function, (2) activate *ras* function independent

---

and Schlessinger, 1990). Thus, one pathway of *let-23 RTK* stimulates vulval induction via *let-60 ras* but another pathway of *let-23 RTK* might inhibit vulval induction via a negative feedback loop. An alternative intracellular inhibition model is given in the text. In terms of intercellular inhibition (2b), *let-23 RTK* might result in an inhibitory signal sent to a neighboring VPC. In fact, such a signal is consistent with the lateral signal discussed previously since one aspect of the lateral signal might be to inhibit the intrinsic pathway in a neighboring cell. (B) Hyperinduction, associated with reduction-of-function *let-23 RTK* alleles, might result from a loss of the inhibitory pathway, but not the stimulatory pathway, so inhibition (intracellular or intercellular) is lacking. As a result, an overall increase occurs in the levels of induction. This loss of only the inhibitory pathway could occur because activation of the inhibitory pathway (e.g., via PLC-$_{\gamma}$ or lateral signal) is more sensitive to an initial reduction in *let-23 RTK* levels than is activation of the stimulatory pathway (via *let-60 ras*). Alternatively, loss of only the inhibitory pathway could occur because the hyperinduced mutations perturb the interactions of *let-23 RTK* with a substrate necessary for inhibition (e.g., PLC-$_{\gamma}$) and not another necessary for activation. (See text for another model of this kind.) (C) Finally, complete loss of *let-23 RTK* function eliminates both pathways. The end result is no induction.

of upstream genes such as *let-23 RTK*, and (3) poison *ras* function in *trans*. Sequences of all these types of mutations have been analyzed (Beitel *et al.*, 1990; Han and Sternberg, 1991). Many of the recessive *let-60 ras* mutations affect codons highly conserved in the Ras family. These residues might be important for activation of, but not for activity of, Ras (Beitel *et al.*, 1990). One recessive reduction-of-function allele alters a residue in the effector domain (Beitel *et al.* 1990). The dominant gain-of-function alleles all alter the conserved glycine at position 13 to glutamic acid (Beitel *et al.*, 1990). Alterations of this glycine in other systems has been associated with oncogenic activation of Ras. Thus, the effects of *let-60 ras* mutations on vulval development highly parallel our knowledge of Ras mutations in other systems and, with the *let-23 RTK* data presented here, suggest that *C. elegans* vulval development provides a good system for studying important processes of more complex organisms.

The residues altered by many *let-60(dn)* mutations and the properties of these mutations also have been studied (Han and Sternberg, 1991). These data suggest that the *let-60(dn)* mutations likely influence guanine nucleotide binding and that the *let-60(dn)* protein interacts nonproductively with some positive factor required by the wild-type *let-60* protein.

B. Mutations that Affect Execution, Not Determination, of Vulval Fate

The phenotype of Muv/Vul double mutants suggests that the Vul genes discussed in this chapter are defective in determination, but not execution, of 1° or 2° vulval fate (Sternberg and Horvitz, 1989). For example, in a *let-23(lf)* background, *let-60 ras* gain-of-function mutations still result in selection of all 1° and 2° fates by the VPCs (Han *et al.*, 1990). Thus, even if *let-23 RTK* activity is eliminated, the VPCs are still competent to execute 1° and 2° fate subprograms.

Three genes are known to affect execution and not determination of vulval fate: *lin-11, lin-17*, and *lin-18* (Fig. 16.26; Ferguson and Horvitz, 1985; Ferguson *et al.*, 1987; Sternberg and Horvitz, 1988). Mutations in *lin-11* result in abnormal LLLL 2° lineages (Fig. 16.26; Ferguson *et al.*, 1987). Considering its molecular structure, the *lin-11* gene may encode a transcription factor that is activated by *lin-12* and is required in VPC daughter cells for execution of the NT portion of 2° lineages. The genes *lin-17* and *lin-18* mainly affect P7.p. Although mutations in these genes cause variable effects, one important effect is to reverse the polarity of P7.p 2° lineage occasionally from NTLL to LLTN (Fig. 16.27; Ferguson *et al.*, 1987; W. Katz and P. W. Sternberg, unpublished observation). P5.p polarity is unaffected, suggesting that some mechanism, operating via *lin-17* and *lin-18*, is actively required to make P7.p polarity the mirror image of P5.p. Mutant *lin-11* hermaphrodites display a single ventral protrusion and are Egl, whereas mutant *lin-17* and *lin-18* hermaphrodites

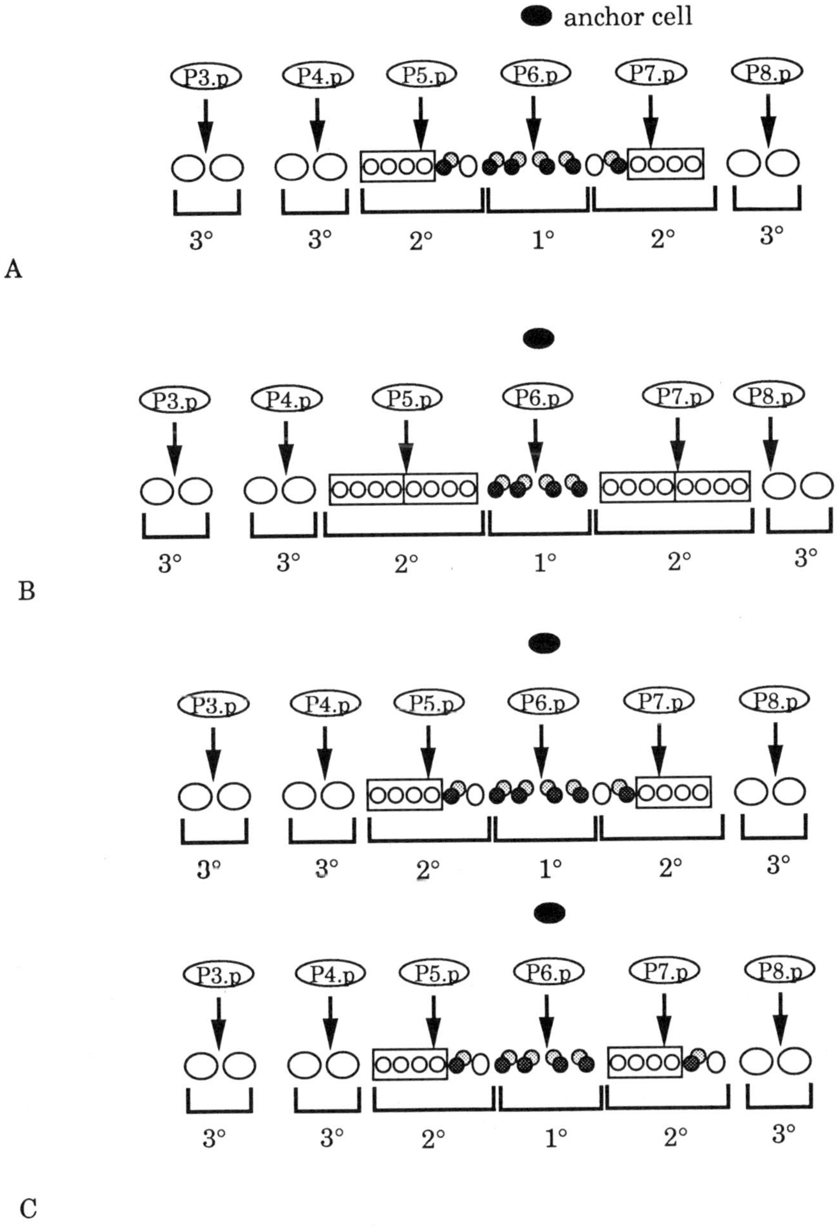

Figure 16.26 Effects of execution mutants. (A) Wild-type induction. (B) In a *lin-11* loss-of-function mutant [*lin-11(lf)*], the TN portion of 2° lineages is lacking and the LL portion is reiterated. (C) One of the effects of *lin-17(lf)* and *lin-18(lf)* mutants is to randomize the polarity of P7.p. Sometimes the polarity is wild type (NTLL) and sometimes the polarity is reversed (LLTN). Data from Ferguson *et al.* (1987) and W. Katz and P. W. Sternberg, unpublished results.

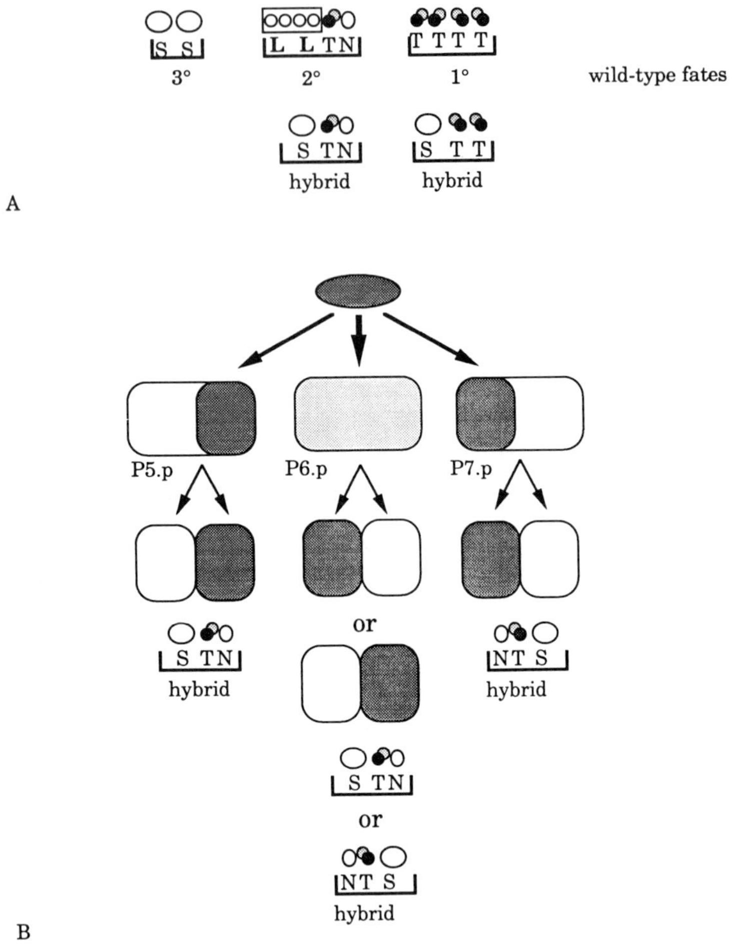

Figure 16.27 Hybrid lineages. (A) Hybrid lineages result from a partial reduction of function in any of the *vul* genes (e.g., *let-23 RTK* and *let-60 ras*). They are most often half 3°–half 2° or half 3°–half 1° as shown. The half-vulval portion of the lineage coincides with the location of the anchor cell relative to the hybrid lineage, so for P5.p the vulval portion faces posterior, for P7.p the vulval portion faces anterior, and for P6.p the vulval portion faces either anterior or posterior. (B) These data are interpreted to mean that some factor necessary for execution of vulval lineages is distributed asymmetrically in each VPC (shaded part of the cells) based on the direction of the anchor cell signal. In a VPC with lower signal transduction levels, this factor is limiting and results in only the anchor cell-proximal half of a VPC receiving enough of it to divide past one round. See Aroian and Sternberg (1991) for data and further discussion.

sometimes show two small ventral protrusions, one at and one posterior to the vulva (Ferguson and Horvitz, 1985; W. Katz and P. W. Sternberg, unpublished observation).

C. *lin-25*

Mutations in the gene *lin-25* do not fit nicely into any one category because they seem to fit into many categories. Most *lin-25* mutant hermaphrodites are Egl, although some are sterile (Ferguson and Horvitz, 1985). The *lin-25* mutation often results in extra VPCs and the precocious division of VPCs, suggesting that *lin-25* is involved in proper generation of the VPCs (Ferguson *et al.*, 1987). Further, in those mutant animals in which the VPCs divide at their normal time, no 2° lineages occur; only 3° lineages and an abnormal 1° lineage at P6.p are formed (Ferguson *et al.*, 1987). Therefore, in conjunction with *lin-12, lin-25* is needed for proper determination of 2° fate. Also, the temperature-sensitive period of *lin-25* extends beyond the first division of the VPCs, past the time when determination is thought to occur (Ferguson *et al.*, 1987). Thus *lin-25* also is likely to affect execution of vulval lineages. Consistent with this hypothesis, *lin-25* mutations result in abnormal 1° lineages and are epistatic to *lin-1* (i.e., the *lin-25, lin-1* double mutant is Egl not Muv; E. Ferguson, P. W. Sternberg, and H. R. Horvitz, unpublished observations), suggesting that *lin-25* performs a downstream function.

D. Nonstandard Vulval Lineages—A Possible Link between Determination and Execution

Hybrid lineages associated with mutations in the Vul determination genes and in *lin-12* may provide evidence for a link between determination and execution. Hybrid lineages are not uncommon and are characterized by selection of half nonvulval and half vulval (Fig. 16.27A). Although they are associated with reduction-of-function mutations in the Vul determination genes, these hybrids are explained most simply as occurring because of defects in execution of vulval fate. In the case of *let-23 RTK* mutants, hybrid lineages show a polarity bias centered toward the anchor cell so the vulval half of the lineage is always closest to the anchor cell (Aroian and Sternberg, 1991). This result suggests that the anchor cell polarizes these lineages. Further, no obvious correlation exists between selection of hybrid fate and the fate executed by neighboring VPCs, suggesting that perhaps hybrids are determined cell autonomously (Aroian and Sternberg, 1991).

Such lineages have been proposed to reflect a defect in execution of vulval fate in one of the daughters of a VPC (Sternberg and Horvitz, 1986; Aroian and Sternberg, 1991). In one simple model, determination of 1° and 2° vulval fate by the Vul determination genes *lin-2, lin-3, lin-7*, and so on results in the production of a Factor X that is segregated to the daughters of the VPCs and is required for those daughters to execute a vulval lineage (Aroian and Sternberg, 1991). Further, the anchor-cell signal biases the distribution of Factor X so more Factor X is localized in the part of a VPC closest to the anchor cell (Fig. 16.27B). VPCs in which the vulval determination pathway is reduced (but not eliminated) may undergo proper determination but improper execution because not enough Factor X is produced and the amounts that are produced are localized preferentially to the anchor-cell proximal part of a VPC (Fig. 16.27B). Such a model is consistent with the hybrid data but is not the only model possible. For example, hybrid lineages may represent truly novel lineages that result from defective determination, not execution.

VII. Prospects

The development of the *C. elegans* vulva provides a powerful system in which to study signal transduction via cell biology, genetics, and molecular biology. The ability to analyze the vulva at the cellular level and to perturb the surroundings makes it possible to understand in some detail the processes that are occurring and how various mutations affect those processes. Further, the ease of genetic manipulation makes it possible to isolate many mutations in the pathways involved and to probe the relationships among these genes. Finally, molecular analysis now shows that these signal transduction processes share many elements with other important signal transduction processes in other organisms. The combination of this molecular information with the other tools of studying the vulval determination system are verifying and extending our current knowledge of these elements. In addition, as more genes in the pathway are characterized and cloned, novel relationships and processes in these pathways might be uncovered that can shed new light on related pathways and processes in other systems.

Acknowledgments

We thank Helen Chamberlin, Min Han, Mike Herman, Russell Hill, Linda Huang, Gregg Jongeward, and Wendy Katz for unpublished results. We thank Nancy Bonini, Andy Golden, Linda Huang, Paul Kayne, Katharine Liu, Wendy Katz, Lloyd Martinez, Anna Newman, and Erich Schwarz for helpful comments and critical reading of this manuscript, and we thank members of our laboratory for discussions. P. W. Sternberg is an investigator of the Howard Hughes Medical Institute. Research conducted in our laboratory that was reviewed here has been supported by the Howard Hughes Medical Institute and grants to P. W. Sternberg from the U.S. Public Health Service, March of Dimes Birth Defects Foundation, and Searle Scholar's Program.

References

Aroian, R. V., and Sternberg, P. W. (1991). Multiple functions of *let-23*, a *Caenorhabditis elegans* receptor tyrosine kinase gene required for vulval induction. *Genetics* **128**, 251–267.

Aroian, R. V., Koga, M., Mendel, J. E., Ohshima, Y., and Sternberg, P. W. (1990). The *let-23* gene necessary for *Caenorhabditis elegans* vulval induction encodes a tyrosine kinase of the EGF receptor subfamily. *Nature (London)* **348**, 693–699.

Austin, J., and Kimble, J. (1987). *glp-1* is required in the germ line for regulation of the decision between mitosis and meiosis in *C. elegans. Cell* **51**, 589–599.

Austin, J., Maine, E. M., and Kimble, J. (1989). Genetics of intercellular signalling in *C. elegans. Dev. Suppl.*, 53–57.

Beitel, G., Clark, S., and Horvitz, H. R. (1990). The *Caenorhabditis elegans ras* gene *let-60* acts as a switch in the pathway of vulval induction. *Nature (London)* **348**, 503–509.

Bourne, H. R., Sanders, D. A., and McCormick, F. (1991). The GTPase superfamily: conserved structure and molecular mechanism. *Nature (London)* **349**, 117–127.

Brenner, S. (1974). The genetics of *Caenorhabditis elegans*. *Genetics* **77**, 71–94.

Coffman, C., Harris, W., and Kintner, C. (1990). *Xotch*, the *Xenopus* homolog of *Drosophila Notch. Science* **249**, 1438–1441.

Collins, J., Saari, B., and Anderson, P. (1987). Activation of a transposable element in the germ line but not the soma of *C. elegans. Nature (London)* **328**, 726–728.

Coulson, A. R., Sulston, J., Brenner, S., and Karn, J. (1986). Toward a physical map of the genome of the nematode *Caenorhabditis elegans. Proc. Natl. Acad. Sci. U.S.A.* **83**, 7821–7825.

Coulson, A., Waterston, R., Kiff, J., Sulston, J., and Kohara, Y. (1988). Genome linking with yeast artificial chromosomes. *Nature (London)* **335**, 184–186.

Davidson, E. H. (1990). How embryos work: A comparative view of diverse modes of cell fate specification. *Dev.* **108**, 365–389.

Desai, C., and Horvitz, H. R. (1989). *Caenorhabditis elegans* mutants defective in the functioning of the motor neurons responsible for egg laying. *Genetics* **121**, 703–721.

Ellisen, L. W., Bird, J., West, D. C., Soreng, A. L., Reynolds, T. C., Smith, S. D., and Sklar, J. (1991). *TAN-1*, the human homolog of the *Drosophila Notch* gene, is broken by chromosomal translocation in T lymphobalstic neoplasms. *Cell* **66**, 649–661.

Ferguson, E., and Horvitz, H. R. (1985). Identification and characterization of 22 genes that affect the vulval cell lineages of *Caenorhabditis elegans*. *Genetics* **110**, 17–72.

Ferguson, E., and Horvitz, H. R. (1989). The multivulva phenotype of certain *C. elegans* mutants results from defects in two functionally-redundant pathways. *Genetics* **123**, 109–121.

Ferguson, E. L., Sternberg, P. W., and Horvitz, H. R. (1987). A genetic pathway for the specification of the vulval cell lineages of *Caenorhabditis elegans. Nature (London)* **326**, 259–267.

Fire, A. (1986). Integrative transformation of *Caenorhabditis elegans. EMBO J.* **5**, 2675–2680.

Fire, A., White-Harrison, S., and Dixon, D. (1990). A modular set of *lacZ* fusion vectors for studying gene expression in *Caenorhabditis elegans. Gene* **93**, 189–198.

Freyd, G., Kim, S. K., and Horvitz, H. R. (1990). Novel cysteine-rich motif and homeodomain in the product of the *Caenorhabditis elegans* cell lineage gene *lin-11. Nature (London)* **344**, 876–879.

Greenwald, I. S. (1985). *lin-12*, a nematode homeotic gene, is homologous to a set of mammalian proteins that includes epidermal growth factor. *Cell* **43**, 583–590.

Greenwald, I. S., Sternberg, P. W., and Horvitz, H. R. (1983). The *lin-12* locus specifies cell fates in *Caenorhabditis elegans. Cell* **34**, 435–444.

Han, M., and Sternberg, P. W. (1990). *let-60*, a gene that specifies cell fates during *C. elegans* vulval induction, encodes a ras protein. *Cell* **63**, 921–931.

Han, M., and Sternberg, P. W. (1991). Analysis of dominant negative mutations of the *C. elegans let-60 ras* gene. *Genes Dev.* **5**, 2188–2198.

Han, M., Aroian, R., and Sternberg, P. W. (1990). The *let-60* locus controls the switch between vulval and non-vulval cell types in *C. elegans. Genetics* **126**, 899–913.

Herman, R. K. (1989). Mosaic analysis in the nematode *Caenorhabditis elegans. J. Neurogenetics* **5**, 1–24.

Herman, R. K., and Hedgecock, E. M. (1990). The size of the *C. elegans* vulval primordium is limited by *lin-15* expression in surrounding hypodermis. *Nature (London)* **348**, 169–171.

Horvitz, H. R., and Sternberg, P. W. (1991). Multiple intercellular signalling systems control the development of the *Caenorhabditis elegans* vulva. *Nature (London)* **351**, 535–541.

Horvitz, H. R., and Sulston, J. E. (1980). Isolation and genetic characterization of cell-lineage mutants of the nematode *Caenorhabditis elegans. Genetics* **96**, 435–454.

Kim, S. K., and Horvitz, H. R. (1990). The *Caenorhabditis elegans* gene *lin-10* is broadly expressed while required specifically for the determination of vulval cell fates. *Genes Dev.* **4**, 357–371.

Kimble, J. (1981). Lineage alterations after ablation of cells in the somatic gonad of *Caenorhabditis elegans*. *Dev. Biol.* **87**, 286–300.

Kimble, J. E., and White, J. G. (1981). On the control of germ cell development in *Caenorhabditis elegans. Dev. Biol.* **81**, 208–219.

Li, C., and Chalfie, M. (1990). Organogenesis in *C. elegans*: Positioning of neurons and muscles in the egg-laying system. *Neuron* **4**, 681–695.

Mello, C. C., Kramer, J. M., Stinchcomb, D., and Ambros, V. (1991). Efficient gene transfer in *C. elegans* Extrachromosomal maintenance and integration of transforming sequences. *EMBO. J.* **10**, 3959–3970.

Mulcahy, L. S., Smith, M. R., and Stacey, D. W. (1985). Requirement for *ras* proto-oncogene function during serum-stimulated growth of NIH3T3 cells. *Nature (London)* **313**, 241–243.

Price, J. V., Clifford, R. J., and Schupbach, T. (1989). The maternal ventralizing locus *torpedo* is allelic to *faint little ball*, an embryonic lethal, and encodes the *Drosophila* EGF receptor homolog. *Cell* **56**, 1085–1092.

Priess, J. R., and Thomson, J. N. (1987). Cellular interactions in early *Caenorhabditis elegans* embryos. *Cell* **48**, 241–250.

Ruvkun, G., and Giusto, J. (1989). The *Caenorhabditis elegans* heterochronic gene *lin-14* encodes a nuclear protein that forms a temporal development switch. *Nature (London)* **338**, 313–319.

Schejter, E. D., and Shilo, B. (1989). The *Drosophila* EGF receptor homolog (DER) gene is allelic to *faint little ball*, a locus essential for embryonic development. *Cell* **56**, 1093–1104.

Seydoux, G., and Greenwald, I. (1989). Cell autonomy of *lin-12* function in a cell fate decision in *C. elegans. Cell* **57**, 1237–1245.

Simpson, P. (1990). *Notch* and the choice of cell fate in *Drosophila* neuroepithelium. *Trends Genet.* **6**, 343–345.

Sporn, M. B., and Roberts, A.B. (1988). Peptide growth factors are multifunctional. *Nature (London)* **332**, 217–219.

Sternberg, P. W. (1988a). Control of cell fates in equivalence groups in *C. elegans. Trends Neurosci.* **11**, 259–264.

Sternberg, P. W. (1988b). Lateral inhibition during vulval induction in *Caenorhabditis elegans. Nature (London)* **335**, 551–554.

Sternberg, P. W., and Horvitz, H. R. (1986). Pattern formation during vulval development in *Caenorhabditis elegans. Cell* **44**, 761–772.

Sternberg, P. W., and Horvitz, H. R. (1988). *lin-17* mutations of *C. elegans* disrupt asymmetric cell divisions. *Dev. Biol.* **130**, 67–73.

Sternberg, P. W., and Horvitz, H. R. (1989). The combined action of two intercellular signalling pathways specifies three cell fates during vulval induction in *C. elegans. Cell* **58**, 679–693.

Sulston, J., and Horvitz, H. R. (1977). Postembryonic cell lineages of the nematode *Caenorhabditis elegans. Dev. Biol.* **56**, 110–156.

Sulston, J. E., and Horvitz, H. R. (1981). Abnormal cell lineages in mutants of the nematode *Caenorhabditis elegans. Dev. Biol.* **82**, 41–55.

Sulston, J. E., and White, J. G. (1980). Regulation and cell autonomy during postembryonic development of *Caenorhabditis elegans. Dev. Biol.* **78**, 577–597.

Thomas, J. H., Stern, M. J., and Horvitz, H. R. (1990). Cell interactions coordinate the development of the *C. elegans* egg-laying system. *Cell* **62**, 1041–1052.

Thompson, C. C., Brown, T. A., and McKnight, S. L. (1991). Convergence of *Ets*- and *Notch*-related structural motifs in a heteromeric DNA binding complex. *Science* **253**, 762–768.

Trent, C., Tsung, N., and Horvitz, H. R. (1983). Egg-laying defective mutants of the nematode *Caenorhabditis elegans. Genetics* **104**, 619–647.

Ullrich, A., and Schlessinger, J. (1990). Signal transduction by receptors with tyrosine kinase activity. *Cell* **61**, 203–212.

Waterston, R. H., Thomson, J. N., and Brenner, S. (1980). Mutants with altered muscle structure in *Caenorhabditis elegans. Dev. Biol.* **77**, 271–302.

Way, J. C., and Chalfie, M. (1989). The *mec-3* gene of *Caenorhabditis elegans* requires its own product for maintained expression and is expressed in three neuronal cell types. *Genes Dev.* **3**, 1823–1833.

Weissman, B. E., and Aaronson, S. A. (1983). BALB and Kirsten murine sarcoma viruses alter growth and differentiation of EGF-dependent BALB/c mouse epidermal keratonocyte lines. *Cell* **32**, 599–606.

Yochem, J., and Greenwald, I. (1989). *glp-1* and *lin-12*, genes implicated in distinct cell–cell interactions in *C. elegans* encode similar transmembrane proteins. *Cell* **58**, 553–563.

Yochem, J., Weston, K., and Greenwald, I. (1988). *C. elegans lin-12* encodes a transmembrane protein similar to *Drosophila Notch* and yeast cell cycle gene products. *Nature (London)* **335**, 547–550.

Index